Lecture Notes in Computer Science 8207

Commenced Publication in 1973
Founding and Former Series Editors:
Gerhard Goos, Juris Hartmanis, and Jan van Leeuwen

Allan Tucker   Frank Höppner
Arno Siebes   Stephen Swift (Eds.)

Advances in Intelligent Data Analysis XII

12th International Symposium, IDA 2013
London, UK, October 17-19, 2013
Proceedings

 Springer

Volume Editors

Allan Tucker
Brunel University, School of Information Systems, Computing and Mathematics
Uxbridge, Middlesex, UK
E-mail: allan.tucker@brunel.ac.uk

Frank Höppner
Ostfalia University of Applied Sciences, Faculty of Computer Science/IT
Wolfenbüttel, Germany
E-mail: f.hoeppner@ostfalia.de

Arno Siebes
Universiteit Utrecht, Faculty of Science
Utrecht, The Netherlands
E-mail: a.p.j.m.siebes@uu.nl

Stephen Swift
Brunel University, School of Information Systems, Computing and Mathematics
Uxbridge, Middlesex, UK
E-mail: stephen.swift@brunel.ac.uk

ISSN 0302-9743   e-ISSN 1611-3349
ISBN 978-3-642-41397-1   e-ISBN 978-3-642-41398-8
DOI 10.1007/978-3-642-41398-8
Springer Heidelberg New York Dordrecht London

Library of Congress Control Number: 2013949601

CR Subject Classification (1998): H.3, H.4, I.2, F.1, H.2.8, J.3, I.4-5

LNCS Sublibrary: SL 3 – Information Systems and Application,
incl. Internet/Web and HCI

Typesetting: Camera-ready by author, data conversion by Scientific Publishing Services, Chennai, India

Printed on acid-free paper

Springer is part of Springer Science+Business Media (www.springer.com)

Preface

We are proud to present the proceedings of the 12th International Symposium on Intelligent Data Analysis, which was held during October 17–19 in London, UK. The series started in 1995 and was held biennially until 2009. In 2010, the symposium re-focussed to support papers that go beyond established technology and offer genuinely novel and game-changing ideas, while not always being as fully realized as papers submitted to other conferences.

IDA 2013 continued this approach and sought first-look papers that might elsewhere be considered preliminary, but contain potentially high-impact research. The IDA Symposium is open to all kinds of modelling and analysis methods, irrespective of discipline. It is an interdisciplinary meeting that seeks abstractions that cut across domains. IDA solicits papers on all aspects of intelligent data analysis, including papers on intelligent support for modelling and analyzing data from complex, dynamical systems.

Intelligent support for data analysis goes beyond the usual algorithmic offerings in the literature. Papers about established technology were only accepted if the technology was embedded in intelligent data analysis systems, or was applied in novel ways to analyzing and/or modelling complex systems. The conventional reviewing process, which favours incremental advances on established work, can discourage the kinds of papers that IDA 2013 has published. The reviewing process addressed this issue explicitly: referees evaluated papers against the stated goals of the symposium, and any paper for which at least one program chair advisor wrote an informed, thoughtful, positive review was accepted, irrespective of other reviews. Indeed, it was noted that this had notable impact on some of the papers included in the program.

We were pleased to have a very strong program. We received 84 submissions from 215 different authors from at least 23 different countries on 6 continents. As in IDA 2012, we included a poster session for PhD students to promote their work and also introduced the use of a 2-minute video slot for all PhD posters and standard posters with a prize for the best.

We were honored to have distinguished invited speakers at IDA 2013:

- David Hand from Imperial College, London talked about the opportunities in big and open data, examining the statistical, data mining, and machine learning tools which are currently in use, and the potential that lies ahead.
- John Quinn from Makerere University, Kampala, Uganda discussed data analysis in developing countries, focussing on techniques for rapid, accurate and low-cost crop disease monitoring in Uganda.
- Tijl de Bie from Bristol University, UK talked about subjective interestingness in exploratory data mining, and highlighted the importance of focussing on the user and their notion of interestingness.

The conference was held at the Royal Statistical Society in London (a stone's throw from Reverend Thomas Bayes' grave). We wish to express our gratitude to all authors of submitted papers for their intellectual contributions; to the Program Committee members and the additional reviewers for their effort in reviewing, discussing, and commenting on the submitted papers; to the members of the IDA Steering Committee for their ongoing guidance and support; and to the Senior Program Committee for their active involvement. We thank Richard van de Stadt for running the submission website and handling the production of the proceedings. Special thanks go to the poster chair, Matthijs van Leeuwen, and the frontier prize chairs, Jaakko Hollmén and Frank Klawonn. We gratefully acknowledge those who were involved in the local organization of the symposium: Valeria Bo, Neda Trifonova, Yuanxi Li, Stelios Pavlidis, Mahir Arzoky, Samy Ayed and Stefano Ceccon. We are grateful for our sponsors: Brunel University, London; The Heilbronn Institute, Bristol; The Artificial Intelligence Journal; and SMESH. We are especially indebted to KNIME, who funded the IDA frontier prize for the most visionary contribution presenting a novel and surprising approach to data analysis in the understanding of complex systems.

August 2013

Allan Tucker
Frank Höppner
Arno Siebes
Stephen Swift

Organization

General Chair

Allan Tucker   Brunel University, UK

Local Chair

Stephen Swift   Brunel University, UK

Program Chairs

Frank Höppner   Ostfalia University, Germany
Arno Siebes   Utrecht University, The Netherlands

Poster Chair

Matthijs van Leeuwen   KU Leuven, Belgium

Frontier Prize Chairs

Jaakko Hollmén   Aalto University, Finland
Frank Klawonn   Ostfalia University, Germany

Advisory Chairs

Xiaohui Liu   Brunel University, UK
Hannu Toivonen   University of Helsinki, Finland

Local Organization Committee

Stelios Pavlidis   Brunel University, UK
Neda Trifonova   Brunel University, UK
Valeria Bo   Brunel University, UK
Yuanxi Li   Brunel University, UK
Stefano Ceccon   Brunel University, UK
Mahir Arzoky   Brunel University, UK
Samy Ayed   Brunel University, UK

Program Chair Advisors

Niall Adams	Imperial College, UK
Michael R. Berthold	University of Konstanz, Germany
Liz Bradley	University of Colorado, USA
João Gama	University of Porto, Portugal
Jaakko Hollmén	Aalto University, Finland
Frank Klawonn	Ostfalia University, Germany
Joost Kok	Leiden University, The Netherlands
Xiaohui Liu	Brunel University, UK
Hannu Toivonen	University of Helsinki, Finland
Allan Tucker	Brunel University, UK
David Weston	University of London, UK

Program Committee

Fabrizio Angiulli	University of Calabria, Italy
Alexandre Aussem	University of Lyon 1, France
Barbro Back	Åbo Akademi University, Finland
Christian Borgelt	European Centre for Soft Computing, Spain
Henrik Boström	Stockholm University, Sweden
Bruno Crémilleux	University of Caen, France
Tijl De Bie	University of Bristol, UK
Andre de Carvalho	University of São Paulo, Brazil
Jose del Campo	University of Malaga, Spain
Sašo Džeroski	Jožef Stefan Institute, Slovenia
Fazel Famili	IIT - National Research Council Canada, Canada
Ad Feelders	Utrecht University, The Netherlands
Ingrid Fischer	University of Konstanz, Germany
Doug Fisher	Vanderbilt University, USA
Élisa Fromont	University of Jean Monnet, France
Johannes Fürnkranz	Technische Universität Darmstadt, Germany
Ricard Gavaldà	Technical University of Catalonia (UPC), Spain
Kenny Gruchalla	National Renewable Energy Lab, USA
Lawrence Hall	University of South Florida, USA
Eyke Hüllermeier	University of Marburg, Germany
Alípio Jorge	University of Porto, Portugal
Nada Lavrač	Jožef Stefan Institute, Slovenia
Matthijs van Leeuwen	KU Leuven, Belgium
Jose Lozano	University of the Basque Country, Spain
George Magoulas	University of London, UK
Maria-Carolina Monard	University of São Paulo, Brazil
Giovanni Montana	Imperial College, UK
Mohamed Nadif	Descartes University of Paris, France
Andreas Nürnberger	University of Magdeburg, Germany

Panagiotis Papapetrou	Aalto University, Finland
Stelios Pavlidis	Brunel University, UK
Mykola Pechenizkiy	Eindhoven University of Technology, The Netherlands
José-Maria Peña	Polytechnic University of Madrid, Spain
Ruggero Pensa	University of Turin, Italy
Alexandra Poulovassilis	University of London, UK
Miguel A. Prada	University of Leon, Spain
Ronaldo Prati	Universidade Federal do ABC, Brazil
Fabrizio Riguzzi	University of Ferrara, Italy
Stefan Rüping	Fraunhofer IAIS, Germany
Antonio Salmerón Cerdán	University of Almeria, Spain
Vítor Santos Costa	University of Porto, Portugal
Roberta Siciliano	University of Naples Federico II, Italy
Myra Spiliopoulou	Otto-von-Guericke University Magdeburg, Germany
Maarten van Someren	Universiteit van Amsterdam, The Netherlands
Stephen Swift	Brunel University, UK
Maguelonne Teisseire	UMR TETIS - Irstea & LIRMM, France
Antony Unwin	University Augsburg, Germany
Veronica Vinciotti	Brunel University, UK
Zidong Wang	Brunel University, UK

Additional Referees

Darko Aleksovski
Nicolas Bechet
Elena Bellodi
Hanen Borchani
Igor Braga
Alireza Chakeri
Everton Cherman
Marcos Cintra
Mickaël Fabrègue
Fabio Fassetti
Gianluca Frasso
Maxime Gasse
Tatiana Gossen
Samuel Hawkins
Thanh Lam Hoang
Dino Ienco

Elena Ikonomovska
Baptiste Jeudy
Jurica Levatic
Aritz Pérez Martínez
João Mendes-Moreira
Jean Metz
Luigi Palopoli
Matic Perovšek
Vid Podpečan
Mathieu Roche
Hugo Alatrista Salas
Ivica Slavkov
Jovan Tanevski
Nejc Trdin
Bernard Ženko

Table of Contents

Data, Not Dogma: Big Data, Open Data, and the Opportunities Ahead

David J. Hand[1,2]

[1] Imperial College, London
[2] Winton Capital Management

Abstract. Big data and open data promise tremendous advances. But the media hype ignores the difficulties and the risks associated with this promise. Beginning with the observation that people want *answers to questions*, not simply data, I explore some of the difficulties and risks which lie in the path of realising the opportunities.

1 Introduction

Everyone at this meeting will be fully aware of the power conferred by the ability to extract meaning from data: you've all been doing just that for years. Perhaps, like me, for many years. But suddenly the world seems to have woken up to the potential as well. The phrases 'big data' and 'open data', two particular manifestations of this sudden dawning of awareness, are just two terms which appear to be cropping up everywhere.

Big data, obviously enough, refers to massive data sets, but quite what is meant by 'big' depends on the context. 'Big', in any case, is likely to grow over time. However, one quick definition is that it is a data set which is too large to fit into the computer's memory in one go.

Open data are data which are released to the public. This will often be official data, collected, for example, by a National Statistical Institute, but the term can also refer to scientific data. The open data movement in science is being parallelled by other similar initiatives, such as open access publishing.

I could give countless examples of the wonderful things that big data and open data are said to promise, but here are just two.

From McKinsey an example on big data: 'we are on the cusp of a tremendous wave of innovation, productivity, and growth, as well as new modes of competition and value capture - all driven by big data as consumers, companies, and economic sectors exploit its potential' [10].

And from Stephan Shakespeare's review of public sector information in the UK, published earlier this year [14] a comment on open data: 'from data we will get the cure for cancer as well as better hospitals; schools that adapt to children's needs making them happier and smarter; better policing and safer homes; and of course jobs. Data allows us to adapt and improve public services and businesses and enhance our whole way of life, bringing economic growth, wide-ranging social benefits and improvements in how government works ... the

A. Tucker et al. (Eds.): IDA 2013, LNCS 8207, pp. 1–12, 2013.

new world of data is good for government, good for business, and above all good for citizens'.

From these two examples we see a wonderful vision of the future. Unfortunately, I suggest, we also see a vision rather short on content and reality.

My aim in this paper is to look more closely at that vision. To suggest that, perhaps, all is not quite as wonderful as the quotations above suggest.

2   No-One Wants Data

I think a good place to start is to point out the awkward fact that no-one, or to be more precise, very few people, want data. In general, *people are not interested in data*. What people want are *answers*. Data are of value only to the extent that they can lead to answers, just as iron ore is of value only if we can extract the iron from it.

I think that's quite a nice analogy: data are the raw material from which we aim to extract information, meaning, and answers. I almost said, the raw material which we have to *process* to turn into something useful. But 'data processing' has rather different overtones. It generally refers to the manipulation - the sorting, searching, and juggling with data. All rather trivial exercises compared with the *analysis* through which information is extracted. Analysis - as in the phrase *intelligent data analysis* - is what we are really concerned with. The extraction of information from data generally requires more than trivial exercises of data manipulation. It requires us to apply the blast furnaces of statistical inference, machine learning, pattern recognition, data mining, and so on.

This distinction between what I mean by processing and by analysis is illustrated by the fact that, for most problems, we are not really interested in describing or summarising the data we have available. What we usually want to do is to use the data as the basis for an inference: perhaps about the future, perhaps about a larger population from which those data have been taken, perhaps about a counterfactual which might have happened, or perhaps for some other reason. Such inferences will increase our understanding and enable us to make better decisions. Such inferences go far beyond the mere 'processing' of the data. Inference from the data we have to the data we *might have had* or *might have in the future* is a non-trivial exercise which requires very deep theory.

3   Big Data and Statistical Models

Some have suggested that the new world of big data will sweep away the need for clever theory. In 2008 Chris Anderson, then editor of *Wired* magazine wrote an article called 'The end of theory: the data deluge makes scientific method obsolete'. It wasn't even phrased as a question. He began his article with the famous comment by the late George Box: *all models are wrong, but some are useful*. He also quoted Peter Norvig, Google's research director, extending Box's comment: *all models are wrong, and increasingly you can succeed without them*. Anderson wrote 'Out with every theory of human behavior, from linguistics to sociology. Forget taxonomy, ontology, and psychology. Who knows why people do what they do? The point is they do it, and we can track and measure it with unprecedented fidelity. With enough data, the numbers speak for themselves.'

But I'm afraid that's wrong. The numbers don't speak for themselves. They only speak if they have a model which can, metaphorically, turn them into sounds.

Underlying this misunderstanding is the fact that there are two kinds of model. They go under various names, but here I shall call them *substantive* and *empirical*.

Substantive models are based on substantive theory. That is, they are based on a theory about the mechanism or process underlying the phenomenon. My guess is that this is what Anderson means when he thinks of a 'model'. A familiar example is the physics model relating the distance a dropped stone has fallen to the time it has been falling: the inverse square law of gravitational attraction tells us that the distance is proportional to the square of the time. Substantive models are simplifying summaries of empirical relationships - stripping out the superfluous - in this case air resistance, for example. Statistical methods are needed when building substantive models so that the values of parameters can be determined (the value of g in the falling stone example), and so that models can be tested, evaluated, and chosen.

In contrast, empirical models are based solely on data. They are statistical summaries, which capture the main features of a data set. Like substantive models, they strip out the irrelevant detail, but they are not based on an underlying substantive theory relating the variables in the model. Typically empirical models are constructed with a particular aim in mind. A good example would be a logistic regression model to predict which applicants are likely to default on a bank loan. We'll feed in whatever variables we think might be relevant, and construct our model based on a data set consisting of these variables and the observed outcomes - default or not default - of previous customers. It's unnecessary to have any psychological theory in order to construct the model. We simply know that, in the past, people with certain combinations of characteristics have had a higher or lower propensity to default, and we assume that similar people will behave in a similar way in the future - and our logistic regression model aims to capture this. Of course, there are subtleties. Even if we have a very large data set it does not mean we can ignore issues of overfitting (just how many parameters are there in the model?) or nonstationarity (has a dramatic slump in the economy meant that people are behaving differently or that different types of people are applying for loans?). In fact, as we will see below, very large data sets may make us vulnerable to additional dangers.

So what I think Anderson really meant when he said 'we don't have to settle for models at all' is that to make predictions we don't have to have substantive models.

Now it's not as if empirical models are new. I picked credit scoring as my empirical modelling example above partly because it's an area I know a lot about, and partly because there is a long history of highly successful empirical modelling in this area, going back to the 1960s. Bruce Hoadley [8] gives a very nice description of how Fair Isaac Corporation developed highly sophisticated empirical credit scoring models decades ago. But with the increasing availability of large data sets the possibilities of empirical modelling have suddenly been thrust to the fore. You will doubtless be familiar with the following examples:

- Nate Silver's correct prediction of the winning candidate in 49 of the 50 states in the US presidential election in November 2008, and his correct prediction of all 50 states in the 2012 US presidential election;
- Google Translate, which does not base its translation on grammatical parsing and deconstructing sentences, but rather on statistical matching between massive corpora from the two languages - the originals were based on UN documents, since it publishes in six official languages;

It is certainly clear from these examples that we can do wonderful things with empirical models. But that does not mean that substantive models are obsolete. In particular, while empirical models are very useful for prediction, they are essentially useless for enlightening us about underlying mechanisms. They give no insight into why variables interact in the way they do. This is in stark contrast to substantive models, based on some underlying theory. Substantive models are the core of understanding.

Substantive models also permit generalisation to other contexts in a way which empirical models do not. Having devised our model saying that the force between two objects is inversely proportional to their distance, we can model the moon's orbit, the apple's fall, and the shape of the galaxy. In fact I picked that last example deliberately, since the curious pattern of rotation of stars in galaxies tells us there must be more to the universe than merely the matter and energy we can see. Simply modelling the rotation empirically will not tell you about the existence of dark matter. If you do model the rotational data, all you will produce is an equation which describes how the galaxies behave, and gives no insights into why they behave that way or whether it clashes with what you might have expected.

I have previously described this weakness of empirical models in the context of credit scoring. Empirical models, by definition, are built only on the available data. If we are building a model to predict future default probability of loan applicants, and our data were collected during a benign economic period, we might expect its performance to deteriorate if the economic situation were to change. In fact, there is a risk of a 'cliff-edge' effect [2,7], in which changing circumstances provoke a catastrophic change in performance. Substantive models are less vulnerable to such changes (which does not mean they are unaffected, or that any effect is necessarily minor) because they are modelling at a deeper level: at the relationships between variables. Others have also pointed out this danger of empirical models, though often in different terms (see for example [9]).

All of this means that, when Anderson says 'This is a world where massive amounts of data and applied mathematics replace every other tool that might be brought to bear', he is simply wrong. Massive data provides additional tools and additional approaches, but that does not mean the old tools are redundant or that (substantive) models are unnecessary.

Empirical models typically have other characteristics which can make them attractive. They often take the form of multiple applications of relatively simple component models. In the credit scoring example, the large population of applicants is divided into segments (as in a classification tree) and separate models (typically models linear in the characteristics describing the applicants) are constructed for each segment. Such a model is very easy to describe, even to the statistically unsophisticated. The whole structure is vastly more accessible than an elaborate model based on (for example) a belief network summarising the relationships between multiple interacting characteristics.

Random forest classifiers provide another example. Here many very simple component classifiers are added to give overall estimated probabilities of class membership. And in Google Translate, the simple operation of matching text strings lies at its core.

I have a basic theorem about statistical analysis. This is that *sophisticated and sensitive use of simple models, applied with deep understanding of their properties and limitations, is more effective than blind application of advanced methods*. Empirical models of the kind I have just described, involving very large numbers of very simple component models, are an illustration of this.

I'd like to conclude this section by coming back to the comment by George Box, that all models are wrong, but some are useful, and note the fact that, *by definition*, all models are wrong. The point is that all models must be simplifications, and hence wrong. You will have noticed that, in my definitions of both substantive and empirical models above I commented that they stripped out the superfluous and removed the detail. As Rosenblueth and Wiener put it in 1945: 'No substantial part of the universe is so simple that it can be grasped and controlled without abstraction. Abstraction consists in replacing the part of the universe under consideration by a model of similar but simpler structure.' Abstraction is a necessary and valuable aspect of modelling, whether it takes the form of ignoring minor fluctuations in data, as in empirical modelling, or of ignoring minor influences and higher order effects, as in substantive modelling.

I am reminded of the one paragraph story *On Exactitude in Science* by Jorge Luis Borges. This describes how cartographers, dissatisfied with the accuracy of successively larger scale maps, eventually constructed one which was of the same size as the empire, matching it point for point. Later generations recognised that this map was useless and let it be destroyed by the elements: '...in the Deserts of the West, still today, there are Tattered Ruins of that Map....'.

This illustration suggests we might go so far as to say that models *must be wrong if they are to be of any use*. But they have to be wrong in the right way, and that's why data analysis is a skilled profession.

4 Big Data

In the opening, I suggested that one definition of big data is that it is a data set which is too large to fit into the computer's memory in one go. There are others. Big data might be

- data which are too extensive to permit iterative methods: one pass analysis is necessary (streaming data would often fall into this category);
- 'big data refers to datasets whose size is beyond the ability of typical database software tools to capture, store, manage, and analyze' [10];
- 'high volume, high velocity, and/or high variety information assets that require new forms of processing to enable enhanced decision making, insight discovery and process optimization' [3];
- 'a data set is large if it exceeds 20% of the RAM on a given machine, and massive if it exceeds 50%' [1].

I had planned to give examples of big data, but I realised that this is redundant. Countless examples are readily at hand by googling 'big data', and the Wikipedia entry lists many application areas[1].

It might seem pretty obvious from these examples, and from the quotes I opened with, that big data is a new phenomenon. Or, at least, so one would think. But one would be mistaken. Some thirteen years ago, Hand et al listed [6]:

- Wal-Mart, with over 7 billion transactions annually;
- AT&T carrying over 70 billion long distance calls per year;
- Mobil Oil aiming to store over 100 terabytes of data;
- the NASA Earth Observing System, which was projected to generate around 50 gigabytes of data per hour.

Now it may be true that modern big data sets are even larger than those, but surely those would be large by anybody's standards?

The uncomfortable fact is that the phrase 'big data' is just a media rebranding of what was formerly known as 'data mining': the application of data analytic tools to extract meaningful or valuable information from large data sets. Other phrases describing very similar ideas, though often in different contexts, have also come and sometimes gone. 'Business analytics' and 'business intelligence' are examples.

At first glance, it might look as if big data will require novel analytic tools. This is partly true. Some problems (e.g. anomaly detection) require elaborate searches, and large enough data sets make this infeasible. Other problems are concerned with deeper inferential issues. With a large enough data set, any slight quirk in the underlying data generating mechanism, no matter how small and inconsequential in practical terms, is likely to be statistically significant. Large data sets also invite multiple testing - and as you all know, if you carry out enough tests you can guarantee finding some significant values. This problem was the motivation behind the development of false discovery rate: the strategy here, and one which can be adopted more widely, is to say 'if we cannot solve this problem, is there a related one which we can solve and which will shed light on what we want to know?'

In contrast, as well as motivating deep thought about inferential issues, the tools used for analysing very large data sets are often very simple (as my opening

---

[1] http://en.wikipedia.org/wiki/Big_data

examples illustrated). This is partly of necessity - a complex nonlinear model would take too long to fit, but a collection of linear models (each merely involving inverting a matrix) would be feasible. But it is also partly because the data are first split into subsets. The Fair Isaac credit card case illustrates this. As does MapReduce. In both cases a divide and conquer strategy is being applied: the biggest army can be defeated if you tackle them one at a time.

Sometimes this splitting goes further, and takes the form of screening. That is, what at first appears to be a truly massive data set is reduced to manageable proportions by a process of discarding much of it. For example:

- Graham (2012) gives the example of using Twitter data to understand the London riots from 2011. He says 1/3 of UK internet users have a Twitter profile, a small proportion of those produce the bulk of the content, and about 1% of that group geocode their tweets.
- the Large Hadron Collider at CERN produces about one petabyte of data per second. An online filter system reduces this by a factor of about 10,000, and then further selection steps reduce it by a further factor of 100.

It is true that big data provide us with analytic opportunities. These come in various shapes and forms. One is the possibility of detecting deviations, anomalies, or structural characteristics which are so small they would be imperceptible, or swamped by noise in anything but a large data set. The financial markets illustrate this. The efficient market hypothesis essentially says that it is not possible to predict the way they will move. In fact, as the success of CTA hedge funds shows, this 'hypothesis' is not true - but it is *very nearly true*. The ratio of signal to noise is tiny, so that large data sets are needed to detect a signal with any degree of reliability. The search for the Higgs boson, already mentioned, is another example.

Much big data is generated as a consequence of automatic data capture arising as a side effect of other human activity, such as credit card transaction data, supermarket purchase pattern data, web search data, phone call data, and social network data, using tools such as Facebook and Twitter. Such data sets allow novel insights into how people behave. But all of these examples also have another characteristic in common - they are *streaming* data.

Streaming data are data which simply keep on coming. Certainly batches of the data can be stored for later analysis, perhaps for exploration or model building, but often this is also associated with an additional requirement for on-line analysis. Credit card fraud and cyber-security illustrate this: the models can be built off-line, but they must be applied on-line, to detect fraud as it is attempted, and the models really need to automatically update and evolve if they are to be useful.

As well as novel analytic challenges, big data can also pose novel data quality issues. Close manual examination of a billion data points is impossible, so that a computer is a necessary intermediary. One can look at the data through the lens of a particular summary statistic or plot, but no finite number of these can cover all possible ways in which errors can occur. Particularly pernicious in the big data context is the possibility of selection bias, because it is often disguised. As will be evident, big data most commonly arise as observational data, not collected from a deliberately designed experiment. Observational data can all too easily fall prey to all sorts of selection effects. One sometimes hears the phrase 'natural experiment' used to describe a situation where the researcher has been unable to specify the conditions under which the observations are made, but where these conditions are determined by the natural course of events. If you hear this phrase you should be very suspicious!

Here are four examples of selection problems:

- synchronisation issues arise in financial market data and other areas, and the effects can often be very subtle. Even for assets traded on the same day problems can arise. Suppose that two assets are being traded, one of which trades much more frequently than the other. Then the last trade price of the one which trades more frequently is likely to be taken nearer the end of the day than the other, so that its price encapsulates more information (about anything that happened between the last trade prices of the two assets);
- self-selected samples in surveys. This is especially problematic with web surveys, where you may not know who responded, or even how often they responded, let alone what group of people are more likely to respond - other than that those with a vested interest may do so. My favourite example in this area is the (imaginary!) case of the magazine survey which asks the one question 'do you reply to magazine surveys' and, receiving no negative responses, believes that all its readers reply to such things;
- between June and December 2002 a smoking ban was imposed in all public spaces in Helena, Montana. During these six months, the heart attack rate dropped by 60%, increasing again when the bad was lifted. It looks like a clear argument for banning smoking, but one can only tell if that is the case if background changes in rates are taken into account.
- the effectiveness of speed cameras is another illustration of the dangers of relying on purely observational data, free from careful experimental design. For obvious reasons, such cameras are placed at accident black spots, where the rates are highest. But doing that invites the phenomenon of regression to the mean to come into effect: we are likely to see an apparent reduction in accident rates at those sites, purely by chance, even if the cameras are totally ineffective. (In fact, careful statistical analysis shows that cameras do have a beneficial effect, but much smaller than the raw figures would suggest.)

5   Open Data

In [5], on the topic of open data, I wrote: 'But the word open is a bit of a wriggler. It is a word which has a huge number of synonyms in different contexts. In some contexts, the words transparent, evident, explicit, fair, honest, public, disclosed, unconcealed, and sincere are synonyms for open, and presumably these are the sorts of things the government had in mind. On the other hand, the words vacant and empty are synonyms in other contexts, as are indefensible, unprotected, unrestricted, artless, and blatant.' Given that, it is hardly surprising that, like big data, open data has various definitions. One is: 'a piece of data is open if anyone is free to use, reuse, and redistribute it - subject only, at most, to the requirement to attribute'.

I have already mentioned the UK government's open data initiative. There are various claims made for it, including:

- that it enables accountability. That must surely be true. If people can actually look at the raw data, they can see whether people are doing the job they are claiming to do, and if they are doing it properly;
- that it empowers communities. The argument here is that it enables people to see where actions are needed, and how effective they are;
- that it drives economic growth. This is more subtle. The argument derives from the old adage that knowledge is power. Information, data, can be used to produce new products and services, and to make older systems more efficient. A McKinsey report estimated that, if used effectively, public data assets would benefit the European economy to the tune of some quarter of a trillion Euros per year ([10], p8).

To make this concrete, here are a couple of examples of open data sets which are already available for you to download from the web. (You can download similar data sets for the US from http://www.data.gov/ and for the UK from http://data.gov.uk.)

- *the COINS database.* COINS stands for Combined Online Information System, and it is a database of UK Government expenditure. These data are widely used in government, to produce reports for Parliament, and also by the Office for National Statistics. COINS raw data were initially released in June 2010, and further releases have been made since then. To give you an idea of the amount of data released, this data set alone is of the order of 44Gb.
- *crime maps.* Originally trialled in Chicago, the advent of the web has permitted the creation of regularly updated maps showing the location and date of crimes. From the police perspective, crime maps enable better decisions, better targeting of resources, and improved tactics. From the public's perspective, the maps enable citizens to identify risky areas to avoid, and to demand more police action if necessary. From May of 2012, the UK public will also be able to see what action or outcome has occurred after a crime has been reported

If all of this looks as if it's too good to be true, then it might well be. The unfortunate truth is that some data has the potential to do harm. There are privacy and confidentiality issues. Individual medical or banking records should certainly not be released. Recent high profile cases of confidential data relating to US national security being released to the public illustrate the concern.

Academic and other researchers usually require statistical data rather than the individual records, so the data are condensed and summarised. Unfortunately, this does not mean that individual information cannot be obtained from them. One strategy is to make multiple separate queries of a statistical database, cross-classifying the results. In other cases, individuals might be identified because they are the only people with a certain combination of attributes. Even more seriously is the jigsaw effect, whereby different databases are cross referenced. This can be extraordinarily powerful: if you know just the sex, date and year of birth, and the city of someone in the US, then 53% of the US population can be uniquely identified. There have been some recent widely publicised examples of the power of cross-matching databases:

- on 3rd August 2006, to stimulate research, and following the principle of crowdsourcing, AOL posted 20 million search queries from some 650,000 users of its search engine. From the perspective of a data geek like myself, this was a wonderful resource. But the point about being a data geek is that you are interested in finding things in the data. Although AOL had attempted to anonymise the release, some people were quickly identified.
- on 2nd October 2006, the online film rental service Netflix released 100 million records of film ratings, again after seeking to anonymise the records. In contrast to AOL, here the aim was not simply to give geeks something to play with, but to serve as data for a competition (with a million dollar prize), in which competitors were invited to develop a superior film recommendation algorithm. Once again, however, researchers soon found that individuals could be identified on the basis of very little extra information (from other sources) about them [13].

Other problems arise from lost data: almost every week it seems that the media describe a disc, memory stick, or laptop containing confidential data has gone astray.

Lurking beneath all of this is the disturbing fact that *the data will be wrong*. I don't mean all of it, of course, but merely some of it: no large collection of data should be expected to be free of errors. If the data describe people, then such errors can be critically important.

To conclude this section, I thought I would give an example of how open data, while a wonderful aspiration, can create problems of its own. My example refers to the Crime mapping data mentioned above. Direct Line Insurance conducted a survey which showed that recording and publishing crime statistics, coupled with the creation of crime maps, may be making people less willing to report antisocial behaviour in their neighbourhood, fearing it could have a detrimental impact on local house prices [12]. The survey found that 11% of respondents claim to have seen but not reported an incident 'because they were scared it would drive away potential purchasers or renters'. I should make a disclaimer here: this was an internet poll, and we have already seen these may not be entirely reliable.

6   Conclusion

Big data, open data, the promise they bring, and the awareness shown by our politicians and the public are indeed very exciting developments. The opportunities are very great - not least for the delegates at this conference, who understand the promise and have the skills to take advantage of the ore that the data represent. But all advanced technologies have risks as well as benefits, and those based on sophisticated data analysis are no exception.

Particular risks here are

- extracting useful information from large data sets requires considerable expertise. It is all too easy to draw incorrect conclusions. This is especially so if the data are subject to unquantified selection pressures and are of unknown quality;
- the need for skilled analysts presents its own problems. Although recent figures suggest the number with appropriate skills being trained at our universities is beginning to increase, decades of under-supply, during which the number of competing job opportunities for skilled data analysts has increased dramatically means that there is considerable catching up to do;
- data, by itself, no matter how much of it there is, cannot necessarily answer your question. Data collected for a different reason, with a different question in mind, may be useless for answering your question;
- often a small, targetted, quality controlled, and carefully designed experiment will generate a small data set which is better for answering your question than gigabytes of irrelevant material.

Big data and open data have great promise, but we should approach them with our eyes open.

References

1. Emerson, J.W., Kane, M.J.: Don't drown in data. Significance 9(4), 38–39 (2012)
2. Hand, D.J.: Mining the past to determine the future: problems and possibilities. International Journal of Forecasting 25, 441–451 (2008)
3. Gartner (2012), http://www.gartner.com/DisplayDocument?id=2057415&ref=clientFriendlyUrl
4. Graham, M.: Big data and the end of theory (2012), http://www.guardian.co.uk/news/datablog/2012/mar/09/big-data-theory?INTCMP=SRCH
5. Hand, D.J.: The dilemmas of open data. In: Herzberg, A.M. (ed.) Statistics, Science, and Public Policy XVII: Democracy, Danger, and Dilemmas. Queen's University, Canada, pp. 67–74 (2013)
6. Hand, D.J., Blunt, G., Kelly, M.G., Adams, N.M.: Data mining for fun and profit. Statistical Science 15, 111–131 (2000)
7. Hand, D.J., Brentnall, A., Crowder, M.J.: Credit scoring: a future beyond empirical models. Journal of Financial Transformation 23, 121–128 (2008)
8. Hoadley, B.: Statistical Modeling: The Two Cultures: Comment. Statistical Science 16, 220–224 (2001)

9. Lucas, R.: Econometric policy evaluation: a critique. Carnegie-Rochester Conference Series on Public Policy 1, 19–46 (1976)
10. Manyika, J., Chui, M., Brwon, B., Bughin, J., Dobbs, R., Roxburgh, C., Byers, R.H.: Big data: the next frontier for innovation, competition, and productivity (2011), http://www.mckinsey.com/insights/business_technology/big_data_the_next_frontier_for_innovation
11. Rosenblueth, A., Winer, N.: The role of models in science. Philosophy of Science 12, 316–321 (1945)
12. Timmins, N.: Crime maps 'hit reporting of crime'. Financial Times (July 13, 2011)
13. Narayanan, A., Shmatikov, V.: How to break anonymity of the netflix prize dataset. Computing Research Repository cs/0610105 (2006), http://arxiv.org/abs/cs/0610105
14. Shakespeare, S.: Shakespeare Review: An independent review of public sector information (2013), https://www.gov.uk/government/publications/shakespeare-review-of-public-sector-information

Computational Techniques for Crop Disease Monitoring in the Developing World

John Quinn

Department of Computer Science, Makerere University
P.O. Box 7062, Kampala Uganda

Abstract. Tracking the spread of viral crop diseases is critically important in developing countries. It is also a problem in which several data analysis techniques can be applied in order to get more reliable information more quickly and at lower cost. This paper describes some novel ways in which computer vision, spatial modelling, active learning and optimisation can be applied in this setting, based on experiences of surveying viral diseases affecting cassava and banana crops in Uganda.

1 Introduction

The problem of monitoring the spread of infectious disease among crops in developing regions is interesting in two regards. First, it is of critical practical significance, as the effects of crop disease can be devastating in areas where one of the main forms of livelihood is subsistence farming. It is therefore important to monitor the spread of crop disease, allowing the planning of interventions and early warning of famine risk. Second, it provides an example of the scope of opportunity for applying novel data analysis methods in under-resourced parts of the world.

The standard practice currently in a country such as Uganda is for teams of trained agriculturalists to be sent to visit areas of cultivation and make assessments of crop health. A combination of factors conspire to make this process expensive, untimely and inadequate, including the scarcity of suitably trained staff, the logistical difficulty of transport, and the time required to coordinate paper reports. Although computers remain a rarity in much of the developing world, smartphones are increasingly available: for example they account for 15-20% of all phones in Kenya, projected to be at 50% by the end of 2015 [2], and there are 8 million mobile internet subscribers [6] in a country with population of 41 million. Among other benefits, the prevalence of mobile computing devices and mobile internet makes it easy to collect different types of data, and in new ways such as crowdsourcing. Once data is collected electronically, this opens up opportunities to apply computational techniques which allow the process of crop disease survey in such an environment to be reinvented entirely.

We outline here three ways in which novel data analysis techniques can be used to improve the speed, accuracy and cost-efficiency of crop disease survey, using examples of cassava and banana crops in Uganda. After briefly discussing the mobile data collection platform we have implemented for this purpose (Section 2),

A. Tucker et al. (Eds.): IDA 2013, LNCS 8207, pp. 13–18, 2013.

we describe automated diagnosis of diseases, and image-based measurement of disease symptoms (Section 3), possibilities for incorporating spatial and spatio-temporal models for mapping (Section 4), and ways in which survey resources can be used optimally by prioritising data collection at the locations that the spatial model determines to be most informative (Section 5). These methods are currently being trialled with collaborators in the Ugandan National Crop Resources Research Institute, which specialises in cassava disease, and the Kawanda Agricultural Research Institute, which specialises in banana disease.

2   Mobile Data Collection

We implemented a system for collecting crop disease survey information with low cost (under 100 USD) Android phones, based on the Open Data Kit [3]. This provides a convenient interface for digitising the existing forms used by surveyors, with the ability to also collect richer data including images and GPS coordinates. Data collected on this system can be plotted on a map in real-time, see for example `http://cropmonitoring.appspot.com`. Clearly there are a number of immediate benefits from simply collecting data on a phone instead of paper, in that costs are reduced since the time needed to do data-entry and print paper forms far outweighs the costs of the phones and data, and results are immediately available. It also means that the survey can be conducted without experts being required to travel to the field; images can be collected and assessed remotely. More importantly to the purposes of this discussion, however, it allows data analysis methods to be applied which have the potential to fundamentally change the way in which the survey is conducted.

3   Automated Diagnosis and Symptom Measurement

Since the collection of survey data with mobile devices can include images of crops, removing the requirement for experts to be physically present to carry out inspection, we next focus on automating the judgements that those experts make based on images of leaves and roots. A typical national-scale survey of cassava disease in Uganda, for example, would include judgements about disease status and levels of symptoms on around 20,000 plants. The automation of judgements on this quantity of images constitutes a considerable saving of time and resources. With sufficient labelled training data this is a feasible problem for many diseases with clearly visible symptoms, and we can therefore collect data more rapidly and at lower cost. Automatic image-based diagnosis of crop diseases from leaf images is an active field [10,11,7], though little previous work has focused on crops grown primarily in developing countries.

Symptoms which need to be assessed for cassava include the extent of necrosis of the roots. It is also useful to count the number of whiteflies found on the leaves, as these are the vectors for multiple viruses. Figure 1 shows the ways in which we can carry out these measurements using computer vision. Assessment of roots is currently done in the manual survey by assigning root samples to

Fig. 1. Automated symptom measurement. Left: cassava root with necrotisation caused by cassava brown streak disease; center: classification of pixels to measure proportion of necrotisation; right: whitely count on cassava leaf.

Fig. 2. Banana leaf image patches. Left: healthy leaf; center: banana bacterial wilt; right: black sitagoka disease.

one of five categories, from completely healthy to completely necrotised. The main problem with this process is that the intermediate grades are easily confused; automating the process with image processing leads to more accurate and standardised results, removing the variability caused by different surveyors. Counting whiteflies on leaves is an infuriating and slow task for surveyors. The underside of a cassava leaf might have hundreds of these small, mobile insects, hence accurate counts are not feasible. In image processing terms, however, this is not a difficult problem, being essentially a form of blob detection.

Identification of viral diseases from leaf images is also possible given labelled data for training a classification model. Figure 2 shows examples of a healthy leaf surface and two diseases common in Uganda, banana bacterial wilt and black sitagoka disease. We have found that classification based on colour histogram features gives good results, though the incorporation of texture features is likely to improve this further. We have had similar experiences with diagnosis of cassava diseases from leaf images [1].

We have also found that with such straightforward classification techniques, it is possible to implement this process directly on the phone being used for the survey for real-time feedback. Figure 3 shows how the system works when these elements are combined. Capturing a cassava leaf image on the phone allows us to obtain an immediate diagnosis, which is uploaded to a server and plotted on a map online.

4   Incorporating a Spatial Model

Models of crop disease are used for understanding the spread or severity of an epidemic, predicting the future spread of infection, and choosing disease manage-

Fig. 3. Phone based survey with automated diagnosis. Left: mobile-phone based survey of cassava field; center: software on the phone detects cassava mosaic disease from leaf appearance; right: data collected with the phone is instantly uploaded to the web.

ment strategies. Common to all of these problems is the notion of spatial interpolation. Observations are made at a few sample sites, and from these we infer the distribution across the entire spatial field of interest. Standard approaches to this problem (reviewed in [9]) include the use of spatial autocorrelation, or Gaussian process regression [5]. Often the extent to which each plant is affected by disease is quantified in ordinal categories, in which case a spatial model which makes efficient use of the available data is Gaussian process ordinal regression [8]. Temporal dynamics can be added to these models, allowing forecasts to be made.

4.1   Combining Diagnosis and Mapping

The above tasks of estimating the density of an infectious disease in space and diagnosing that disease in individual cases (as in Section 3) are generally done separately. Informally, a surveyor may be aware of outbreaks of a disease in particular places or seasonal variations in disease risk, and they may interpret test results accordingly. But the diagnosis is not usually formally coupled with estimates of disease risk from the emerging spatial model.

The tasks of mapping disease density over space and time and of diagnosing individual cases are complementary, however. A "risk map" can be used to give a prior in diagnosis of an individual plant with a known location. In turn, the results of individual diagnoses can be used to update the map in a more effective way than simply making hard decisions about infection statuses and using summary count data for the update. The potential for combining maps and diagnosis in this way comes about with the possibility of performing diagnosis with networked location aware devices that can carry out the necessary calculations, as discussed in Section 3. In practice, this combined inference of spatial disease density and diagnosis in individual cases can be done with multi-scale Bayesian models, as described in [4]. By selecting an appropriate model structure, this can be done tractably even for very large numbers of individual plants as in the case of a national survey. This can improve both the accuracy of the risk map and of individual diagnoses, since the uncertainty in both tasks is jointly modeled.

5 Optimising Survey Resources

A probabilistic spatial or spatio-temporal model is useful not just in building up a picture of the disease map, but in knowing which locations would be most informative for collecting new data. While this was impossible in the traditional paper-based survey system, in which data entry would happen after the return of surveyors, the methodology described in this paper allows models to be learned in real-time as data is collected in the field. Therefore our models can be used to guide surveyors to collect more valuable data, holding fixed their budgeted number of samples.

This problem is essentially active learning, in which we prefer to collect data from locations in which the model has the lowest confidence. For example, in a Gaussian process model, we prefer to sample from locations where the density estimate has the highest covariance with the data already collected. This approach would be suitable for example in a crowd sourcing setting: if phones were given to agricultural extension workers across the country, and micro-payments are made to those workers in return for sending image data, it would be possible to adjust the levels of those payments based on location in order to use the budget optimally with respect to building an informative model.

When we attempt to direct the progress of a survey in which data collection teams are sent to travel around the country, the situation is a little different. There is a fixed travel budget, e.g. for fuel, and we cannot simply collect data from arbitrary locations on the map. Considering the constraints of being able to travel along a given road network with some budget, this optimisation problem is in general very complex. However, we can simplify this constraint somewhat by considering that in rural parts of the developing world, the road network is often sparse. This makes it reasonable to assume that survey teams will follow a set route, corresponding to a one dimensional manifold $\mathcal{R}$ within the spatial field. With a survey budget allowing k stops, we are interested in finding a set of points along $\mathcal{R}$ that maximise the informativeness of the survey. Under this constraint, optimisation is tractable with a Monte Carlo algorithm [8], where we recompute after each stop the optimal next sample location based on the spatial model given the most recent observation. This can also be done for multiple groups of surveyors simultaneously traveling along different routes.

6 Discussion

This paper has outlined various ways in which computational techniques can make crop disease survey more effective given tight resource constraints. It is an illustration of one of the ways in which data analysis can be used to address problems in the developing world, where we often wish to automate the judgements of experts who are in short supply, collect intelligence about socio-economic or environmental conditions from different, noisy data sources, or optimise the allocation of some scarce resource. Similar methods can be directly applied to the survey and diagnosis of human disease, for example, another active area of current work.

References

1. Aduwo, J.R., Mwebaze, E., Quinn, J.A.: Automated vision based diagnosis of cassava mosaic disease. In: Proceedings of the ICDM Workshop on Data Mining in Agriculture (2010)
2. Anonymous. The Arrival of Smartphones and the Great Scramble for Data. The East African (May 25, 2013)
3. Hartung, C., Lerer, A., Anokwa, Y., Tseng, C., Brunette, W., Borriello, G.: Open Data Kit: Tools to build information services for developing regions. In: Proceedings of the 4th ACM/IEEE International Conference on Information and Communication Technologies and Development (2010)
4. Mubangizi, M., Ikae, C., Spiliopoulou, A., Quinn, J.A.: Coupling spatiotemporal disease modeling with diagnosis. In: Proceedings of the International Conference on Artificial Intelligence, AAAI (2012)
5. Nelson, M.R., Orum, T.V., Jaime-Garcia, R.: Applications of geographic information systems and geostatistics in plant disease epidemiology and management. Plant Disease 83, 308–319 (1999)
6. Communications Commission of Kenya. Quarterly sector statistics report (July-September 2012)
7. Perez, A.J., Lopeza, F., Benlloch, J.V., Christensen, S.: Colour and shape analysis techniques for weed detection in cereal fields. Computers and Electronics in Agriculture 25(3), 197–212 (2000)
8. Quinn, J.A., Leyton-Brown, K., Mwebaze, E.: Modeling and monitoring crop disease in developing countries. In: Proceedings of the International Conference on Artificial Intelligence, AAAI (2011)
9. van Maanen, A., Xu, X.-M.: Modelling plant disease epidemics. European Journal of Plant Pathology 109, 669–682 (2003)
10. Wang, L., Yang, T., Tian, Y.: Crop disease leaf image segmentation method based on color features. In: Computer And Computing Technologies in Agriculture. IFIP, vol. 258, pp. 713–717. Springer, Boston (2008)
11. Zhihua, Z., Guo, X., Zhao, C., Lu, S., Wen, W.: An algorithm for segmenting spots in crop leaf disease image with complicated background. Sensor Letters 8, 61–65 (2010)

Subjective Interestingness
in Exploratory Data Mining

Tijl De Bie

Intelligent Systems Lab, University of Bristol, UK
tijl.debie@gmail.com

Abstract. Exploratory data mining has as its aim to assist a user in improving their understanding about the data. Considering this aim, it seems self-evident that in optimizing this process the data as well as the user need to be considered. Yet, the vast majority of exploratory data mining methods (including most methods for clustering, itemset and association rule mining, subgroup discovery, dimensionality reduction, etc) formalize interestingness of patterns in an objective manner, disregarding the user altogether. More often than not this leads to subjectively uninteresting patterns being reported.

Here I will discuss a general mathematical framework for formalizing interestingness in a subjective manner. I will further demonstrate how it can be successfully instantiated for a variety of exploratory data mining problems. Finally, I will highlight some connections to other work, and outline some of the challenges and research opportunities ahead.

1 Introduction

Exploratory Data Analysis (EDA), introduced and championed by John Tukey in the seventies, has greatly affected the productivity of statistical analysis. With EDA Tukey stressed the importance of data for the *generation* of hypotheses through interactive exploration of the data by a human user, alongside its use for hypothesis *testing*. Since then EDA has only gained in importance: Data has become a primary source for new scientific discoveries (dubbed the *4th scientific paradigm*), insights in businesses and their customers, assisting government workers in the analysis of security threats, and satisfying the interests of end-users and consumers more generally.

1.1 The Concept of Interestingness in Data Exploration Tasks

EDA aims to provide insights to users by presenting them with human-digestible pieces of 'interesting' information about the data. While initially this was limited to simple statistics (such as the 'five number summary'), this was soon complemented with sophisticated methods such as Projection Pursuit (PP) [3], which aims to present the user with interesting low-dimensional projections of the data. Research into PP focused on the identification of an 'index' quantifying how interesting a projection is—this index was sometimes referred to as the *interestingness* of the projection [9].

A. Tucker et al. (Eds.): IDA 2013, LNCS 8207, pp. 19–31, 2013.

Today, the term *interestingness* and *Interestingness Measure* (IM) is still in use in certain areas of Exploratory Data Mining (EDM), in particular in the context of frequent pattern mining (and notably frequent itemset and association rule mining—see [5] for an excellent survey, and [13] for a survey of novelty-based IMs). However, under different names and guises (e.g. 'objective function', 'quality function', 'score function', or 'utility function'), the concept of interestingness remains central to all EDM prototypes, such as clustering, dimensionality reduction, community detection in networks, subgroup discovery, local pattern mining in multi-relational databases, and more.

1.2 Flavours of Interestingness

The early work by Friedman and Tukey was informed by actual experiments on human users and how they manually explore different possible data projections [3]. In a seminal paper on PP, Huber explicitly stated that "We cannot expect universal agreement on what constitutes an 'interesting' projection" [9].

This early focus on the user makes it remarkable that the vast majority of contemporary research tacitly attempts to quantify interestingness in an 'objective' way, ignoring variations among users. That is, for a specific EDM task, e.g. clustering, one attempts to propose a mathematical function of the clustering pattern—the IM—that quantifies how good the clustering is deemed to be. To be able to do this, one has no other choice than to design the IM for a problem setting faced or imagined, with a particular type of intended use or user in mind. It should be no surprise that this has led to an explosion in the number of possible IMs for each of the common EDM tasks, including association rule and frequent itemset mining, frequent pattern mining more generally, subgroup discovery and variants (e.g. exceptional model mining), dimensionality reduction, clustering, community detection, and multi-relational data mining.

It is not until Tuzhilin, Silberschatz, Padmanabhan, and colleagues that the term *subjective IM* was proposed. Their research was centred on association rule mining, formalizing subjective interestingness of a rule as a quantification of 'unexpectedness' or 'novelty' (see e.g. [18,16]). Their focus on association rule mining is not surprising given that the void between available IMs and practical needs was probably the widest for such types of pattern. Their approach relied on the concept of a 'belief system', formalizing the beliefs of the data miner, to which association rules can then be contrasted to determine their interestingness. Though conceptually groundbreaking in our view, the approach was still somewhat ad hoc and specific to association rules and related patterns, limiting the impact of this work.

In the last few years, a number of research groups have emphasized the need to assess or validate EDM patterns (see e.g. [6,8]). Often, this is done by comparing the found patterns with a 'background model' that represents 'random data'—any pattern that can thus be explained by this model can be dismissed as a fluke. Due to the required complexity of these background models, they are often specified implicitly by specifying permutation invariants, which allows one to randomize the data and to use empirical hypothesis testing to quantify to which extent the patterns can be explained by random chance. Much of this work was done (like the work by Tuzhilin et al.) in the area of frequent itemset and association rule mining.

This more recent line of work can be interpreted as being concerned with the quantification of subjective interestingness. Indeed, if the background model is related to the user's prior knowledge, then the assessment of a pattern is similar to the quantification of its subjective interestingness. The techniques also appear more flexible than those from Tuzhilin et al. Yet, their reliance on empirical hypothesis testing and background models specified in terms of permutation invariants has important disadvantages, such as lack of resolution, the inability to incorporate anything but relatively simple types of prior knowledge into the background model, the intractability of the empirical hypothesis testing approach on large datasets, and computational and conceptual difficulties in updating the background model with new information. As a result, this approach for assessing data mining results seems limited to their actual assessment with respect to relatively simple background models, rather than as a generically applicable way to derive subjective IMs.

Finally, another related area of research attempts to quantify interestingness in terms of *actionability*. The actionability of a pattern is meant to quantify (often in monetary terms) the 'use' of a pattern for a particular purpose (e.g., a marketeer can be interested in identifying optimal package deals—the goal is to identify the best package optimizing profits). Clearly, if the purpose is known, this strategy is the right one. However, we consider such problems outside the scope of EDM, where the user is interested in exploring without a clear anticipation of what to expect or what to do with the patterns found.

1.3 Formalizing Subjective Interestingness

It thus appears that the notion of subjective interestingness, while its importance has been acknowledged, has been elusive to rigorous and comprehensive formal study. The aim of our research surveyed in this paper is thus to provide a positive and constructive answer to the following question:

> *Can we develop a theoretically rigorous while practically applicable framework for the quantification of subjective interestingness of patterns in EDM?*

We argue that this question is well-posed and can be approached in a mathematically rigorous way, with minimal and plausible assumptions on how human users interact with a data mining system. However, this can only be done if the data miner (the user) is an integral part of the analysis, considered as much as the data and the patterns themselves. More specifically, to understand what is interesting to a user, we will need to understand how to model the beliefs of that user, how to contrast a pattern with this model, and how that model evolves upon presentation of a pattern to that user.

2   A Mathematical Framework for Subjective Interestingness

In this section we will provide a high-level overview of the formal foundations of the framework as it has been developed so far.

2.1   Notation and Terminology

We first need to establish some notation and formally define a few concepts. Arguably the most important concept is the *data*, which we assume to be drawn from a set of possible values, to which we refer as the *data space*. We formally define and denote them as follows.

Definition 1 (Data and data space). *Given a set Ω called the data space, the data is an element $x \in \Omega$.*

Common examples of data are: a set of vectors in a vector space; a time series; a network; a binary matrix representing a market-basket database; a multi-relational database.

   We aim to define the notion of a pattern as generally as possible, as any kind of information that restricts the range of possible values the data may have. To avoid having to refer to the form and way in which this information is provided, we thus choose to define a pattern by specifying this restricted set of values the data may have, i.e. by specifying a subset Ω' of the data space: $\Omega' \subseteq \Omega$.

   More precisely, we will define patterns in terms of elements from a sigma algebra $\mathcal{F}$ over the data space Ω, thus defining a measurable space $(\Omega, \mathcal{F})$. Each measurable set $\Omega' \in \mathcal{F}$ then defines a pattern as follows.

Definition 2 (Pattern). *Let $\mathcal{F}$ be a sigma algebra over Ω, and thus $(\Omega, \mathcal{F})$ a measurable space. We say that a pattern defined by a measurable set $\Omega' \in \mathcal{F}$ is present in the data x iff $x \in \Omega'$.*

Common examples of a pattern are the specification of: a low-dimensional projection of the data; a clustering in the data; an itemset and the set of transactions in which it occurs (also known as a tile); a clique in a network.

Remark 1. The definition of a pattern is such that a *set of* patterns, defined by the sets $\Omega'_i \in \mathcal{F}$, is a pattern itself, defined by the set $\Omega' = \bigcap_i \Omega'_i \in \mathcal{F}$.

Remark 2. It is common to specify a pattern by specifying the value of a measurable function f (called a statistic) when evaluated on the data x, i.e. by stating that $f(x) = f_x$. Then Ω' defining the pattern is $\Omega' = f^{-1}(f_x) = \{x \in \Omega | f(x) = f_x\}$. In such cases we may also denote a pattern by the triple (Ω, f, f_x).

   Our intention is to define IMs as a subjective concept—i.e. depending on the user. To achieve this it seems inevitable to let IMs depend also on the a representation of the 'state-of-mind' of the human user. A promising avenue to do this, which we will pursue here, is to let it depend on the degree of belief the user attaches to each possible value for the data from the data space. We will approximate this belief state by means of a so-called background distribution:

Definition 3 (Background distribution). *A background distribution is a probability measure P defined over the measurable space* $(\Omega, \mathcal{F})$*, such that for a set* $F \in \mathcal{F}$*,* $P(F)$ *approximates the belief that the data miner would attach to the statement* $x \in F$*. The set of possible background distributions is denoted as* $\mathcal{P}$*.*

We can now formally define an IM as follows:

Definition 4 (Interestingness Measure (IM)). *An IM* I *is a real-valued function of a background distribution* P *and a pattern defined by* Ω'*, i.e.*

$$I : \mathcal{F} \times \mathcal{P} \to \mathbb{R}.$$

An objective IM is an IM I that is independent of the background model P, i.e. for which $I(P_1, \Omega') = I(P_2, \Omega')$ for any $P_1, P_2 \in \mathcal{P}$. A subjective IM is an IM that is not objective.

2.2 Summary of Theoretical Research Results

In our recent research we have laid a number of foundations for the quantification of subjective interestingness, roughly relying on the definitions above. We summarize the key results here.

Modelling the Data Mining Process. Our intended framework, so far most formally presented in [1], is based on a view of the data mining process illustrated in Fig. 1. In this model, the user is assumed to have a belief state about the data, which starts in an initial state, and evolves during the mining process.

When a pattern is revealed to the user, this reduces the set of possible values the data may have. This is reflected in an update of the user's belief state. We assume no more than that the update in their belief state is such that the user attaches no belief to values of the data excluded by the revealed patterns. This is illustrated in the top row of Fig. 1, for two patterns Ω' and Ω'' shown in the bottom row of Fig. 1: upon presentation of these patterns the belief the user attaches to values outside Ω' and subsequently also Ω'' becomes zero. Clearly, patterns are subjectively more informative to a user if they exclude regions of the data space to which the user attached a higher belief.

Continuing this process, the part of the data space to which a non-zero belief is attached shrinks, and the data miner would ultimately end up knowing the exact value of the data. Of course, in practice the user would stop the mining process as soon as they are satisfied with their understanding of the data, or as soon as their resources (e.g. time, or mental capacity to remember newly presented information) have run out.

While we postulate the existence of the user's evolving belief state, we acknowledge that in reality the data mining system will have incomplete knowledge of it at best. Thus, it will need to rely on an evolving proxy. This evolving proxy is the *background distribution* P, illustrated in the middle row of Fig. 1. Shortly below we will discuss how this can initially be obtained based on limited available information about the prior belief state, and how it can be updated after the presentation of a pattern to the user.

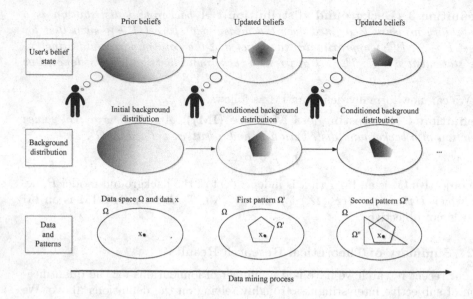

Fig. 1. The first row of figures illustrates diagrammatically how the belief state of the user evolves during the data mining process, the second row illustrates how our model of this belief state (i.e. the background model) evolves, and the third row illustrates the data space, the data, and the patterns revealed. Shading illustrates the degree of belief the user attaches to the data (first row) and the probability under the background model (second row). The three columns show: (1) the prior beliefs and initial background model, (2) the effect of revealing a first pattern Ω' on the user's beliefs (all we assume here is that the user will attach 0 belief to data values excluded by the shown patterns) and on the background distribution used as a proxy for the belief state (3) the same after revealing a second pattern Ω''.

Optimizing the Data Mining Process. Crucial in this model for data mining is that *the user is an integral part* of it, *as much as the data itself.* By doing this we can try and optimize its efficiency from a subjective user perspective. Indeed, the goal should be to pick those patterns that will result in the best updates of the user's belief state, while presenting a minimal strain on the user's resources. This means that the IM should be a trade-off between two aspects: the pattern's *information content* (which should quantify how useful it is to the user in updating her beliefs about the data), and its *descriptional complexity* (which should quantify how much resources are used to assimilate it).

In the following paragraphs we briefly discuss how each of these concepts can be usefully formalized.

Determination of the Initial Background Distribution. In an ideal world, the process is started by the user doing a 'brain dump' to inform the data mining system about all their prior beliefs about the data, formalized in the form of a background distribution. Of course, in reality this is impossible, and a lighter touch approach is needed to extract as much information as possible. The approach we have advocated is to let the user specify a number of expectations on the data, which can be formalized as:

$$\mathbb{E}_{x \sim P} \{f(x)\} = \hat{f},$$

where f is a statistic computing a property, of which the user expects the value to be equal to $\hat{f}$.

This type of expectations can be formulated at a specific detailed level, though in practice it is more useful to formulate large numbers of them in a more generic way. For example, in a binary matrix the user may have certain expectations on all row sums and column sums. In a network they may have expectations on the degrees of the nodes.

This information represents a constraint on the background distribution P, thus limiting the set of possible background distributions. Only in contrived cases will this suffice to uniquely determine it though. Thus, among the possible choices we need to select one. In [1] we laid out two arguments in favour of picking the distribution of maximum entropy among those not ruled out by the constraints. The most intuitive argument goes that the maximum entropy distribution is the most unbiased one, such that no undue information is injected into the process.

The Information Content of a Pattern. Recall that a pattern is defined in terms of the measurable subset Ω' from the data space to which it says the data belongs. The probability $P(\Omega')$ represents the probability for the event $x \in \Omega'$, i.e. the degree of belief the user attaches to this pattern being present in the data x. Clearly, if $P(\Omega')$ is small, this pattern is subjectively surprising and informative to the user. Reflecting this, and using formal arguments, in [1] we argued that

$$\text{InformationContent}(\Omega', P) = -\log\left(P(\Omega')\right)$$

is a meaningful and robust way to quantify the information content embodied by that pattern. Note that it depends on the pattern itself as well as on the background distribution, i.e. it is a subjective quantity. As pointed out in [1], in special cases this information content reduces to a p-value—i.e. it is strictly more general than statistical hypothesis testing based interestingness measures.

Updating the Background Distribution. Upon being presented with a pattern, the user will consciously or unconsciously update their beliefs about the data. Again in [1], we argued that a robust way of doing this is to simply condition the background distribution on the new domain Ω' (see Fig. 1, middle row). As a result, the beliefs attached to any value for the data outside Ω' become zero, and the probability of all values within Ω' (including the true value x for the data) is divided by $P(\Omega')$.

This provides one of the formal arguments in favour of quantifying the information content as $-\log\left(P(\Omega')\right)$: the larger this quantity, the larger the increase in probability of the data x under the background distribution.

The Descriptional Complexity of a Pattern. The final building block is the descriptional complexity of a pattern. While it is harder to give generic advice on how to quantify the complexity of a pattern, in practice it is often fairly intuitive to specify it, perhaps in a parametric way. For example, the specification of a set of items (e.g. in frequent itemset mining) will typically be more complex for a human user to assimilate if it contains more items. Similarly, the specification of the members of a community in a network is more complex if the community is larger. The dependency on the number of items or on the size of the community can be parameterized such that the user can adapt it to approximate her own notion of complexity best, and also allowing the user to zoom in or out to smaller or larger patterns.[1]

Trading Off Information Content with Descriptional Complexity. Assuming that the user is interested in gaining as much information about the data as possible within a bounded amount of resources, the data mining problem can be formalized as an optimization problem: maximize the total amount of information carried by the conveyed patterns, subject to an upper bound constraint on their cumulative descriptional complexity.

As we showed in [1], this problem can be reduced to a Weighted Budgeted Set Coverage (WBSC) problem, which is NP-hard but can be approximated to a constant factor $1 - \frac{1}{e}$ using a greedy method. Indeed, a revealed pattern defined by Ω' excludes ('covers') a part $\Omega \setminus \Omega'$ of the data space Ω from the set of possible values for the data x. The larger the probability $P(\Omega \setminus \Omega')$ (the 'weight' of the excluded subset), the larger the pattern's information content $-\log(P(\Omega'))$. Thus, to find a set of patterns that have maximal information content, one has to find patterns that jointly exclude a subset from Ω with maximal probability under the initial background distribution. This is done within a finite 'budget' of cumulative descriptional complexity.

An appealing feature of this greedy approach is that it is not only approximately optimal at the end but also any intermediate solution is approximately optimal (with the same approximation factor) given the budget already consumed. This continuous near-optimality of the greedy approach is a strong argument in favour of the iterative nature of the framework's view of the data mining process (Fig. 1), in which patterns are revealed one by one. Indeed, not only is iterative data mining convenient from a user's perspective, it is also as close to optimal as any other computationally efficient one-shot pattern *set* mining approach could be.

Conveniently, the greedy criterion to be used in selecting the next pattern is simply the ratio of the information content (given the current background distribution) and the descriptional complexity. Thus, it is this ratio that we put forward as a general subjective IM:

---

[1] A separate rationale for limiting the descriptional complexity of patterns is available for the special case when the information content is equivalent to a *p*-value. In those cases, limiting the complexity will effectively limit the number of hypotheses considered, controlling the adverse effects of multiplicity.

$$I(\Omega', P) = \frac{\text{InformationContent}(\Omega', P)}{\text{DescriptionalComplexity}(\Omega')}.$$

Remark 3. Although it is good news that the problem of finding the best set of patterns can be approximated well using a greedy algorithm, it should be noted that each greedy step can still be computationally expensive. Whether this is the case depends on the particular problem setting.

3   Instantiations of the Framework

While abstract, the developed framework is highly practical, and we believe it has the potential of altering the way EDM is being conducted. To illustrate this, we survey a few instantiations of this framework we have developed so far.

3.1   Tile Mining

The first instantiation concerns the tile mining problem. Assume we are given a rectangular binary matrix, representing e.g. a binary database or an attribute-value database. Then, a tile is defined as a set of rows along with a set of columns in this matrix such that all matrix elements in the intersection of these rows and columns are equal to one [4]. Tiles are similar to frequent itemsets, with as a difference the fact that also a set of supporting transactions is specified along with the itemset itself.

In the terminology of the current paper, a tile-pattern is therefore the specification that a tile is present in the data (the given binary matrix). This limits the set of possible values of this matrix to only those matrices that have ones in the intersection of the rows and columns defining the tile.

In practice, the interestingness of a tile is probably a trade-off between its dimensions as well as other properties such as the density of the rows and columns covered by the tile. The interestingness of a tile could thus be formalized in a multitude of ways, each of which could make sense in specific situations.

Our framework allows one to let the IM be dictated by the prior beliefs of the user—rather than having to design it directly. To demonstrate this we have taken a number of steps. In a first paper on this topic, we demonstrated the basic idea of the framework by deriving the IM subject to prior beliefs on the density of the rows and columns (the so-called row and column marginals) [2], soon afterward generalized toward noisy tiles [10]. More recently, we expanded the set of prior belief types to include also prior beliefs on the density of certain subsets of matrix entries (including rows, columns, and tile areas), on a cluster structure on the rows of the data, as well as itemset frequencies [11]. This allows for iterative data mining in a similar vain to [8].

3.2 Tile Mining in Real-Valued Data, and Iterative Bi-clustering

Our tile-mining work was generalised in [14,15], to include iterative mining of subjectively interesting bi-clusters from real-valued data matrices. In [14] the background distribution for real-valued data is also applied for the purpose of quantifying subjective interestingness of subgroups.

3.3 Clustering and Alternative Clustering

Another EDM task for which the number of IMs available is large and growing is data clustering. A few examples are the average squared distance to the nearest cluster centre in K-means clustering, a particular eigenvalue in spectral clustering, and the likelihood in mixture of Gaussians clustering.

After introducing a simple new way of formalizing a cluster pattern, we were able to apply our theoretical ideas to formalize the interestingness of a cluster pattern with respect to a specific type of prior beliefs, namely expectations on the mean and covariance of the data set [12]. Importantly, we demonstrated how the background distribution can be updated to account for the knowledge of a certain clustering pattern, allowing one to use the resulting method in an alternative clustering setup. In further work we intend to expand the set of prior belief types that can be used, and explore other clustering pattern syntaxes.

3.4 Multi-relational Data Mining

A last instantiation we wish to highlight is the development of a pattern syntax for multi-relational data (e.g. data in relational databases, RDF data and the semantic web, etc.). This work builds on the insights from Sec. 3.1, considerably broadening its applicability and developing sophisticated algorithms for mining multi-relational patterns.

The basic ideas were introduced in [19], in [20] the mining algorithm was substantially improved, and in [21] the pattern syntax, algorithm, and the IM were generalised to allow for also n-ary relations in the database.

4 Discussion

The novelty of the proposed framework lies mainly in its interpretation and novel use of pre-existing mathematical tools, including information theory, exponential family models, and convex optimization theory and duality. Although it has a number of striking links with other frameworks for data mining and machine learning, it is unique in its insistence on the central role of the data miner and subjectivity. Here we point out some similarities and differences with related frameworks.

The *Minimum Description Length (MDL)* principle [7] aims to detect regularities in data that allow one to describe it in the most compact way. Such regularities are roughly equivalent to what we call patterns in this paper.

The MDL principle has been successfully applied for the development of objective IMs for various data mining problems, notably for frequent itemset mining [17]. While mathematically the language of MDL is related to the language used in our framework (especially to the presentation we used in [1]), the emphasis and interpretation are very different. For example, MDL has inductive inference as its main goal, and human interpretability of the code is not required in principle. In contrast, the proposed framework is not concerned with inference at all, and interpretability is key (i.e. the descriptional complexity should be a reflection of a perceived complexity by the user). More crucially, the concept of a background model for the prior beliefs of a user is absent in MDL, such that IMs derived by the MDL principle are objective ones. Finally, as far as we know the reduction of the iterative EDM process to a WBSC problem, providing a solid algorithmic basis to the proposed framework, is absent from the MDL theory.

In the Introduction we already mentioned *hypothesis testing* based approaches to the assessment of EDM results, and the proposed use of (empirical) p-values as IMs, and we have highlighted some practical disadvantages of this approach. Other pioneering work that relies on exact hypothesis testing but that seems more restricted in terms of null hypotheses used includes [22,23]. The similarity of such approaches with the proposed framework is clear from the fact that for certain pattern types, the information content is equivalent to a p-value. However, it differs in being more general (the information content is not always equivalent to a p-value). More crucially, the null hypothesis (or more generally the background distribution) is given a precise interpretation by our framework, as the model for the prior beliefs of the user, and a mechanism for obtaining it and for understanding its evolution during the mining process is provided.

Mathematically, the proposed framework also seems related to *Bayesian inference*. However, an important conceptual difference is that the goal of Bayesian inference is to infer a model for the stochastic source of the data, whereas our goal is to help the user understand the data itself by unveiling interesting parts or aspects of it.

We thus believe the proposed framework may serve an important role as an alternative framework, with a unique focus on subjectivity and bringing the user into the equation. The first data mining methods that were inspired by it appear to support this belief.

5   Challenges and Opportunities Ahead

Our ultimate goal is to further develop these initial ideas with the ultimate aim of transforming the way EDM is done: from an expert-driven process requiring a thorough understanding of data mining techniques and processes, to a process where the data mining system thoroughly understands the needs and wishes of the user, presenting only what is interesting, in a subjective sense.

The instantiations developed so far illustrate the breadth of applicability as well as the practical usefulness of the simple theoretical foundations we proposed to this end. Nevertheless, significantly more work is needed to ensure a wide impact of this way of formalizing subjective interestingness: conceptually, algorithmically, as well as in terms of instantiating the ideas for actual data mining problems with the aim of making them more user-oriented.

Conceptually, fruitful research avenues may include considering alternative definitions of a pattern (to include more probabilistic notions), as well as alternative kinds of background models and models for how they evolve in response to the presentation of a pattern. Consideration of cognitive aspects may become important here as well.

Algorithmically, a characterization of problem types that lead to efficient algorithms is currently beyond reach but would be extremely valuable.

In terms of developing new instantiations: expanding the range of prior belief types that can be dealt with (along the way expanding the types of data), as well as developing it for new pattern types and syntaxes, are major challenges to which we invite the broader research community.

Finally, designing empirical evaluations (short of expensive user studies) for subjective IMs (or frameworks for designing them) is non-trivial by their very nature. The associated difficulty in publishing research on this topic thus poses a serious stumbleblock to this line of research. We therefore believe a broad discussion on how to evaluate research on subjective IMs is overdue.

Acknowledgements. I am heavily indebted to Kleanthis-Nikolaos Kontonasios and Eirini Spyropoulou, as well as to Jilles Vreeken and Mario Boley, who have made significant contributions to the development of these ideas. When I said 'we' or 'our' in this paper, this was usually meant refer to one or several of them as co-authors on a paper. This work was partially supported by the EPSRC project EP/G056447/1. Finally, I wish to express my gratitude to the organizers of IDA 2013 for inviting me to talk about this work.

References

1. De Bie, T.: An information-theoretic framework for data mining. In: Proc. of the 17th ACM SIGKDD International Conference on Knowledge Discovery and Data Mining (KDD) (2011)
2. De Bie, T.: Maximum entropy models and subjective interestingness: an application to tiles in binary databases. Data Mining and Knowledge Discovery 23(3), 407–446 (2011)
3. Friedman, J., Tukey, J.: A projection pursuit algorithm for exploratory data analysis. IEEE Transactions on Computers 100(9), 881–890 (1974)
4. Geerts, F., Goethals, B., Mielikäinen, T.: Tiling databases. In: Suzuki, E., Arikawa, S. (eds.) DS 2004. LNCS (LNAI), vol. 3245, pp. 278–289. Springer, Heidelberg (2004)
5. Geng, L., Hamilton, H.J.: Interestingness measures for data mining: A survey. ACM Computing Surveys 38(3), 9 (2006)
6. Gionis, A., Mannila, H., Mielikäinen, T., Tsaparas, P.: Assessing data mining results via swap randomization. ACM Transactions on Knowledge Discovery from Data 1(3), 14 (2007)

7. Grünwald, P.: The Minimum Description Length Principle. MIT Press (2007)

8. Hanhijarvi, S., Ojala, M., Vuokko, N., Puolamäki, K., Tatti, N., Mannila, H.: Tell me something I don't know: Randomization strategies for iterative data mining. In: Proc. of the 15th ACM SIGKDD International Conference on Knowledge Discovery and Data Mining (KDD), pp. 379–388 (2009)

9. Huber, P.: Projection pursuit. The Annals of Statistics, 435–475 (1985)

10. Kontonasios, K.-N., De Bie, T.: An information-theoretic approach to finding informative noisy tiles in binary databases. In: Proc. of the 2010 SIAM International Conference on Data Mining (SDM) (2010)

11. Kontonasios, K.-N., DeBie, T.: Formalizing complex prior information to quantify subjective interestingness of frequent pattern sets. In: Hollmén, J., Klawonn, F., Tucker, A. (eds.) IDA 2012. LNCS, vol. 7619, pp. 161–171. Springer, Heidelberg (2012)

12. Kontonasios, K.-N., De Bie, T.: Subjectively interesting alternative clusterings. Machine Learning (2013)

13. Kontonasios, K.-N., Spyropoulou, E., De Bie, T.: Knowledge discovery interestingness measures based on unexpectedness. WIREs Data Mining and Knowledge Discovery 2(5), 386–399 (2012)

14. Kontonasios, K.-N., Vreeken, J., De Bie, T.: Maximum entropy modelling for assessing results on real-valued data. In: Proc. of the IEEE International Conference on Data Mining (ICDM) (2011)

15. Kontonasios, K.-N., Vreeken, J., De Bie, T.: Maximum entropy models for iteratively identifying subjectively interesting structure in real-valued data. In: Proc. of the European Conference on Machine Learning and Principles and Practice of Knowledge Discovery from Databases (ECML-PKDD) (2013)

16. Padmanabhan, B., Tuzhilin, A.: A belief-driven method for discovering unexpected patterns. In: Proc. of the 4th ACM SIGKDD International Conference on Knowledge Discovery and Data Mining (KDD), pp. 94–100 (1998)

17. Siebes, A., Vreeken, J., van Leeuwen, M.: Item sets that compress. In: Proc. of the 2006 SIAM International Conference on Data Mining (SDM) (2006)

18. Silberschatz, A., Tuzhilin, A.: On subjective measures of interestingness in knowledge discovery. In: Proc. of the 1st ACM SIGKDD International Conference on Knowledge Discovery and Data Mining (KDD), pp. 275–281 (1995)

19. Spyropoulou, E., De Bie, T.: Interesting multi-relational patterns. In: Proc. of the IEEE International Conference on Data Mining (ICDM) (2011)

20. Spyropoulou, E., De Bie, T., Boley, M.: Interesting pattern mining in multi-relational data. Data Mining and Knowledge Discovery (2013)

21. Spyropoulou, E., De Bie, T., Boley, M.: Mining interesting patterns in multi-relational data with n-ary relationships. In: Discovery Science (DS) (2013)

22. Webb, G.: Discovering significant patterns. Machine Learning 68(1), 1–33 (2007)

23. Webb, G.: Filtered-top-k association discovery. WIREs Data Mining and Knowledge Discovery 1(3), 183–192 (2011)

Time Point Estimation of a Single Sample from High Throughput Experiments Based on Time-Resolved Data and Robust Correlation Measures

Nada Abidi[1], Frank Klawonn[1,2], and Jörg Oliver Thumfart[3]

[1] Cellular Proteomics
Helmholtz Centre for Infection Research
Inhoffenstr. 7, D-38124 Braunschweig, Germany
[2] Department of Computer Science
Ostfalia University of Applied Sciences
Salzdahlumer Str. 46/48, D-38302 Wolfenbuettel, Germany
[3] Institute for Clinical Chemistry, Universitätsmedizin Mannheim
Medical Faculty Mannheim, Heidelberg University
Theodor-Kutzer-Ufer 1-3, D-68167 Mannheim, Germany

Abstract. Recent advances of modern high-throughput technologies such as mass spectrometry and microarrays allow the measurement of cell products like proteins, peptides and mRNA under different conditions over time. Therefore, researchers have to deal with a vast amount of available measurements gained from accomplished experiments using the above techniques.

In this paper, we set our focus on methods that analyze consistency of time-resolved replicates by using similarity patterns between measured cell products over time. This fact led us to develop and evaluate a method for time points estimation of a single sample using independent replicate sets taking the existing noise in the measurements and biological perturbations into account. Moreover, the established approach can be applied to assess the preanalytical quality of biobank samples used in further biomarker research.

1 Introduction

Microarray-based genomic surveys and other high-throughput approaches ranging from genomics to combinatorial chemistry are becoming increasingly important in biology and chemistry [1]. Therefore, biologists need data analysis approaches to deal with the complexity of biological systems in order to extract useful and relevant information from measurements.

To understand and model biological systems, it is necessary to consider their dynamics over time. However, due to the restrictions imposed by experimental settings and measurement devices, in many cases only snapshots of two different states or time points are taken and compared. The advancement of technologies in recent years allows to carry out more experiments in which not only two conditions can be compared, but in which measurements are taken at a number of time points (see for instance [2,3]). Nevertheless, the number of time points at which measurements can be taken is usually still very limited and therefore the time points are chosen very carefully, mostly in such

A. Tucker et al. (Eds.): IDA 2013, LNCS 8207, pp. 32–43, 2013.
© Springer-Verlag Berlin Heidelberg 2013

a way that the most interesting time points are covered. This means that we have to deal with a small number of time points with varying intervals between them. Therefore, the term time-resolved data instead of time series data is used.

We are mainly interested in two questions here. Given a number of replicate data from time-resolved experiments and measurements at a single unknown time point. Can we make a good estimation of the unknown time point? The second question concerns data consistency, i.e. do the replicates show a consistent behaviour or do they highly deviate from each other?

During this work, we studied the relationship between cell products at different time points using data from more than 30 patients. The correlation derived from the patients gave a reasonably accurate estimation of time points coming from other independent samples.

Several realisations of time-resolved experiments often contain systematic variations of measured values. Furthermore, we need to take the existence of random perturbations of the biological and medical data into consideration. Therefore, we used the robust rank correlation measures that are freely available as an R package named Rococo [4] to derive the relationship between the replicated sets at different time points. This family of rank correlation measures shows a smoother behaviour with respect to the noisy numerical data and offers more robustness to noise for small samples [5]. As a result, the relationship derived from the replicates allow us to gain accurate estimation of a single sample time course.

The paper is organized into six sections. The next section reviews briefly some traditional correlation methods as well as the robust rank correlation measures employed in statistical data analysis. Section 3 presents the problem in abstract terms. In Section 4, a data consistency check is introduced using different correlation methods. While Section 5 develops an approach used to determine the status of a single sample, section 6 provides an evaluation of our method using real data. Finally, a discussion and some concluding remarks are given.

2 Rank Correlation Measures

Correlation measures have contributed significantly in order to analyse and understand biological systems. The coefficient of correlation evaluates the similarity of two sets of measurements (i.e., two variables obtained in the same observation)[6]. Correlation measures are applied to pairs of observations

$$(x_i, y_i)_{i=1}^n \tag{1}$$

with ($n \geq 2$), to measure to which extent the two observations comply with a certain model. According to [5], the most common approaches of rank correlation measure are *Pearson's correlation coefficient* that assumes a linear relationship as the underlying model, *Spearman's rank correlation coefficient* [7,8] and *Kendall's tau (rank correlation coefficient)* [9,10,11]. The basic variant of Kendall's tau is defined as

$$\tau = \frac{C - D}{\frac{1}{2}n(n-1)},$$

where C and D are the numbers of concordant and discordant pairs, respectively:

$$C = |\{(i,j) \mid x_i < x_j \text{ and } y_i < y_j\}| \qquad D = |\{(i,j) \mid x_i < x_j \text{ and } y_i > y_j\}|$$

Gaussian scoring: (with parameter $\sigma > 0$)

$$R_\sigma^{\text{Gauss}}(\alpha, \beta) = \begin{cases} 1 - \exp(-\frac{1}{2\sigma^2}(\alpha - \beta)^2) & \text{if } \beta > \alpha \\ 0 & \text{otherwise} \end{cases}$$

with $R(\alpha, \beta) \in [0, 1]$ [12]. For a given scoring function $R : \mathbb{R} \to [0, 1]$, an operator $E : \mathbb{R} \to [0, 1]$ was defined in [5] as:

$$E(\alpha, \beta) = 1 - \max(R(\alpha, \beta), R(\alpha, \beta)).$$

Gamma rank correlation coefficient: We can compute the degree to which (i, j) is a concordant pair as

$$\tilde{C}(i, j) = \min(R_X(x_i, x_j), R_Y(y_i, y_j))$$

and the degree to which (i, j) is a discordant pair as

$$\tilde{D}(i, j) = \min(R_X(x_i, x_j), R_Y(y_j, y_i)),$$

Then we can compute the overall score of concordant pairs $\tilde{C}$ and the overall score of discordant pairs $\tilde{D}$, respectively, as sums of the following scores:

$$\tilde{C} = \sum_{i=1}^{n} \sum_{j=1}^{n} \tilde{C}(i, j), \qquad \tilde{D} = \sum_{i=1}^{n} \sum_{j=1}^{n} \tilde{D}(i, j).$$

Consequently, we can define the *generalized gamma rank correlation measure* $\tilde{\gamma}$ as

$$\tilde{\gamma} = \frac{\tilde{C} - \tilde{D}}{\tilde{C} + \tilde{D}}.$$

With the use of correlation coefficients, it is possible to indicate the amount of common cell products behaviour that exists between two time points from different observation sets. It enables us to measure the monotonic relationship between cell products of two different time points. Since we are dealing with noisy replicated sets which may obscure monotonic associations, in our experiments, the robust gamma rank correlation coefficient designed for dealing with noisy numerical data lead to the best results.

3   Formal Problem Statement

Replicated time-resolved data are characterized by non-constant intervals between measured time points. Additionally, replicates could also have different measured time courses due to encountered technical difficulties to obtain measurements at exact times while retrieving samples.

Table 1. General structure of the data

Time point	Replicate1		...	ReplicateR	
Cell product	$s_1^{(1)}$ ...	$s_1^{(T_1)}$	...	$s_R^{(1)}$ ...	$s_R^{(T_R)}$
i_1	$x_{1,1}^{(1)}$ ...	$x_{1,1}^{(T_1)}$	...	$x_{1,R}^{(1)}$ ...	$x_{1,R}^{(T_R)}$
$\vdots$	$\vdots$	$\vdots$	$\vdots$	$\vdots$	$\vdots$
i_N	$x_{N,1}^{(1)}$ ...	$x_{N,1}^{(T_1)}$	...	$x_{N,R}^{(1)}$ ...	$x_{N,R}^{(T_R)}$

Table 1 shows the principle structure of the data we intend to analyse. We consider N cell products (genes, proteins, peptides or metabolites) that are measured in a replicate r at T_r different time points $s_r^{(1)}, \ldots, s_r^{(T_r)}$ and the intervals between time points might vary [13]. For the N cell products, we have R replicate sets, which would have identical values in the ideal case. But this usually applies neither to the time points nor to the measurement values. But we would expect that at similar time points, replicates would have roughly the same values. It should be noted that there can also be missing values.

The main objective of this work is to develop a method in order to estimate the time point for a single sample using an independent data set of repeated replicates from time-resolved data, measured from high-throughput experiments. To achieve reliable results, two aspects must be considered. Firstly, a measurement error at one time point of one replicate could affect its trustworthiness and corrupt our estimation. Secondly, for data consistency, the existing noise in the measurements must be taken into account by finding out similarity patterns of replicates coming from time-resolved samples.

4 Data Consistency Check

When observing multiple time series generated by a noisy and stochastic process, we often encounter large systematic sources of variability. For example within a set of replicate biological time series, the time axis can be variously shifted, compressed and expanded in complex non-linear ways. Additionally, in some circumstances, the scale of the measured data can vary from one replicate to another, or even within the same replicate [14]. Suitable normalisation can sometimes amend this effect. In order to establish the relationship between different replicate sets we computed various traditional correlation measures as well as the robust gamma rank correlation coefficient between all available time points.The tolerance argument of the robust gamma rank correlation coefficient was chosen as 10 percent of the interquartile range of the data.

Now we consider N cell products that are measured in each replicate set at possibly different time points. As we mentioned above, time points might vary from one replicate to another. We consider two replicates x and y. For replicate x each cell product is measured at K different time points ($k \in \{1, \ldots, K\}$). For the second time-resolved replicate y each cell product is measured at L different time points ($l \in \{1, \ldots, L\}$). For the measured cell products at time point t_k of replicate x the maximum degree of similarity with cell products of replicate y measured at t_l is obtained in the following way:

$$t_{\text{opt}}(t_k; x, y) = \text{argmax}_{l \in \{1,\dots,L\}} \{R(x^{(t_k)}, y^{(t_l)})\}$$

with R being the chosen rank correlation function.

The basic idea behind the calculation of the robust gamma rank correlation between replicates at all available time points was to detect replicates with consistent behaviour over time. We expect that two replicates of the same biological process have different biological speeds, due to the metabolism of organisms. Which means, that cell products measured at two different time points in one replicate are best correlating with those measured at one time point of the second replication set.

We consider t_l and $t_l + 1$, as the most suitable time points from the replicate set y for time points t_k, and $t_k + 1$ respectively. These time points represent the highest obtained correlation coefficients. We assume that inconsistent behaviour between the replicate sets x and y does exist if:

$$t_l > t_{l+1}$$

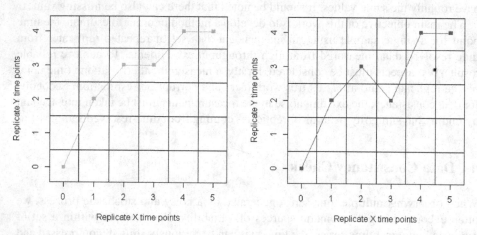

Fig. 1. Resulting correlation curve: each time point of replicate x is plotted against the best correlating time point from replicate y

Figure 1 (left) illustrates two replicates x and y with consistent behaviour over time. Horizontal or vertical segments show that the time point of one replicate highly correlates with multiple time points of the other replicated set due to differences in metabolism speed of both organisms. While in Figure 1 (right) an inconsistent behaviour between both replicates is clearly detected. A consistent behaviour corresponds to a non-decreasing graph.

This method of checking consistency is not suitable when cyclic behaviour is observed within replicates, for instance when the biological system is assumed to return to the original state, e.g. measurements taken before infection, during infection and after recocery. The speed of the reaction might vary from one organism to another. However the initial status before the infection and the final status after the treatment are expected to deliver the same measurements.

5   Time Point Estimation

The main target of this work is to establish a method to estimate time points for human plasma and serum time-resolved samples. The repeated measurements of one cell product at different time points under the same condition are expected to deliver a similar behaviour for all replicate sets. A high correlation between cell product measurements could lead us to the exact time of the measurements. However some of the identified cell products might do nothing or something completely unrelated to the specific condition or process of interest. Therefore, they could deteriorate the correlation coefficient between cell products at a given time point. In order to reduce this effect, we adopted the solution presented in [15] by not considering the correlation with respect to all cell products. For each pair of time points, we are allowed to remove a fixed small number p of cell products that lead to the highest increase of the rank correlation coefficient. For a given single sample x, N cell products are measured at an unknown time point t_k. To assess the time point t_k, the following steps were carried out:

1. By means of the previously mentioned consistency check we can gain insight into the behavior of the replicate set having the structure mentioned in Table 1. Therefore, the data set Y will be considered as a training set,

$$Y = \{y_1, \ldots, y_R\}$$

   where y_r is a replicate set with N cell products that are measured at L_r time points $(l_r \in \{1, \ldots, L_r\}$
2. Remove replicates that raise suspicious behaviour.
3. For measured cell products at the estimation time point t_k, we compute the robust gamma rank correlation coefficient at all available time points of the training data set.
4. For each replicate of the training data set we choose the time point $t^{(r)}$ that delivers the highest obtained rank correlation coefficient with measured cell products at t_k, so that:

$$t_{\mathrm{opt}}(t_k; x, y) = \mathrm{argmax}_{l_r \in \{1, \ldots, L_r\}} \{R(x^{(t_k)}, y_r^{(l_r)})\}$$

5. We build the mean value from all obtained time points to assess t_k:

$$t_k = \frac{\sum_{r=1}^{R} \left(t^{(r)}\right)}{r} \tag{2}$$

   A weighted mean or median could also be applied at this level to determine the time value to be estimated from the different replicates.

6   Evaluation

Our approach is demonstrated using experimental data from human serum samples. Peptides in human serum samples were measured by nHPLC-MS/MS under the same condition. However, each replicate set has a specific time course due to the encountered

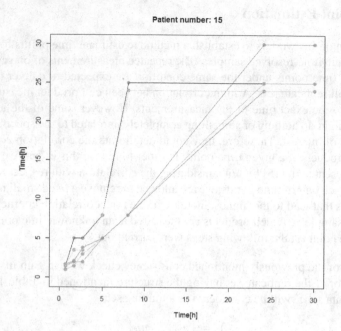

Fig. 2. Patient 15

technical difficulties to obtain measurements at exact times. The available serum sets were collected from three independent patient cohorts. Applying our method described in Section 5 should asses the sample age in order to judge the quality of the serum samples and consequently to decide whether to eliminate it or not from further biomedical studies for instance biomarkers discovery research.

During the evaluation step, the training set was composed of independent donor cohorts 1 and 2, which were collected from 30 patients. About 60 peptides were measured at different time points for each sample. As mentioned above, the replicates might have different time courses. The test set consisted of a third serum sample collected from more than 20 patients and with different time courses between replicates.

In order to visualize the data consistency check results, we generated a plot for each replicate set where $(R - 1)$ curves are presented. Each curve presents the highest correlation coefficient between one replicated measured cell products at each of its time points and one of the $(R - 1)$ replicate time courses. The number of detected inconsistent curves was almost reduced by half when using the robust gamma rank correlation in comparison to traditional rank correlation methods.

Figure 2 contains 5 out of 29 computed curves plotted between all time points of patient 15 and all other replicates' time points. Most of the obtained curves build a consistent behaviour between the mentioned patient and all other replicates over time. Patient 11 illustrated in Figure 3 shows an inconsistent behaviour with all other replicate sets at time point 30 (here only 5 curves are presented). Therefore, patient 11 has been removed from the training set to avoid the wrong effect on time point estimation of the test set.

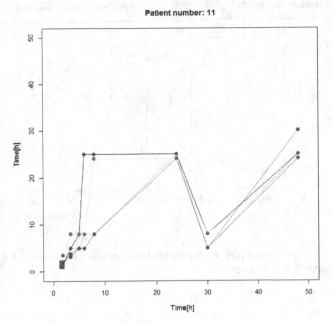

Fig. 3. Patient 11

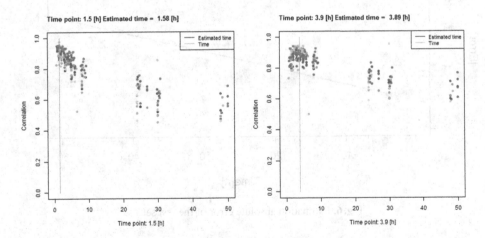

Fig. 4. Time Point estimation examples: both presented time points were measured at early stage with use of three independent data sets

After testing the training set for data consistency and removing potential outliers like patient 11, we can apply our approach to estimate time courses of a single sample using the training set already mentioned. For each time point estimation, we compute the correlation coefficient with all available replicate time points in the training set after

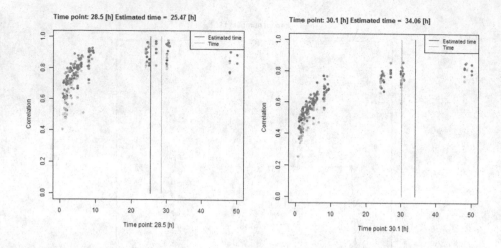

Fig. 5. Time Point estimation examples: both presented time points were measured at late stage with use of three independent data sets

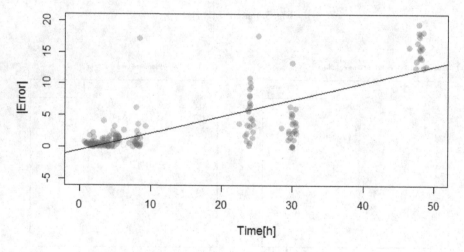

Fig. 6. Estimation absolute error of the test set

removing the three peptides that allow the highest increase of the correlation coefficient during each calculation. Therefore, we obtain from each replicate one time point that represents the highest obtained correlation coefficient. By computing the mean of the 29 resulting time points, we determined the time point at which the measurement of the 58 peptides was performed.

To visualize our approach, we plotted all computed correlation coefficients with all available replicates time points in the training set for each time point where each patient is illustrated with a different colour. The plotted correlation points provide us with a primary idea about the behavior of the computed correlation coefficient depending on the time point we intend to estimate. The average of the selected time points with the highest correlation from each patient of the training set is displayed as the estimated time point. The illustration of both estimated time point and real time point, allow an intuitive visualization and identification of the sufficiency of our presented aging method.

As illustrated in Figure 4, the presented approach demonstrates a relatively reliable result at an initial time interval located between 0 and 10 hours, where the biological process is still active. The lines for estimated time point and the the true time point are almost identical. At the same time, an estimation inaccuracy occurs at late time span (after 15 hours) as shown in Figure 5. This inaccuracy might be caused by the end status that arises at a later stage of the biological process and also by sparsity measurements of later time points in the data set. At this point, we have to note that the relevant biological period for biologist is the intitial time span, where reliable estimation results are required to make further decisions. While at later stage cell products are anyway too inaccurate to be processed in additional analysis.

In Figure 6, the computed absolute error was plotted against the real time point. The observed error is increasing in a time dependent manner, so that the assessment inaccuracy is showing a higher deviation of the estimated time point from the real time point at later time span.

To judge the reliability of our proposed approach, we intend to determine the standard deviation of the absolute error. As we mentioned before, an increasing trend is observed so that fitting a linear regression model to the obtained data can denote the relationship between the computed error and the real time. Let X denotes the error. We assume that X_t follows a normal distribution with mean μ equal zero and variance σ_t^2. Therefore the expected value of the absolute error can be written as follows:

$$E(|X_t|) = \sqrt{\frac{2}{\pi}}\sigma_t \approx at + b$$

and consequently, the standard deviation of the absolute error noted σ_t can be abtained by: $\sigma_t = \sqrt{\frac{\pi}{2}}(at + b)$. Based on the resulting coefficient from the linear regression, a and b were consecutively equal to 0.33 and -0.64.

7   Discussion

During this work, our goal was to develop an approach to determine the time course of one single sample by using an independent time-resolved data set of repeated replicates from high-throughput experiments as they are common in genomics, proteomics and metabolomics.

The provided technique can be used as a quality tool to estimate the analytical quality of the samples e.g. biobank samples used in biomarker research. With use of this presented method, the estimation of time points which belong to the initial time span, considered as the biological relevant period, delivered reliable results that can be considered as sufficient for the quality judgment of the sample of interest. The data consistency check approach provides an intuitive way to judge the replicates consistency and therefore enable biologist to discard unsuitable replicates from additional analysis. Furthermore the resulting consistency curves can be used for cluster analysis in order to find groups of replicate sets with similar cell product behaviour over time.

Nevertheless, the presented work is in primary stage and requires more evaluation with different biological data sets with enlarged numbers of replicates. Besides, the elaborated algorithm should be adapted to meet requirements of samples with cyclic behaviour used for instance in infection research studies.

Acknowledgments. The research leading to these results has received support from the Innovative Medicines Initiative Joint Undertaking under grant agreement n° 115523-2 – COMBACTE, resources of which are composed of financial contribution from the European Union's Seventh Framework Programme (FP7/2007-2013) and EFPIA companies in kind contribution.

References

1. Eisen, M.B., Spellman, P.T., Brown, P.O., Botstein, D.: Cluster analysis and display of genome-wide expression patterns. Proceedings of the National Academy of Sciences 95 (1998)
2. Kleemann, R., van Erk, M., Verschuren, L., van den Hoek, A.M., Koek, M., Wielinga, P.Y., Jie, A., Pellis, L., Bobeldijk-Pastorova, I., Kelder, T., Toet, K., Wopereis, S., Cnubben, N., Evelo, C., van Ommen, B., Kooistra, T.: Time-Resolved and Tissue-Specific Systems Analysis of the Pathogenesis of Insulin Resistance. PLoS ONE 5 (2010)
3. Blom, E.J., Ridder, A.N.J.A., Lulko, A.T., Roerdink, J.B.T.M., Kuipers, O.P.: Time-resolved transcriptomics and bioinformatic analyses reveal intrinsic stress responses during batch culture of bacillus subtilis. PLoS ONE 6 (2011)
4. Bodenhofer, U., Krone, M.: RoCoCo: an R package implementing a robust rank correlation coefficient and a corresponding test (2011), Software available at http://www.bioinf.jku.at/software/rococo/
5. Bodenhofer, U., Krone, M., Klawonn, F.: Testing noisy numerical data for monotonic association. Inform. Sci. 245, 21–37 (2013)
6. Abdi, H.: Coefficients of correlation, alienation and determination. In: Salkind, N.J. (ed.) Encyclopedia of Measurement and Statistics, Sage, Thousand Oaks (2007)
7. Spearman, C.: The proof and measurement of association between two things. Am. J. Psychol. 15, 72–101 (1904)
8. Spearman, C.: Demonstration of formulae for true measurement of correlation. Am. J. Psychol. 18, 161–169 (1907)
9. Abdi, H.: The Kendall rank correlation coefficient. In: Salkind, N.J. (ed.) Encyclopedia of Measurement and Statistics, Sage, Thousand Oaks (2007)
10. Kendall, M.G.: A new measure of rank correlation. Biometrika 30, 81–93 (1938)
11. Kendall, M.G.: Rank Correlation Methods, 3rd edn. Charles Griffin & Co., London (1962)
12. Bodenhofer, U., Demirci, M.: Strict fuzzy orderings with a given context of similarity. Internat. J. Uncertain. Fuzziness Knowledge-Based Systems 16, 147–178 (2008)
13. Klawonn, F., Abidi, N., Berger, E., Jänsch, L.: Curve fitting for short time series data from high throughput experiments with correction for biological variation. In: Hollmén, J., Klawonn, F., Tucker, A. (eds.) IDA 2012. LNCS, vol. 7619, pp. 150–160. Springer, Heidelberg (2012)

14. Listgarten, J., Neal, R.M., Roweis, S.T., Emili, A.: Multiple alignment of continuous time series. In: Advances in Neural Information Processing Systems, pp. 817–824. MIT Press (2005)
15. Krone, M., Klawonn, F.: Rank correlation coefficient correction by removing worst cases. In: Hüllermeier, E., Kruse, R., Hoffmann, F. (eds.) IPMU 2010. Part I. CCIS, vol. 80, pp. 356–364. Springer, Heidelberg (2010)

Detecting Events
in Molecular Dynamics Simulations

Iris Adä and Michael R. Berthold

Nycomed-Chair for Bioinformatics and Information Mining
Dept. of Computer and Information Science
University of Konstanz
first.last@uni-konstanz.de

Abstract. We describe the application of a recently published general
event detection framework, called EVE to the challenging task of molecular event detection, that is, the automatic detection of structural changes
of a molecule over time. Different types of molecular events can be of interest which have, in the past, been addressed by specialized methods.
The framework used here allows different types of molecular events to be
systematically investigated. In this paper, we summarize existing molecular event detection methods and demonstrate how EVE can be configured
for a number of molecular event types.

1 Introduction

Research in Chemistry/Chemical Biology has been interested in understanding
molecular dynamics simulations for around 25 years [14,19]. There are many different ways of modeling molecules; here we are primarily interested in molecules
(as in Figure 1) represented as a connected group of multiple atoms, ranging

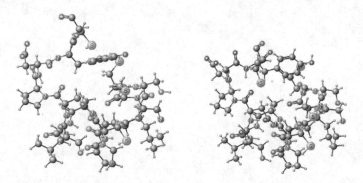

Fig. 1. Two conformations of alpha-conotoxin pnib (1AKG). The molecule's atoms
are connected by bonds and the entire arrangement in space represents the molecule's
conformation.

A. Tucker et al. (Eds.): IDA 2013, LNCS 8207, pp. 44–55, 2013.
© Springer-Verlag Berlin Heidelberg 2013

from tens to several hundreds of atoms and more. The geometrical arrangement of atoms to each other (known as the *conformation* of the molecule) under normal conditions changes continuously, even in a vacuum. These changes range from small oscillations to drastic changes in the overall molecular shape. It is important to differentiate between small, unimportant conformational changes and more important ones, which are relevant e.g. for the biological function of a molecule. The automatic detection of unexpected or irregular transformations of the molecule's conformation is of particular interest.

There are two main reasons why molecular dynamics (MD) is concerned with simulating data. First of all, monitoring atom positions and hence the molecular conformation at a sufficiently high resolution (both in terms of time and location) is complicated and second the influence of the molecule's surroundings has a substantial effect on the molecule's conformations. Using simulated movements over time helps to better understand the behavior of a molecule. However, actually processing the vasts amount of data generated by these simulations poses enormous problems.

The core idea behind molecular dynamics simulations is to simulate the behavior of molecules, mostly of proteins, over time by using "simple potential-energy functions" [2]. Hence this is an artificially, but not randomly generated data set, which depicts the true behavior of the molecule. The forces and reciprocal effects between the molecule's atoms and bonds can be explained with fairly simple mathematical functions. Basically every atom can influence all other atoms with the bonds between the atoms adding to this effect. The dynamics are calculated by solving equation systems based on all these functions together [2]. By iteratively calculating the steps of the molecule's internal positions a sequence of conformations is generated. This sequence captures the movement of the molecule, represented by all of the individual atom positions in three dimensional coordinates over time. The interest of molecular event detection lies in finding unexpected movements among atom positions of a molecule. Such movements can e.g. refer to a conformational change or a folding of the molecule.

The internal relations between molecule's atoms and bonds are investigated for other aspects (e.g. kinetic and thermodynamic information [2]) as well in molecular mechanics [8]. However, in this work we are interested only in the changes of atom positions over time, hence, in the sequence of molecule conformations. There are various types of molecular events. The key point of interest is that these structural changes are relevant for the chemical state or mechanism of the molecule.

In this work we concentrate on the problem of general molecular event detection as an application for change and event detection in high dimension. We begin by discussing current approaches, starting with an overview of feature based methods before discussing more recent methods. Afterwards we briefly summarize the concepts of EVE [1], an event detection framework we use to formalize the underlying event detection problem. We conclude by demonstrating how EVE can be used to detect molecular events of interest for a real molecular sequence.

2   Related Work

In the area of data mining and statistics considerable work has already been invested in the detection of irregularities in series data. Event detection [12], drift detection [24] and anomaly detection [9] are vibrant research directions.

In this section we focus on how events are detected in molecular dynamics. One of the main difficulties is the high dimensionality of the feature vector. Following the structure of a recent survey paper [6] we will first discuss traditional analysis methods, which are mainly concerned with the extraction of features. Recent methods regard the molecule's atoms as the nodes in a graph and apply different strategies to introduce edges and then monitor changes of the resulting graph over time.

2.1   Traditional Analysis

A large number of atoms – easily hundreds – is encountered when analyzing a molecular times series. Using their coordinates and other properties as one huge feature vector is time consuming and often yields uninterpretable results. For this reason quite a few applications of molecular event detection are first concerned with preprocessing the features of the molecule's conformation.

H Atom Filtering: As already mentioned, molecules consist of many atoms like carbonate (C), hydrogen (H), or oxygen (O), to name just some of the common ones. Smaller atoms (like hydrogen), for example, are known to be more prone to move. Common practice therefore ignores the movements of hydrogens as they are not related to an overall structural change of the molecule and can easily be derived from the remainder of the molecular structure anyway.

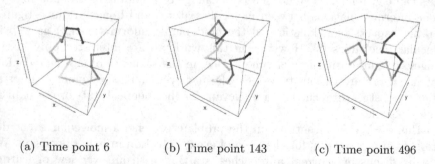

(a) Time point 6   (b) Time point 143   (c) Time point 496

Fig. 2. The C^α trace of alpha-conotoxin pnib. The C^α atoms are connected to show the overall structure of the molecule. This reduces the complexity tremendously, as only 16 of the 213 atoms in the molecule remain. The figures show three conformational states of the molecule, at time points 6, 143 and 496. At step 143 the molecule is folded and opened again in a last step (496). The three time points were chosen because they demonstrate the movement of the molecule nicely.

C^α Atom Extraction: A more compact representation retains only the so called C^α trace [11], where a C^α-carbon is the central carbon of an amino acid [21]. Put simply, each protein consists of multiple amino acids, which are "substructures" of the molecule. In each amino acid the C^α-carbon can be uniquely determined and all of these carbons are combined to the C^α trace, which is an abstraction of the overall structural appearance. Therefore only the C^α carbons are used as features and all other atoms are filtered. This reduction naturally reduces the dimension immensely. It is questionable whether it also filters a possible movement from the data. In Figure 2 the C^α trace of 1AKG is visualized for three consecutive time points.

Distances: One problem when monitoring molecular dynamics is that only relations within the molecule's atoms are of interest and not an overall movement. One solution for this problem is to calculate relative distances (e.g. Euclidean or absolute) between each pair of atoms inside the molecule. The resulting feature space is quadratic in the number of atoms.

Angles: A second solution for filtering an overall movement from the data is to use angles between the position of individual atoms. The feature vector then consists of the angles in three dimensional space.

RMSD: A "well-known and most widely accepted" [15] method for measuring the similarity between two confirmations of a molecule is the root-mean-square deviation (RMSD). It is calculated by using the average distance between all atom positions.

$$\mathrm{RMSD}\left(M^{(1)}, M^{(2)}\right) = \sqrt{\frac{1}{n}\sum_{i=1}^{n}|m_i^{(1)} - m_i^{(2)}|^2}$$

$M^{(t)}$ is the conformation of molecule M at time point t and $m_i^{(t)} \in \mathbb{R}^3$ is the atom position of atom i.

The disadvantage is that the RMSD measure also reacts to movements and rotations of the molecule. The problem of movements can be resolved by shifting one molecule and rotation can be solved by pre-applying a rotation. The Kabsch Algorithm [13] is one of the most popular solutions for this optimization problem and uses a singular value decomposition to find the minimizing transformation matrix. A more recent approach uses Quaternions to solve the problem [10].

Visualization: The detection of an event can also be determined by visual inspection. In a line plot of the RMSD measure, molecular events are mapped to peaks in the dissimilarity. More recently heatmaps of the complete distance matrix between all time points are used as well. Areas showing a small in-between dissimilarity and a large dissimilarity to neighboring times are then further investigated.

2.2 Non-traditional Analysis

During the last few years, new methods for analyzing MD simulations came up. In contrast to traditional analysis methods they are designed to deal with much larger simulations.

In flexibility analysis [22,4] each atom is individually investigated using a principal component analysis (PCA). The results enable atoms with fast vibrations or small movements to be filtered and summarize the major states of the molecule. Finally the flexibility vectors, generated from the PCA, are plotted on a mean structure of the molecule for further analysis [6]. In the wavelet analysis of MD simulations [3,7] the atoms are individually analyzed as well. Using a continuous wavelet transform Benson and Daggett [7] are able to find trajectories of different proteins that show similar structural movements.

The most recent research direction uses graphs to model the overall dependencies in the molecule. Wriggers et al. [23] presented probably one of the earliest approaches. In this work a graph is generated from the atoms' positions. They propose multiple methods for generating the graph using a distance cut-off, whereby atoms are connected in the graph if their distance is below a certain threshold, or the generalized masked delaunay tetrahedralization. Afterwards changes in the consequent graphs are tracked over time, enabling the number of appearing and disappearing edges to be counted. Finally they apply different filters to the achieved series of graph changes to detect the event. The Dynamic Tensor Analysis [17,18] applies tensor analysis to identify conformational substates of the molecular sequence. More recently graphs are generated using additional expert knowledge of the chemical structures [5].

3 Goals of Molecular Event Detection

As described previously molecular event detection is concerned with the analysis of movements of molecules over time. The molecule's movement is simulated in three dimensional space and every few picoseconds, or even less, a snap shot/conformation of the current positions is calculated. The molecule M is represented as list of its three dimensional atom positions $M = (m_1, \ldots, m_n)^t$, where $m_i = (x, y, z)^t \in \mathbb{R}^3$. The molecules are generated at time points $t = 1 \ldots m$. The molecule at time point t is entitled $M^{(t)}$.

The goal of molecular event detection is to find interesting changes in the consequent states of a molecule. Next we assume that a dissimilarity function $d(\cdot, \cdot)$ is provided, which is able to calculate the dissimilarity between two conformations of a molecule. For example RMSD can be used.

Constant: A molecular dynamics is called constant in a time interval $[t_0, t_1]$ if the relations in-between the atom positions do not change
$(\exists \epsilon \geq 0 \; \forall i, j \in [t_0, t_1] : |d(M^{(i)}, M^{(j)})| < \epsilon)$.

Changing: Molecular dynamics change if the dissimilarities increase over consequent time steps. A molecule is changing in the time interval $[t_0, t_p] = (t_0, \ldots, t_p)$

Fig. 3. The process of molecular folding/refolding, illustrated by showing four exemplary significant atoms during the process

if the dissimilarity to predecessors increases, hence
$$\forall i \in \{1, \ldots, p-1\} : d(M^{(t_0)}, M^{(t_i)}) < d(M^{(0)}, M^{(t_p)}).$$

Reoccurring Change: A change is regarded as reoccurring if the atoms return to their previous relations. With respect to the definition of changing, there would be a second interval where the dissimilarities to predecessors decrease. The interval $[t_0, t_1]$ is called a reoccurring change if $\exists t_2, t_3 : [t_2, t_0] \cup [t_1, t_3]$ is constant. Note that this type of change is mostly tripartite, starting with a change, followed by a constant state and finishing with the reoccurrence to the beginning structure.

Outlier: An outlier is defined as a single data point (or only a few) within a constant state where the molecule shows high dissimilarity to the previous and following steps. More formally: t is an outlier, if $\exists t_0, t_1, t \in [t_0, t_1] : [t_0, t_1] \setminus t$ not changing. Note that if the step size of the molecular simulation is fine enough, one would not expect to find such outliers.

A special kind of a reoccurring change is a folding and refolding of the molecule, which is a particularly interesting molecular behavior. During this event the molecule changes its shape into a new conformation (where it usually has a different biological function) and, after a certain time, refolds into the original conformation. Of key interest here is the identification of the exact start and end of the process. Figure 3 shows the visualization of a folding with subsequent refolding. We will now focus on illustrating below how these exemplary molecular event types can be detected using EVE.

4   EVE for the Detection of Molecular Events

This section presents an exemplary setup to demonstrate how EVE is configured for the application of molecular event detection. In the following the main aspects of the EVE framework are summarized: for a more detailed discussion see our previous work [1]. We consequently analyze 1AKG as a relatively small but well understood example of a molecule.

4.1   The EVE Framework

EVE [1] is a general framework for event detection. The main intention behind the framework is twofold. The first goal is to model the detection of events in a general setting. Using EVE it is straightforward to classify existing approaches as well as emphasizing dissimilarities and similarities between individual event detection techniques. Secondly, the framework is intended to provide a fast entry point to examine new, also complex data types. This fits perfectly for the application of molecular event detection.

EVE structures any event detection algorithm into three key components: window configuration, dissimilarity measure and detection mechanism.

1. *Window configuration*

   Two windows need to be defined, the past and the current window. The past window models behavior which is assumed to be normal at this point in time. The current window on the other hand is tested to see if it contains a possible event. The process of a window over consequent time steps is defined by two terms. The first refers to the start position which can be sliding or fixed (short: S or F) and the second one to the window size which is either constant or growing (short: C or G). Two such windows, one for past and one for current (e.g. SC ↪SC), determine the process of a window over consequent time steps as a *window combination*.

   - FC ↪SC (Fixed Constant to Sliding Constant): Comparing a window at the start of the series with a sliding window ending at the most recent data point.
   - FC ↪FG (Fixed Constant to Fixed Growing): Comparing the start window with the rest of the following data.
   - FG ↪SC (Fixed Growing to Sliding Constant): The idea behind this concept is to extract statistics or models out of all past information and compare it to the most recent data points.
   - SC ↪SC (Sliding Constant to Sliding Constant): Comparing two consequent sliding windows.

2. *Dissimilarity measure*

   The goal of the dissimilarity function is to indicate the probability of an event being detected. By previously building an abstracted model on the window, multiple possibilities of calculating this dissimilarity are possible. The Kullback-Leibler distance [16], the Euclidean distance (L_2), other L-norms or classification measures (e.g. the false positive rate) are obvious examples.

3. *Detection mechanism*

   Detection is performed to identify events by evaluating the previously calculated dissimilarity measures, often by simply applying a threshold function or using a control chart [20].

After introducing the main ingredients of EVE, we now demonstrate how it can be applied to the specific application of molecular dynamics and illustrate how relevant events of different type can be detected.

4.2 Experimental Setup

For the setup of the EVE framework the representation, dissimilarity function, and threshold are chosen as follows.

Data Model: When a window contains multiple conformations the positions of the atoms are averaged to filter out small oscillations of the atoms. This average structure is calculated as follows:

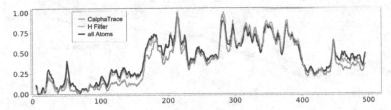

Fig. 4. The 0-1-normalized distance of three representation of one molecule (1AKG) with the same EVE setup (**FG** $\hookrightarrow$ **SC** and window size 5)
(**Legend:** x axis: time (ns), y axis: dissimilarity)

$$\text{avg}\left(M^{(1)},\dots,M^{(c)}\right) = \frac{1}{c}\left(\sum_{i=1}^{c} m_1^{(i)},\dots,\sum_{i=1}^{c} m_n^{(i)}\right).$$

The Root-mean-square deviation (as introduced in Section 2.1) is used as **dissimilarity measure**.

As **detection** mechanism a control chart [20] is used. An event is reported if the dissimilarity measure exceeds the sum of mean μ and 1.5 fold std. dev. σ.

We investigated four different window combinations:

$$\text{FC} \hookrightarrow \text{SC},\ \text{FC} \hookrightarrow \text{FG},\ \text{FG} \hookrightarrow \text{SC},\ \text{SC} \hookrightarrow \text{SC}.$$

In the experiments we used three window sizes: $c \in \{1, 10, 20\}$.

We here previously filtered the H atoms of the molecule. In Figure 4 the **FG** $\hookrightarrow$ **SC** with window size 5 is applied on three representations, C^{α}-trace, H atoms filtering and all atoms. We normalized the distance measures to 0-1. However the visualization shows that the results are very similar and especially filtering the H atoms only had small effects to the overall error measure.

4.3   Molecular Data Set

The analysis is demonstrated on 1AKG, alpha-conotoxin pnib[1] (see Fig. 1), which is a relatively small molecule, actually a protein, containing 16 amino acids and 213 atoms. We chose this molecule because it is already well studied and hence the ground truth, e.g. the interesting events in the series are known.

The behavior of the molecule's conformation is as follows: During the first 70 time steps it does not change significantly, only small movements inside and a small overall rotation occur. Afterwards the molecule starts to fold. At time point ~ 230 folding reaches a maximum and the folding angles start to decrease again until time point ~ 400, where a constant conformation is reached.

In Figure 6 to 8 three windowing concepts are applied to the alpha-conotoxin pnib over time. The x axis of each plot represents the time and the y axis the calculated dissimilarity. In all of these visualizations, the dissimilarity measure is shown in green, the lower bound of the control chart is depicted in blue and the upper bound is orange.

---

[1] Detailed information can be found in the protein data bank:
http://www.rcsb.org/pdb/explore.do?structureId=1akg

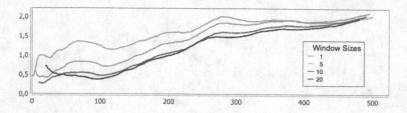

Fig. 5. FC↪FG (Fixed Constant to Fixed Growing): When comparing a start window to the rest of the data there are two assumptions. First the past window has to represent the data nicely. The model of the current window should be able to smoothen events over time. This is not the case here, as the constant increase of all four windows does not reveal any insights.

(**Legend:** x axis: time (ns), y axis: dissimilarity)

4.4 Discussion of Experiments

In this section we discuss the analysis results of the molecular event detection using EVE.

The first type of window combination investigated is the FC↪FG because it is the one with the most irrelevant result. Figure 5 shows the error calculated for the FC↪FG on four different window sizes. The constant phase in the beginning of the molecular series is determined. However, afterwards the mean structure of the current window is not representative as it contains the change and the base line.

In addition to the changes of interest, the molecules' conformation changes steadily as the atoms are constantly see-sawing and there is always a lot of movement in smaller regions as well. This is clearly visible e.g. in the SC↪SC of window size $c = 1$ (Figure 6a). There is so much movement in the dissimilarity function that the number of false positives is much too high. These oscillations can be filtered by adjusting the window sizes accordingly. However using a window size that is too big can smooth out possible events. In FC↪FG (Figure 5)

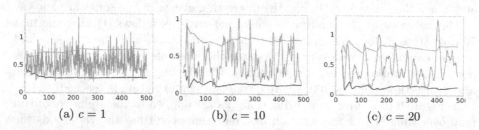

(a) $c = 1$ (b) $c = 10$ (c) $c = 20$

Fig. 6. SC↪SC (Sliding Constant to Sliding Constant): Comparing two sliding windows worked out very nicely for 1AKG. While smaller window sizes contain much noise, $c = 20$ clearly show 3 very significant events.

(**Legend:** x axis: time (ns), y axis: dissimilarity, green: dissimilarity measured, blue: $\mu - 1.5 * \sigma$, orange: $\mu + 1.5 * \sigma$)

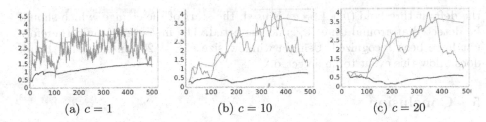

(a) $c = 1$ (b) $c = 10$ (c) $c = 20$

Fig. 7. FC↪SC (Fixed Constant to Sliding Constant): In this analysis one can see, that the conformation of the molecule is less similar to the original state in the middle of the series and in the end returns to the initial conformation.
(**Legend:** x axis: time (ns), y axis: dissimilarity, green: dissimilarity measured, blue: $\mu - 1.5 * \sigma$, orange: $\mu + 1.5 * \sigma$)

an increasing line only is visible for all combinations, due to the fact that the second window is too big to produce meaningful results. However the opposite concept FG↪SC (as can be seen in Figure 8 works well. The outlier is observed with the window size $c = 1$ (Fig. 8a).

The second part of our analysis was concerned with the question whether a reoccurring change can also be detected in the data. Two insights were provided by the FC↪SC combination. The first part of the window up to time point 100 does not change very much, however, there is an outlier at time point 52 indicating a high dissimilarity to the baseline window. Although the outlier was smoothed out by bigger windows, they showed the overall movement trend in the conformational states of the molecule much more clearly. Considering the dissimilarity for FC↪SC and window size $c = 20$, the constant phase in the beginning, the change, which was previously marked as a folding; and finally the return to the initial state can be seen clearly (Figure 7). The reoccurring change also becomes clear by inspecting Figure 8b where the constant behavior is reached again after time step 400.

The last observation is provided by the plots of the SC↪SC combination (Figure 6). A window size of $c = 10$ yields good results on the first view. When

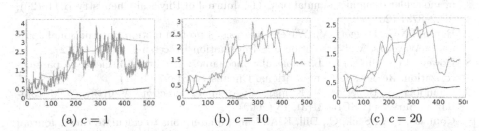

(a) $c = 1$ (b) $c = 10$ (c) $c = 20$

Fig. 8. FG↪SC (Fixed Growing to Sliding Constant): The beginning and the end of the change can be seen in all three error plots. However the recorring can not be detected with the control chart.
(**Legend:** x axis: time (ns), y axis: dissimilarity, green: dissimilarity measured, blue: $\mu - 1.5 * \sigma$, orange: $\mu + 1.5 * \sigma$)

the default threshold $(\mu + 1.5 * \sigma)$ is used, the start of the change, which should be detected at around time 70, is not detected. The first event would therefore not have been recognized. Using a window size of $c = 20$, on the other hand, does allow this event to be detected.

5   Conclusion

In this paper we demonstrated how the EVE framework can be applied to a complex problem, namely molecular event detection. Using this framework we were able to identify the points of interest and demonstrated that EVE can be used to detect events in challenging real world problems in a structure manner.

Acknowledgements. This research was partly supported by the DFG under grant GRK 1042 (Research Training Group "Explorative Analysis and Visualization of Large Information Spaces").

We would like to thank Dr. Thomas Exner and Fredrick Robin Devadoss for providing the data and explaining many chemical details of molecular dynamics.

References

1. Adä, I., Berthold, M.R.: Eve: a framework for event detection. Evolving Systems 4(1), 61–70 (2013)
2. Adcock, S.A., McCammon, J.A.: Molecular dynamics: survey of methods for simulating the activity of proteins. Chemical Reviews 106(5), 1589 (2006)
3. Askar, A., Cetin, A.E., Rabitz, H.: Wavelet transform for analysis of molecular dynamics. The Journal of Physical Chemistry 100(49), 19165–19173 (1996)
4. Benson, N.C., Daggett, V.: Dynameomics: Large-scale assessment of native protein flexibility. Protein Science 17(12), 2038–2050 (2008)
5. Benson, N.C., Daggett, V.: A chemical group graph representation for efficient high-throughput analysis of atomistic protein simulations. Journal of Bioinformatics and Computational Biology 10(04) (2012)
6. Benson, N.C., Daggett, V.: A comparison of multiscale methods for the analysis of molecular dynamics simulations. The Journal of Physical Chemistry B 116(29), 8722–8731 (2012)
7. Benson, N.C., Daggett, V.: Wavelet analysis of protein motion. International Journal of Wavelets, Multiresolution and Information Processing 10(04) (2012)
8. Bowen, J.P., Allinger, N.L.: Molecular mechanics: The art and science of parameterization. Reviews in Computational Chemistry, 81–97 (1991)
9. Chandola, V., Banerjee, A., Kumar, V.: Anomaly detection: A survey. ACM Computing Surveys (CSUR) 41(3), 15 (2009)
10. Coutsias, E.A., Seok, C., Dill, K.A.: Using quaternions to calculate rmsd. Journal of Computational Chemistry 25(15), 1849–1857 (2004)
11. Flocco, M.M., Mowbray, S.L.: C^{α}-based torsion angles: A simple tool to analyze protein conformational changes. Protein Science 4(10), 2118–2122 (1995)
12. Guralnik, V., Srivastava, J.: Event detection from time series data. In: Proceedings of the Fifth ACM SIGKDD International Conference on Knowledge Discovery and Data Mining, pp. 33–42. ACM (1999)

13. Kabsch, W.: A solution for the best rotation to relate two sets of vectors. Acta Crystallographica Section A 32(5), 922–923 (1976)
14. Karplus, M., McCammon, J.A.: Molecular dynamics simulations of biomolecules. Nature Structural & Molecular Biology 9(9), 646–652 (2002)
15. Kavraki, L.: Molecular distance measures [connexions web site] (June 2007), http://cnx.org/content/m11608/1.23/
16. Kullback, S.: The Kullback-Leibler distance. The American Statistician (1987)
17. Ramanathan, A., Agarwal, P.K., Kurnikova, M., Langmead, C.J.: An online approach for mining collective behaviors from molecular dynamics simulations. Journal of Computational Biology 17(3), 309–324 (2010)
18. Ramanathan, A., Yoo, J.O., Langmead, C.J.: On-the-fly identification of conformational substates from molecular dynamics simulations. Journal of Chemical Theory and Computation (2011)
19. Rapaport, D.C.: The art of molecular dynamics simulation. Cambridge Univ. Pr. (2004)
20. Shewhart, W.A.: Economic control of quality of manufactured product, vol. 509. American Society for Qualit (1980)
21. Smith, A., Datta, S.P., Smith, G.H., Campbell, P.N., Bentley, R., McKenzie, H.A., et al.: Oxford dictionary of biochemistry and molecular biology. Oxford University Press (OUP) (2000)
22. Teodoro, M.L., Phillips Jr., G.N., Kavraki, L.E.: Understanding protein flexibility through dimensionality reduction. Journal of Computational Biology 10(3-4), 617–634 (2003)
23. Wriggers, W., Stafford, K.A., Shan, Y., Piana, S., Maragakis, P., Lindorff-Larsen, K., Miller, P.J., Gullingsrud, J., Rendleman, C.A., Eastwood, M.P., et al.: Automated event detection and activity monitoring in long molecular dynamics simulations. Journal of Chemical Theory and Computation 5(10), 2595–2605 (2009)
24. Zliobaite, I.: Learning under concept drift: an overview. Technical report, Technical report, Faculty of Mathematics and Informatics, Vilnius University: Vilnius, Lithuania (2009)

Graph Clustering by Maximizing Statistical Association Measures

Julien Ah-Pine

University of Lyon, ERIC Lab
5, avenue Pierre Mendès France
69676 Bron Cedex, France
julien.ah-pine@eric.univ-lyon2.fr

Abstract. We are interested in objective functions for clustering undirected and unweighted graphs. Our goal is to define alternatives to the popular modularity measure. To this end, we propose to adapt statistical association coefficients, which traditionally measure the proximity between partitions, for graph clustering. Our approach relies on the representation of statistical association measures in a relational formulation which uses the adjacency matrices of the equivalence relations underlying the partitions. We show that graph clustering can then be solved by fitting the graph with an equivalence relation *via* the maximization of a statistical association coefficient. We underline the connections between the proposed framework and the modularity model. Our theoretical work comes with an empirical study on computer-generated graphs. Our results show that the proposed methods can recover the community structure of a graph similarly or better than the modularity.

Keywords: Graph clustering, Community detection, Statistical association measures, Modularity.

1 Introduction

Many real-world problems can be designed using graphs where entities of the studied system are represented as nodes and their relationships as edges between nodes. In many domains such as biology, ecology, social network analysis ..., graph theory tools are employed as means for representing complex systems. In this context, graph clustering consists in partitioning nodes into groups such that vertices belonging to the same group are better interconnected to each other than to vertices outside of the group. Discovering such clusters can lead to new and important insights. In biology for example, clustering a protein-protein interaction network helps to find proteins with the same biological function. Another example is in social network analysis, where graph clustering leads to the detection of community structures [1]. Such knowledge can help to better understand the social system and its related phenomenons.

There exist many graphs clustering techniques. We particularly focus on methods that optimize an objective function. The benefit criterion aims at reflecting

A. Tucker et al. (Eds.): IDA 2013, LNCS 8207, pp. 56–67, 2013.
© Springer-Verlag Berlin Heidelberg 2013

the quality of a clustering. In this context, density-based objective functions are well-known approaches. In these cases, clusters are defined as sub-graphs with high densities of edges. The modularity measure proposed by Newman and Girvan in [2] is a popular density-based objective function. It assumes that two nodes belong to the same community if the number of edges between them is greater than the expected number of edges under a null random model.

We address the graph clustering task from a viewpoint different from the one underlying the modularity. We suppose that an undirected and unweighted graph can be seen as a perturbed equivalence relation and finding groups of nodes can be interpreted as fitting the graph with a partition. To this end, we need to quantify the proximity between two partitions. In the statistical literature there are numerous coefficients addressing this exact problem. These criteria are known as statistical association measures (SAM) between categorical variables or partitions. Our proposal is thus to fit a given graph with a partition by maximizing a SAM. However, using such measures in this context is not straightforward. Indeed, these coefficients are typically defined by using contingency tables over the set of categories of the two partitions. Yet, the contingency table between a given graph (which is not an equivalence relation) and a partition does not exist. To overcome this drawback, we review the research works of Marcotorchino who showed in [3,4], that many SAM can be equivalently expressed through the adjacency matrices of the equivalence relations underlying the categorical variables. Based on this approach, we show how we can convert SAM to define new density-based quality functions for graph clustering.

In section 2, we recall some density-based objective functions for graph clustering. We particularly emphasize the modularity concept. Then, in section 3, we introduce our framework. We recall SAM both in their contingency and their relational formulations. Then we show how these measures can lead to graph clustering methods. Moreover, we study the relationships between the modularity and SAM. Next, in section 4, we empirically examine the behaviors of the proposed objective functions on artificial graphs and we compare their results with the ones provided by the modularity. We finally conclude and sketch some future works in section 5.

2   Related Work: Modularity Optimization

There are several types of density-based benefit functions for graph clustering [5,6]. One first family is based on graph cuts measures which iteratively split the set of nodes of a graph into two, providing that the density of edges between the two clusters is low. To apply such methods, one can generally use any max-flow/min-cut algorithm such as the Ford-Fulkerson one. Another method is spectral clustering which computes the Fiedler eigenvector of the Laplacian of the graph. Edges cuts criteria and the aforementioned algorithms are particularly used to tackle graph partitioning problems. These tasks are slightly distinct from graph clustering problems. In graph partitioning, the number of clusters and their sizes are known and one has to recover the correct partition given these pieces of information. In contrast, in graph clustering, we do not assume any information about the number nor the shape of the communities.

In order to better deal with the graph clustering task, Newman and Girvan proposed the modularity concept [2]. Their approach has the advantage to better formalize the concept of community and to avoid setting the number of clusters manually. This measure is denoted by Q and it can be expressed as follows [7] :
Q = Number of edges within communities − Expected number of such edges.

More formally, let us assume that we are given a graph with n vertices and m edges. Its adjacency matrix is denoted by A. Let us denote by P the pairwise matrix of general term P_{ij} which is the expected number of edges between nodes i and j. Since we are concerned with undirected and unweighted graph, P_{ij} can be interpreted as the probability to have an edge between i and j. The modularity can be formulated by the equation below [7] :

$$Q(A, \delta) = \frac{1}{2m} \sum_{i=1}^{n} \sum_{j=1}^{n} (A_{ij} - P_{ij}) \, \delta(g_i, g_j) \ . \tag{1}$$

where g_i is the cluster of i and $\delta(g_i, g_j) = 1$ if $g_i = g_j$ and 0 otherwise.

From this general formulation, Newman adopted different assumptions which led to the definition of a specific null random model. His hypothesis are the following ones [7] : (i) since the graph is undirected then P should satisfy the relation, $\forall i, j : P_{ij} = P_{ji}$; (ii) Q should be null when all vertices are in a single group and thus[1] $\sum_{i,j} A_{ij} = \sum_{i,j} P_{ij} = 2m$; (iii) the degrees distribution of the random model should be approximately the same as the one of the given graph which leads to the following constraint, $\forall i : \sum_j P_{ij} = k_i$ where $k_i = \sum_j A_{ij}$ is the observed degree of node i; (iv) edges should be placed at random meaning that the probability of observing an edge between i and j should be independent from the probability of observing an edge involving i and the probability of observing and edge involving j. Under these assumptions, the simplest null random model is when $\forall i, j : P_{ij} = k_i k_j / 2m$ [7]. Accordingly, the following modularity formulation is the one which is commonly used in the literature :

$$Q(A, \delta) = \frac{1}{2m} \sum_{i=1}^{n} \sum_{j=1}^{n} \left(A_{ij} - \frac{k_i k_j}{2m} \right) \delta(g_i, g_j) \ . \tag{2}$$

It is worthwhile to mention that apart from (2), other coefficients relying on the modularity concept could be designed from (1). In that perspective, Newman suggested that the assumptions (i) and (ii) are fundamental and should be considered as axioms of the modularity framework unlike (iii) and (iv) [7].

Adopting (2), one can optimally solve the graph clustering problem *via* modularity maximization with the following integer linear program (see for e.g. [8]) :

$$\max_{Y} \frac{1}{2m} \sum_{i,j} \left(A_{ij} - \frac{k_i k_j}{2m} \right) (1 - Y_{ij}) \text{ subject to : } \begin{cases} Y_{ik} \leq Y_{ij} + Y_{jk} \ \forall i, j, k \\ Y_{ij} \in \{0, 1\} \quad \forall i, j \end{cases} .$$
$$\tag{3}$$

---

[1] In order to lighten the notations we write $\sum_{i,j}$ as a shortcut for $\sum_{i=1}^{n} \sum_{j=1}^{n}$.

where $Y_{ij} = 1$ if i and j are not in the same cluster and 0 otherwise.

However, optimizing the modularity (and any other objective functions) over the set of possible partitions is an NP-hard problem. As a result, many research works have been devoted to approximately maximize the modularity with different strategies and heuristics [5,6].

The application of modularity to the graph clustering task has demonstrated good performances both on artificial and real-world networks. However, some recent works have highlighted certain limits of this method [9]. In particular, optimizing the modularity tends to split large groups while small communities below a certain threshold are not correctly detected.

In the next section, we introduce new quality functions for graph clustering which provide alternatives to the modularity criterion given in (2).

3   The Proposed Approach: Statistical Association Measures (SAM) Optimization

Density-based techniques typically rely on the definition of a community and use heuristics to discover sub-graphs satisfying this definition. From our viewpoint, since clustering a given graph consists in detecting a hidden community structure, we can interpret the graph as an equivalence relation perturbed by noise. Thereby, we argue that graph clustering can be thought of as recovering the real community structure and this can be achieved by fitting the graph with a partition. This approach assumes there is a way to assess the proximity between the graph and a partition. In what follows, we introduce some statistical association measures which aim at measuring the similarity between two partitions by means of contingency tables. Then, we recall the relational formulation of these coefficients due to Marcotorchino. Using the latter expressions of SAM, we show how to use these coefficients as benefit functions for the graph clustering task. In that perspective, we underline some theoretical links between SAM and the modularity concept in order to bring into light some conceptual similarities between these two frameworks in the context of graph clustering.

3.1   SAM and Their Relational Representation

Let us assume two categorical variables V^k and V^l with respectively p_k and p_l categories. Note that a categorical variable infers a set of disjoint groups of items which in turn can be interpreted as a partition or a clustering or an equivalence relation[2]. In categorical data analysis, in order to analyse the relationship between two categorical variables, we use the contingency table of dimensions $(p_k \times p_l)$ denoted by $\mathbf{N}$ whose general term is defined by : $\mathbf{N}_{uv}$ = Number of items belonging to both category u of V^k and category v of V^l.

Then, a core concept in categorical data analysis is the deviation from the statistical independence situation which occurs when for all pairs of categories

---

[2] Therefore, we will use these different terms interchangeably.

(u, v), the probability of jointly observing u and v equals the product between the probability of observing u and the probability of observing v. Using $\mathbf{N}$, this principle translates into the following formula[3] : $\forall (u, v) : \mathbf{N}_{uv}/n = (\mathbf{N}_{u.}\mathbf{N}_{.v})/n^2$ where $\mathbf{N}_{u.} = \sum_v \mathbf{N}_{uv}$. In this context, the greater the difference between $\mathbf{N}_{uv}$ and $(\mathbf{N}_{u.}\mathbf{N}_{.v})/n$ (for all pairs (u, v)), the stronger the relationship between the categorical variables.

Accordingly, we propose to study the following coefficients :

$$B(V^k, V^l) = \sum_{u=1}^{p_k} \sum_{v=1}^{p_l} \left(\mathbf{N}_{uv} - \frac{\mathbf{N}_{u.}\mathbf{N}_{.v}}{n} \right)^2 . \tag{4}$$

$$E(V^k, V^l) = \sum_{u=1}^{p_k} \sum_{v=1}^{p_l} \left(\mathbf{N}_{uv}^2 - \frac{\mathbf{N}_{u.}^2 \mathbf{N}_{.v}^2}{n^2} \right) . \tag{5}$$

$$J(V^k, V^l) = \frac{1}{n} \sum_{u=1}^{p_k} \sum_{v=1}^{p_l} \left(\mathbf{N}_{uv} \left(\mathbf{N}_{uv} - \frac{\mathbf{N}_{u.}\mathbf{N}_{.v}}{n} \right) \right) . \tag{6}$$

$$LM(V^k, V^l) = \sum_{u=1}^{p_k} \sum_{v=1}^{p_l} \frac{\mathbf{N}_{uv}^2}{\mathbf{N}_{u.}} - \frac{1}{n} \sum_{v=1}^{p_l} \mathbf{N}_{.v}^2 . \tag{7}$$

The SAM B, E, J and LM are respectively the Belson [10], Marcotorchino's square independence deviation [3], the Jordan[4] [11] and the Light-Margolin [12] criteria. They are all null in case of statistical independence. B and LM can only have positive values while E and J can be either positive or negative [3]. Given V^k, these coefficients achieve their maxima when V^l is exactly the same partition as V^k [3].

The contingency representation is the usual way to introduce SAM. However, there exists an equivalent representation of these coefficients which emphasizes the relational nature of categorical variables. Indeed, as we mentioned beforehand, categorical variables are equivalence relations and such algebraic structures can be represented by graphs. This point of view was adopted by Marcotorchino and enabled him to formulate SAM with adjacency matrices[5] [3,4]. Let us denote by C^k the adjacency matrix[6] associated to V^k. Its general term is defined by $C_{ij}^k = 1$ if i and j belong to the same category and 0 otherwise. Marcotorchino provided correspondence formulas between the contingency table $\mathbf{N}$ on the one hand and the relational representations C^k and C^l on the other hand [3,4]. Here are some of these transformation formulas : (i) $\sum_{u=1}^{p_k} \sum_{v=1}^{p_l} \mathbf{N}_{uv}^2 = \sum_{i=1}^{n} \sum_{j=1}^{n} C_{ij}^k C_{ij}^l$; (ii) $\sum_u \mathbf{N}_{u.}^2 = \sum_{i,j} C_{ij}^k$; (iii) $\sum_{u,v} \mathbf{N}_{uv}\mathbf{N}_{u.}\mathbf{N}_{.v} = \sum_{i,j}((C_{i.}^k +$

---

[3] In order to lighten the notations we write $\sum_u$ as a shortcut for $\sum_{u=1}^{p_k}$.

[4] It is actually an interpretation of Jordan's measure given by Marcotorchino in [3].

[5] The study of association and aggregation of binary relations using graph theory and mathematical programming led to the Relational Analysis method developed by Marcotorchino and which has many applications in statistics, data-mining and multiple-criteria decision making (see for e.g. [13] and references therein).

[6] Also called relational matrix in the Relational Analysis framework.

$C^k_{.j})/2)C^l_{ij}$; (iv) $\sum_{u,v}(N^2_{uv}/N_{u.}) = \sum_{i,j}(2C^k_{ij}/(C^k_{i.} + C^k_{.j}))C^l_{ij}$ where $C^k_{i.} = \sum_j C^k_{ij}$ and $C^k_{i.} = C^k_{.i}$ since C^k is symmetric.

Applying these correspondence formulas enables the following expressions of SAM in terms of C^k and C^l :

$$B(C^k, C^l) = \sum_{i=1}^{n}\sum_{j=1}^{n}\left(C^k_{ij} - \frac{C^k_{i.}}{n} - \frac{C^k_{.j}}{n} + \frac{C^k_{..}}{n^2}\right)C^l_{ij} . \tag{8}$$

$$E(C^k, C^l) = \sum_{i=1}^{n}\sum_{j=1}^{n}\left(C^k_{ij} - \frac{C^k_{..}}{n^2}\right)C^l_{ij} . \tag{9}$$

$$J(C^k, C^l) = \frac{1}{n}\sum_{i=1}^{n}\sum_{j=1}^{n}\left(C^k_{ij} - \frac{1}{2}\left(\frac{C^k_{i.}}{n} + \frac{C^k_{.j}}{n}\right)\right)C^l_{ij} . \tag{10}$$

$$LM(C^k, C^l) = \sum_{i=1}^{n}\sum_{j=1}^{n}\left(\frac{2C^k_{ij}}{C^k_{i.} + C^k_{.j}} - \frac{1}{n}\right)C^l_{ij} . \tag{11}$$

It is noteworthy that the different formulations of the statistical independence deviation with contingency tables in (4), (5), (6) and (7), translate into different types of deviation concepts in the relational representation (8), (9), (10) and (11). Such properties were examined in [14] and led to the formalization of the central tendency deviation principle in cluster analysis. Indeed, one can observe the following central tendencies : in (9) $C^k_{..}/n^2$ is the mean average over all the terms of C^k; in (10) $(C^k_{i.} + C^k_{.j})/2n$ is the arithmetic mean of $C^k_{i.}/n$ and $C^k_{.j}/n$ and in (11) $1/n$ is the mean average over all terms of the matrix of general term $2C^k_{ij}/(C^k_{i.} + C^k_{.j})$ (which is equivalent to $C^k_{ij}/C^k_{i.}$). Regarding (8), the central tendency concept is of geometrical nature. Since C^k is a dot product matrix (or Gram matrix) the transformation of C^k_{ij} into $C^k_{ij} - C^k_{i.}/n - C^k_{.j}/n + C^k_{..}/n^2$ is known as the double centering (or Torgerson) transformation. This operation results in dots products between vectors centered with respect to the mean vector.

Now that we have provided the expression of SAM using the graph relations underlying partitions, we show in the next paragraph how to employ such criteria for clustering graphs.

3.2   Graph Clustering by Maximizing SAM

We interpret a given undirected and unweighted graph as a perturbed equivalence relation and clustering the graph can be seen as attempting to recover the real partition. To solve the graph clustering task, we thus propose to fit the graph encoded by its adjacency matrix A with an equivalence relation by maximizing one of the SAM introduced previously. In other words, we want to find the partition that is the most similar to A according to a given SAM. More formally, we introduce the following benefit functions for clustering graphs :

$$B(A, X) = \sum_{i=1}^{n}\sum_{j=1}^{n}\left(A_{ij} - \left(\frac{A_{i.}}{n} + \frac{A_{.j}}{n} - \frac{A_{..}}{n^2}\right)\right)X_{ij} . \tag{12}$$

$$E(A, X) = \sum_{i=1}^{n} \sum_{j=1}^{n} \left(A_{ij} - \frac{A_{..}}{n^2} \right) X_{ij} . \tag{13}$$

$$J(A, X) = \frac{1}{n} \sum_{i=1}^{n} \sum_{j=1}^{n} \left(A_{ij} - \frac{1}{2} \left(\frac{A_{i.}}{n} + \frac{A_{.j}}{n} \right) \right) X_{ij} . \tag{14}$$

$$LM(A, X) = \sum_{i=1}^{n} \sum_{j=1}^{n} \left(\frac{2A_{ij}}{A_{i.} + A_{.j}} - \frac{1}{n} \right) X_{ij} . \tag{15}$$

where X is the adjacency matrix of the partition we want to recover and whose general term is $X_{ij} = 1$ if nodes i and j are in the same cluster and 0 otherwise.

X represents an equivalence relation which, from an algebraic standpoint, is a binary relation with the following properties : (i) reflexivity ($X_{ii} = 1, \forall i$); (ii) symmetry ($X_{ij} = 1 \Leftrightarrow X_{ji} = 1, \forall i, j$) and (iii) transitivity ($X_{ij} = 1 \wedge X_{jk} = 1 \Rightarrow X_{ik} = 1, \forall i, j, k$). Marcotorchino and Michaud showed that these relational properties can be formulated as linear constraints through X [15] and this finding allowed them to model the clustering problem as an integer linear program :

$$\max_{X} \Delta(A, X) \text{ subject to :} \begin{cases} X_{ii} = 1 & \forall i \\ X_{ij} - X_{ji} = 0 & \forall i, j \\ X_{ij} + X_{jk} - X_{ik} \leq 1 & \forall i, j, k \\ X_{ij} \in \{0, 1\} & \forall i, j \end{cases} \tag{16}$$

where, in our case, $\Delta(A, X)$ is either (12) or (13) or (14) or (15).

It is important to mention that this integer linear program allowed Marcotorchino to design the maximal association model for clustering data described by categorical variables. In Marcotorchino's works, A was considered as an equivalence relation [15] or the sum over several equivalence relations [16]. Our proposal can thus be understood as the extension of the maximal association framework to graph clustering by considering A to be a general adjacency matrix without any particular property (except being undirected).

Before moving to the section dedicated to the experiments, we establish some interesting relationships between the modularity framework and our proposal based on SAM.

3.3   Some Relationships between Modularity and SAM

Firstly, using the notations introduced previously, the standard modularity defined in (2) can be reformulated as below :

$$Q(A, X) = \frac{1}{A_{..}} \sum_{i=1}^{n} \sum_{j=1}^{n} \left(A_{ij} - \frac{A_{i.} A_{.j}}{A_{..}} \right) X_{ij} . \tag{17}$$

In addition to the correspondence formulas given in paragraph 3.1, let us introduce the following identity : (v) $\sum_{v}(\sum_{u} \mathbf{N}_{u.} \mathbf{N}_{uv})^2 = \sum_{i,j} C_{i.}^{k} C_{.j}^{k} C_{ij}^{l}$ [4]. From this equation, if we identify C^k and C^l to A and X respectively and if we assume

A and X to be partitions associated to two categorical variables V^k and V^l, we can easily show that the modularity given in (17) can be expressed by means of a contingency table as follows :

$$Q(V^k, V^l) = \frac{1}{\sum_{u=1}^{p_k} \mathbf{N}_{u.}^2} \left(\sum_{u=1}^{p_k} \sum_{v=1}^{p_l} \mathbf{N}_{uv}^2 - \frac{1}{\sum_{u=1}^{p_k} \mathbf{N}_{u.}^2} \left(\sum_{v=1}^{p_l} \left(\sum_{u=1}^{p_k} \mathbf{N}_{u.} \mathbf{N}_{uv} \right)^2 \right) \right).$$
(18)

Then, one can easily check that $Q(V^k, V^l)$ is null in case of statistical independence between V^k and V^l. This outcome shows the potential application of the modularity measure in categorical data analysis.

Let us now place the SAM in the context of the modularity concept developed by Newman. Let us formally introduce the following central tendencies : $\mu_{ij}^Q = A_{i.} A_{.j} / A_{..}$; $\mu_{ij}^B = A_{i.}/n + A_{.j}/n - A_{..}/n^2$; $\mu_{ij}^E = A_{..}/n^2$; $\mu_{ij}^J = A_{i.}/(2n) + A_{.j}/(2n)$ and $\mu_{ij}^{LM} = 1/n$. In that case, (17), (12), (13), (14) and (15) can all be reformulated as : $\alpha \sum_{i=1}^n \sum_{j=1}^n (A_{ij} - \mu_{ij}^Z)$ with $Z \in \{Q, B, E, J, LM\}$, $\alpha = 1$ when $Z \in \{B, E, LM\}$, $\alpha = 1/A_{..}$ when $Z = Q$, $\alpha = 1/n$ when $Z = J$ and by substituting A_{ij} with $\hat{A}_{ij} = 2A_{ij}/(A_{i.} + A_{.j})$ when $Z = LM$. This expression of SAM better underlines the connections between the general modularity framework given in (1) and cluster analysis methods based on the central tendency deviation principle [14]. Furthermore, one can easily check that Newman's axioms we recalled in section 2 are both satisfied by all SAM under study except the LM method : (i) $\forall i, j : \mu_{ij}^Z = \mu_{ji}^Z$ for $Z \in \{B, E, J, LM\}$; (ii) $\sum_{i,j} A_{ij} = \sum_{i,j} \mu_{ij}^Z$ for $Z \in \{B, E, J\}$. As a result, B, E and J fit in the modularity model. However, hypothesis (iii) and (iv) are not satisfied by any SAM under study except B for which we have (iii) $\forall i : \sum_j \mu_{ij}^B = k_i = A_{i.}$.

In such a context, it is also interesting to notice that the SAM E given in (13) corresponds to another suggested modularity model which assumes a Bernoulli distribution for P_{ij} in (1) and which boils down to the following constant[7], $P_{ij} = A_{..}/n^2$, $\forall i, j$ [7].

After having introduced the proposed objective functions and some properties about the relationships between the modularity concept and SAM, we examine in the next section if our proposals lead to interesting graph clustering methods from an empirical standpoint. Another goal of these experiments is to enable us to initiate a comparison between the modularity framework and SAM based optimization with regard to the hypothesis underlying each method.

4   Experiments

Our experiments are based on computer-generated graphs of different sizes. We give below the details about the algorithm we used to maximize the different density-based objective functions presented previously. We explain the tool we

---

[7] Note that in the case of E, we assume that the graph is reflexive unlike Q. In the latter case, the constant is $A_{..}/(n(n-1))$.

employed and the parameters we set to generate the artificial graphs. We then analyse the quality of the graph clustering results obtained with the different techniques.

4.1 Greedy Optimization by Agglomerative Hierarchical Clustering

The optimization problems given in (16) and (3) are NP-hard[8], and thus, numerous heuristics attempting to provide sub-optimal solutions have been proposed (see for e.g. the surveys [5,6,1]). In our experiments, we used the greedy optimization algorithm proposed by Newman in [17] in order to maximize the modularity criterion given in (2).

This heuristic is based on a simple agglomerative hierarchical clustering strategy. It starts with n distinct clusters and at each iteration it merges the two clusters that allow the best improvement of the modularity value. The merging process goes on until there is no pair of clusters whose fusion enables increasing the modularity value. This heuristic has the advantage to avoid fixing the number of clusters as a parameter.

This algorithm can be applied to other kinds of quality measures and in order to provide a fair comparison between the different objective functions, we thus used this technique to maximize (12), (13), (14) and (15) as well.

4.2 LFR Benchmark Graphs

The computer-generated graphs we analyzed in our experiments rely on the LFR benchmarks proposed by Lancichinetti, Fortunato and Radicchi in [18,19]. These benchmarks aim at providing the research community with graphs whose properties reflect real-world cases. Indeed, observed complex networks are characterized by heterogeneous distributions both for node degrees and cluster sizes. As a consequence, the authors developed a model that generates graphs which satisfy these features. They also implemented a freely available tool[9] that we used to conduct our empirical work.

Their approach is based on the planted l-partition model in which node degrees follow a power law distribution with exponent τ_1 and clusters size a power law distribution with exponent τ_2. Overall the parameters of their model are : (i) n the number of nodes; (ii) the average degree of nodes; (iii) the maximum degree of nodes; (iv) τ_1; (v) τ_2 and (vi) $\mu \in [0,1]$ the mixing parameter. The latter parameter μ is the one that allows gradually monitoring the presence or the absence of a community structure in the graph. It represents the percentage of edges that a node shares with vertices that do not belong to its group. Therefore, as μ grows, the community structure progressively degrades and the

---

[8] Note that the constraints in (16) and in (3) are equivalent : the former models the properties of an equivalence relation while the latter models the properties of the complementary of an equivalence relation which is a distance relation satisfying the triangle inequality constraint.

[9] http://santo.fortunato.googlepages.com/inthepress2

limit case $\mu = 1$ corresponds to the situation where edges are totally placed randomly. Typically, we assume that there is a strong community structure within the graph as long as $\mu < 0.5$.

4.3   Experiments Settings and Results

We studied graphs of different sizes : 500, 1000 and 2000 nodes. We used the same parameters values as in [19] : the average degree was set to 20; the maximum degree to 50; $\tau_1 = 2$; $\tau_2 = 1$ and we vary μ from 0 to 0.7 with a 0.1 step.

The LFR benchmark graphs also provide a built-in community structure which allowed us to compare the clustering results with the real partition. To assess the proximity between the found partition and the ground-truth, the normalized mutual information (NMI) measure was used. Let us denote by V^k the clustering output of our algorithm with p_k clusters and by V^l the real clustering with p_l groups. Let $P(V^k = u, V^l = v) = P(u,v) = \mathbf{N}_{uv}/n$ be the probability of jointly observing u and v and let $P(u) = \mathbf{N}_{u.}/n$ and $P(v) = \mathbf{N}_{.v}/n$ be the probability of observing u and v respectively. The mutual information measure between V^k and V^l denoted by $I(V^k, V^l)$ is defined as follows : $I(V^k, V^l) = \sum_{u=1}^{p_k} \sum_{v=1}^{p_l} P(u,v) \log(P(u,v)/(P(u)P(v)))$. Its normalized version denoted by $NMI(V^k, V^l)$ is then given by $NMI(V^k, V^l) = 2I(V^k, V^l)/(H(V^k) + H(V^l))$ where $H(V^k) = -\sum_u P(u) \log(P(u))$ is the entropy of V^k. The NMI measure ranges from 0 and 1. It equals 1 when $V^k = V^l$ and it is null when V^k and V^l are statistically independent. Note that we used the NMI coefficient to assess our clustering models because this measure is often used in the graph clustering literature (see for e.g. [19]). In that way, we can also position our contributions with respect to other papers and graph clustering techniques.

To have a better estimation of the performances, we generated 5 different graphs for each distinct parameter setting and we took the median value. The experimental results obtained for NMI measures are shown in the first row of Fig. 1. We also computed the number of clusters found by each method in order to examine if the different techniques are able to recover the right number of clusters. In this case, we also took the median over the 5 trials. These results are shown in the second row of Fig. 1.

The first row of Figure 1 allows us to compare the quality of the clustering outputs found by the different methods. We claim the following outcomes : (i) as expected, all the methods have their quality diminishing as μ grows; (ii) the LM coefficient dominates the modularity Q and other methods and this superiority seems to grow with the size of the graphs; (ii) B and J perform similarly than Q whatever the size of the graphs; (iii) E is the less good approach and it particularly performs the worst when $\mu \in [0.3, 0.7]$.

Concerning the number of clusters found, we can make the following observations from the second row of Fig. 1 : (i) the number of real communities provided by the LFR benchmarks is stable and varies around 20, 40 and 80 for graphs of size 500, 1000, 2000 respectively; (ii) except for E with graphs of size 500, all the techniques produce less clusters than the correct number of groups; (iii) as μ grows the number of clusters decreases for all the methods and beyond a certain

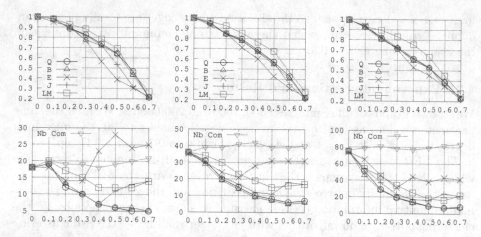

Fig. 1. First row : NMI values (vertical axis) *versus* mixing parameters μ (horizontal axis). Second row : Number of clusters found (vertical axis) *versus* mixing parameters μ (horizontal axis). We show the curves for each objective function. From left to right, the plots correspond to graphs with 500, 1000 and 2000 nodes respectively.

point (approximately 0.6 for Q, B, LM; 0.5 for J and 0.3 for E), it tends to either grow or stabilize; (iv) the LM criterion tends to produce more clusters than Q, B and J.

Overall, B and J are comparable to Q while LM is clearly a better objective function than the other ones. One reason that could explain the superiority of LM is the fact that it implicitly transforms the binary matrix A into a non negative one, $\widehat{A}$ whose general term is $\widehat{A}_{ij} = 2A_{ij}/(A_{i.} + A_{.j})$. Then its related central tendency scheme, $\mu_{ij}^{LM} = 1/n$, gives the same value for all pairs of nodes (i, j). Such an approach is indeed different from the other quality functions we examined, since they all keep the binary matrix but what changes from one function to the other is the underlying central tendency scheme μ^Z, $Z \in \{Q, B, J, E\}$ as we underlined in paragraph 3.3.

Moreover, our experimental results invite us to further analyze the hypothesis underlying the different objective functions. Regarding Newman's assumptions for the modularity given in (2), it is interesting to notice that B and J perform similarly than Q despite the fact they do not satisfy all the hypothesis assumed by the latter criterion. More importantly, LM violates most of Newman's hypothesis but outperforms all other methods including Q.

5   Conclusion

We have proposed new objective functions for clustering undirected and unweighted graphs. Our method consists in maximizing SAM represented in their relational representation, in order to fit the given graph with a partition. Our empirical study on artificial graphs has shown encouraging results. Most of the proposed SAM perform equivalently or better than the modularity criterion. In particular, the Light-Margolin coefficient dominates the latter approach. As for future work, we plan to develop the analysis provided in paragraph 3.3 and leverage the empirical results presented previously by further comparing the modularity concept and SAM from a theoretical viewpoint. We also plan to extend our experiments on larger graphs both for computer-generated cases and for real-world networks in order to further validate our findings.

References

1. Papadopoulos, S., Kompatsiaris, Y., Vakali, A., Spyridonos, P.: Community detection in social media. Data Min. Knowl. Discov. 24(3), 515–554 (2012)
2. Newman, M.E.J., Girvan, M.: Finding and evaluating community structure in networks. Phys. Rev. E 69(2), 026113 (2004)
3. Marcotorchino, J.F.: Utilisation des comparaisons par paires en statistique des contingences partie I. Technical Report F069, IBM (1984)
4. Marcotorchino, J.F.: Utilisation des comparaisons par paires en statistique des contingences partie II. Technical Report F071, IBM (1984)
5. Schaeffer, S.E.: Graph clustering. Computer Science Review 1(1), 27–64 (2007)
6. Fortunato, S.: Community detection in graphs. Physics Reports 486(35), 75–174 (2010)
7. Newman, M.E.J.: Finding community structure in networks using the eigenvectors of matrices. Physical Review E 74(3), 036104 (2006)
8. Agarwal, G., Kempe, D.: Modularity-maximizing graph communities via mathematical programming. The European Physical Journal B 66(3), 409–418 (2008)
9. Lancichinetti, A., Fortunato, S.: Limits of modularity maximization in community detection. Phys. Rev. E Stat. Nonlin. Soft Matter Phys. 84(6 Pt. 2), 066122 (2011)
10. Belson, W.: Matching and prediction on the principle of biological classification. Applied Statistics 7 (1959)
11. Jordan, C.: Les coefficients d'intensité relative de Korosy. Revue de la société hongroise de statistique 5 (1927)
12. Light, R.J., Margolin, B.H.: An analysis of variance for categorical data. Journal of the American Statistical Association 66(335), 534–544 (1971)
13. Ah-Pine, J., Marcotorchino, J.F.: Overview of the relational analysis approach in data-mining and multi-criteria decision making. In: Usmani, Z.U.H. (ed.) Web Intelligence and Intelligent Agents (2010)
14. Ah-Pine, J.: Cluster analysis based on the central tendency deviation principle. In: Huang, R., Yang, Q., Pei, J., Gama, J., Meng, X., Li, X. (eds.) ADMA 2009. LNCS (LNAI), vol. 5678, pp. 5–18. Springer, Heidelberg (2009)
15. Marcotorchino, J.F., Michaud, P.: Heuristic approach of the similarity aggregation problem. Methods of Operations Research 43, 395–404 (1981)
16. Marcotorchino, J.F.: Maximal association theory as a tool of research. In: Classification as a Tool of Research, pp. 275–288. North-Holland, Amsterdam (1986)
17. Newman, M.: Fast algorithm for detecting community structure in networks. Physical Review E 69 (September 2003)
18. Lancichinetti, A., Fortunato, S., Radicchi, F.: Benchmark graphs for testing community detection algorithms. Phys. Rev. E 78(4),0 46110 (2008)
19. Fortunato, S., Lancichinetti, A.: Community detection algorithms: a comparative analysis: invited presentation, extended abstract. In: Stea, G., Mairesse, J., Mendes, J. (eds.) VALUETOOLS, vol. 27. ACM (2009)

Evaluation of Association Rule Quality Measures through Feature Extraction[*]

José L. Balcázar[1] and Francis Dogbey[2]

[1] Departament de Llenguatges i Sistemes Informàtics
Universitat Politècnica de Catalunya
Barcelona, Spain
jose.luis.balcazar@upc.edu
[2] Advanced Information Technology Institute
Ghana-India Kofi Annan Centre of Excellence in ICT
Accra, Ghana
francisd@aiti-kace.com.gh

Abstract. The practical success of association rule mining depends heavily on the criterion to choose among the many rules often mined. Many rule quality measures exist in the literature. We propose a protocol to evaluate the evaluation measures themselves. For each association rule, we measure the improvement in accuracy that a commonly used predictor can obtain from an additional feature, constructed according to the exceptions to the rule. We select a reference set of rules that are helpful in this sense. Then, our evaluation method takes into account both how many of these helpful rules are found near the top rules for a given quality measure, and how near the top they are. We focus on seven association rule quality measures. Our experiments indicate that multiplicative improvement and (to a lesser extent) support and leverage (a.k.a. weighted relative accuracy) tend to obtain better results than the other measures.

Keywords: Association rules, Feature Extraction, Prediction, Support, Confidence, Lift, Leverage, Improvement.

1 Introduction and Preliminary Definitions

Association rules have the syntactic form of implications, $X \rightarrow A$, where X is a set of items and A is an item. Although several works allow for several items in the right-hand side (often under the term *partial implications*, e.g. [20]), here, like in many other references, we follow [2,6] and restrict ourselves to single-item consequents. Semantically, association rule $X \rightarrow A$ is expected to mean that A tends to appear when X appears. However, there are many different formalizations for such a relaxed implication; this issue applies both to supervised learning and in the context of associations, see [13,15,17,18,24].

[*] This work has been partially supported by project BASMATI (TIN2011-27479-C04) of Programa Nacional de Investigación, Ministerio de Ciencia e Innovación (MICINN), Spain, and by the Pascal-2 Network of the European Union.

A. Tucker et al. (Eds.): IDA 2013, LNCS 8207, pp. 68–79, 2013.

We aim at finding ways to objectively evaluate the rule quality measures themselves. One existing approach consists of generating algorithmically datasets where certain itemsets are to be found later [27]. We explore an alternative approach to the same end. This paper is based on the following informal working hypothesis: since different rule quality measures select different association rules, we might be able to compare objectively these sets of "top" rules by combining them with other Data Mining decision processes, and evaluating their contribution. Specifically, here we relate associations with feature extraction and with predictors.

In classification problems of Machine Learning, feature selection and extraction are processes, often considered as preprocessing, through which better accuracies can be sometimes obtained. Along these processes, every observation acquires, explicitly or implicitly, additional columns, that may even fully replace the original features. It is a wide area with a large body of references: see [16] and the references there.

In our previous paper [4], we explored the average accuracy improvements (which can be negative) of predictors as new features are added to the observations. Each of these added features is Boolean, and flags exceptions to association rules. In turn, these association rules are picked as the "top-N", according to each of several popular rule quality measures. In that paper, our results indicated, essentially, very little average positive accuracy increment or, in many cases, none at all. Also, the differences among the average accuracy increments for several quality measures were rather marginal.

Hence, we focus now on individual rules that actually do improve accuracy. We propose to evaluate rule quality measures according to how many of these helpful rules are found among the top-quality rules for the measure, and how close to the top they appear. We study two variants, depending on whether it is desired to employ the new feature for actual classification tasks. We demonstrate our new score on seven quality measures, and find that multiplicative improvement and, to a lesser extent, support and leverage (a.k.a. weighted relative accuracy) tend to obtain better scores.

We must point out here that we do not wish to argue, nor even implicitly suggest, that helpfulness for classifiers, at the current status of our understanding, is to be equated with subjective interest for the user. However, our proposal combines, in an interesting, novel way a precise formal definition, leading to numeric scores, with other data mining processes that could take place on the same data.

2   Setting

Our notation is standard. The reader is assumed to be familiar with association rules (see the survey [9]), and with the C5.0 tree-based predictor, an improved version of C4.5 [23]. See also e.g. [25] and the references there.

For our development, we need our datasets both in relational and transactional formats. In transactional datasets, each observation ("row" or "transaction") consists of a set of items. In the relational case, each observation ("row") consists of an observed value for each of a fixed set of attributes ("columns"). We use a relational and a transactional variant for each dataset. Association rules are computed on the transactional variant, then used to add extra features (new columns) to the relational variant before passing it on to the predictor. Transactional data can be seen as relational in several non-equivalent ways; here, we consider one binary column per item. Thus the association rules only use *positive* appearances from the relational table. Relational datasets are casted in transactional form through the fully standard approach of adding one item for each existing pair ⟨attribute,value⟩.

The support of an itemset in a transactional dataset is the number of transactions in which it is contained. It is used often normalized, by dividing by the total number of transactions, and then it is akin to a frequentist estimator of the probability of the itemset. Accordingly, we will denote by $p(X)$ the normalized support of itemset X. This notation is naturally extended to conditional probabilities $p(A|X) = p(AX)/p(X)$.

2.1   Rule Quality Measures

There are literally dozens of measures proposed in the literature for choosing association rules. They assign a real number as a value to each association rule. This value can be used either by thresholding, thus discarding rules that receive low value, or by priorization, by sorting the rules obtained according to their value. There are many studies of the properties of the different measures; see [15,17,18,24]. To evaluate association rule $X \to A$, most measures involve an arithmetic combination of the quantities $p(X)$, $p(A)$, and $p(XA)$. We define now the measures we are interested in.

Definition 1. *[1] The* support *of association rule $X \to A$ is $s(X \to A) = p(XA)$.*

Definition 2. *[1,20] The* confidence *of association rule $X \to A$ is $c(X \to A) = p(A|X) = p(XA)/p(X)$.*

The most basic and widely applied scheme for selection of association rules consists in imposing thresholds for the support (thereby reducing the exponentially growing exploration space of itemsets) and for the confidence.

Definition 3. *[17] The* relative confidence *of association rule $X \to A$, also called* centered confidence *or* relative accuracy*, is $r(X \to A) = p(A|X) - p(A) = c(X \to A) - c(\emptyset \to A)$.*

The relative confidence is, therefore, measuring additively the effect of "adding the condition" or antecedent X on the support of the consequent A.

Definition 4. *[8] The* lift *of association rule $X \to A$ is $\ell(X \to A) = \frac{p(XA)}{p(X) \times p(A)}$.*

Definition 5. *[17,22] The* leverage *of association rule $X \to A$ is $v(X \to A) = p(XA) - p(X) \times p(A)$.*

Both in lift and in leverage, if supports are unnormalized, extra factors n are necessary. A widespread criticism of lift and leverage is that they are symmetric, that is, give the same value to the rule obtained by permuting antecedent and consequent, and, therefore, fail to support the intuitive directionality of the syntax $X \rightarrow A$.

In case of independence of both sides of the rule at hand, $p(XA) = p(X)p(A)$; therefore, both lift and leverage are measuring deviation from independence: lift measures it multiplicatively, and leverage does it additively. By multiplying and dividing by $p(X)$, it is easy to check that leverage can be rewritten as $v(X \rightarrow A) = p(X)(p(A|X) - p(A)) = p(X)r(X \rightarrow A)$, and is therefore called also *weighted relative accuracy* [17].

One of the few published measures that do not depend only on the supports $p(X), p(A), p(XA)$, and similar quantities, but requires instead exploring a larger space, is:

Definition 6. *[5,19] The improvement of association rule $X \rightarrow A$, where $X \neq \emptyset$, is $i(X \rightarrow A) = \min\{c(X \rightarrow A) - c(Y \rightarrow A)|Y \subset X\}$.*

Improvement generalizes relative confidence by considering not only the alternative rule $\emptyset \rightarrow A$ but all rules where the left-hand side is some proper subset of X. The same process can be applied to lift, which is, actually, a multiplicative, instead of additive, version of relative confidence: $\ell(X \rightarrow A) = \frac{p(XA)}{p(X) \times p(A)} = \frac{c(X \rightarrow A)}{p(A)} = \frac{c(X \rightarrow A)}{c(\emptyset \rightarrow A)}$. Taking inspiration in this correspondence, we introduced in [4] a multiplicative variant of improvement that generalizes lift, exactly in the same way as improvement generalizes relative confidence:

Definition 7. *The multiplicative improvement of association rule $X \rightarrow A$, where $X \neq \emptyset$, is $m(X \rightarrow A) = \min\{c(X \rightarrow A)/c(Y \rightarrow A)|Y \subset X\}$.*

2.2   Datasets

A key condition we have imposed on our empirical study is to employ datasets that allow for sensible relational and transactional formulations without resorting to further algorithmics. In this sense, numeric attributes may need a discretization process and, then, it would be far from clear how to tell which empirical effects were due to the interaction between the associator and the classifier, as we wish to study, and which ones were introduced by the discretization phase. We restrict ourselves to datasets that allow for direct run of a standard associator.

We introduce now the datasets on which we run our tests. They are among the far most common benchmark datasets in association rule quality studies. All of them are publicly available [11].

CMC is short for Contraceptive Method Choice, and contains data from an Indonesian survey of demographic features, socio-economic features and contraceptives choices (predicted attribute) among married women. The dataset is made up of 1473 observations and 10 attributes. ADULT contains census information, extracted from questionnaire data. It has 48842 observations and 11

Table 1. Size, items, support, rules, and initial accuracies of C5.0

Dataset	Size	Nb. of items	Attrib. Predict	Supp.	Nb of rules	C5.0 Accur.
CMC	1473	74	Contracep. Meth.	1%	19067	50.17%
ADULT	48842	272	Income	1.18%	19840	83.49%
GERMAN	1000	1077	Customer class	11.75%	19762	72.69%
VOTES	435	50	Party	22.5%	19544	95.40%
MUSHROOM	8124	119	Cap surface	15.9%	19831	53.30%

attributes. As usual for this dataset, we predict whether the annual income of individuals is over a certain threshold (50K). We prepared the adult dataset by merging the known train and test sets. We removed the four numeric attributes fnlwgt (final weight), education-num (which is redundant), capital-gain, and capital-loss. GERMAN is a dataset on credit scoring from Germany. There, the data from clients must be used to predict whether the client is a good candidate to receive a loan. It has been suggested that prediction should be made by weighting differently (by a factor of 5) mistakes in predicting good than mistakes in predicting bad; however, in order to keep fair comparisons with all our other datasets, here we did not weigh differently the mistakes. VOTES records votes of representatives from the US Congress to a number of law proposals in 1984; the attribute to be predicted is the party the representative belongs to. Finally, MUSHROOM reports characteristics that can appear together in a sample of certain mushroom species. The usual prediction task for this dataset is whether the exemplar is edible or poisonous. However, the predictor attains full accuracy directly on the original data, which renders the task useless to evaluate association rule quality measures according to our proposal. Therefore, we only report on results predicting the sort of "cap surface" of the exemplar, given the rest of the attributes. The attributes to be predicted are two-valued for all the datasets, except for CMC and MUSHROOM, where predictions are respectively three-valued and four-valued.

Table 1 indicates the main characteristics of the chosen datasets. The *size* is the number of observations; the *number of items* is the total of different values existing for all the attributes, and, therefore, coincides with the number of items in the transactional version of the dataset. For each dataset, we impose a support constraint and compute association rules with no confidence constraint with a standard Apriori associator [6]. Since our purposes are essentially conceptual, here we are not (yet) after particularly fast algorithmics, and that associator was often fast enough. The support is set at a value appropriate for Apriori to yield between 19000 and 20000 rules in these conditions. These supports were identified manually, and are also reported in the table, together with the number of rules. We also report in the table the baseline accuracies of C5.0, for each dataset, before any extra features are added.

The feature extraction and accuracy test is developed as follows. For a given association rule, an enlarged relational version of the dataset is fed to the predictor: it contains one additional binary column, which indicates whether each row is an exception of the association rule, that is, fulfills the antecedent of the rule but not the consequent. Only rules of confidence higher than 50% and strictly lower than 100% (in order to ensure the presence of exceptions) are recorded; also, only rules with nonempty antecedent are employed: otherwise, the additional feature would be redundant, flagging just failure of the consequent. The parameter settings of C5.0 are left at their default values. The accuracy of the predictor on the enlarged dataset is compared to the accuracy obtained on the original unexpanded dataset. All accuracies are computed by 10-fold cross-validation.

In [4], for each rule quality measure under study, the top N rules were recorded, and the accuracy test performed as just described; then, the process continued by averaging, separately by measure, the resulting accuracy changes. The average accuracy, however, changed little if we compare among them the different rule quality measures, and its ability to evaluate rule quality measures is, thus, debatable. Thus, a finer analysis is needed. This is the aim of the present paper.

3   A Score for Rule Quality Measures

This section describes our major contribution: a score for association rule quality measures in terms of usefulness for predictive tasks. Our proposal for a finer evaluation of the rule quality measures is as follows: first, one identifies a number of "good", helpful, rules, that is, rules whose corresponding new features do improve clearly the accuracy of the predictors. Then, we can evaluate how many of these good rules actually appear among the top N rules according to each of the measures. Thus, we ask ourselves which rule quality measure is able to pick these "good" rules.

Let us move on into the details. First of all, we need to fix a predictor. We use one of the most popular options, namely C5.0. Other commonly employed predictors might be studied in later work; but we note here that [4] included Naïve Bayes predictors (see our discussion of its inappropriateness there), and that numerically-based Support Vector Machines are difficult to conciliate with the categorical attributes that we look for in order to have sensible transactional versions of our datasets without resorting to additional discretizations.

Then, we require to work with a set of rules that, through our feature addition process, increase the C5.0 accuracy by a certain amount. We denote the set of selected rules as G. For each dataset, we select for G those association rules that lead to noticeable accuracy improvements, after adding the new feature flagging exceptions to the rule, as indicated above. In order to be somewhat fair, these rules are taken from the joint pool of all the top-N rule sets provided by all our measures. In our case, we will use $N = 50$. More precisely, from all these rules, 50 from each measure (with a handful of duplications), in turn,

1. the dataset is expanded with one further feature flagging the exceptions, as already indicated,
2. the predictor is run both on the expanded dataset and on the original dataset,
3. the respective accuracies are evaluated via 10-fold cross-validation, and

4. the relative improvement of accuracy (ratio among both accuracies) is determined.

This allows us to identify the set G of helpful rules: those that increase by ϵ the accuracy of the predictor upon having the new feature available. We set $\epsilon = 0.5\%$. To each quality measure, we can assign now a score related to how many of these rules from G do actually appear among the top N rules.

We observe that it is better if helpful rules are captured near the top. Hence, we wish to assign a higher score for a measure both if more helpful rules from G are captured, and if they are captured near the top. More precisely, assume that any given measure has, among its top N rules, k rules from G, and that they are in positions $a_1, \ldots, a_k$, for $1 \le a_i \le N$. Then, we assign to it the score

$$\frac{1}{Z_D} \sum_{i=1}^{k} (N + 1 - a_i) = \frac{1}{Z_D} \left(k(N+1) - \sum_{i=1}^{k} a_i \right)$$

In this way, hitting a helpful rule as first one ($a_i = 1$) contributes N units to the score; $a_i = 2$ contributes $N - 1$, and so on. The value $N + 1$ ensures that even the rule at the N-th place, $a_k = N$, would contribute something, if little. The value Z_D is a dataset-dependent normalization factor, defined as the highest score reached by any of the measures for that dataset. Its aim is to provide us with a way of comparing the outcomes of the experiments on different datasets along the same scale: Z_D is set in such a way that the rule quality measure that reaches highest score is scored exactly at 1. (A different option for normalization could be the maximum reachable value. We prefer this Z_D as the other option leads to very small values, harder to undertand, for all measures tested.) Then, the presence of values substantially less than 1 means that some rule quality measures are substantially worse than the best one, whereas, if all scores are near 1, it means that the dataset is such that the different measures score relatively similarly.

In a sense, this proposal corresponds to the natural "tuning" of an Area-Under-Curve (AUC) approach [7,10]. In the usual computation of AUC, binary predictions are ranked, and low-ranked predictions are expected to correlate with negative labels, and high-ranked predictions with positive labels. On the other hand, here, many of the rules in the set G may not be "ranked" at all by a given measure, if they do not appear among the top-N. Of course, we must take this fact into account. Except for this, our score can be seen actually as an area-under-curve assessment.

4   Empirical Results

We report the scores for our datasets, and provide below also the outcome of a variant of the process, in which we will disallow the access to some attributes at the time of computing association rules. In the first variant, though, the associator has no such limit.

Table 2. Scores and other parameters at 0.5% accuracy increase

Dataset	Conf.	Lift	R. conf.	Impr.	M. impr.	Supp.	Lev.	$|G|$	Max.	Z_D
ADULT	*0*	0.25	*0*	0.24	0.28	**1**	*0.02*	34	20	507
CMC	0.95	0.98	*0.93*	**1**	0.97	0.98	0.94	253	47	1220
VOTES	*0.22*	0.52	0.35	0.64	0.82	0.67	**1**	71	17	425
GERMAN	*0.17*	0.56	0.46	0.79	**1**	0.35	0.21	72	27	745
MUSHROOM	0.41	*0*	*0.22*	0.49	**1**	0.56	0.26	18	4	148
(predict cap surface)										

4.1 Associator Accesses All Attributes

Our experiments use, as indicated, a very mild threshold of at least $\epsilon = 0.5\%$ accuracy increase in determining the helpfulness of a rule; see Table 2. We also ran experiments on a quite large dataset with census data of elderly people (about 300000 transactions) but, after considerable computational effort, no rule at all was found helpful for that dataset.

Both in this table and in the next, different typefaces mark the highest and the lowest nonzero scores per dataset: boldface marks the measure scoring highest and italics the measure(s) scoring lowest; if some are zero, the lowest nonzero is also marked. We also report the value of Z_D to show how the unnormalized-sum score varies considerable among the datasets. We report as well the size of the set of helpful rules and the maximum amount of helpful rules, "Max.", found among the top 50 rules as all measures are considered. Except for one dataset, this column shows that only a small part of the top rules for each measure are actually helpful.

Generally speaking, we see that multiplicative improvement tends to attain higher values of our score, whereas both confidence and relative confidence tend to perform rather poorly; support, leverage, and, to a lesser extent, also lift and improvement offer relatively good scores. Altogether, however, each measure only catches a handful of the set of good rules.

4.2 Disallow the Associator Access to the Predicted Class

We explore now the following issue: besides the usage for ranking rule quality measures, do we intend to actually employ the new features obtained from the helpful rules to improve the performance of the classifiers?

This question is relevant because, in the affirmative case, we cannot afford to make use of the predicted attribute for the computation of the new features: upon predicting, it will be unavailable. Thus, this attribute is to be excluded from the computation of association rules.

The price is that we rely on hypothetical correlations between the class attribute and the exceptions to rules that do not involve the class attribute; these correlations may not exist. In contrast, in the previous section, exceptions to all rules, whether they involve or not the class attribute, are available. The results

Table 3. Scores at 0.5% accuracy increase (class disallowed)

Dataset	Conf.	Lift	R. conf.	Impr.	M. impr.	Supp.	Lev.	$\|G\|$	Max.	Z_D
CMC	0.88	0.90	0.89	0.91	0.96	0.99	1	261	48	1236
VOTES	0.87	0.53	0.22	0.90	0.53	0.76	1	47	9	264
GERMAN	1	0.59	0.34	0.70	0.68	0.61	0.34	39	8	250
MUSHROOM (predict cap surface)	0.18	0.41	0.15	0.56	1	0.57	0.57	20	9	216

in this section were therefore obtained by excluding the predicted attribute from the computation of the association rules, all the rest being the same. The results are given in Table 3. There was only one helpful rule for the ADULT dataset: $|G| = 1$; hence, it is omitted from this table.

The figures support the candidacy of leverage, multiplicative improvement, and support as occassionally good measures; however, all show some cases of mediocre performance. The outcomes are also less supportive of lift. The unreliability of relative confidence is confirmed, and the behavior of confidence is more erratic than in the previous case: top or close to the top for some cases, low and close to the minimum scores in others.

A fact we must mention is the direct effect of allowing the class attribute to appear in the associations. Comparing both tables, we see that several datasets provide a larger $|G|$ if the class is available, but not substantially larger; and it is in fact smaller for other datasets. In general, with or without access to the predicted class attribute, in most datasets, few helpful rules are captured by each measure ("Max." columns in both tables). The access to the class attribute is not, therefore, as key a point as it could be intuitively expected.

5   Conclusions and Further Work

Along a wide study of proposals to measure the relevance of association rules, we have added a novel approach. We have deployed a framework that allows us to evaluate the rule quality evaluation measures themselves, in terms of their usefulnes for subsequent predictive tasks.

We have focused on potential accuracy improvements of predictors on given, public, standard benchmark datasets, if one more Boolean column is added, namely, one that is true exactly for those observations that are exceptions to one association rule: the antecedent holds but the consequent does not. In a sense, we use the association rule as a "hint of outliers", but, instead of removing them, we simply offer direct access to this label to the predictor, through the extra column.

Of course, in general this may lead astray the predictor instead of helping it. In our previous work [4] we presented an initial analysis of the average change of accuracy, and saw that it may well be negative, and that, generally speaking, it does not distinguish well enough among various rule quality measures. Therefore, we have concentrated on an analysis based on selected rules that actually provided accuracy increases. We defined a score for rule quality measures that represents how well a given measure is able to bring, near the top, rules that are helpful to the predictor. We account both for the number of helpful rules among the top-k, and for how close to the top they appear, by means of a mechanism akin to the AUC measure for predictor evaluation.

Our experiments suggest that leverage, support, and our recently proposed measure, multiplicative improvement, tend to be better than the other measures with respect to this evaluation score. Leverage and support are indeed known to offer good results in practice. Possibly further empirical analysis of the multiplicative improvement measure may confirm or disprove whether it has similar potential.

5.1 Related Work

We hasten to point out the important differences of association rules versus classification tasks: they differ in the consideration of locally applicable patterns versus modeling for prediction in the global dataset; in association tasks, no particular attribute is a "class" to be predicted, and the aim is closer to providing the user with descriptive intuitions, rather than to foresee future labels. An interesting related discussion is [12].

As a token consequence of the difference, we must observe that, in a context of associations, "perfect" rules of confidence 1 (that is, full implications without exceptions) are, in the vast majority of cases, useless in practice—a surprising fact for those habitued to using rules for classification, where finding a predictive rule without exceptions is often considered progress. See additional discussion in [3].

That said, the idea of evaluating associators through the predictive capabilities of the rules found has been put forward e.g. in [21]. The usage of association rules for direct prediction (where the "class" attribute is forced to occur in the consequent) has been widely studied (e.g. [26]). In [21], two different associators are employed to find rules with the "class" as consequent, and they are compared in terms of predictive accuracy. Predictive Apriori turns out to be better than plain Apriori, but neither compares too well to other rule-based predictor. (We must point out as well that this reference employs datasets with numeric attributes that are discretized, see our discussion in Section 2.2.) Our work can be seen as a natural next step, by decoupling the associator from the predictor, as they do not need to be the same sort of model, and, more importantly, conceptually decoupling the rule quality measure employed to rank the output associations from the algorithm that is actually used to construct the rules.

5.2 Future Work

A number of natural ideas to explore appear. We have evaluated only a handful of rule quality measures; many others exist. Also, other related explorations remain open: we limited, somewhat arbitrarily, to 50 the top rules considered per measure, but smaller or larger figures may provide different intuitions. Our choice was dictated by the consideration that the understanding of 50 association rules by a human expert would require some time, and one hardly could expect this human attention span to reach much further.

Also, we only tried one popular predictor, C5.0, and we could study others; one could run as well heavy, exhaustive searches in order to find the best set of helpful rules G on which to base the score. The alternative normalization mentioned in the text, instead of Z_D, could be explored, and might end up in figures offering more clear cross-comparison among different datasets. Other evaluations of the predictor could substitute accuracy. Yet other variations worth exploring would be to flag conformance to a rule instead of being an exception to it, that is, marking those observations that support both the antecedent and the consequent of the rule; to explore partial implications, that is, association rules with more than one item in the consequent, where multiplicative improvement would partially correspond to confidence boost [3]; and to allow addition of more than one feature at a time. We hope to explore some of these avenues soon.

Additionally, we consider that our approach might be, at some point, of interest in subgroup discovery (see [14] for its connection to association rules via the common notion of closure spaces, and the further references in that paper). The new features added in our approach, in fact, do identify regions of the space (identified by the antecedent of the association rule, hence having a geometric form of Boolean hypersubcubes) where the consequent of the rule behaves quite differently than in the general population. How this relates to the other existing proposals of subgroup discovery, and whether it is efficient at all in that sense, remains an interesting topic for further research.

References

1. Agrawal, R., Imielinski, T., Swami, A.N.: Mining association rules between sets of items in large databases. In: Buneman, P., Jajodia, S. (eds.) SIGMOD Conference, pp. 207–216. ACM Press (1993)
2. grawal, R., Mannila, H., Srikant, R., Toivonen, H., Verkamo, A.I.: Fast discovery of association rules. In: Advances in Knowledge Discovery and Data Mining, pp. 307–328. AAAI/MIT Press (1996)
3. Balcázar, J.L.: Formal and computational properties of the confidence boost in association rules. To appear in ACM Transactions on KDD (2013),
   http://www.lsi.upc.edu/~balqui/
4. Balcázar, J.L., Dogbey, F.K.: Feature extraction from top association rules: Effect on average predictive accuracy. In: 3rd EUCogIII Members Conference and Final Pascal Review Meeting (2013), http://www.lsi.upc.edu/~balqui/
5. Bayardo, R., Agrawal, R., Gunopulos, D.: Constraint-based rule mining in large, dense databases. In: ICDE, pp. 188–197 (1999)
6. Borgelt, C.: Efficient implementations of Apriori and Eclat. In: Goethals, B., Zaki, M.J. (eds.) FIMI, CEUR Workshop Proceedings, vol. 90. CEUR-WS.org (2003)
7. Bradley, A.P.: The use of the area under the ROC curve in the evaluation of machine learning algorithms. Pattern Recognition 30(7), 1145–1159 (1997)

8. Brin, S., Motwani, R., Ullman, J.D., Tsur, S.: Dynamic itemset counting and implication rules for market basket data. In: Peckham, J. (ed.) SIGMOD Conference, pp. 255–264. ACM Press (1997)
9. Ceglar, A., Roddick, J.F.: Association mining. ACM Comput. Surv. 38(2) (2006)
10. Fawcett, T.: ROC graphs: Notes and practical considerations for researchers. Pattern Recognition Letters 27(8), 882–891 (2004)
11. Frank, A., Asuncion, A.: UCI machine learning repository (2010), http://archive.ics.uci.edu/ml
12. Freitas, A.A.: Understanding the crucial differences between classification and discovery of association rules - a position paper. SIGKDD Explorations 2(1), 65–69 (2000)
13. Fürnkranz, J., Flach, P.A.: ROC 'n' rule learning—towards a better understanding of covering algorithms. Machine Learning 58(1), 39–77 (2005)
14. Garriga, G.C., Kralj, P., Lavrac, N.: Closed sets for labeled data. Journal of Machine Learning Research 9, 559–580 (2008)
15. Geng, L., Hamilton, H.J.: Interestingness measures for data mining: A survey. ACM Comput. Surv. 38(3) (2006)
16. Guyon, I., Elisseeff, A.: An introduction to variable and feature selection. Journal of Machine Learning Research 3, 1157–1182 (2003)
17. Lavrač, N., Flach, P.A., Zupan, B.: Rule evaluation measures: A unifying view. In: Džeroski, S., Flach, P.A. (eds.) ILP 1999. LNCS (LNAI), vol. 1634, pp. 174–185. Springer, Heidelberg (1999)
18. Lenca, P., Meyer, P., Vaillant, B., Lallich, S.: On selecting interestingness measures for association rules: User oriented description and multiple criteria decision aid. European Journal of Operational Research 184(2), 610–626 (2008)
19. Liu, B., Hsu, W., Ma, Y.: Pruning and summarizing the discovered associations. In: Proc. Knowledge Discovery in Databases, pp. 125–134 (1999)
20. Luxenburger, M.: Implications partielles dans un contexte. Mathématiques et Sciences Humaines 29, 35–55 (1991)
21. Mutter, S., Hall, M., Frank, E.: Using classification to evaluate the output of confidence-based association rule mining. In: Webb, G.I., Yu, X. (eds.) AI 2004. LNCS (LNAI), vol. 3339, pp. 538–549. Springer, Heidelberg (2004)
22. Piatetsky-Shapiro, G.: Discovery, analysis, and presentation of strong rules. In: Proc. Knowledge Discovery in Databases, pp. 229–248 (1991)
23. Quinlan, J.R.: C4.5: Programs for Machine Learning. Morgan Kaufmann (1993)
24. Tan, P.N., Kumar, V., Srivastava, J.: Selecting the right objective measure for association analysis. Information Systems 29(4), 293–313 (2004)
25. Wu, X., Kumar, V., Quinlan, J.R., Ghosh, J., Yang, Q., Motoda, H., McLachlan, G.J., Ng, A.F.M., Liu, B., Yu, P.S., Zhou, Z.H., Steinbach, M., Hand, D.J., Steinberg, D.: Top 10 algorithms in data mining. Knowl. Inf. Syst. 14(1), 1–37 (2008)
26. Yin, X., Han, J.: CPAR: Classification based on predictive association rules. In: Barbará, D., Kamath, C. (eds.) SDM. SIAM (2003)
27. Zimmermann, A.: Objectively evaluating interestingness measures for frequent itemset mining. In: Li, J., Cao, L., Wang, C., Tan, K.C., Liu, B., Pei, J., Tseng, V.S. (eds.) PAKDD 2013 Workshops. LNCS, vol. 7867, pp. 354–366. Springer, Heidelberg (2013)

Towards Comprehensive Concept Description Based on Association Rules

Petr Berka

University of Economics
W. Churchill Sq. 3, 130 67 Prague, Czech Republic
berka@vse.cz

Abstract. The paper presents two approaches to post-processing of association rules that are used for concept description. The first approach is based on the idea of meta-learning; a subsequent association rule mining step is applied to the results of "standard" association rule mining. We thus obtain "rules about rules" that in a condensed form represent the knowledge found using association rules generated in the first step. The second approach finds a "core" part of the association rules that can be used to derive the confidence of every rule created in the first step. Again, the core part is substantially smaller than the set of all association rules. We experimentally evaluate the proposed methods on some benchmark data taken from the UCI repository. The system LISp-Miner has been used to carry out the experiments.

Keywords: concept description, association rules, meta-learning.

1 Introduction

Concept description is one of the typical data mining tasks. According to CRISP-DM methodology concept description "aimes at understandable descriptions of concepts or classes" [9]. Concept description is thus similar to classification as there are predefined classes (given by values of the target attribute) we are interested in. But unlike to classification, the focus of concept description is on understandability not on classification accuracy. So association rules, decision rules or decision trees are preferred to model the concepts.

In our paper we will focus on concept description using association rules. Association rules have been proposed by R. Agrawal in the early 90th as a tool for so called market basket analysis [2]. An association rule has the form of an implication

$$X \Longrightarrow Y$$

where X and Y are sets of items and $X \cap Y = \emptyset$. An association rule expresses that transactions containing items of set X tend to contain items of set Y, so e.g. a rule $\{A, B\} \Longrightarrow \{C\}$ says, that customers who buy products A and B also often buy product C. Such statements can be used to guide the placement of goods in a store, for cross-selling or to promote new products. This idea of association

A. Tucker et al. (Eds.): IDA 2013, LNCS 8207, pp. 80–91, 2013.

rules can be applied to any data in the tabular, attribute-value form. So data describing values of attributes can be analyzed in order to find associations between conjunctions of attribute-value pairs (categories). Let us denote these conjunctions as *Ant* (antecedent) and *Suc* (succedent) and the association rule as

$$Ant \implies Suc.$$

When using association rules for concept description, *Suc* will be a category of the target attribute.

The two basic characteristics of an association rule are *support* and *confidence*. Support is the estimate of the probability $P(Ant \wedge Suc)$, (the frequency of $Ant \wedge Suc$ is the *absolute support*), confidence is the estimate of the probability $P(Suc|Ant)$.

In association rule discovery the task is to find all rules with support and confidence above the userdefined thresholds *minconf* and *minsup*. There is a number of algorithms, that perform this task. The probably best-known algorithm `apriori` proceeds in two steps. All frequent itemsets are found in the first step during breath-first search in the space of all frequent itemsets. Then, association rules with a confidence of at least *minconf* are generated in the second step [2]. Another well known algorithm is `FP-Growth`. This algorithm uses FP-tree to generate frequent itemsets. This way of representing the itemsets reduces the computational costs because (unlike apriori) it requires only two scans of the whole data [12]. The found frequent itemsets are then again splitted into antecedent and succedent to create a rule. The search space of all possible itemsets (or conjunctions of categories) can be very huge. For K items, there is 2^K itemsets, for K categorial attributes $A_1, A_2, ...A_K$, having $v_1, v_2, ...v_K$ distinct values, the number of all possible conjunctions is

$$\prod_{i=1}^{K}(1 + v_i) - 1. \tag{1}$$

The main problem when using association rules for data mining is their interpretation. Usually we end up with a huge number of associations and each of them might be interesting for the domain expert or end-user. So some automatic support for the interpretation in the form of association rules post-processing would be of a great help. We present some ideas in this direction and show their experimental evaluation using `LISp-Miner`, a data mining toolbox for mining different types of rules, that is under development at the University of Economics, Prague [16,18].

The rest of the paper is organized as follows. Section 2 reviews work related to the problem of post-processing of association rules, section 3 gives an overview of GUHA method and the 4FT rules, section 4 introduces the concept of association meta-rules and describes how they can be obtained using `4FT-Miner` procedure, section 5 presents `KEX`, another procedure of the `LISp-Miner` system and shows how the resulting rules can be understand as condensed representation of association rules, section 6 summarizes the experimental evaluation of the proposed approach on some benchmark datasets from the UCI Machine learning repository, and section 7 concludes the paper.

2 Related Work

Various approaches have been proposed in the past to post-process the huge list of found associations. These approaches can be divided into several groups. One group are methods for visualization, filtering or selection of the created rules. This are the standard options in most systems.

Second group contains methods that use some algorithms to further process the rules: clustering, grouping or using some inference methods fits into this group as well as our approach. An application of deduction rules to post-process the results of GUHA method is described in [15]; these rules allow to remove association rules that are logical consequences of another association rules. Similar idea, but applied to "Agrawal-like" association rules can be found in [19]. This paper also describes clustering of association rules that have the same succedent; the distance between two rules is defined "semantically", i.e. as the number of examples covered only by one of the rules. Both semantical and syntactical (i.e. based on the lists of attribute-value pairs that occur in the rules) clustering of association rules can be found e.g. in [17].

The third possibility is to post-process the rules using some domain knowledge. So e.g. An et all use expert-supplied taxonomy of items for clustering the discovered association rules with respect to the taxonomic similarity ([1]), or Domingues and Rezende ([10]) iteratively scan the itemset rules and updates a taxonomy that is then used to generalize the association mining results.

An additional possibility is to filter out consequences of domain knowledge via application of logic of association rules [15].

3 GUHA Method and 4FT Rules

GUHA is an original Czech method of exploratory data analysis developed since 1960s. Its principle is to offer all interesting facts following from the given data to the given problem. A milestone in the GUHA method development was the monograph [11], which introduces the general theory of mechanized hypothesis formation based on mathematical logic and statistics. Association rules defined and studied in this book are relations between two general Boolean attributes derived from the columns of an analyzed data table. Various types of relations of Boolean attributes are used including relations corresponding to statistical hypothesis tests.

Within the GUHA framework, we understand the association rule as the expression

$$Ant \approx Suc/Cond,$$

where *Ant*, *Suc* and *Cond* are conjunctions of literals called antecedent, succedent and conjunction, and $\approx$ denotes a relation between *Ant* and *Suc* for the examples from the analyzed data table, that fulfill the condition *Cond*. The rule $Ant \approx Suc/Cond$ is true in the analyzed data table, if the condition associated with $\approx$ is satisfied for the frequencies a, b, c, d of the corresponding contingency table. Here a denotes the number of examples, that are covered both by *Ant* and *Suc*, b denotes the number of examples, that are covered by *Ant* but not covered by *Suc*, c denotes the number of examples, that are covered by *Suc* but not covered by *Ant* and d denotes the number of examples that are covered neither by *Ant*, nor by *Suc*. When comparing this notion of association rules (we will call them 4FT rules) with the "standard" understanding, we will find, that:

— 4FT rules offer more types of relations between *Ant* and *Suc*; we can search not only for implications (based on standard definitions of support and confidence of a rule), but also for equivalences or statistically based relations. In the sense of 4FT rules "classical" association rules can be considered as founded implications (for examples of various 4FT relations see Table 1.

Table 1. Examples of 4FT-relations

4ft relation $\approx$		
Name	Symbol	$\approx (a, b, c, d) = 1$ iff
Founded implication	$\Rightarrow_{p,B}$	$\frac{a}{a+b} \geq p \wedge a \geq B$
Founded double implication	$\Leftrightarrow_{p,B}$	$\frac{a}{a+b+c} \geq p \wedge a \geq B$
Founded equivalence	$\equiv_{p,B}$	$\frac{a+d}{a+b+c+d} \geq p \wedge a \geq B$
Simple deviation	$\sim_{\delta,B}$	$\frac{ad}{bc} > e^{\delta} \wedge a \geq B$
χ^2 quantifier	$\sim^2_{\alpha,B}$	$\frac{(ad-bc)^2}{rkls} n \geq \chi^2_{\alpha} \wedge a \geq B$
Above average dependence	$\sim^+_{q,B}$	$\frac{a}{a+b} \geq (1+q)\frac{a+c}{a+b+c+d} \wedge a \geq B$

— 4FT rules offer more expressive syntax of *Ant* and *Suc*; *Ant* and *Suc* are conjunctions of literals (i.e. expressions in the form $A(coef)$ or $\neg A(coef)$, where A is an attribute and *coef* is a subset of possible values), not only of attribute-value pairs (which are literals as well). If e.g. the analyzed data contain attribute A with values a, b, c, attribute B with values x, y, z, and attribute C with values k, l, m, n, then a 4ft-rule can be e.g.

$$A(b) \wedge B(x \vee y) \Rightarrow_{p,B} \neg C(k)$$

— a 4FT rule consists not only of *Ant* and *Suc* but can contain also a condition *Cond*; this condition is generated during the rule learning process as well.

There is also an important difference in the process of generating the rules itself. Unlike `apriori` or `FP-Growth` where the whole frequent itemsets (or conjunctions of categories) are created first and these itemsets are then splitted into antecedent and succedent of a rule, in `4FT-Miner`, antecedent, succedent (and eventually condition) are generated separately, so we can easily set the target attribute as the only one that can occur in the succedents. Moreover, we can control the complexity of the searched rule space by determining (using parameters *maxlenA*, *maxlenS* and *maxlenC*) the maximal number of literals that can occur in conjunctions of antecedent, succedent or condition of a rule. So 4FT-Miner is better suited for concept description tasks than apriori or FP-Growth.

4 Association Meta-rules

We propose to apply association rule mining algorithm to the set of original association rules obtained as a result of a particular data mining task. This idea thus follows the stacking concept that is used to combine classifiers, but that has not been presented yet for descriptive tasks. The input to the proposed meta-learning step will be association rules encoded in a way suitable for association rule mining algorithm; the result will be a set of association meta-rules uncovering relations between various characteristics of the original set of rules.

In [7] we proposed several types of association meta-rules: *qualitative* and *quantitative*, and *frequent cedents*. Out of them, qualitative meta-rules and frequent cedents can be adopted to find condensed concept description. Qualitative meta-rules will in general represent the meta-knowledge in the form "if original association rules contain a conjunction *Ant*, then they also contain the conjunction *Suc*", i.e. qualitative rules have the form

$$Ant \implies Suc.$$

Here these rules represent the knowledge about what conjunctions occur frequently together in the concept description.

By *frequent cedents* we understand conjunctions of categories, that frequently occur in the list of original association rules. With respect to the concept description task, these cedents represent a meta-knowledge about frequent co-occurrence of specific categories in the concept description.

We used LISp-Miner to find association meta-rules and frequent cedents we used, but any other association rule mining algorithm can be used as well. Encoding of the original rules is thus the key problem of this approach. *Ant* and *Suc* can be encoded either (1) using binary attributes, where each attribute represents one possible literal or (2) using the attributes from the original data set. Another open question concerning the representation of a rule is whether categories not occurring in the rule should be treated as missing or as negative ones. In the first approach, attributes not used in the rule will be encoded using missing value code. In the second approach, when using the binary representation, categories not used in the rule will get the value false, and when using the original attributes, categories not used in the rule will get a new special value interpreted as not used.

In the experiments reported in this paper, we:

- represent a rule using the original attributes. This leads formally to the same structure of data table as for original data (i.e. the table representing the association rules has the same columns as the table representing the data).

– used missing value code to represent categories not occurring in a rule. This option is more suitable as it will prevent the meta-learning step to generate a great number of meta-rules about non-occurrence of literals in the original rules, this option also corresponds to the original notion of association rules where only items that do occur in the market baskets are taken into consideration.

This approach is not limited to binary classification problems where the data can be considered as examples or counter-examples of a single concept. We can use this approach to a data containing examples of arbitrary number of concepts (in our experiments reported in section 6 such data are represented by the Iris dataset). In such a case, when creating the association meta-rules using LISp-Miner we can set the condition $Cond$ to take the values of the target attribute and thus obtain meta-rules only from original rules covering a specific concept.

We will illustrate the concept of association meta-rules using the Monk1 dataset from the UCI repository [20]. This dataset consists of 123 examples and following six attributes: head_shape, body_shape, smile, holding, jacket_color, tie (these attributes are the input ones), and class (this is a binary target attribute). Running 4FT-Miner with the input parameters $maxlenA = 6$, $minsup = 5\%$, $minconf = 0.9$, we obtain 34 association rules, some of them listed in Tab 2. These rules have been turned into examples for the subsequent run of 4FT-Miner (Table 3, shows the representation of rules from Table 2). Table 4 shows some meta-rules we obtained (we used the same settings for input parameters as before), and table 5 shows some obtained frequent cedents. Notice, that among the meta-rules and cedents, there are rules with similar syntax as have the association rules. But their interpretation is of course different. Compare the association rule

$$\text{Body(o)} \wedge \text{Head(o)} \Longrightarrow \text{Class}(+)$$

the meta-rule

$$\text{Body(o)} \Longrightarrow \text{Head(o)}$$

and the frequent cedent

$$\Longrightarrow \text{Body(o)} \wedge \text{Head(o)}$$

The association rule says, that the concept Class(+) can be described using the conjunction Body(o) $\wedge$ Head(o), the meta-rule says, that whenever the concept is described using Body(o), it is also described by Head(o), and the frequent cedent says, that the conjunction Body(o) $\wedge$ Head(o) frequently occurs in the concept description. We can also see a significant "compression" of the list of the build association rules; while we obtained 34 4FT rules, we have only 14 4FT meta-rules (this makes the reduction to 41 % of the number of association rules) and 13 frequent cedents (this makes the reduction to 38 % of the number of association rules). We use these numbers to evaluate the results for other data as presented in section 6 as well.

Table 2. Example 4FT rules for Monk1 data

$$Body(o) \implies Class(+) \ (0.1951, 1)$$
$$Body(o) \wedge Head(o) \implies Class(+) \ (0.1301, 1)$$
$$Body(o) \wedge Head(o) \wedge Holding(b) \implies Class(+) \ (0.0569, 1)$$
$$Body(o) \wedge Head(o) \wedge Holding(s) \implies Class(+) \ (0.0569, 1)$$
$$Body(o) \wedge Head(o) \wedge Smile(n) \implies Class(+) \ (0.0569, 1)$$
$$Body(o) \wedge Head(o) \wedge Smile(y) \implies Class(+) \ (0.0732, 1)$$
$$Body(o) \wedge Head(o) \wedge Tie(n) \implies Class(+) \ (0.0894, 1)$$
$$\cdots$$

Table 3. Example 4FT rules for Monk1 data encoded as data

id	Head	Body	Smile	Holding	Jacket	Tie	Class	support	confidence
1	?	o	?	?	?	?	+	0.1951	1
2	o	o	?	?	?	?	+	0.1301	1
3	o	o	?	b	?	?	+	0.0569	1
4	o	o	?	s	?	?	+	0.0569	1
5	o	o	n	?	?	?	+	0.0569	1
6	o	o	y	?	?	?	+	0.0732	1
7	o	o	?	?	?	n	+	0.0894	1

. . .

Table 4. Example 4FT meta-rules for Monk1 data

$$Body(o) \implies Head(o)$$
$$Body(s) \implies Head(s)$$
$$Head(o) \implies Body(o)$$
$$Head(o) \implies Jacket(r)$$
$$\cdots$$

Table 5. Example 4FT frequent cedents for Monk1 data

$$\implies Body(o)$$
$$\implies Body(s)$$
$$\implies Body(o) \wedge Head(o)$$
$$\implies Head(s)$$
$$\cdots$$

5 KEX Rules

KEX is an algorithm that learns rules in the form

$$Ant \Rightarrow C(w)$$

where Ant is a conjunction of categories, C is a category of target attribute and w (called weight) expresses the uncertainty of the rule [5,6].

KEX performs a heuristic top-down search in the space of candidate rules. In this algorithm the covered examples are not removed during learning, so an example can be covered by more rules. Thus more rules can be used during classification each contributing to the final assignment of an example. KEX uses a pseudo-bayesian combination function borrowed from the expert system PROSPECTOR [8] to combine contributions of more rules:

$$w_1 \oplus w_2 = \frac{w_1 \times w_2}{w_1 \times w_2 + (1 - w_1) \times (1 - w_2)}. \tag{2}$$

KEX works in an iterative way, testing and expanding an implication $Ant \Rightarrow C$ in each iteration. This process starts with a default rule weighted with the relative frequency of class C and stops after testing all implications created according to user defined criteria. The induction algorithm inserts only such rules into the knowledge base, for which the confidence (defined in the same way as the confidence of association rules) cannot be inferred (using formula 2) from weights of applicable rules found so far. A sketch of the algorithm is shown in Fig. 1.

KEX algorithm

Initialization
  1. forall category (attribute-value pair) A add $A \Rightarrow C$ to $OPEN$
  2. add empty rule to the rule set KB

Main loop
while $OPEN$ is not empty
  1. **select** the first implication $Ant \Rightarrow C$ from $OPEN$
  2. **test** if this implication significantly improves the set of rules KB built so far (using the χ^2 test, we test the difference between the rule validity and the result of classification of an example covered by Ant) then add it as a new rule to KB
  3. for all possible categories A
     (a) **expand** the implication $Ant \Rightarrow C$ by adding A to Ant
     (b) **add** $Ant \wedge A \Rightarrow C$ to $OPEN$ so that $OPEN$ remains ordered according to decreasing frequency of the condition of rules
  4. **remove** $Ant \Rightarrow C$ from $OPEN$

Fig. 1. Simplified sketch of the KEX rule learning algorithm

The KEX procedure has been designed to create a set of classification rules. So for a new example, all rules that cover this example are used to compute the composed weight of the target class. But we can also use KEX to "compress" the set of association rules to get more condensed representation. The rules generated by KEX can be understood as a "core" set of association rules in such a sense that for each association rule created using the same settings for $minconf$, $maxlengthA$ and $minsup$:

- either the composed weight exactly corresponds to the confidence of the association rule (if this association rule is part of the core set),
- or the composed weight doesn't significantly differ from the confidence of the association rule (if this rule is not a part of the core set).

So we can use the rule set created by KEX to "query" about confidences of all association rules created (with the same settings for $minconf$, $maxlengthA$ and $minsup$) to describe the concept C.

6 Experiments

The experiments were carried out on several benchmark data sets from the UCI Machine Learning Repository [20]. The characteristics of the data (number of examples, number of attributes and number of concepts) are shown in Table 6. The Agaricus, Breast, Kr-vs-kp, Monk1, Tic-tac-toe and Vote datasets contain only categorial attributes, the numeric attributes occurring in the remaining datasets were discretized prior to using LISp-Miner and KEX.

Table 7 summarizes the results. The first column in this table shows the numbers 4FT (association) rules, that were created in the first step. Here we were looking for strong concept descriptions, so we set the parameters $minsup = 5\%$ and $minconf = 0.9$. Then the number of KEX rules, 4FT qualitative meta-rules and 4FT frequent cedents is shown together with the relative size of the corresponding rule set with respect to the 4FT rules created in the first step. So the relative size is computed as number of rules in the respective rule set divided by the number of rules in the 4FT rule set. To make the numbers comparable we set $minconf = 0.9$ and $minsup$ between 1% and 5% for the second step (i.e. for creating KEX rules, 4FT meta-rules and 4FT cedents). With the exception of Tic-tac-toe data and 4FT meta-rules for Iris data, we always reduced the number of original 4TF rules to less than one half.

There is no straightforward relationship between the size of analyzed data (number of objects, number of attributes), the number of 4FT rules and the number of 4FT meta-rules. High relative size of the set of 4FT meta-rules just means low redundancy in the set of 4FT rules and low relative size of the set of 4FT meta-rules means high redundancy in the set of 4FT rules. And low redundancy in the set of 4FT rules is usually related to smaller size of this set of rules.

Table 6. Data description

data set	examples	attributes	concepts
Agaricus	8124	22	2
Australi	690	14	2
Breast	289	9	2
Diab	769	8	2
Iris	150	4	3
JapCred	125	10	2
Kr-vs-kp	3196	36	2
Monk1	123	6	2
Tic-tac-toe	958	9	2
Vote	435	16	2

Table 7. Summary of results

data set	4FT rules	KEX rules		4FT meta-rules		4FT cedents	
	rules	rules	size	rules	size	rules	size
Agaricus	2072	883	43%	12	0.6%	22	1%
Australi	1734	274	16%	226	13%	25	1%
Breast	93	24	26%	16	17%	8	9%
Diab	106	50	47%	44	42%	19	18%
Iris	58	21	36%	36	62%	16	28%
JapCred	777	55	7%	23	3%	14	2%
Kr-vs-kp	972	61	6%	74	8%	47	5%
Monk1	34	14	41%	14	41%	13	38%
Tic-tac-toe	18	18	100%	12	67%	9	50%
Vote	5554	967	17%	25	0.5%	9	0.2%

7   Conclusions

We present two methods of post-processing of association rules that were created
for concept description. The first approach is based on the idea of meta-learning;
a subsequent association rule mining step is applied to the results of "standard"
association rule mining. We thus obtain "rules about rules" that in a condensed
form represent the knowledge found using association rules generated in the first
step. The second approach finds a "core" part of the association rules that can
be used to derive the confidence of every rule created in the first step.

The reported experiments support our working hypothesis, that the number of
KEX rules as well as the meta-rules will be significantly smaller than the number
of original rules. Thus the interpretation of meta-rules by domain expert will be
significantly less time consuming and less difficult compared to the interpretation
of the original association rules. Beside this, the meta-rules can give the user a
summarized interpretation of the original rules. The KEX rules on the contrary

should be used to automated answering the questions about the confidence of any of the original association rules.

Anyway, the interpretation of the found meta-rules by domain experts in respective problem area is necessary to validate the usefulness of the proposed methods.

References

1. An, A., Khan, S., Huang, X.: Objective and Subjective Algorithms for Grouping Association Rules. In: Third IEEE Conference on Data Mining (ICDM 2003), pp. 477–480 (2003)
2. Agrawal, R., Imielinski, T., Swami, A.: Mining Association Rules Between Sets of Items in Large Databases. In: SIGMOD Conference, pp. 207–216 (1993)
3. Baesens, B., Viaene, S., Vanthienen, J.: Post-processing of association rules. In: The Sixth ACM SIGKDD International Conference on Knowledge Discovery and Data Mining (KDD 2000), Boston, Massachusetts (2000)
4. Bauer, E., Kohavi, R.: An Empirical Comparison of Voting Classification Algorithms: Bagging, Boosting, and Variants. Machine Learning 36(1/2), 105–139 (1999)
5. Berka, P., Ivánek, J.: Automated Knowledge Acquisition for PROSPECTOR-like Expert Systems. In: Bergadano, F., De Raedt, L. (eds.) ECML 1994. LNCS (LNAI), vol. 784, pp. 339–342. Springer, Heidelberg (1994)
6. Berka, P.: Learning compositional decision rules using the KEX algorithm. Intelligent Data Analysis 16(4), 665–681 (2012)
7. Berka, P., Rauch, J.: Meta-learning for Post-processing of Association Rules. In: Bach Pedersen, T., Mohania, M.K., Tjoa, A.M. (eds.) DAWAK 2010. LNCS, vol. 6263, pp. 251–262. Springer, Heidelberg (2010)
8. Duda, R.O., Gasching, J.E.: Model design in the Prospector consultant system for mineral exploration. In: Michie, D. (ed.) Expert Systems in the Micro Electronic Age. Edinburgh University Press, UK (1979)
9. Chapman, P., Clinton, J., Kerber, R., Khabaza, T., Reinartz, T., Shearer, C., Wirth, R.: CRISP-DM 1.0 Step-by-step data mining guide. SPSS Inc. (2000)
10. Domingues, M.A., Rezende, S.O.: Using Taxonomies to Faciliate the Analysis of the Association Rules. In: Second International Workshop on Knowledge Discovery and Ontologies (KDO 2005), ECML/PKDD, Porto (2005)
11. Hájek, P., Havránek, T.: Mechanising Hypothesis Formation – Mathematical Foundations for a General Theory. Springer (1978)
12. Han, J., Pei, J., Yin, Y.: Mining Frequent Patterns without Candidate Generation. In: Proc. ACM-SIGMOD Int. Conf. on Management of Data (2000)
13. Jorge, A., Poas, J., Azevedo, P.J.: Post-processing Operators for Browsing Large Sets of Association Rules. Discovery Science, 414–421 (2002)
14. The LISp-Miner Project, http://lispminer.vse.cz/
15. Rauch, J.: Logic of association rules. Applied Intelligence 22, 9–28 (2005)
16. Rauch, J., Šimuånek, M.: An Alternative Approach to Mining Association Rules. In: Lin, T.Y., Ohsuga, S., Liau, C.J., Tsumoto, S. (eds.) Foundations of Data Mining and knowledge Discovery. SCI, vol. 6, pp. 211–231. Springer, Heidelberg (2005)
17. Sigal, S.: Exploring interestingness through clustering. In: Proc. of the IEEE Int. Conf. on Data Mining (ICDM 2002). Maebashi City (2002)

18. Šimuánek, M.: Academic KDD Project LISp-Miner. In: Abraham, A., Franke, K., Koppen, K. (eds.) Advances in Soft Computing Intelligent Systems Design and Applications, pp. 263–272. Springer, Heidelberg (2003)
19. Toivonen, H., Klementinen, M., Roikainen, P., Hatonen, K., Mannila, H.: Pruning and grouping discovered association rules. In: Workshop Notes of the ECML 1995 Workshop on Statistics, Machine Learning and Knowledge Discovery in Databases, Heraklion, pp.47–52 (1995)
20. UCI Machine Learning Repository, http://archive.ics.uci.edu/ml/

CD-MOA: Change Detection Framework for Massive Online Analysis

Albert Bifet[1], Jesse Read[2], Bernhard Pfahringer[3],
Geoff Holmes[3], and Indrė Žliobaitė[4]

[1] Yahoo! Research Barcelona, Spain
abifet@yahoo-inc.com
[2] Universidad Carlos III, Spain
jesse@tsc.uc3m.es
[3] University of Waikato, New Zealand
{bernhard,geoff}@waikato.ac.nz
[4] Aalto University and HIIT, Finland
indre.zliobaite@aalto.fi

Abstract. Analysis of data from networked digital information systems such as mobile devices, remote sensors, and streaming applications, needs to deal with two challenges: the size of data and the capacity to be adaptive to changes in concept in real-time. Many approaches meet the challenge by using an explicit change detector alongside a classification algorithm and then evaluate performance using classification accuracy. However, there is an unexpected connection between change detectors and classification methods that needs to be acknowledged. The phenomenon has been observed previously, connecting high classification performance with high false positive rates. The implication is that we need to be careful to evaluate systems against intended outcomes–high classification rates, low false alarm rates, compromises between the two and so forth. This paper proposes a new experimental framework for evaluating change detection methods against intended outcomes. The framework is general in the sense that it can be used with other data mining tasks such as frequent item and pattern mining, clustering etc. Included in the framework is a new measure of performance of a change detector that monitors the compromise between fast detection and false alarms. Using this new experimental framework we conduct an evaluation study on synthetic and real-world datasets to show that classification performance is indeed a poor proxy for change detection performance and provide further evidence that classification performance is correlated strongly with the use of change detectors that produce high false positive rates.

Keywords: data streams, incremental, dynamic, evolving, online.

1 Introduction

Real-time analytics is a term used to identify analytics performed taking into account recent data that is being generated in real time. The analytical models

A. Tucker et al. (Eds.): IDA 2013, LNCS 8207, pp. 92–103, 2013.

should be up-to-date to match the current distribution of data. To be able to do that, models should be able to adapt quickly. Drift detection is a very important component in adaptive modeling, detecting a change gives a signal about when to adapt models. Typically, the streaming error of predictive models is monitored and when the detector raises a change alarm, then the model is updated or replaced by a new one.

Currently, drift detection methods are typically evaluated by the final classification accuracy [1,2]. For example, in [2], the authors notice that for a real dataset, the Electricity Market Dataset [3], performance increases when there is a large number of false positives (or low ARL_0): "*Interestingly, the fact that the best performance is achieved with a low ARL_0 suggests that changes are occurring quite frequently.*" Thus, evaluating drift detection methods only using classifiers may not be informative enough, since the adaptation strategy occludes change detector performance. Further, given that classification is not the only context for change detection we need to match our drift detection evaluation methodologies with what it is that we want to achieve from the task as a whole.

This paper investigates change detection for real time predictive modeling, and presents the following contributions:

1. CD-MOA, a new experimental framework for evaluating concept drift detection methods,
2. MTR, a new measure of performance of a concept drift detection method.

It is important to note that the framework generalises to other tasks but in this paper our focus is on change detection in the context of classification. The proposed framework is intended to serve as a tool for the research community and industry data analysts for experimentally comparing and benchmarking change detection techniques on synthetic data where ground truth changes are known. On real data, the framework allows to find changes in time series, and monitor the error in classification tasks. The framework and the proposed techniques are implemented in the open source data stream analysis software MOA and are available online[1].

In Section 2 we present the new change detection framework and propose a new evaluation measure for change detection. Section 3 presents the results of our experimental evaluation. We conclude the study in Section 4.

2   Experimental Framework

CD-MOA is a new framework for comparing change detection methods. It is built as an extension of MOA. Massive Online Analysis (MOA) [4] is a software environment for implementing algorithms and running experiments for online learning from data streams.

CD-MOA contains a graphical user interface where experiments can be run. Figure 1 shows the GUI. It contains three different components. The first is the

---

[1] http://moa.cs.waikato.ac.nz/

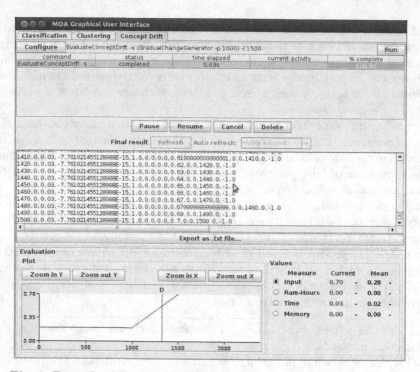

Fig. 1. Example of the new CD-MOA framework graphical user interface

panel where the user can specify the experiment they would like to run. Another panel in the middle shows the numeric results of the experiment, and the panel at the bottom displays a plot of the experiment, showing graphically where the change has been detected.

CD-MOA contains a Java API for easier customization, and implementation of new methods and experiments. The main components of CD-MOA are:

- Tasks: experiments to run combining change detectors and streams
- Methods: change detection algorithms used to detect change
- Streams: time series used to run the experiment. If they have been artificially generated and have ground truth, then the system will output the statistics about detection time.

CD-MOA is connected to MOA, and it is easy to use the change detector methods in CD-MOA to evaluate classifiers, checking how their accuracy evolves. For evaluation purposes, all the methods in CD-MOA have a measure of the resources consumed: time, memory, and RAM-Hours, a measure of the cost of the mining process that merges time and memory into a single measure.

2.1 Evaluation of Change Detection

Change detection is a challenging task due to a fundamental limitation [5]: the design of a change detector is a compromise between detecting true changes and avoiding false alarms.

When designing a change detection algorithm one needs to balance false and true alarms and minimize the time from the change actually happening to detection. The following existing criteria [5,6] formally capture these properties for evaluating change detection methods.

Mean Time between False Alarms (MTFA) characterizes how often we get false alarms when there is no change. The false alarm rate FAR is defined as 1/MTFA. A good change detector would have high MTFA.

Mean Time to Detection (MTD) characterizes the reactivity of the system to changes after they occur. A good change detector would have small MTD.

Missed Detection Rate (MDR) gives the probability of not receiving an alarm when there has been a change. It is the fraction of non-detected changes in all the changes. A good detector would have small or zero MDR.

Average Run Length (ARL(θ)) generalizes over MTFA and MTD. It quantifies how long we have to wait before we detect a change of size θ in the variable that we are monitoring.

$$ARL(\theta = 0) = MTFA, \quad ARL(\theta \neq 0) = MTD$$

Our framework needs to know ground truth changes in the data for evaluation of change detection algorithms. Thus, we generate synthetic datasets with ground truth. Before a true change happens, all the alarms are considered as false alarms. After a true change occurs, the first detection that is flagged is considered as the true alarm. After that and before a new true change occurs, the consequent detections are considered as false alarms. If no detection is flagged between two true changes, then it is considered a missed detection. These concepts are graphically illustrated in Figure 2.

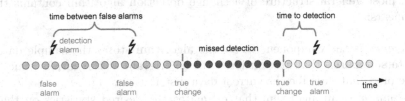

Fig. 2. The setting of change detection evaluation

We propose a new quality evaluation measure that monitors the compromise between fast detection and false alarms:

$$MTR(\theta) = \frac{MTFA}{MTD} \times (1 - MDR) = \frac{ARL(0)}{ARL(\theta)} \times (1 - MDR). \qquad (1)$$

This measure MTR (Mean Time Ratio) is the ratio between the mean time between false alarms and the mean time to detection, multiplied by the probability of detecting an alarm. An ideal change detection algorithm would have a low false positive rate (which means a high mean time between false alarms), a low mean time to detection, and a low missed detection rate.

Comparing two change detectors for a specific change θ is easy with this new measure: the algorithm that has the highest $MTR(\theta)$ value is to be preferred.

2.2   Change Detectors

A *change detector* or *drift detector* is an algorithm that takes a stream of instances as input and outputs an alarm if it detects a change in the distribution of the data. A detector may often be combined with a predictive model to output a prediction of the next instance to come. In general, the input to a change detection algorithm is a sequence $x_1, x_2, \ldots, x_t, \ldots$ of data points whose distribution varies over time in an unknown way. At each time step the algorithm outputs:

1. an estimate of the parameters of the input distribution, and
2. an alarm signal indicating whether a change in this distribution has occurred.

We consider a specific, but very frequent case, of this setting with all x_t being real values. The desired estimate is usually the current expected value of x_t, and sometimes other statistics of the distribution such as, for instance, variance. The only assumption about the distribution of x is that each x_t is drawn independently from each other. This assumption may be not satisfied if x_t is an error produced by a classifier that updates itself incrementally, because the update depends on the performance, and the next performance depends on whether we updated it correctly. In practice, however, this effect is negligible, so treating them independently is a reasonable approach.

The most general structure of a change detection algorithm contains three components:

1. *Memory* is the component where the algorithm stores the sample data or data summaries that are considered to be relevant at the current time, i.e., the ones that describe the current data distribution.
2. *Estimator* is an algorithm that estimates the desired statistics on the input data, which may change over time. The algorithm may or may not use the data contained in Memory. One of the simplest Estimator algorithms is the *linear estimator*, which simply returns the average of the data items contained in Memory. Other examples of run-time efficient estimators are Auto-Regressive, Auto Regressive Moving Average, and Kalman filters [7].
3. *Change detector* (hypothesis testing) outputs an alarm signal when it detects a change in the input data distribution. It uses the output of the Estimator, and may or may not in addition use the contents of Memory.

There are many different algorithms to detect change in time series. Our new framework contains the classical ones used in statistical quality control [6], time series analysis [8], statistical methods and more recent ones such as ADWIN[9].

2.3   Statistical Tests with Stopping Rules

These tests decide between the hypothesis that there is change and the hypothesis that there is no change, using a stopping rule. When this stopping rule is achieved, then the change detector method signals a change. The following methods differ in their stopping rule.

The CUSUM Test. The cumulative sum (CUSUM algorithm), which was first proposed in [10], is a change detection algorithm that raises an alarm when the mean of the input data is significantly different from zero. The CUSUM input ϵ_t can be any filter residual, for instance the prediction error from a Kalman filter.
The stopping rule of the CUSUM test is as follows:

$$g_0 = 0, \qquad g_t = \max(0, g_{t-1} + \epsilon_t - v), \qquad \text{if } g_t > h \text{ then alarm and } g_t = 0$$

The CUSUM test is memoryless, and its accuracy depends on the choice of parameters v and h. Note that CUSUM is a one sided, or *asymmetric* test. It assumes that changes can happen only in one direction of the statistics, detecting only increases.

The Page Hinckley Test. The Page Hinckley Test [10] stopping rule is as follows, when the signal is increasing:

$$g_0 = 0, \qquad g_t = g_{t-1} + (\epsilon_t - v), \qquad G_t = \min(g_t, G_{t-1})$$

$$\text{if } g_t - G_t > h \text{ then alarm and } g_t = 0$$

When the signal is decreasing, instead of $G_t = \min(g_t, G_{t-1})$, we should use $G_t = \max(g_t, G_{t-1})$ and $G_t - g_t > h$ as the stopping rule. Like the CUSUM test, the Page Hinckley test is memoryless, and its accuracy depends on the choice of parameters v and h.

2.4   Drift Detection Method

The drift detection method (DDM) proposed by Gama et al. [1] controls the number of errors produced by the learning model during prediction. It compares the statistics of two windows: the first contains all the data, and the second contains only the data from the beginning until the number of errors increases. Their method doesn't store these windows in memory. It keeps only statistics and a window of recent errors data.

The number of errors in a sample of n examples is modelled by a binomial distribution. For each point t in the sequence that is being sampled, the error rate is the probability of misclassifying (p_t), with standard deviation given by $s_t = \sqrt{p_t(1 - p_t)/t}$. They assume that the error rate of the learning algorithm (p_t) will decrease while the number of examples increases if the distribution of the examples is stationary. A significant increase in the error of the algorithm, suggests that the class distribution is changing and, hence, the actual decision model is supposed to be inappropriate. Thus, they store the values of p_t and s_t when $p_t + s_t$ reaches its minimum value during the process (obtaining p_{min} and s_{min}). DDM then checks if the following conditions trigger:

- $p_t + s_t \geq p_{min} + 2 \cdot s_{min}$ for the warning level. Beyond this level, the examples are stored in anticipation of a possible change of context.
- $p_t + s_t \geq p_{min} + 3 \cdot s_{min}$ for the drift level. Beyond this level the concept drift is supposed to be true, the model induced by the learning method is reset and a new model is learnt using the examples stored since the warning level triggered. The values for p_{min} and s_{min} are reset.

In the standard notation, they have two hypothesis tests h_w for warning and h_d for detection:

- $g_t = p_t + s_t$, if $g_t \geq h_w$ then alarm warning, if $g_t \geq h_d$ then alarm detection, where $h_w = p_{min} + 2s_{min}$ and $h_d = p_{min} + 3s_{min}$.

The test is nearly memoryless, it only needs to store the statistics p_t and s_t, as well as switch on some memory to store an extra model of data from the time of warning until the time of detection.

This approach works well for detecting abrupt changes and reasonably fast changes, but it has difficulties detecting slow gradual changes. In the latter case, examples will be stored for long periods of time, the drift level can take too much time to trigger and the examples in memory may overflow.

Baena-García et al. proposed a new method EDDM (Early Drift Detection Method) [11] in order to improve DDM. It is based on the estimated distribution of the distances between classification errors. The window resize procedure is governed by the same heuristics.

2.5 EWMA Drift Detection Method

A new drift detection method based on an EWMA (Exponential Weighted Moving Average) chart, was presented by Ross et al. in [2]. It is similar to the drift detection method (DDM) described previously, but uses an exponentially weighted moving average chart to update the estimate of error faster.

This method updates the following statistics for each point t in the sequence:

$$p_t = p_{t-1}(t-1)/t + \epsilon_t/t, \qquad s_t = \sqrt{p_t(1-p_t)}$$

$$g_0 = p_0, \qquad g_t = (1-\lambda)g_{t-1} + \lambda\epsilon_t, \qquad s_t^{(g)} = s_t\sqrt{\lambda(1-(1-2\lambda)^{2t})/(2-\lambda)}$$

EWMA uses the following trigger conditions:

- $g_t > h_w$ for the warning level, where $h_w = p_t + 0.5L_ts_t^{(g)}$.
- $g_t > h_d$ for the drift level, where $h_d = p_t + L_ts_t^{(g)}$.

The values of L_t are computed using a different polynomial for each choice of $MTFA$ of the form $L(p_t) = c_0 + c_1p_t + \cdots + c_mp_t^m$ using a Monte Carlo approach. A value of $\lambda = 0.2$ is recommended by the authors of this method.

2.6 ADWIN: ADaptive Sliding WINdow Algorithm

ADWIN[12] is a change detector and estimator that solves in a well-specified way the problem of tracking the average of a stream of bits or real-valued numbers. ADWIN keeps a variable-length window of recently seen items, with the property that the window has the maximal length statistically consistent with the hypothesis "there has been no change in the average value inside the window".

More precisely, an older fragment of the window is dropped if and only if there is enough evidence that its average value differs from that of the rest of the window. This has two consequences: one, that change can reliably be declared whenever the window shrinks; and two, that at any time the average over the existing window can be reliably taken as an estimate of the current average in the stream (barring a very small or very recent change that is still not statistically visible). These two points appears in [12] in a formal theorem.

ADWIN is data parameter- and assumption-free in the sense that it automatically detects and adapts to the current rate of change. Its only parameter is a confidence bound δ, indicating how confident we want to be in the algorithm's output, inherent to all algorithms dealing with random processes.

Table 1. Evaluation results for an experiment simulating the error of a classifier, that after t_c instances with a probability of having an error of 0.2, this probability is increased linearly by a value of $\alpha = 0.0001$ for each instance

Method	Measure	No Change	$t_c = 1,000$	$t_c = 10,000$	$t_c = 100,000$	$t_c = 1,000,000$
ADWIN	1-MDR		0.13	1.00	1.00	1.00
	MTD		111.26	1,062.54	1,044.96	1,044.96
	MTFA	5,315,789				
	MTR		6,150	5,003	5,087	5,087
CUSUM(h=50)	1-MDR		0.41	1.00	1.00	1.00
	MTD		344.50	902.04	915.71	917.34
	MTFA	59,133				
	MTR		70	66	65	64
DDM	1-MDR		0.44	1.00	1.00	1.00
	MTD		297.60	2,557.43	7,124.65	42,150.39
	MTFA	1,905,660				
	MTR		2,790	745	267	45
Page-Hinckley(h=50)	1-MDR		0.17	1.00	1.00	1.00
	MTD		137.10	1,320.46	1,403.49	1,431.88
	MTFA	3,884,615				
	MTR		4,769	2,942	2,768	2,713
EDDM	1-MDR		0.95	1.00	1.00	1.00
	MTD		216.95	1,317.68	6,964.75	43,409.92
	MTFA	37,146				
	MTR		163	28	5	1
EWMA Chart	1-MDR		1.00	1.00	1.00	1.00
	MTD		226.82	225.51	210.29	216.70
	MTFA	375				
	MTR		2	2	2	2

Table 2. Evaluation results for an experiment simulating the error of a classifier, that after $t_c = 10,000$ instances with a probability of having an error of 0.2, this probability is increased linearly by a value of α for each instance

Method	Measure	No Change	$\alpha = 0.00001$	$\alpha = 0.0001$	$\alpha = 0.001$
ADWIN	1-MDR		1.00	1.00	1.00
	MTD		4,919.34	1,062.54	261.59
	MTFA	5,315,789.47			
	MTR		1,080.59	5,002.89	20,320.76
CUSUM	1-MDR		1.00	1.00	1.00
	MTD		3,018.62	902.04	277.76
	MTFA	59,133.49			
	MTR		19.59	65.56	212.89
DDM	1-MDR		0.55	1.00	1.00
	MTD		3,055.48	2,557.43	779.20
	MTFA	1,905,660.38			
	MTR		345.81	745.15	2,445.67
Page-Hinckley	1-MDR		1.00	1.00	1.00
	MTD		4,659.20	1,320.46	405.50
	MTFA	3,884,615.38			
	MTR		833.75	2,941.88	9,579.70
EDDM	1-MDR		0.99	1.00	1.00
	MTD		4,608.01	1,317.68	472.47
	MTFA	37,146.01			
	MTR		7.98	28.19	78.62
EWMA Chart	1-MDR		1.00	1.00	1.00
	MTD		297.03	225.51	105.57
	MTFA	374.70			
	MTR		1.26	1.66	3.55

ADWIN does not maintain the window explicitly, but compresses it using a variant of the exponential histogram technique storing a window of length W using only $O(\log W)$ memory and $O(\log W)$ processing time per item.

3   Comparative Experimental Evaluation

We performed a comparison using the following methods: DDM, ADWIN, EWMA Chart for Drift Detection, EDDM, Page-Hinckley Test, and CUSUM Test. The two last methods were used with $v = 0.005$ and $h = 50$ by default.

The experiments were performed simulating the error of a classifier system with a binary output 0 or 1. The probability of having an error is maintained as 0.2 during the first t_c instances, and then it changes gradually, linearly increasing by a value of α for each instance. The results were averaged over 100 runs.

Tables 1 and 2 show the results. Every single row represents an experiment where four different drifts occur at different times in Table 1, and four different drifts with different incremental values in Table 2. Note that MTFA values come

Table 3. Evaluation results of a prequential evaluation using an adaptive Naive Bayes classifier on Electricity and Forest Covertype datasets: accuracy, κ, and number of changes detected

Change Detector	Warning	Forest Covertype			Electricity		
		Accuracy	κ	Changes	Accuracy	κ	Changes
ADWIN	No	83.24	73.25	1,151	81.03	60.79	88
CUSUM	No	81.55	70.66	286	79.21	56.83	28
DDM	Yes	88.03	80.78	4,634	81.18	61.14	143
Page-Hinckley	No	80.06	68.40	117	78.04	54.43	10
EDDM	Yes	86.08	77.67	2,416	84.83	68.96	203
EWMA Chart	Yes	**90.16**	**84.20**	6,435	**86.76**	**72.93**	426

from the no-change scenario. We observe the tradeoff between faster detection and smaller number of false alarms. Page Hinckley with $h = 50$ and ADWIN are the methods with fewer false positives, however CUSUM is faster at detecting change for some change values. Using the new measure MTR, ADWIN seems to be the algorithm with the best results.

We use the EWMA Chart for Drift Detection with L_t values computed for a MTFA of 400. However it seems that this is a very low value compared with other change detectors. EDDM has a high number of false positives, and performs worse than DDM using the new measure MTR.

This type of test, has the property that by increasing h we can reduce the number of false positives, at the expense of increasing the detection delay.

Finally, we use the change detector algorithms inside the MOA Framework in a real data classification task. The methodology is similar to the one in [1]: a classifier is built with a change detector monitoring its error. If this change detector detects a warning, the classifier begins to store instances. After the change detector detects change, the classifier is replaced with a new classifier built with the instances stored. Note that this classifier is in fact similar to a data stream classification algorithm that exploits a window model. The size of this window model is not fixed and depends on the change detection mechanism. We test the change detectors with a Naive Bayes classifier, and the following datasets:

Forest Covertype Contains the forest cover type for 30 x 30 meter cells obtained from US Forest Service (USFS) Region 2 Resource Information System (RIS) data. It contains $581,012$ instances and 54 attributes. It has been used before, for example in [13,14].

Electricity Contains $45,312$ instances describing electricity demand. A class label identifies the change of the price relative to a moving average of the last 24 hours. It was described by [3] and analysed also in [1].

The accuracy and κ statistic results using prequential evaluation are shown in Table 3. The classifier that uses the EWMA Chart detection method is the method with the best performance on the two datasets. It seems that having a large amount of false positives, detecting more changes, and rebuilding the classifier more often with more recent data helps to improve accuracy for these datasets. For the classification setting, the fact that the detector has a warning signal detection helps to improve results. Also, the success of EWMA Chart may be due to the fact that Naive Bayes is a high-bias algorithm, which can attain good performance from smaller batches of data.

However, as we have seen, the fact that a change detection algorithm produces high accuracy figures in a classification setting does not necessarily imply a low false alarm rate or a high MTR value for that algorithm. It should be possible, for example, to tune all the detectors in the framework to produce better classification results by deliberately raising their false positive rates. Additionally, it should be possible to demonstrate that under normal circumstances, low-bias algorithms suffer high false positive rates.

4   Conclusions

Change detection is an important component of systems that need to adapt to changes in their input data. We have presented a new framework for evaluation of change detection methods, and a new quality measure for change detection. Using this new experimental framework we demonstrated that classification performance is a poor proxy for change detection performance and provide further evidence that if high classification performance is a requirement then using a change detector that produces a high false positive rate can be beneficial for some datasets. We hope that the new framework presented here will help the research community and industry data analysts to experimentally compare and benchmark change detection techniques.

Acknowledgements. I. Žliobaitė's research has been supported by the Academy of Finland grant 118653 (ALGODAN).

References

1. Gama, J., Medas, P., Castillo, G., Rodrigues, P.: Learning with drift detection. In: Bazzan, A.L.C., Labidi, S. (eds.) SBIA 2004. LNCS (LNAI), vol. 3171, pp. 286–295. Springer, Heidelberg (2004)
2. Ross, G.J., Adams, N.M., Tasoulis, D.K., Hand, D.J.: Exponentially weighted moving average charts for detecting concept drift. Pattern Recognition Letters 33(2), 191–198 (2012)
3. Harries, M.: Splice-2 comparative evaluation: Electricity pricing. Technical report, The University of South Wales (1999)
4. Bifet, A., Holmes, G., Kirkby, R., Pfahringer, B.: MOA: Massive online analysis. Journal of Machine Learning Research 11, 1601–1604 (2010)
5. Gustafsson, F.: Adaptive Filtering and Change Detection. Wiley (2000)
6. Basseville, M., Nikiforov, I.V.: Detection of abrupt changes: theory and application. Prentice-Hall, Inc., Upper Saddle River (1993)

7. Kobayashi, H., Mark, B.L., Turin, W.: Probability, Random Processes, and Statistical Analysis. Cambridge University Press (2011)

8. Takeuchi, J., Yamanishi, K.: A unifying framework for detecting outliers and change points from time series. IEEE Transactions on Knowledge and Data Engineering 18(4), 482–492 (2006)

9. Bifet, A., Gavaldà, R.: Adaptive learning from evolving data streams. In: Adams, N.M., Robardet, C., Siebes, A., Boulicaut, J.-F. (eds.) IDA 2009. LNCS, vol. 5772, pp. 249–260. Springer, Heidelberg (2009)

10. Page, E.S.: Continuous inspection schemes. Biometrika 41(1/2), 100–115 (1954)

11. Baena-García, M., del Campo-Ávila, J., Fidalgo, R., Bifet, A., Gavaldá, R., Morales-Bueno, R.: Early drift detection method. In: Fourth International Workshop on Knowledge Discovery from Data Streams (2006)

12. Bifet, A., Gavaldà, R.: Learning from time-changing data with adaptive windowing. In: SIAM International Conference on Data Mining (2007)

13. Gama, J., Rocha, R., Medas, P.: Accurate decision trees for mining high-speed data streams. In: ACM SIGKDD International Conference on Knowledge Discovery and Data Mining, pp. 523–528 (2003)

14. Oza, N.C., Russell, S.J.: Experimental comparisons of online and batch versions of bagging and boosting. In: ACM SIGKDD International Conference on Knowledge Discovery and Data Mining, pp. 359–364 (2001)

Integrating Multiple Studies of Wheat Microarray Data to Identify Treatment-Specific Regulatory Networks

Valeria Bo[1], Artem Lysenko[2], Mansoor Saqi[2],
Dimah Habash[3], and Allan Tucker[1]

[1] Department of Information System and Computing, Brunel University, London, UK
[2] Department of Computational and Systems Biology, Rothamsted Research,
Harpenden, UK
[3] Securewheat Ltd, St Albans, UK

Abstract. Microarrays have allowed biologists to better understand gene regulatory mechanisms. Wheat microarray data analysis is a complex and challenging topic and knowledge of gene regulation in wheat is still very superficial. However, understanding key mechanisms in this plant holds much potential for food security, especially with a changing climate. The purpose of this paper is to combine multiple microarray studies to automatically identify subnetworks that are distinctive to specific experimental conditions. For example, identifying a regulatory network of genes that only exists under certain types of experimental conditions will assist in understanding the nature of the mechanisms. We derive unique networks from multiple independent networks to better understand key mechanisms and how they change under different conditions. We compare the results with biclustering, detect the most predictive genes and validate the results based upon known biological mechanisms. We also explore how this pipeline performs on yeast microarray data.

1 Introduction

Microarray data measures the simultaneous expression of thousands of genes allowing the modelling of the underlying mechanisms of gene regulation through Gene Regulatory Networks (GRNs). In this paper we explore the use of GRNs to analyse wheat microarray data from a number of independent studies. Understanding of gene regulation in wheat is challenging because of the size and complexity of its genome and gaining a deeper knowledge of this plant may lead to improved food security under different climate scenarios. Because the knowledge of wheat gene regulation is still very young, we aim to automatically identify some basic regulatory mechanisms as well as how they differ between a number of independent studies in order to better understand those mechanisms. Microarray data consists of many thousands of genes and generally only a few samples. What is more, the integration of data collected from different studies is an ongoing problem but with some reported successes [4]. In [2] more robust

A. Tucker et al. (Eds.): IDA 2013, LNCS 8207, pp. 104–115, 2013.

models are built from multiple datasets by ordering them based on the level of noise and informativeness and using different Bayesian classifiers to select the informative genes. Steele et.al [16] combine various microarray datasets using post-learning aggregation to build robust regulatory networks. We adopt a similar approach here but focus less on generating the consensus of all datasets and more on identifying mechanisms that are specific to a subset of studies. [17] also explore consensus approaches but use a clustering technique coupled with a statistically based gene functional analysis for the identification of novel genes. Often the sheer number of genes makes the understanding of GRNs difficult and sometimes modules are created by grouping genes that perform some similar function [15]. Networks of these modules can then be discovered to identify mechanisms at a more general level. Clustering helps to preserve all information but might increase noise and bias. In [19] two cancer datasets are compared (case and control). For each dataset gene-pair expression correlation is computed and then used to build a frequency table whose values are used to build a weighted gene co-expression frequency network. After this they identify sub-networks with similar members and iteratively merge them together to generate the final network for both cancer and healthy tissue. Alaakwaa et.al [1] instead explore the biclustering technique [3] which aims to cluster both genes and samples simultaneously. They apply six different biclustering methods and use the resulting biclusters to build Bayesian Networks for each and finally merge the results in one single network which captures the overall mechanism. In this paper, instead of focusing only on consensus networks/clusters we explore a method to "home in" on both the similarities and differences of GRNs generated from different studies by using a combination of clustering, network discovery and graph theory. What is more, we go beyond the simple pairwise correlations between genes which is common in many studies e.g. [19]. We build independent networks for each study using the R package *glasso* [5] which identifies the inverse covariance matrix using the lasso penalty to make it as sparse as possible. We compare the networks obtained for each study using graph similarity techniques and cluster studies with similar regulatory behaviour and network structures. At this point we detect the glasso networks that are unique for each "*study-cluster*". Furthermore, we exploit these results to build Bayesian Networks for each study-cluster to identify the most predictive group of genes. As a validation of our pipeline we compare the results with the ones obtained with a popular state-of-the-art technique known as Biclustering which aims to simultaneously cluster genes and samples. We also explore our approach in yeast to demonstrate the generalisation of the approach to different microarray data.

2 Methods

In this paper we analyse a number of publicly available wheat transcriptome datasets derived from multiple experiments from plants subjected to a range of treatments: stress, development, cultivar, etc. The number of common genes to all studies is 61290 and 595 samples that can be grouped into 16 different studies representing different treatments the plant has undergone as shown in Table 1. Each study contains data derived from treated and non-treated samples. Studies 1-6, 12, and 13 were considered stress-enriched, the remaining as non-stressed treatments. Labels and details for each study are available http://www.ebi.ac.uk/arrayexpress/.

Table 1. Study numbers, labels, number of samples and descriptions

Study	Label	Number samples	Description
1	E-MEXP-971	60	Salt stress
2	E-MEXP-1415	36	S and N deficient conditions
3	E-MEXP-1193	32	Heat and Drought Stress
4	E-MEXP-1694	6	Re-supply of sulfate
5	E-MEXP-1523	30	Heat stress
6	E-MEXP-1669	72	Different nitrogen fertiliser levels
7	E-GEOD-4929	4	Study parental genotypes 2
8	E-GEOD-4935	78	Study 39 genotypes 2
9	E-GEOD-6027	21	Meiosis and microsporogenesis in hexaploid bread wheat
10	E-GEOD-9767	16	Genotypic differences in water soluble carbohydrate metabolism
11	E-GEOD-12508	39	Wheat development
12	E-GEOD-12936	12	Effect of silicon
13	E-GEOD-11774	42	Cold treatment
14	E-GEOD-5937	4	Parental genotypes 2 biological replicates from SB location
15	E-GEOD-5939	72	36 genotypes 2 biological replicates from SB location
16	E-GEOD-5942	76	Parental and progenies from SB location

In this work, to demonstrate the generalisation of the approach, we also analyse *Saccharomyces cerevisiae* (82 different studies). We applied to both types of data the following general steps:

1. Select the most informative genes to reduce the number of variables,
2. Build a network for each study using *glasso*,
3. Create consensus network by identifying links common to a proportion of the discovered GRNs,
4. Apply graph similarity analysis in order to identify similarities between the different studies and therefore cluster them,
5. Identify unique networks,
6. Build Bayesian networks for each study-cluster
7. Identify the most predictive genes

2.1 Selection of Informative Genes and Networks Construction

For computational and practical reasons we need, first, to reduce the number of the variables. To prevent noise and bias, we choose to avoid clustering but discard all the non informative genes. Firstly, we discard those genes that are currently not in the Gene Ontology (GO) database, meaning we can focus on genes that have some annotation. We then discard those genes for which the standard deviation of expression values across all experiments was less than 2. After this step still many genes survive in each study. Thus, to further reduce their number we select only those that survived in at least 4 studies (25% of the studies in wheat).

GRNs have two main components: nodes which represent the variables/genes and edges that encode the joint probability distribution by representing conditional independences between variables. In this paper we want to identify networks that go beyond simple pairwise relationships so our procedure is centred around using *glasso*. The problem of identifying the structure of the network can be solved by estimating the relationships between variables. In the case of undirected graphs it is the same as learning the structure of the conditional independence graph (CIG), which in the case of Gaussian random variables, means to identify the zeros of the inverse covariance matrix (also called precision or concentration matrix). Given a p-dimensional normally distributed random variable X, assuming that the covariance matrix is non-singular, the conditional independence structure of the distribution can be represented by the graphical model G = (N, E) where N = (1,..,p) is the set of nodes and E is the set of edges in N×N. If a pair of variables is not in the set E it means that the two variables are conditionally independent given the other variables. This corresponds to a zero in the inverse covariance matrix. Therefore it imposes an L1 penalty in the estimation of the inverse covariance matrix in order to increase its sparsity [5].

The *glasso* package in R estimates a sparse inverse covariance matrix using a lasso (L_1) penalty. Suppose, we have N multivariate normal observations of dimension p, with mean μ and covariance Σ. The problem is to maximizes the penalized log likelihood $\log det\Theta - tr(S\Theta) - \rho \|\Theta\|_1$ where $\Theta = \Sigma^{-1}$, S is the empirical covariance matrix and $\|\Theta\|_1$ is the L1 norm the sum of the absolute values of the elements of Σ^{-1} and ρ is the regularization parameter. The parameter ρ can be a scalar (typical situation) or a p×p matrix. $\rho = 0$ means no regularization. As stated earlier, in this paper we apply the *glasso* package to build one network for every single study.

2.2   Graph Similarity and Unique Networks

This pipeline automatically identifies subnetworks distinctive to specific studies. The first important step, then, is to be able to automatically detect groups of similar studies. Wheat datasets were therefore selected which have 595 samples derived from 16 different studies listed in Table 1. But different studies still have some network paths in common, the more they have, the more similar they might be. Because we want to identify mechanisms common to similar studies we cluster them using the sensitivity metric. We consider the connections in common between two study-networks and build a contingency table. To verify the reliability of the clusters, we compare the results with the description of the studies in Table 1.

We explored a number of clustering techniques but found that k-means (R function based on [7]) generated the most convincing, based on Table 1, study-clusters. Table 1 demonstrates that the studies can be grouped in two: stress-enriched and non-stress conditions. We explored different values of k but found that 3 clusters were the most revealing: The resulting clusters are: {2, 5, 6, 10, 12}, {1, 3, 4, 9, 11, 13} and {7, 8, 14, 15, 16} based upon the studies numbering from Table 1. While the third cluster clearly groups together all the non-stress studies, the other two reflect studies that are stress enriched. In the process of identifying unique networks we first build the consensus network [16] for each study-cluster as a representative of the general mechanism for that group of studies. Then we select only those edges that exist in the consensus-study network in consideration but not in the remaining ones. The resulting list of nodes involved in the unique connections are used to build the unique networks as explained in details in the next section. We also compare our results (the discovered clusters and their associated networks) with *Biclustering* which aims to cluster samples and genes simultaneously.

2.3   Bayesian Networks, Prediction and Biological Validation

We choose to validate the networks through prediction and Bayesian Networks (BNs) which naturally perform this given the graphical structure obtained using the gene in the unique networks provided by glasso. BNs [8,6] are a class of graphical models that represent the probabilistic dependencies between a given set of random variables. A Bayesian network has a set of variables called nodes and a set of directed edges between variables called arcs. The nodes and arcs together form a *directed acyclic graph* (DAG) G=(V,A). Each variable in the network has attached a conditional probability table of itself given the parents. The reason why we chose BNs at this specific point of the work is because this technique uses probability to measure the available knowledge which means that the more the knowledge we have the lesser the uncertainty. Given the conditional probability table BNs allow to work from parent to child but also backwards from child to parent (cause → effect and cause ← effect). BNs treat uncertainty explicitly and allow to include experts' knowledge in the form of a prior distribution. Furthermore, the directed graph aspect of this representation simplifies the identification of important paths. Having reduced the number of variables and samples by identifying the unique networks, we build one BN for each of the study-clusters previously identified focusing on the genes involved in the unique connections detected with the glasso networks. To do this we used the R package *bnlearn* [13,14]. After this we are interested in finding the most predictive and predictable genes within and outside the study-clusters using the leave one out cross validation technique. Given the m studies and n genes within each studies-cluster we use $m-1$ studies as a training set and the remaining one as test set. Then we use the $n-1$ genes to predict the one left out. We do this within all the studies-clusters and for all the combinations of training and test set of studies and genes. The prediction is made with the R package *gRain* [9]. Finally, we need to make sure that the subnetworks we identified principally with glasso have some biological meaning. To do this we exploit two pieces of software:

1. *Mapman* [18] which explores gene-by-gene the functions related to it and returns a list of functions and a graph of connections
2. The software described in [11] returns the highlighted biological functions in Table 2 which further reinforce the validity of the discovered genes returned by Mapman.

3   Results

In the figures below we show the unique-networks, learnt with *bnlearn*, for wheat in one of the two study-clusters of stress-enriched conditions and the unique network for the non-stress conditions cluster. The numbers identify the genes and the black circles represent in both the genes that are involved in biotic (caused by living organisms, such as insects and microbes) and abiotic (caused by non-alive factors such as heat, drought, cold and agronomy practices) stress response. In both networks we clearly see specific paths and groups of genes that are highly connected. Using Mapman we were able to associate a function to each gene. Focusing on the stress-enriched conditions network, the procedure has managed to identify a relatively small number (58) of well connected nodes. In fact only one node is isolated while the others form a distinctive path. We see that genes involved in both kinds of stress response (biotic and abiotic stress) are involved in the network. Specifically the first four genes that start the network pathway (29 - 47 - 17 - 30) are all involved in biotic stress. The remaining highlighted genes instead are mostly involved in heat stress. A good number of photosynthesis related genes are also involved, in particular (18 - 27 - 21 - 28 - 6 - 22). On the non-stress network, we have again identified a reasonable number of genes though these are less connected. However, one very well defined pathway exists that consists mainly of photosynthesis-related genes (not highlighted). Fewer genes are found that are related to stress response and those that do appear are much less connected, except for the path formed by (46 - 57 - 26 - 50) nodes. The software described in [11] returns the following (see Table 2) highlighted biological functions which go to reinforce the results from Mapman. Higher values of Information Content (IC) are associated with more informative terms. Values greater than 3 are generally considered to be biologically informative. In Figures 3 and 4 we show the predictive accuracy for each gene. What we expect is a better prediction within the study-clusters and a weaker one outside the clusters. Each boxplot represents the percentage of how many times the gene has been predicted correctly among all the different given samples. The chance of correctly predicting the genes randomly is one in three (there are three possible states for each gene: *under-regulated, normal, over-regulated*). Values above this can be considered better than random. In the figures we clearly see that the *internal predictions* (predictions made by cross validating within a study-cluster) are quite high for most of the genes with little variations. For the *external predictions* (predictions on data outside of the study-cluster), however, the mean prediction values are mostly not better than chance as one would expect, and the standard deviations are very high making them not reliable. This implies that the identified subnetworks are indeed specific to their study cluster, making them easier to characterise.

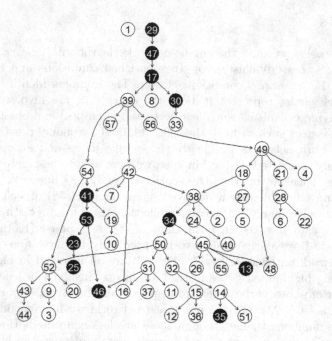

Fig. 1. Unique-Network for wheat under stress-enriched conditions

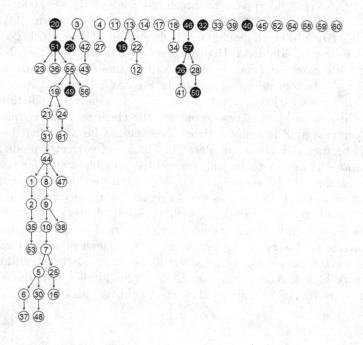

Fig. 2. Unique-Network for wheat under non-stress conditions

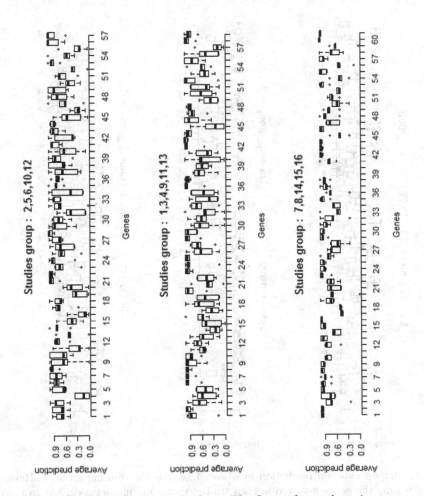

Fig. 3. Boxplot internal prediction in wheat. The figure shows that given n genes in the network, using n-1 genes as predictor, the probability of correctly predicting the value of the remaining gene is almost always better than chance (0.3). This means that for each unique network, considering only the genes involved in it gives a better understanding of the behaviour under specific conditions.

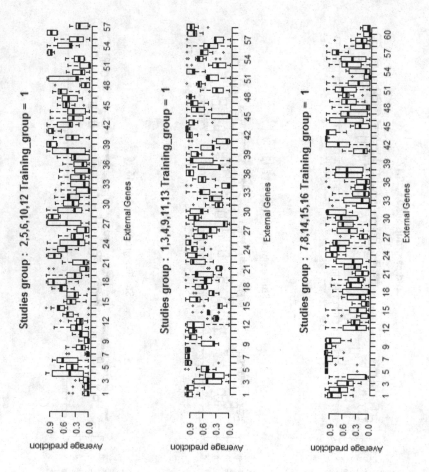

Fig. 4. Boxplot external prediction in wheat. Example for training group 1. Here, we consider only the genes involved in each studies group unique network, but to predict we used data of the genes from studies outside the studies group. The figure shows that the external prediction ability for the first studies group has a weaker mean and a larger range meaning instability. This proves that the mechanisms discovered in each studies group unique network is very specific to describe what is happening inside each studies group and is not for all the other studies.

Table 2. Unique networks biological process functions from Gene Ontology as described in [11]

Unique network	GO Id	GO Name	IC Term
1	GO:0019538	protein metabolic process	3.19
1	GO:0006950	response to stress	3.96
1	GO:0071840	cellular component organization or biogenesis	3.98
2	GO:0006950	response to stress	3.96
2	GO:0071840	cellular component organization or biogenesis	3.98
2	GO:0019684	photosynthesis, light reaction	8.32
2	GO:0044267	cellular protein metabolic process	3.45
3	GO:0006950	response to stress	3.96
3	GO:0015979	photosynthesis	7.13
3	GO:0071840	cellular component organization or biogenesis	3.98
3	GO:0009628	response to abiotic stimulus	4.97
3	GO:0042221	response to chemical stimulus	4.12
3	GO:0006091	generation of precursor metabolites and energy	5.14
3	GO:0044267	cellular protein metabolic process	3.45

3.1   Comparison with Bicluster

Finally we compare the results obtained with our algorithm in wheat with the one obtained using Bicluster [3]. The biclustering technique identifies a subset of genes and a subset of conditions simultaneously. There are various methods in the literature for biclustering [12] but for this work we specifically chose the BCS method implemented in the R package *biclust*. This is a state-of-the-art method that normalizes the data matrix and looks for checkerboard structures using the well known technique of singular value decomposition in eigenvectors applied to both rows and columns [10]. It is important to highlight that biclustering works on each sample and not on the studies. The method, after appropriately tuning the parameters, identifies 17 biclusters. On the wheat data each resulting bicluster highlights a different set of samples but the same set of six genes, 5 of which are related to abiotic heat stress. The genes highlighted by biclustering are also in the list of genes detected by the algorithm described in this paper, specifically we can see five of these genes also highlighted in Figure 1 (23 -25 - 41 - 46 - 53). This discovery points out the importance of these 5 stress-related and 1 protein-degradation-related genes but unfortunately biclustering fails at identifying other equally important stress-related genes identified by our algorithm. In addition the six genes that are identified do not seem to be associated with a specific subset of samples. Rather each of them have been detected in all of the biclusters. Regarding the samples, about half of the biclusters manage to group together samples of stress-enriched studies but split samples from the same study. Unfortunately, none of the biclusters group the non-stress studies accurately enough to identify specific non-stress clusters. Furthermore, considering that each study consists of both actual treatment samples and a small number of controls it might be that biclustering merges together the control samples of the stress-conditions with non-stress samples but this union occurs too often and with too many samples for this to be considered the case. In conclusion, we have found that the resulting biclusters do not properly cluster the samples together, even ones belonging to the same study. Every bicluster highlights the same group of genes preventing any discovery of differences between treatments. It still discovers some important genes but much less than the ones we are able to find with the method proposed in this paper.

3.2   Results in Yeast

For the yeast data, following the same steps as in the wheat dataset, we found
9 study-clusters. Here most internal predictions are good though one cluster results in poor accuracies that are not much better than chance. Table 3 shows
the internal and external mean and variance prediction for each study-cluster.
Within study prediction is high. Unlike the wheat, it seems that the gene networks identified are a little better at predicting between different studies. This
might be because of the larger variety of available studies (82), that cluster
together supplying the networks with more data resulting in reasonably good
prediction even outside the study-cluster.

Table 3. Internal and External prediction accuracy for Yeast

Studies-group	Internal Mean	Internal Variance	External Mean	External Variance
1	0.33	0.08	0.34	0.12
2	0.72	0.11	0.50	0.10
3	0.71	0.09	0.53	0.09
4	0.53	0.12	0.52	0.08
5	0.47	0.13	0.40	0.09
6	0.62	0.08	0.60	0.08
7	0.63	0.07	0.58	0.08
8	0.48	0.11	0.52	0.09
9	0.62	0.12	0.54	0.08

4   Conclusions

There is a large number of microarray data sets available and the relatively new
interest in wheat renders the exploration of complexity of biological mechanisms
within the plant all the more important. Gene regulatory networks (GRNs)
can help in the exploration and understanding of biological processes operating
under different conditions. The complexity of networks learnt from all studies
can involve a large number of genes, making it very difficult to identify the most
important. In this paper we focus on the most relevant genes to different clusters
of studies. To do this we derive "unique network" using a combination of filtering,
clustering and consensus network modelling. We evaluated the predictive power
of our unique networks using Bayesian networks for each study-cluster. We have
shown that this sequence of steps automatically identifies subnetworks that are
characteristic (and more predictive) of a specific group of studies, in particular
highlighting two pathways involved in two types of stress-conditions. Future work
will involve further exploration of graph theoretic approaches to comparing the
glasso networks, further analysis of the discovered genes and pathways, and a
better understanding of the performance of the algorithm using simulated data.

References

1. Alakwaa, F., Solouma, N., Kadah, Y.: Construction of gene regulatory networks
   using biclustering and bayesian networks. Theoretical Biology and Medical Modelling 8(1), 39 (2011)

2. Anvar, S.Y., Tucker, A., et al.: The identification of informative genes from multiple datasets with increasing complexity. BMC Bioinformatics 11(1), 32 (2010)
3. Cheng, Y., Church, G.M.: Biclustering of expression data. In: Proceedings of the Eighth International Conference on Intelligent Systems for Molecular Biology, vol. 8, pp. 93–103 (2000)
4. Choi, J.K., Yu, U., Kim, S., Yoo, O.J.: Combining multiple microarray studies and modeling interstudy variation. Bioinformatics 19(suppl. 1), i84–i90 (2003)
5. Friedman, J., Hastie, T., Tibshirani, R.: Sparse inverse covariance estimation with the graphical lasso. Biostatistics 9(3), 432–441 (2008)
6. Friedman, N., Linial, M., Nachman, I., Pe'er, D.: Using bayesian networks to analyze expression data. Journal of Computational Biology 7(3-4), 601–620 (2000)
7. Hartigan, J.A., Wong, M.A.: Algorithm as 136: A k-means clustering algorithm. Journal of the Royal Statistical Society. Series C (Applied Statistics) 28(1), 100–108 (1979)
8. Heckerman, D., Geiger, D., Chickering, D.M.: Learning bayesian networks: The combination of knowledge and statistical data. Machine Learning 20(3), 197–243 (1995)
9. Højsgaard, S.: Graphical independence networks with the grain package for r (2012)
10. Kluger, Y., Basri, R., Chang, J.T., Gerstein, M.: Spectral biclustering of microarray data: coclustering genes and conditions. Genome Research 13(4), 703–716 (2003)
11. Lysenko, A., Defoin-Platel, M., Hassani-Pak, K., Taubert, J., Hodgman, C., Rawlings, C.J., Saqi, M.: Assessing the functional coherence of modules found in multiple-evidence networks from arabidopsis. BMC Bioinformatics 12(1), 203 (2011)
12. Madeira, S.C., Oliveira, A.L.: Biclustering algorithms for biological data analysis: a survey. IEEE/ACM Transactions on Computational Biology and Bioinformatics 1(1), 24–45 (2004)
13. Scutari, M.: Learning bayesian networks with the bnlearn r package. arXiv preprint arXiv:0908.3817 (2009)
14. Scutari, M., Scutari, M.M.: Package bnlearn (2012)
15. Segal, E., Shapira, M., Regev, A., Pe'er, D., Botstein, D., Koller, D., Friedman, N.: Module networks: identifying regulatory modules and their condition-specific regulators from gene expression data. Nature Genetics 34(2), 166–176 (2003)
16. Steele, E., Tucker, A.: Consensus and meta-analysis regulatory networks for combining multiple microarray gene expression datasets. Journal of Biomedical Informatics 41(6), 914–926 (2008)
17. Swift, S., Tucker, A., Vinciotti, V., Martin, N., Orengo, C., Liu, X., Kellam, P.: Consensus clustering and functional interpretation of gene-expression data. Genome Biology 5(11), R94 (2004)
18. Thimm, O., Bläsing, O., Gibon, Y., Nagel, A., Meyer, S., Krüger, P., Selbig, J., Müller, L.A., Rhee, S.Y., Stitt, M.: mapman: a user-driven tool to display genomics data sets onto diagrams of metabolic pathways and other biological processes. The Plant Journal 37(6), 914–939 (2004)
19. Zhang, J., Lu, K., Xiang, Y., Islam, M., Kotian, S., Kais, Z., Lee, C., Arora, M., Liu, H., Parvin, J.D.: et al. Weighted frequent gene co-expression network mining to identify genes involved in genome stability. PLoS Computational Biology 8(8), e1002656 (2012)

Finding Frequent Patterns
in Parallel Point Processes

Christian Borgelt and David Picado-Muiño

European Centre for Soft Computing
Gonzalo Gutiérrez Quirós s/n, 33600 Mieres, Spain
{christian.borgelt,david.picado}@softcomputing.es

Abstract. We consider the task of finding frequent patterns in parallel point processes—also known as finding frequent parallel episodes in event sequences. This task can be seen as a generalization of frequent item set mining: the co-occurrence of items (or events) in transactions is replaced by their (imprecise) co-occurrence on a continuous (time) scale, meaning that they occur in a limited (time) span from each other. We define the support of an item set in this setting based on a maximum independent set approach allowing for efficient computation. Furthermore, we show how the enumeration and test of candidate sets can be made efficient by properly reducing the event sequences and exploiting perfect extension pruning. Finally, we study how the resulting frequent item sets/patterns can be filtered for closed and maximal sets.

1 Introduction

We present methodology and algorithms to identify *frequent patterns* in parallel point processes, a task that is also known as finding *frequent parallel episodes* in event sequences (see [7]). This task can be seen as a generalization of frequent item set mining (FIM)—see e.g. [2]. While in FIM items co-occur if they are contained in the same transaction, in our setting a continuous (time) scale underlies the data and items (or events) co-occur if they occur in a (user-defined) limited (time) span from each other. The main problem of this task is that, due to the absence of (natural) transactions, counting the number of co-occurrences (and thus determining what is known as the *support* of an item set in FIM) is not a trivial problem. In this paper we rely on a *maximum independent set* approach, which has the advantage that it renders support *anti-monotone*. This property is decisive for an efficient search for frequent patterns, because it entails the so-called *apriori property*, which allows to prune the search effectively. Although NP-complete in the general case, the maximum independent set problem can be solved efficiently in our case due to the restriction of the problem instances by the underlying one-dimensional domain (i.e., the continuous time scale).

The application domain that motivates our investigation is the analysis of *parallel spike trains* in neurobiology: sequences of points in time, one per neuron, representing the times at which an electrical impulse (*action potential* or *spike*) is emitted. Our objective is to identify *neuronal assemblies*, intuitively

A. Tucker et al. (Eds.): IDA 2013, LNCS 8207, pp. 116–126, 2013.

understood as groups of neurons that tend to exhibit synchronous spiking. Such neuronal assemblies were proposed in [5] as a model for encoding and processing information in biological neural networks. In particular, as a (possibly) first step in the identification of neuronal assemblies, we look for *frequent neuronal patterns* (i.e., groups of neurons that exhibit *frequent synchronous spiking*).

The remainder of this paper is structured as follows: Section 2 covers basic terminology and notation. In Section 3 we compare two characterizations of synchrony: *bin-based* and *continuous*, exposing the challenges presented by them. In Section 4 we present our maximum independent set approach to support counting as well as an efficient algorithm. In Section 5 we employ an enumeration scheme (directly inspired by common FIM approaches) to find all frequent patterns in a set of parallel point processes. In particular, we introduce core techniques that are needed to make the search efficient. In Section 6 we present experimental results, demonstrating the efficiency of our algorithm scheme. Finally, in Section 7 we draw conclusions from our discussion.

2 Event Sequences and Parallel Episodes

We (partially) adopt notation and terminology from [7]. Our data are (finite) *sequences of events* of the form $S = \{\langle e_1, t_1 \rangle, \ldots, \langle e_k, t_k \rangle\}$, for $k \in \mathbb{N}$, where e_i in the *event* $\langle e_i, t_i \rangle$ is the *event type* (taken from a domain set E) and $t_i \in \mathbb{R}$ is the time of occurrence of e_i, for all $i \in \{1, ..., k\}$. We assume that S is ordered with respect to time, that is, $\forall i \in \{1, ..., k-1\}: t_i \leq t_{i+1}$. Such data may be represented as *parallel point processes* $\mathcal{L} = \{(a_1, [t_1^{(1)}, \ldots, t_{k_1}^{(1)}]), \ldots, (a_m, [t_1^{(m)}, \ldots, t_{k_m}^{(m)}])\}$ by grouping events with the same type $a_i \in E$ and listing the times of their occurrences (also sorted with respect to time) for each of them.[1] We employ both representations, based on convenience. In our motivating application (i.e., spike train analysis), the event types are given by the neurons and the corresponding point processes list the times at which spikes were recorded for these neurons.

Episodes (in S) are defined as sets of event types in E embedded with a partial order, usually required to occur in S within a certain time span. *Parallel episodes* have no constraints on the relative order of their elements. An *instance (or occurrence) of a parallel episode* (or a *set of synchronous events*) $A \subseteq E$ in S with respect to a (user-specified) time span $w \in \mathbb{R}^+$ can be defined as a subsequence $\mathcal{R} \subseteq S$, which contains exactly one event per event type $a \in A$ and in which any two events are separated by a time distance at most w.

Event synchrony is formally characterized by means of the operator σ_C:

$$\sigma_C(\mathcal{R}, w) = \begin{cases} 1 & \text{if } \max\{|t_i - t_j| \mid \langle e_i, t_i \rangle, \langle e_j, t_j \rangle \in \mathcal{R}\} \leq w, \\ 0 & \text{otherwise.} \end{cases}$$

In words, $\sigma_C(\mathcal{R}, w) = 1$ if all events in $\mathcal{R}$ lie within a distance at most w from each other; otherwise $\sigma_C(\mathcal{R}, w) = 0$. The set of all instances of parallel episodes

---

[1] We use square brackets (i.e., [. . .]) to denote *lists*.

(or all sets of synchronous events) of a set $A \subseteq E$ of event types is denoted by $\mathcal{E}(A, w)$, which we formally define as

$$\mathcal{E}(A, w) = \{\mathcal{R} \subseteq \mathcal{S} \mid A = \{e \mid \langle e, t \rangle \in \mathcal{R}\} \wedge |\mathcal{R}| = |A| \wedge \sigma_C(\mathcal{R}, w) = 1\}.$$

That is, $\mathcal{E}(A, w)$ contains all event subsets of $\mathcal{S}$ with exactly one event for each event type in A that lie within a maximum distance of w from each other.

3 Event Synchrony

Although the above definitions are clear enough, the intuitive notion of the (total) amount of synchrony of a set A of event types suggested by it poses problems: simply counting the occurrences of parallel episodes of A—that is, defining the *support* of A (i.e. the total synchrony of A) as $s(\mathcal{E}(A, w)) = |\mathcal{E}(A, w)|$—has undesirable properties. The most prominent of these is that the support/total synchrony of a set $B \supset A$ may be larger than that of A (namely if an instance of a parallel episode for A can be combined with multiple instances of a parallel episode for $B \setminus A$), thus rendering the corresponding support measure not *anti-monotone*. That is, such a support measure does not satisfy $\forall B \supset A :$ $s(\mathcal{E}(B, w)) \leq s(\mathcal{E}(A, w))$. However, this property is decisive for an efficient search for frequent parallel episodes, because it entails the so-called *apriori property*: given a user-specified minimum support threshold $s_{\min}$, we have $\forall B \supset A :$ $s(\mathcal{E}(A, w)) \leq s_{\min} \Rightarrow s(\mathcal{E}(B, w)) \leq s_{\min}$ (that is, *no superset of an infrequent set is frequent*). This property allows to prune the search effectively: as soon as an infrequent set is encountered, no supersets need to be considered anymore.

In order to overcome this problem, many approaches to find frequent parallel episodes resort to *time binning* (including [7] and virtually all approaches employed in neurobiology): the time interval covered by the events under consideration is divided into (usually disjoint) time bins of equal length (i.e., the originally continuous time scale is *discretized*). In this way transactions of classical frequent item set mining (FIM) [2] are formed: events that occur in the same time bin are collected in a transaction and are thus seen as synchronous, while events that occur in different time bins are seen as not synchronous. Technically, time binning can be characterized by the synchrony operator σ_B, defined as follows for $\mathcal{R} \subseteq \mathcal{S}$ and w now representing the bin width:

$$\sigma_B(\mathcal{R}, w) = \begin{cases} 1 & \text{if } \exists k \in \mathbb{Z}: \ \forall \langle e, t \rangle \in \mathcal{R}: \ t \in (w(k-1), wk], \\ 0 & \text{otherwise} \end{cases}$$

(implicitly assuming that binning is anchored at 0). Clearly, this solves the problem pointed out above: counting the bins in which all event types of a given set A occur (or, equivalently, counting the number of sets of synchronous events of A— at most one per time bin) yields an anti-monotone support measure. However, this *bin-based* model of synchrony has several disadvantages:

– **Boundary problem.** Two events separated by a time distance (much) smaller than the bin length may end up in different bins and thus be regarded as non-synchronous. Such behavior is certainly undesirable.

- **Bivalence problem.** Two events can be either (fully) synchronous or non-synchronous. Small variations in the time distance between two spikes (possibly moving one of them over a bin boundary) cause a jump from (full) synchrony to non-synchrony and *vice versa*. This may be counter-intuitive.
- **Clipping.** In neurobiology the term *clipping* refers to the fact that in a bin-based model it is usually considered only *whether* a neuron emits a spike in a time bin, not *how many* spikes it emits in it. The same can be observed for general event types in many settings employing time binning.

Some of these disadvantages can be mitigated by using overlapping time bins, but this causes other problems, especially certain anomalies in the counting of parallel episodes if they span intervals of (widely) different lengths: parallel episodes with a short span may be counted more often than parallel episodes with a long span, because they can occur in more (overlapping) time bins.

Due to these problems we prefer a synchrony model that does *not* discretize time into bins, but rather keeps the (time) scale continuous: a *continuous* synchrony model, as it was formalized in Section 2. This model captures the *intended* characterization of synchrony in the bin-based approach, solves the boundary problem and overcomes the effects of clipping, while it keeps the synchrony notion bivalent (for a related continuous model with a graded notion of synchrony see [8]). In order to overcome the anti-monotonicity problem pointed out above, we define the support of a set $A \subseteq E$ of event types as follows (see also [6] for a similar characterization):

$$s(\mathcal{E}(A, w)) = \max \left\{ |H| \mid H \subseteq \mathcal{E}(A, w) \wedge \not\exists \mathcal{R}_1, \mathcal{R}_2 \in H \colon \mathcal{R}_1 \neq \mathcal{R}_2 \wedge \mathcal{R}_1 \cap \mathcal{R}_2 \neq \emptyset \right\}.$$

That is, we define the support (or total synchrony) of a pattern $A \subseteq E$ as the size of a maximum independent set of instances of parallel episodes of A (where by *independent set* we mean a collection of instances that do not share any events, that is, the instances do not overlap). Such an approach has the advantage that the resulting support measure is guaranteed to be anti-monotone, as can be shown generally for maximum independent subset (or, in a graph interpretation, node set) approaches—see, e.g., [3] or [10].

4   Support Computation

A core problem of the support measure defined in the preceding section is that the maximum independent set problem is, in the general case, NP-complete and thus not efficiently solvable (unless $\mathcal{P} = \mathcal{NP}$). However, we are in a special case here, because the domain of the elements of the sets is one-dimensional and the elements of the considered sets are no more than a (user-specified) maximum distance w apart from each other. The resulting constraints of the possible problem instances allow for an efficient solution, as shown in [9].

Intuitively, the constraints allow to show that a maximum independent set can be found by a greedy algorithm that always selects the next (with respect to time) selectable instance of the parallel episode we are considering. The idea of the

type train = (int, list of real);   (∗ pair of identifier and list of points/times ∗)
                                               (∗ points/times are assumed to be sorted ∗)

function support (*L*: set of train, *w*: real) : int;
begin   (∗ *L*: trains to process, *w*: window width ∗)
  s := 0;   (∗ initialize the support counter ∗)
  while ∀(*i*, *l*) ∈ *L* : *l* ≠ [] **do begin**   (∗ while none of the lists is empty ∗)
    $t_{\min}$ = min {head(*l*) | (*i*, *l*) ∈ *L*};   (∗ get smallest and largest head element ∗)
    $t_{\max}$ = max{head(*l*) | (*i*, *l*) ∈ *L*};   (∗ and thus the span of the head elements ∗)
    if $t_{\max} - t_{\min} > w$ (∗ if not synchronous, delete smallest heads ∗)
    then *L* := {(*i*, tail(*l*)) | (*i*, *l*) ∈ *L* ∧ head(*l*) = $t_{\min}$}
          ∪ {(*i*, *l*) | (*i*, *l*) ∈ *L* ∧ head(*l*) ≠ $t_{\min}$};
    else *L* := {(*i*, tail(*l*)) | (*i*, *l*) ∈ *L*};   (∗ if synchronous, delete all heads ∗)
        s := *s* + 1; **end**   (∗ (i.e., delete found synchronous points) ∗)
  end   (∗ and increment the support counter ∗)
  return *s*;   (∗ return the computed support ∗)
end   (∗ support() ∗)

Fig. 1. Pseudo-code of the support computation from a set of trains/point processes

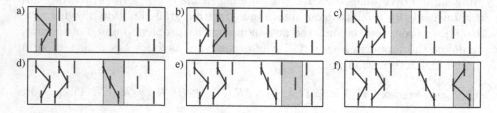

Fig. 2. Illustration of the support computation with a sliding window for three parallel point processes. Blue lines connect selected (i.e. counted) groups of points/times.

proof is that starting from an arbitrary maximum independent set, the selection of the sets can be modified (while keeping the number of sets and thus the maximality property of the selection), so that events occurring at earlier times are chosen. In the end, the first selected set contains the earliest events of each type that together form an instance of the parallel episode *A* under consideration. The second selected set contains the earliest events of each type that form an instance of the parallel episode *A* in an event sequence from which the events of the first instance have been removed, and so on. As a consequence, always selecting greedily the next instance of the parallel episode in time that does not contain events from an already selected one yields a maximum independent set. The details of the proof can be found in [9].

Pseudo-code of the resulting greedy algorithm to compute $s(\mathcal{E}(A, w))$, which works on a representation of the data as *parallel trains* (or parallel point processes) is shown in Figure 1. An illustration in terms of a sliding window that stops at certain points, namely always the next point that is not yet part of a selected set or has already been considered, is shown in Figure 2.

```
function isect (I: list of interval, l: list of real, w: real) : list of interval;
begin                                       (* — intersect interval and point list *)
    J := [];  p := −∞;  q := −∞;            (* init. result list and output interval *)
    while I ≠ [] ∧ x ≠ [] do begin          (* while both lists are not empty *)
        a, b := head(I);  t := head(l);     (* get next interval and next point *)
        if   t < a then l := tail(l);       (* point before interval: skip point *)
        elif t > b then I := tail(I);       (* point after  interval: skip interval *)
        else x := max{a, t − w};            (* if current point is in current interval, *)
             y := min {b, t + w};           (* intersect with interval around point *)
             if x ≤ q then q := y;          (* merge with output interval if possible *)
             else if q > −∞ then J.append([p, q]);  p := x;  q := y;
             l := tail(l);                  (* store pending output interval, *)
        end                                 (* start a new output interval, and *)
    end                                     (* finally skip the processed point t *)
    if q > −∞ then J.append([p, q]); end    (* append the last output interval *)
    return J;                               (* return the created interval list *)
end    (* isect() *)


function recurse (C, L: set of train, I: list of interval, w: real, s_min: int);
begin                                       (* — recursive part of CoCoNAD *)
    while L ≠ ∅ begin                       (* while there are more extensions *)
        choose (i, l) ∈ L;  L := L − {(i, l)};  (* get and remove the next extension *)
        J := isect(I, l);                   (* intersect interval list with extension *)
        D := {(i', [t | t ∈ l' ∧ ∃j ∈ J : t ∈ j]) | (i', l') ∈ C ∪ {(i, l)}};
        s := support(D, w);                 (* filter collected trains with intervals, *)
        if s < s_min then continue;         (* compute support of the train set and *)
        report {i' | (i', l') ∈ C} with support s;(* skip infrequent/report frequent sets *)
        X := {(i', [t | t ∈ l' ∧ ∃j ∈ J : t ∈ j]) | (i', l') ∈ L};
        recurse(D, X, J, w, s_min);         (* filter extensions with interval list and *)
    end                                     (* find frequent patterns recursively *)
end    (* recurse() *)


function coconad (L: set of train, w: real, s_min: int);
begin                                       (* L: list of trains to process *)
    recurse([], L, [[−∞, +∞]], w, s_min);   (* w: window width, s_min: min. support *)
end    (* coconad() *)                      (* initial interval list is whole real line *)
```

Fig. 3. Pseudo-code of the recursive enumeration (support computation see Fig. 1)

5   Finding Frequent Patterns

In order to find all frequent patterns we rely on an enumeration scheme that
is directly inspired by analogous approaches in FIM, especially the well-known
Eclat algorithm [11]. This algorithm uses a *divide-and-conquer* scheme, which
can also be seen as a depth-first search in a tree that results from selecting edges
of the Hasse diagram of the partially ordered set $(2^E, \subseteq)$—see, e.g., [2]. For a
chosen event type a, the problem of finding all frequent patterns is split into two

subproblems: (1) find all frequent patterns containing a and (2) find all frequent patterns *not* containing a. Each subproblem is then further divided based on another event type b: find all frequent patterns containing (1.1) both a and b, (1.2) a but not b, (2.1) b but not a, (2.2) neither a nor b etc. More details of this approach in the context of FIM can be found in [2]. Pseudo-code of this scheme for finding frequent patterns is shown in Figure 3, particularly in the function "recurse": the recursion captures including another event type (first subproblem), the loop excluding it afterwards (second subproblem). Note how the *apriori property* (see above) is used to prune the search.

A core difference to well-known FIM approaches is that we cannot, for example, simply intersect lists of transaction identifiers (as it is done in the Eclat algorithm, see [11] or [2]), because the continuous domain requires keeping all trains (i.e. point processes) of the collected event types in order to be able to compute the support with the function shown in Figure 1. However, simply collecting and evaluating complete trains leads to considerable overhead that renders the search unpleasantly slow (see the experimental results in the next section). In order to speed up the process, we employ a filtering technique based on a list of (time) intervals in which the points/times of the trains have to lie to be able to contribute to instances of a parallel episode under consideration (and its supersets). The core idea is that a point/time in a train that does not have a partner point/time in all other trains already collected (in order to form an instance of a parallel episode) can never contribute to the support of a parallel episode (or any of its supersets) and thus can be removed.

The initial interval list contains only one interval that spans the whole real line (see the main function "coconad" in Figure 3). With each extension of the current set of event types (that is, each split event type in the divide-and-conquer scheme outlined above), the interval list is "intersected" with the train (that is, its list of points/times). Pseudo-code of this intersection is shown in the function "isect" in Figure 3: only spikes lying inside an existing interval are considered. In addition, the intervals are intersected with the intervals $[t - w, t + w]$ around the point/time t in the train, where w is the (user-specified) window width that defines the maximum time distance between events that are to be considered synchronous. The resulting intersections are then merged into a new interval list.

These interval lists are used to filter both the already collected trains before the support is determined (cf. the computation of the set D in the function "recurse" in Figure 3) as well as the potential extensions of the current set of event types (cf. the computation of the set X). Preliminary experiments that we conducted during the development of the algorithm showed that each of these filtering steps actually improves performance (considerably).

A common technique to speed up FIM is so-called *perfect extension pruning*, where an item i is called a *perfect extension* of an item set I if I and $I \cup \{i\}$ have the same support. The core idea is that a subproblem split (as described in the divide-and-conquer scheme above) can be avoided if the chosen split item is a perfect extension. The reason is that in this case the solution of the first subproblem (include the split item) can be constructed easily from the solution of the second (exclude the split item): simply add the perfect extension item to all frequent item sets in the solution of the second subproblem (see [2] for details). However, for this to be possible, it is necessary that the property of being a perfect extension carries over to supersets, that is, if an item i is a perfect extension of an item set I, then it is also a perfect extension of all item sets $J \supset I$. Unfortunately, this does not hold in the continuous case we consider here: there can be patterns $A, B \subseteq E$, with $B \supset A$, and $a \in E$ such that $s(\mathcal{E}(A, w)) = s(\mathcal{E}(A \cup \{a\}, w))$, but $s(\mathcal{E}(B, w)) > s(\mathcal{E}(B \cup \{a\}, w))$. As a consequence, perfect extension pruning cannot be applied directly.

Fortunately, though, we are still able to employ a modified version, which exploits the fact that we can choose the order of the split event types independently in different branches of the search tree. The idea is this: if we find a perfect extension (based on the support criterion mentioned above), we collect it and only solve the second subproblem (exclude the split event type). Whenever we report a pattern A as frequent we check whether this set together with the set B of collected perfect extensions has the same support. If it does, the property of being a perfect extension actually carries over to supersets in this case. Therefore we can proceed as in FIM: we report all sets $A \cup C$ with $C \subseteq B$, using the same support $s(\mathcal{E}(A, w))$. If not, we "restart" the search using the collected perfect extensions as split/extension items again. Note that due to the fact that we know the support of the set $A \cup B$ (computed to check whether A and $A \cup B$ have the same support), we have an additional pruning possibility: as soon as we reach this support in the restarted recursion, the remaining perfect extensions can be treated like "true" perfect extensions. (Note that this technique is not captured in the pseudo-code in Figure 3; for details refer to the source code, which we made available on the internet—see below).

Finally, we consider filtering for closed and maximal frequent patterns, which is a common technique in FIM to reduce the output size. For applications in spike train analysis we are particularly interested in closed frequent patterns, because they capture all frequency information without loss, but (usually) lead to (much) smaller output (while maximal frequent patterns, which can reduce the output even more, lose frequency information and cause certain unpleasant interpretation problems). Because of this filtering for closed sets of neurons that frequently fire together as an indication of assembly activity, we call our algorithm CoCoNAD (for COntinuous-time ClOsed Neuron Assembly Detection—see Figure 3, although the pseudo-code does not capture this filtering).

The main problem of filtering for closed and maximal sets are the "eliminated event types," that is, event types which have been used as split event types and w.r.t. which we are in the second subproblem (exclude item). While event types that have *not* been used on the current path in the search tree are processed in the recursion and hence it can be returned from the recursion whether there exists a superset containing them that has the same support (closed sets) or is frequent (maximal sets), eliminated event types need special treatment. We implemented two approaches: (1) collecting the (filtered trains of the) eliminated event types and explicitly checking the support of patterns that result from adding them and (2) using (conditional) repositories of (closed) frequent sets as suggested for FIM in [4]. (Again these techniques are not captured in Figure 3;

details can be found in the source code). The latter has the disadvantage that it requires extra memory for the (conditional) repositories, but turns out to be significantly faster than the former (see next section).

6   Results

We implemented our algorithm scheme in both Python and C (see below for the sources) and tested it on the task of identifying all frequent patterns in data sets with varying parameters in order to assess its efficiency. Parameters were chosen with a view on our application domain: data sets with a varying number of event types (i.e. neurons in our application domain), chosen based on the number of neurons that can be simultaneously recorded with current technology (around 100, cf. [1]). Event rates were chosen according to typical firing rates observed in spike train recordings (around 20–30Hz), which usually comprise a few seconds. Several minimum support thresholds and window widths were considered as an illustration. Window widths were selected based on typical time bin lengths in applications of the bin-based model of synchrony (1 to 7 milliseconds). Support thresholds were considered down to two sets of synchronous events to demonstrate that highly sensitive detections are possible.

The first three diagrams in Figure 4 show execution times of our algorithm on some of these data sets to give an impression of what impact the parameters

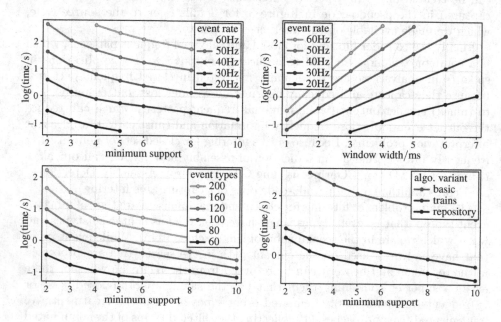

Fig. 4. Execution times in different experimental settings. Default parameters (unless on horizontal axis or in legend) are 5s recording period, 100 event types, 30Hz event rate, $s_{\min} = 2$, $w = 5$ms, closed item sets filtered with (conditional) repositories.

have on the execution time.[2] The last diagram compares the performances of the algorithmic variants described in Section 5. In this diagram "basic" refers to an algorithm without filtering of point processes, that is, as if the pseudo-code in Figure 3 (cf. function "recurse") used the assignments $D := C \cup \{(i, l)\}$ and $X := L$, that is, as if the complete trains were maintained; "trains" means an algorithm with filtering of point processes, where the trains of eliminated items are collected and used to check at reporting time for closed patterns; while "repository" means an algorithm with filtering of point processes, but using (conditional) repositories of already found closed patterns to filter for additional closed patterns. Note that filtering the point processes contributes most to make the search efficient (it reduces the time by about two orders of magnitude).

7   Conclusion

We presented an efficient algorithm scheme aimed at identifying frequent patterns in parallel point processes. This task can be seen as a generalization of frequent item set mining to a continuous (time) scale, where items or events co-occur (that is, are synchronous and thus constitute a set of synchronous events) if they all lie within a certain user-defined (time) span from each other. The main problem of this task is that, due to the absence of natural transactions (on which standard frequent item set mining is based), counting the number of sets of synchronous events (i.e., assessing the support) of a pattern is not a trivial matter. In this paper the support of a pattern is defined as the maximum number of non-overlapping sets of synchronous events that can be identified for that particular set. This has the advantage that it renders support anti-monotone and thus allows to prune the search for frequent patterns effectively. Computing the support thus defined becomes an instance of the maximum independent set problem that, although NP-complete in the general case, can be shown to be efficiently solvable in our case due to the restriction of the problem instances by the underlying one-dimensional domain (i.e., the continuous time scale).

In order to make the search for frequent patterns efficient we introduced several core techniques, such as filtering the point processes to reduce them to the relevant points and using (conditional) repositories to filter for closed (or maximal) patterns. These techniques contribute substantially to speeding up the search, as is demonstrated by the experiments reported in this paper.

Software and Source Code

Python and C implementations of the described algorithm as command line programs as well as a Python extension module that makes the C implementation accessible in Python (2.7.x as well as 3.x) can be found at these URLs:

www.borgelt.net/coconad.html   www.borgelt.net/pycoco.html

---

[2] All tests were run on a standard PC with an Intel Core 2 Quad 9650@3GHz processor, 8GB RAM, Ubuntu Linux 12.10 64bit operating system, using the C implementation of our algorithm compiled with GCC 4.7.2.

Acknowledgments. The work presented in this paper was partially supported by the Spanish Ministry for Economy and Competitiveness (MINECO Grant TIN2012-31372).

References

1. Bhandari, R., Negi, S., Solzbacher, F.: Wafer Scale Fabrication of Penetrating Neural Electrode Arrays. Biomedical Microdevices 12(5), 797–807 (2010)
2. Borgelt, C.: Frequent Item Set Mining. In: Wiley Interdisciplinary Reviews (WIREs): Data Mining and Knowledge Discovery, vol. 2, pp. 437–456. J. Wiley & Sons, Chichester (2012), doi:10.1002/widm.1074
3. Fiedler, M., Borgelt, C.: Subgraph Support in a Single Graph. In: Proc. IEEE Int. Workshop on Mining Graphs and Complex Data, pp. 399–404. IEEE Press, Piscataway (2007)
4. Grahne, G., Zhu, J.: Efficiently Using Prefix-trees in Mining Frequent Itemsets. In: Proc. Workshop Frequent Item Set Mining Implementations (FIMI 2003), Aachen, Germany. CEUR Workshop Proceedings, vol. 90 (2003)
5. Hebb, D.: The Organization of Behavior. J. Wiley & Sons, New York (1949)
6. Laxman, S., Sastry, P.S., Unnikrishnan, K.: Discovering Frequent Episodes and Learning Hidden Markov Models: A Formal Connection. IEEE Trans. on Knowledge and Data Engineering 17(11), 1505–1517 (2005)
7. Mannila, H., Toivonen, H., Verkamo, A.: Discovery of Frequent Episodes in Event Sequences. In: Data Mining and Knowledge Discovery, vol. 1(3), pp. 259–289 (1997)
8. Picado-Muiño, D., Castro-León, I., Borgelt, C.: Fuzzy Characterization of Spike Synchrony in Parallel Spike Trains. Soft Computing (to appear, 2013)
9. Picado-Muino, D., Borgelt, C.: Frequent Itemset Mining for Sequential Data: Synchrony in Neuronal Spike Trains. Intelligent Data Analysis (to appear, 2013)
10. Vanetik, N., Gudes, E., Shimony, S.E.: Computing Frequent Graph Patterns from Semistructured Data. In: Proc. IEEE Int. Conf. on Data Mining, pp. 458–465. IEEE Press, Piscataway (2002)
11. Zaki, M.J., Parthasarathy, S., Ogihara, M., Li, W.: New Algorithms for Fast Discovery of Association Rules. In: Proc. 3rd Int. Conf. on Knowledge Discovery and Data Mining (KDD 1997), pp. 283–296. AAAI Press, Menlo Park (1997)

Behavioral Clustering for Point Processes

Christian Braune[1], Christian Borgelt[2], and Rudolf Kruse[1]

[1] Otto-von-Guericke-University of Magdeburg
Universitätsplatz 2, D-39106 Magdeburg, Germany
[2] European Centre for Soft Computing
Calle Gonzalo Gutiérrez Quirós s/n, E-33600 Mieres (Asturias), Spain
christian.braune@ovgu.de, christian@borgelt.net,
kruse@iws.cs.uni-magdeburg.de

Abstract. Groups of (parallel) point processes may be analyzed with a variety of different goals. Here we consider the case in which one has a special interest in finding subgroups of processes showing a behavior that differs significantly from the other processes. In particular, we are interested in finding subgroups that exhibit an increased synchrony. Finding such groups of processes poses a difficult problem as its naïve solution requires enumerating the power set of all processes involved, which is a costly procedure. In this paper we propose a method that allows us to efficiently filter the process set for candidate subgroups. We pay special attention to the possibilities of *temporal imprecision*, meaning that the synchrony is not exact, and *selective participation*, meaning that only a subset of the related processes participates in each synchronous event.

Keywords: point processes, clustering, spike train analysis.

1 Introduction

Point processes occur in many different situations, such as arrivals of customers or phone calls, accidents on highways or firing of neurons in artificial or natural neural networks [8]. They generate a series of points in time or space and can be used to describe different kinds of event sequences. The work we report about in this paper is motivated by the analysis of (parallel) spike trains in neurobiology [14]: each train refers to a neuron, the associated point process records the times at which the neuron emitted an electrical impulse (*action potential* or *spike*).

The mechanisms by which a single neuron is activated by the release of neurotransmitters and emits electrical impulses are fairly well understood. However, how groups of neurons interact with each other and collectively encode and process information (like stimuli) is still the subject of ongoing research and intense debate in the neuroscience community. Several competing theories have been proposed to describe this processing. In particular, neuron assemblies have been suggested by Hebb [15] as the key elements of information processing in the cortex. Hebb suggested that such assemblies should reveal themselves by increased synchronous activity, i.e. they tend to produce (roughly) coincident spikes.

A. Tucker et al. (Eds.): IDA 2013, LNCS 8207, pp. 127–137, 2013.

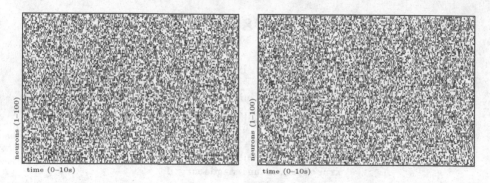

Fig. 1. Two sets of (artificial) parallel spike trains. Right: independent trains, generated as Poisson processes; left: coincident spiking events of 20 neurons (randomly selected) injected into an independent background (see also: [13,12,3]).

Since today the recording of several hundred(s) of spike trains in parallel is possible, there is an increased need for efficiently analyzing the data and to test the assembly hypothesis accordingly. The objective is to detect those groups of neurons (or spike trains) that show more synchronous activity than we would expect to see under the assumption that they are all independent.

The main obstacles we need to deal with in this task are threefold: in the first place, the possible combinations of neurons that may form an assembly increases exponentially with the number of neurons recorded. Furthermore we have to cope with *temporal imprecision* and *selective participation*. Temporal imprecision means that we cannot expect two events that originate from the same underlying coincident event to actually appear at (exactly) the same point in time in the spike trains. This may be due to the underlying biological process that generates the spikes but also due to the measurement procedure in which one probe records the electrical potential of (possibly) several neurons in parallel which then have to be separated in a process called *spike sorting*.

Selective participation means that not each and every neuron that belongs to an assembly actually takes part in every coincidence; rather some neurons may miss some of the coincidences. This is quite likely to occur in real neural networks as neurons need some time (so-called *refractory period*) after they emitted a spike to regenerate and be able to emit the next spike. Selective participation may even lead to situations where we do not even see a single coincidence in which all neurons forming the assembly took part.

In this paper we analyze the task of distinguishing between groups of spike trains that contain just random noise (independent spike trains) and groups that exhibit increased synchronous firing, allowing for temporal imprecision as well as selective participation, but without actually identifying the assemblies themselves. Figure 1 shows two samples of parallel spike trains where in the left case 20 neurons are firing with higher synchrony while the right picture shows independent trains. This is to emphasize the difficulty of the problem posed.

The remainder of this paper is organized as follows: in Section 2 we briefly review related work on methods for the analysis of parallel spike trains. In Section 3 we describe how our method works and evaluate it on artificially generated train sets in Section 4. Section 5 concludes the paper with a discussion of the results and an outlook on future research on this topic.

2   Related Work

The problem of finding cell assemblies in parallel spike train data has been the subject of research for quite some time. Early algorithmic attempts to detect assemblies date back at least to [11] where assemblies are detected by performing several pairwise χ^2-tests for independence on the time-discretized spike trains. The pair of spike trains yielding the lowest p-value is then merged into a single spike train, containing only the coincidences. The result of this merger is then added to the pool of spike trains (keeping the originals for further tests) and the tests are repeated until no further significant pairs can be found.

Generally, attempts to identify assemblies in parallel spike train data can be (roughly) categorized in three classes: (1) finding out whether there is (at least) one assembly present in the data (e.g. [18,21,22]), (2) answering for each neuron whether it belongs to such an assembly (e.g. [3]) and (3) actually identifying the assemblies (e.g. [10] or for selective participation [2]). The approach we present in this paper belongs to the second category, which is particularly useful for preprocessing, and results from previous work we did on the generation of prototypes for the analysis of spike trains on a continuous time domain [6].

Methods that test whether a neuron belongs to an assembly or not (like [3]) sometimes rely on the generation of surrogate data. Such surrogates are spike trains that retain some (desirably: most) of the statistical properties of the original spike trains while other properties (for instance, synchronous spiking) are destroyed on purpose in order to be able to test for this property. Simple examples are the generation of a spike train that contains the same number of spikes but at different points in time or a spike train that has the same distribution of inter-spike intervals. One may then calculate some statistics for each of the surrogates and if the behavior seen in the original spike train does not occur (or occurs only very rarely) in the surrogates one can assume that it is not the product of a random process. However, though fairly simple and statistically sound, generating surrogates for a large set of spike trains is a very time-intensive procedure that can quickly become infeasible if a large number of surrogate data sets need to be generated to meet a chosen significance level. Methods that allow for faster decision on whether a neuron belongs into an assembly or not are desireable and are very useful as a preprocessing step, and then later (computationally more expensive) analysis can be focused on promising subsets.

In this paper we introduce a method that allows for such quick preprocessing of a data set of parallel spike trains by analyzing the overall behavior of the spike trains, especially w.r.t. synchronous events, which groups them by their behavior rather than by their actual coincident events. In this sense it may be seen as a classification algorithm [16].

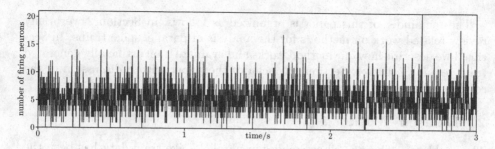

Fig. 2. Spike profile for a set of 100 spike trains with an injected assembly of size 10

3   Behavioral Clustering

In this section we describe how to compute a clustering of neurons into potential assembly neurons and background neurons from parallel spike train data. The method we propose here circumvents an interpretability problem that occurs when calculating metric representatives for the spike trains (see [5]) in the sense that we do not cluster the spike trains directly but rather their behavior: spike trains that belong to no assembly should behave essentially randomly, while the other spike trains should show a different, more organized behavior.

A spike train is essentially a point process, that is, a set of events that are identified by a point in time. We denote such a set by T, every event (or rather the point in time at which it occurs) by $t_i \in T$. Spike trains are often discretized to form an n-dimensional binary vector, each component of which describes one time bin and records whether the neuron emitted a spike in the corresponding time interval or not. As this approach suffers from various problems (especially the boundary problem, which results from how the bin boundaries fall relative to possible synchronous events), we choose a dynamic window placement. That is, by placing windows of a user-specified length $2\delta t$ around each event (spike), we model the events not as single points in time but as intervals during which they may be considered as coincident with other events. Formally we define:

$$f_T(t) = \begin{cases} 1 & \text{if } \exists t_i \in T : t_i - \delta t \le t \le t_i + \delta t, \\ 0 & \text{otherwise.} \end{cases}$$

Spike trains are thus effectively encoded as interval lists that describe the combined influence of all contained spikes. Note that the above definition merges overlapping intervals into a single, longer interval and thus the interval list may contain fewer intervals than the original point process contains points.

A set of parallel recorded spike trains $S = \{T_1, \ldots, T_n\}$ can now be represented by its spike profile which is simply the sum of all individual influence functions (cf. Figure 2):

$$f_S(t) = \sum_{T \in S} f_T(t).$$

To distinguish spike trains that form an assembly from those that do not we need a representation of the spike trains that allows us to study their behavior. As we pointed out in the motivation, we are looking for a subgroup of processes that exhibits higher synchronous activity than we would expect under independence. Synchronous activity should show itself by several spike trains containing events that occur (roughly, in the presence of temporal imprecision) at the same time. As a pairwise comparison of the interval data would be too costly we can use the profile to identify the behavior of each spike train individually with respect to the other spike trains.

To extract what may be called a "behavior profile", we create different interval lists from the spike profile by using a flooding-like approach. That is, we extract from the spike profile all intervals where $f_S(t) > 0$ holds. This, as we may say, "prototype" interval list is then compared to each individual spike train (or rather its representation as an interval list). To this end we calculate the overlap between the interval lists. Formally, we compute $P_x = \{t \mid f_S(t) \geq x\}$ as a "prototype" interval list and then calculate $s_T(x) = d(T, P_x) \; \forall T \in S$, $\forall x \in \{0, 1, \ldots, \max f_S(t)\}$, where d is the overlap of the two interval lists T and P_x. More technically, we define the cut level function (for level x)

$$f_{S,x}(t) = \begin{cases} 1 & \text{if } f_S(t) \geq x, \\ 0 & \text{otherwise} \end{cases} \quad \text{and then} \quad s_T(x) = \int f_{S,x}(t) \cdot f_T(t) \, dt.$$

That is, the function $s_T(x)$ describes the total length of the time intervals in which both the cut level function (for level x) and the spike train function are 1. For a sample set of 100 spike trains with an injected assembly of size 10 the resulting curves can be seen in Figure 3 (left diagram, $\delta = 3ms$).

The profile curves of the assembly can already be distinguished by visual inspection on this leftmost graph as they descend slightly later. To enhance the distinction, we exploit the plausible argument that higher levels are more important to detect synchronous activity. Therefore we weight each point of the behavior profile with the square of the level, i.e., the number of participating spike trains. Formally, we have $\forall T \in S \colon \forall x \in \{0, 1, \ldots, \max f_S(t)\}$:

$$s_T'(x) = x^2 \cdot s_T(x).$$

The resulting curves are shown in the middle diagram of Figure 3.

Finally, in order to make the assembly stand out even more, we normalize the curves by subtracting for each point the minimum value over the spike trains. Formally, we have $\forall T \in S \colon \forall x \in \{0, 1, \ldots, \max f_S(t)\}$:

$$s_T''(x) = x^2 \cdot s_T(x) - m_x \quad \text{where} \quad m_x = \min_{T \in S} \left(x^2 \cdot s_T(x) \right).$$

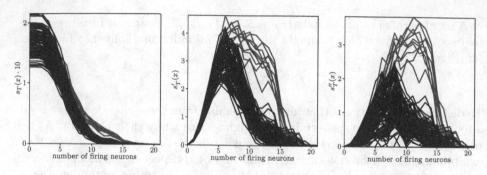

Fig. 3. Profile curves for a set of 100 spike trains with an assembly of size 10 injected. Left: similarity with the prototype/cut level; middle: similarity with the prototype/cut level, weighted with the square of the number of participating neurons (i.e., the level height); right: similarity with the prototype/cut level, weighted with the square of the number of participating neurons (i.e., the level height) and normalized by subtracting the minimum value over all spike trains.

The resulting curves are shown in Figure 3 on the right. Here the assembly clearly stands out from the rest of the spike trains so that we may use the profile curves obtained from the function s'' to describe the behavior of a spike train and perform a clustering on this data, using the vector of values $s''_T(x)$ for $x \in \{0, 1, \dots, \max f_S(t)\}$ as points in a metric space.

To automatically separate the two groups we decided to test both a density-based clustering algorithm (DBSCAN, [9]) and a simple hierarchical clustering with complete linkage. While the former should be able to detect the number of assemblies as well, the latter was set to report only two clusters. Directly reporting the assemblies contained in the set of spike trains would be a nice feature, as it would lift the method from a pure classification of neurons (into assembly and non-assembly neurons) to an assembly detection algorithm. But unfortunately the assembly neurons behave very similar when compared against the rest of the spike trains as can be seen in Figure 4 where two assemblies are shown as red and green lines respectively. Only if the two assemblies differed significantly in size and/or activity the curves would be different enough to become distinguishable. For the time being we consider them indistinguishable and leave better approaches for future work. As we want a procedure that decides without taking too much time if a spike train should be considered noise, we chose to still evaluate DBSCAN as it showed promising results in separating at least noise from assembly spike trains.

As input both algorithms received a similarity matrix, computed from the squared point-wise difference, i.e. the squared Euclidean distance, of the "behavior profiles". The calculation of this matrix is much faster than the calculation of a similarity matrix as we employed it in [5].

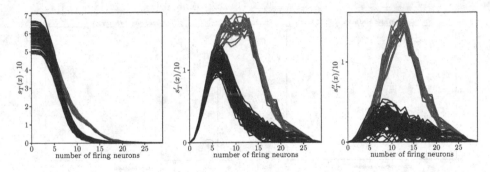

Fig. 4. Profile curves for a set of 100 spike trains with two assemblies injected, each of size 10. Noise is colored in blue, assemblies in red or green. The two disjoint assemblies are nearly indistinguishable.

Both algorithms may report the clusters found in arbitrary order so that we still need a criterion to distinguish them. For that we first calculate the mean profile curve as

$$m_{\mathcal{C}}(x) = \frac{1}{\|\mathcal{C}\|} \sum_{T \in \mathcal{C}} s_T''(x)$$

for each cluster $\mathcal{C}$ found and then calculate the area under the curve (AUC) for each $m_{\mathcal{C}}$. The one that yields the smallest AUC has the smallest overlap with time frames that many spike trains have contributed to. So it is fair to assume that this is the prototype for the behavior of the noise. The remaining spike trains will be labeled as potential assembly candidates and can be further processed with other methods.

As we only need to decide which spike trains should be labeled as noise and which as assembly candidates, we can also justify the choice of restricting the hierarchical clustering to report exactly two clusters. One will be the noise while the other one will contain the assembly spike trains.

4   Evaluation

To evaluate the method we proposed in Section 3 we generated several artificial sets of spike trains and ran our algorithm to report assembly and non-assembly spike trains. As this is a kind of classification, we can use classification quality measures such as the Adjusted Rand Index (ARI, [20]), Adjusted Mutual Information (AMI, e.g. [23]) and others for the evaluation of our method. Both aforementioned measures calculate the agreement between two different clustering results based on the predicted cluster labels but independent of the order of the cluster labels. The first is based on the absolute number of agreements while the latter is based on the mutual information shared by both clusterings.

To generate an artificial spike train, we sample the inter-spike intervals (time between two subsequent events) from an exponential distribution until a specified length of the spike train is reached (i.e., we generate Poisson point processes).

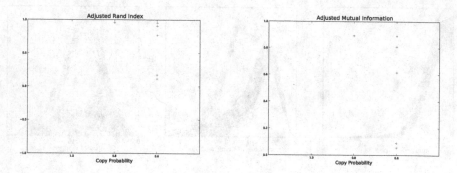

Fig. 5. Results for 1000 sets of parallel spike trains of 10 seconds length with an assembly of 20 spike trains injected, analyzed with DBSCAN

Firing rates were set to 20Hz for the non-assembly spike trains (which is a typical reference in the field). For the assembly spike trains a mother process was generated from which the coincident events were copied into the assembly spike trains with a certain probability ($c = 1.0$ if full participation was to be present, $c < 1.0$ if selective participation was to be modeled). The background firing was adjusted such that the overall firing rate (background and coincidences) was the same as for the noise spike trains (20Hz). Thus the spike trains cannot be distinguished by merely looking at the number of spikes. The temporal imprecision was modeled by shifting each spike after its generation by a certain, specified amount (here: ± 5ms, i.e. $\forall t_i : t_i := t_i + U(-5, 5)$; or $\delta t = 5$ms).

For our experiments we generated 1000 sets of parallel spike trains, consisting of 100 spike trains each. 20 of the spike trains form a single assembly with a coincidence rate of 5Hz embedded. The length of the simulated recording was 10 seconds in the first trials with copy probabilities of 1.0, 0.8 and 0.6. Each of these sets has been analyzed in the same way with either DBSCAN or hierarchical clustering grouping the spike trains together. To show the effects of the assembly size on the detection quality we ran the same number of tests on sets of parallel spike trains with an assembly size of only 10 and only 6 seconds length. The results of these tests can be seen in Figures 5 and 6 respectively for DBSCAN. Please note that the boxplots used seem to disappear in some cases. This is due to the fact that (almost) all values for the quality measures are actually 1.0 which means that the algorithm returned a clustering that perfectly matched the ground truth. Even if we reduce the copy probabilities to 0.6 all non-perfect results have to be considered outliers (i.e. they lie at least 2.698σ / outside the 99% interval from the median).

As the result for shorter spike trains with an average participation probability of 0.8 was significantly worse than we expected, we also used hierarchical clustering to analyze the last set of spike trains (see Figure 7). For copy probabilities of $c = 0.8$ the results are clearly better than when using density-based clustering, but for smaller copy probabilities the results are still bad albeit better. With only 10 spike trains forming an assembly and only six of them taking part in

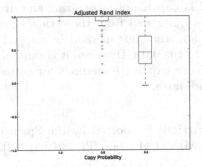
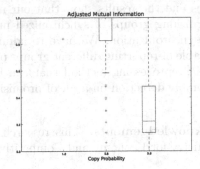

Fig. 6. Results for 1000 sets of parallel spike trains of 6 seconds length with an assembly of 10 spike trains injected, analyzed with DBSCAN

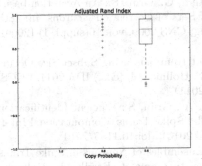
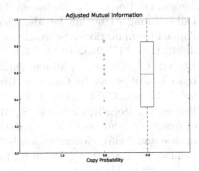

Fig. 7. Results for 1000 sets of parallel spike trains of 6 seconds length with an assembly of 10 spike trains injected, analyzed with hierarchical clustering

a coincidence on average the number of events we can use for the detection of the assembly neurons is already close to the number of coincidences we would expect to see in totally independent processes.

5   Conclusion and Future Work

In this paper we presented a method to group sets of point processes by the similarity of their behavior with respect to the behavior of the other processes by means of clustering algorithms. Synchronization between processes can be detected quite well for processes of different length and under additional obstacles such as selective participation and temporal imprecision. We evaluated our method in several different settings of artificially generated spike trains, i.e. point processes as they commonly appear in neurobiology. The artificial nature of our data allows us to control the experiments and perform a much more restrictive analysis as we can clearly calculate the number of false-positive or false-negative results. We use different measures for classification evaluation to aggregate these rates and the results show that our method is capable of recognizing and distinguishing groups of synchronized processes quite well from those that show no synchronization. We have to admit, though, that our method is not (yet) capable of reporting different groups present in the data. However, it is valuable as a preprocessing method that can focus more expensive methods for actual assembly detection on a set of promising candidates.

Acknowledgements. This research was partially supported by the Spanish Ministry for Economy and Competitiveness (MINECO Grant TIN2012-31372).

References

1. Berger, D., Borgelt, C., Diesmann, M., Gerstein, G., Grün, S.: An Accretion based Data Mining Algorithm for Identification of Sets of Correlated Neurons. In: 18th Annual Computational Neuroscience Meeting, CNS 2009, vol. 10(suppl. 1) (2009), doi:10.1186/1471-2202-10-S1-P254

2. Borgelt, C., Kötter, T.: Mining Fault-tolerant Item Sets using Subset Size Occurrence Distributions. In: Gama, J., Bradley, E., Hollmén, J. (eds.) IDA 2011. LNCS, vol. 7014, pp. 43–54. Springer, Heidelberg (2011)

3. Berger, D., Borgelt, C., Louis, S., Morrison, A., Grün, S.: Efficient Identification of Assembly Neurons within Massively Parallel Spike Trains. Computational Intelligence and Neuroscience, Article ID 439648 (2010), doi:10.1155/2010

4. Braune, C., Borgelt, C., Grün, S.: Finding Ensembles of Neurons in Spike Trains by Non-linear Mapping and Statistical Testing. In: Gama, J., Bradley, E., Hollmén, J. (eds.) IDA 2011. LNCS, vol. 7014, pp. 55–66. Springer, Heidelberg (2011)

5. Braune, C., Borgelt, C., Grün, S.: Assembly Detection in Continuous Neural Spike Train Data. In: Hollmén, J., Klawonn, F., Tucker, A. (eds.) IDA 2012. LNCS, vol. 7619, pp. 78–89. Springer, Heidelberg (2012)

6. Braune, C., Borgelt, C.: Prototype Construction for Clustering of Point Processes based on Imprecise Synchrony. In: 8th Conf. of the European Society for Fuzzy Logic and Technology, EUSFLAT 2013 (submitted, under review, 2013)

7. Brown, E.N., Kass, R.E., Mitra, P.P.: Multiple Neural Spike Train Data Analysis: State-of-the-art and Future Challenges. Nature Neuroscience 7(5), 456–461 (2004), doi:10.1038/nn1228

8. Daley, D.J., Vere-Jones, D.: An Introduction to the Theory of Point Processes. Springer, New York (1988), doi:10.1007/978-0-387-49835-5

9. Ester, M., Kriegel, H.-P., Sander, J., Xu, X.: A Density-based Algorithm for Discovering Clusters in Large Spatial Databases with Noise. In: Proc. 2nd Int. Conf. on Knowledge Discovery and Data Mining (KDD 1996), Portland, Oregon, pp. 226–231. AAAI Press, Menlo Park (1996)

10. Feldt, S., Waddell, J., Hetrick, V.L., Berke, J.D., Ochowski, M.: Functional Clustering Algorithm for the Analysis of Dynamic Network Data. Physical Review E 79, 056104 (2009), doi:10.1103/PhysRevE.79.056104

11. Gerstein, G.L., Perkel, D.H., Subramanian, K.N.: Identification of Functionally Related Neural Assemblies. Brain Research 140(1), 43–62 (1978), doi:10.1016/0006-8993(78)90237-8

12. Grün, S., Abeles, M., Diesmann, M.: Impact of Higher-Order Correlations on Co-
    incidence Distributions of Massively Parallel Data. In: Marinaro, M., Scarpetta,
    S., Yamaguchi, Y. (eds.) Dynamic Brain. LNCS, vol. 5286, pp. 96–114. Springer,
    Heidelberg (2008), doi:10.1007/978-3-540-88853-6_8)
13. Grün, S., Diesmann, M., Aertsen, A.M.: 'Unitary Events' in Multiple Single-neuron
    Spiking Activity. I. Detection and Significance. Neural Computation 14(1), 43–80
    (2002), doi:10.1162/089976602753284455
14. Grün, S., Rotter, S. (eds.): Analysis of Parallel Spike Trains. Springer, Berlin
    (2010), doi:10.1007/978-1-4419-5675-0_10
15. Hebb, D.O.: The Organization of Behavior. J. Wiley & Sons, New York (1949)
16. Kruse, R., Borgelt, C., Klawonn, F., Moewes, C., Steinbrecher, M., Held, P.: Com-
    putational Intelligence. Springer, London (2013), doi:10.1007/978-1-4471-5013-8
17. Lewicki, M.: A Review of Methods for Spike Sorting: The Detection and Classifica-
    tion of Neural Action Potentials. Network 9(4), R53–R78 (1998), doi:10.1088/0954-
    898X_9_4_001
18. Louis, S., Borgelt, C., Grün, S.: Complexity Distribution as a Measure for As-
    sembly Size and Temporal Precision. Neural Networks 23(6), 705–712 (2010),
    doi:10.1016/j.neunet.2010.05.004
19. Picado-Muino, D., Borgelt, C.: Characterization of Spike Synchrony without Dis-
    cretization of Time. Neuroinformatics (submitted)
20. Rand, W.: Objective Criteria for the Evaluation of Clustering Methods. Journal of
    the American Statistical Association 336(66), 846–850 (1971), doi:10.2307/2284239
21. Staude, B., Grün, S., Rotter, S.: Higher-order Correlations in Non-stationary Par-
    allel Spike Trains: Statistical Modeling and Inference. Frontiers in Computational
    Neuroscience 4, 16 (2010), doi:10.3389/fncom.2010.00016
22. Staude, B., Rotter, S., Grün, S.: CuBIC: Cumulant Based Inference of Higher-
    order Correlations in Massively Parallel Spike Trains. Journal of Computational
    Neuroscience 29(1-2), 327–350 (2010), doi:10.1007/s10827-009-0195-x
23. Wells III, W.M., Viola, P., Atsumi, H., Nakajima, S., Kikinis, R.: Multi-modal Vol-
    ume Registration by Maximization of Mutual Information. Medical Image Analy-
    sis 1(1), 35–51 (1996), doi:10.1016/S1361-8415(01)80004-9

Estimating Prediction Certainty
in Decision Trees

Eduardo P. Costa[1], Sicco Verwer[2], and Hendrik Blockeel[1,3]

[1] Department of Computer Science, KU Leuven, Leuven, Belgium
[2] Institute for Computing and Information Sciences, Radboud University Nijmegen,
The Netherlands
[3] Leiden Institute of Advanced Computer Science, Universiteit Leiden, Leiden,
The Netherlands
{eduardo.costa,hendrik.blockeel}@cs.kuleuven.be, siccoverwer@gmail.com

Abstract. Decision trees estimate prediction certainty using the class distribution in the leaf responsible for the prediction. We introduce an alternative method that yields better estimates. For each instance to be predicted, our method inserts the instance to be classified in the training set with one of the possible labels for the target attribute; this procedure is repeated for each one of the labels. Then, by comparing the outcome of the different trees, the method can identify instances that might present some difficulties to be correctly classified, and attribute some uncertainty to their prediction. We perform an extensive evaluation of the proposed method, and show that it is particularly suitable for ranking and reliability estimations. The ideas investigated in this paper may also be applied to other machine learning techniques, as well as combined with other methods for prediction certainty estimation.

Keywords: Decision trees, prediction certainty, soft classifiers.

1 Introduction

In classification, it is often useful to have models that not only have high accuracy, but can also tell us how certain they are about their predictions. Classifiers that output some kind of reliability or likelihood of the quality of predictions are called soft classifiers [1]. A common example of a soft classifier is a probability estimator, which estimates the probability that a data instance belongs to a certain class. Rankers and reliability estimators are other forms of soft classifiers.

The standard way of turning decision trees (DTs) into soft classifiers consists of inferring the certainty of a prediction from the class distribution in the leaf responsible for the prediction. For example, if an instance x is classified in a leaf node with 90% of positive examples, we say that x has 90% probability of being positive. However, it has been shown that these proportions can be misleading: the smaller the leaf is, the more likely the proportion is accidental, and not inherent to the population distribution in the leaf [2,3]; and, as DT learners try to make the leaves as pure as possible, the observed frequencies are systematically shifted towards 0 and 1 [3].

A. Tucker et al. (Eds.): IDA 2013, LNCS 8207, pp. 138–149, 2013.
© Springer-Verlag Berlin Heidelberg 2013

In this paper, we propose an alternative method to estimate prediction certainty in DTs. Assume that, given data set D and k classes $c_1, \ldots, c_k$, we want to make a prediction for an unseen instance x. Suppose that we learn a tree T that predicts with high "certainty" (according to leaf proportions) that x has class c_1. Logically speaking, if we would give the learner the prior knowledge that x has class c_1 (by simply adding (x, c_1) to D), the learner should not return a tree T_{c_1} that predicts with less certainty that the class of x is c_1. If it does, there is a logical contradiction, in the sense that more evidence about some fact being true leads to less certainty about it. Moreover, if we add (x, c_2) to D, and it turns out that the tree learned from the new dataset T_{c_2} makes a different prediction than T_{c_1}, also with high certainty, then we, as observers, know that there is actually high uncertainty about the class.

More specifically, our method works as follows. Given an unseen instance x, we can, for $i = 1, \ldots, k$, add (x, c_i) to D, giving a dataset D_i from which a tree T_{c_i} is learned, and look at the prediction T_i makes for x. If all T_{c_i} predict the same class c, we can be quite certain that c is the correct class. If multiple trees predict different classes, each with high certainty, we must conclude that these predictions are highly uncertain. We also propose a way to combine the predictions of the resulting trees.

We perform an extensive evaluation of the proposed method on 48 randomly selected UCI datasets. We compare our results to those of a standard DT learner and a standard ensemble method. The results show that our method tends to produce better ranking and reliability estimates than the other methods, while producing better probability estimates than standard trees and comparable probability estimates to ensembles. Additionally, compared to a closely related method for reliability estimation, we show that our method produces better estimates.

The remainder of the text is organized as follows. In Section 2 we discuss basic concepts related to prediction certainty in soft classifiers, with special focus on DTs; we also discuss related work and point the main differences w.r.t our method. In Section 3 we describe our new method in detail. In Section 4 we present experiments and results, and in Section 5 we conclude.

2   Background and Related Work

We first discuss three different ways of interpreting prediction certainty and how to evaluate them. Then, we briefly recall DTs and discuss related work.

2.1   Prediction Certainty in Soft Classifiers

The notion of certainty associated to soft classifiers has been defined in different ways in the literature. We discuss three of them: probability, ranking and reliability estimations. For each one of them we present a measure to evaluate it; we use these measures to evaluate our method in the experimental section.

We say that a soft classifier is a probability estimator when it estimates for every possible class the true probability that a random instance with the given attribute values belongs to that class. Probability estimations are usually evaluated with the Brier score [4] (Equation 1), which is also known as mean squared error. In Equation 1, n iterates over all the N predictions, i iterates over all the k classes, $t(c_i|x_n)$ is the true probability that instance x_n belongs to class c_i, and $p(c_i|x_n)$ is the estimated probability. When the true label of a prediction is given but its true probability is unknown, $t(c_i|x_n)$ is defined to be 1 if the true label of x_n is c_i, and 0 otherwise.

$$\text{Brier score} = \frac{\sum_{n=1}^{N} \sum_{i=1}^{k} (t(c_i|x_n) - p(c_i|x_n))^2}{N} \qquad (1)$$

A ranking estimator orders the instances from high to low expectation that the instance belongs to a certain class c. For every pair of instances (x_1, x_2), the ranking defines if x_1 is more likely, equally likely or less likely to belong to c than x_2. As ranking estimation is defined in terms of pairs of sequences, we can say that this is a relative estimation. Probability estimation, on the other hand, is an absolute estimation, since the estimation for each prediction can be interpreted on its own. The ranking ability of a classifier is usually assessed using the area under the ROC curve (AUC). A ROC curve is a two-dimensional plot with the true positive rate in the y-axis and the false positive rate in the x-axis which is obtained by varying the discrimination threshold.

Finally, Kukar and Kononenko [5] use the term "reliability" to define the probability that a prediction is correct. This is, in principle, the same as probability estimation for the predicted class. However, Kukar and Kononenko [5] consider reliability in a more general way. They consider a prediction to be more "reliable" when it is has a higher probability to be correct, and evaluate the reliability skill of a classifier by assessing how well it can distinguish between correct and incorrect predictions based on the calculated reliability for each prediction. This is in fact a ranking evaluation where the predictions define only one rank over all classes together (in terms of correct and incorrect predictions), instead of internally to each class, as for standard ranking evaluation. For this evaluation, we can use the AUC; we call it AUC reliability to avoid confusion with the aforementioned AUC calculation.[1]

2.2   Decision Trees and Certainty Estimates

A decision tree (DT) [6] is a tree-shaped predictive model that assigns to a given instance a prediction by determining the leaf that the instance belongs to, and making a prediction based on this leaf. In a classification context, the predicted class is typically the one that occurs most frequently among the training instances covered by that leaf.

When a certainty estimate is needed, the relative frequency of the predicted class among the training instances covered by the leaf is often used as a probability estimate. However, it is well-known that standard DT learners do not yield

---

[1] In their original evaluation framework, Kukar and Kononenko [5] use information gain to evaluate the rank of reliability scores. The advantage of using AUC is that the evaluation is not dependent on a fixed discrimination threshold.

very good probability estimates [2,3]. Especially for small leaves, the sample class distribution can significantly deviate from the population class distribution, and it typically deviates towards lower class entropy (or higher purity), due to the learning bias of the tree learner. Moreover, DTs assign the same certainty estimates for instances falling into the same leaf, and do not exploit the fact that even in the same leaf there might be prediction certainty differences. For example, it is reasonable to assume that a borderline prediction in a leaf is more likely to be a misclassification than the other predictions in that leaf.

Several methods have been proposed in the literature to improve estimates in DTs. One approach is to apply a smoothing correction (e.g., the Laplace or m-estimate smoothing) to unpruned trees [2,1]. These corrections are used to avoid extreme estimates (i.e, 0 or 1). Other group of methods either modify the decision tree learning (e.g., by learning fuzzy [7] or lazy DTs [8,9]) or the way in which the predictions are made (e.g., by propagating the test instances across multiples branches of the tree and combining estimates from different leaf nodes [10], or by using internal nodes to make predictions [3]). Other methods use alternative probability calculations, e.g., by combining the class distribution from different nodes in the path taken by the test instance [11].

In contrast to these methods, which either develop a new type of DT learner or use different probability estimations, we propose a new way of using the results that can be obtained using any traditional DT learner. We do this by learning multiple trees, and combining their predictions. In contrast to ensemble methods, which also learn multiple trees, we modify the training data in a very restricted and controlled way to obtain different trees. We do this by just complementing the training data with a labeled version of the instance to be classified.

Our method is similar to the method for reliable classification proposed by Kukar and Kononenko [5]. Their method estimates a reliability score for each prediction based on a two-step approach: first an inductive learning step is performed, followed by a transductive step. More specifically, given an unseen instance x, the method makes a prediction for x using a standard machine learning method (e.g., a DT learner). The output of this inductive step is then used as input to the transductive step: x is added to the training data with the predicted label c and a new classifier is learned. Finally, the probability distributions output by both steps are compared in order to infer the reliability of predicting x as belonging to class c. The idea behind this reliability estimation is that the more different the probability distributions are, the less reliable the prediction is. This idea is based on the theory of randomness deficiency [12,13]. Once a reliability score has been calculated for every prediction, the predictions are ranked according to their reliability, and a threshold is defined to separate the predictions into two populations: unreliable and reliable predictions. Kukar and Kononenko [5] propose a supervised procedure to find this threshold automatically.

In contrast to this method, we measure the sensitivity of the learned model w.r.t. all possible labels, instead of only using the label which is believed (predicted) to be the correct one. Our hypothesis is that measuring this sensitivity is crucial to obtaining good certainty estimates for DTs.

Related is also the method of conformal prediction [14], which aims to give confidence bounds on predictions by returning a set of predictions that cover a predetermined (typically 95%) confidence interval. Conformal prediction uses a nonconformity measure to estimate the distance of a new example x from a set of previously seen examples. This estimation is then used to predict a set of one or more labels for x (called the prediction region for x), which is assumed to contain the correct label, given the confidence interval. Intuitively, this only predicts labels that do not result in outliers with the given nonconformity measure. The method of conformal prediction differs from our method in the sense that it may output more than one predicted label for each test instance, while our method aways assign only one label to each prediction.

3   Proposed Method

We start by giving the intuition of our method. Then, we describe its algorithm.

3.1   Intuition of the Proposed Method

We want to investigate the following questions w.r.t. the instance we want to classify: (1) "If we add the test instance to the training set with a different label than the correct one, will the DT learner find a tree that is consistent with this instance according to the wrong label?"; (2) "How certain will the tree be about this prediction?"; (3) "How does this situation compare to the one where the instance is added to the training set with the correct label?".

Our method is therefore based on the idea that the learned DT can be dependent on the label of a single instance. One example of this effect can be seen in Fig. 1, which shows two DTs built to classify an instance x_1 from the Iris dataset [15]. For the left tree, x_1 was included in the training data with the correct label, *Iris-virginica*, while for the right one, x_1 was included with the wrong label *Iris-versicolor*. Observe that the trees make different predictions based on the pre-defined label for x_1. Intuitively, we cannot be very certain about the predicted label when the prediction model itself depends on the label we give to the instance. This uncertainty is not reflected in a certainty measure based only on leaf proportions. This intuition leads to the method we discuss next.

3.2   Description of the Method

The proposed method estimates the prediction certainty by comparing trees generated for the test instance x using different labels. As the correct label of x is not known to the method, we try all possibilities. More specifically, to classify an instance x, the method builds k trees, where k is the number of possible labels for the target attribute. For each label, we insert x in the training set with that label and induce a DT. In the end, we combine the prediction of all the trees.

To combine the predictions, we first calculate the prediction estimates for each tree by applying the Laplace smoothing (with $\alpha = 1$), then average over

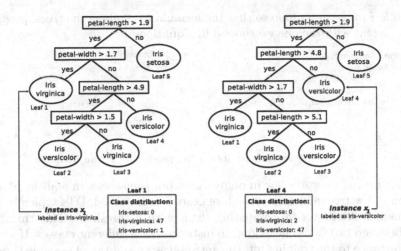

Fig. 1. Two DTs built for the same instance (x_1) of the Iris dataset. Left: x_1 was added in the training data with the correct label, *Iris-virginica*. Right: x_1 was added with label *Iris-versicolor*. For both cases the tree predicts x_1 with the pre-defined label.

the predicted values (Equation 2). In Equation 2, $\mathsf{Pred}(c)$ is the prediction for (probability of) class c, L_i is the leaf node responsible for the prediction in tree T_i, $|T|$ is the number of trees ($|T|$ equals the number of classes), $|L_i(c)|$ is the number of instances belonging to class c within L_i, and $|L_i(\neg c)|$ is the number of instances belonging to a different class within L_i. This strategy was chosen during the fine-tuning of our method on six validation datasets, which were not included in our experimental results.

$$\mathsf{Pred}(c) = \frac{\sum_{i=1}^{|T|} \frac{|L_i(c)|+1}{|L_i(c)|+|L_i(\neg c)|+|T|}}{|T|} \qquad (2)$$

The validation of our method also showed that our combination strategy benefits from pruning. In our strategy, pruning is important to avoid predictions from very small leaves, since such leaves can cause "overfitting" on any label assigned by our method, including the incorrect ones. This conclusion is, however, based on experiments where only one pruning procedure was considered. More specifically, we used the same pruning procedure used by the decision tree learner C4.5 [16]. The investigation of the effect of different pruning procedures on our strategy is an interesting venue for future work.

Pseudocode. The algorithm is described in pseudocode in Fig. 2, where x is the test instance, D is the original training data and C is the set of possible classes. The procedure LearnTree builds a DT for the given training data, and the procedure CombinePred combines the predictions of the resulting trees **T**. Note that, as we have a double loop in the procedure CombinePred, we need two variables (i and j) to iterate over the k possible labels in C. We use the variable j in the outer loop so that the formula to combine the trees' predictions is consistent with the one we showed in Equation 2.

procedure ObtainPrediction (x, D, C)
    for all $c_i \in C$:
        $D' := D \cup \{(x, c_i)\}$
        $T_i :=$ LearnTree(D')
    return CombinePred$(\{T_1, \ldots, T_k\}, x, D, C)$

procedure CombinePred$(\mathbf{T}, x, D, C)$
    for all $c_j \in C$:
        Pred$(c_j) := 0.0$
        for all $T_i \in \mathbf{T}$:
            Pred$(c_j) :=$ Pred$(c_j) + \frac{|L_i(c_j)|+1}{|L_i(c_j)|+|L_i(\neg c_j)|+|T|}$
        Pred$(c_j) := \frac{\text{Pred}(c_j)}{|T|}$
    return $\{$Pred$(c_1), \ldots,$ Pred$(c_k)\}$

Fig. 2. Pseudocode to obtain the prediction for an instance

This procedure works fine in many cases, but sometimes an additional modification of the tree induction procedure LearnTree is needed. DTs typically handle continuous attributes by comparing them with a threshold. This threshold is put between two values belonging to instances with different classes. If we add a test instance to the training set, this introduces an additional possible threshold. This can have undesired effects, as shown in Fig. 3. In this example, the instance x, which is represented in Fig. 3.a as a circle, is a positive instance. This instance is far enough from the negative class (so that a standard DT would not have problems classifying it correctly), but is the positive instance the closest to the negative ones. If we allow our method to use the attribute values of x, it will always choose a decision boundary that perfectly separates the instances depending on the label attributed to x (as shown in Figs. 3.b and 3.c). This would lead our method to conclude that x is a difficult case to be classified, while actually it is not. To avoid this, the attribute values of x are not used when determining the possible split (test) values of the tree. However, they are still used when determining the heuristic value (information gain) of these possible splits.

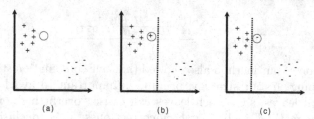

(a)   (b)   (c)

Fig. 3. Undesired effects of using a test instance when constructing decision boundaries

4   Empirical Evaluation

We present an extensive evaluation of our method, which we implemented as an extension of the DT learner Clus (http://dtai.cs.kuleuven.be/clus/); we call it Clus-TPCE (transductive prediction certainty estimation).

4.1 Experimental Setup

In total we use 48 randomly selected UCI datasets [15] in the experiments. For the datasets with no pre-defined test set, we use leave-one-out validation.

With these experiments we want to answer the question: "Does the proposed method yields better prediction certainty estimates than a standard DT?"; for this comparison we use the original version of Clus, which we refer to as Clus-Orig.[2] One could argue, however, that this comparison is not entirely fair since our method uses multiple trees to make the final prediction, and it is known that ensembles tend to yield better estimates than single trees [2]. Therefore, to ensure that an improvement of Clus-TPCE is not simply due to "ensemble effects", we also compare to standard bagging (Clus-Ens) with the same number of trees as we use for Clus-TPCE. For all methods, we use information gain as the splitting criterion.

We evaluate the results as (a) ranking estimates, (b) probability estimates, and (c) reliability estimates. For the reliability estimation evaluation, we include the results for the procedure proposed by Kukar and Konenko [5] (see, Section 2.2). We implemented this procedure to estimate the reliability estimation of the predictions given by Clus-Orig; we call it Clus-K&K.

For each comparison, additionally to the results themselves, we also report the p-value of a two-sided Wilcoxon signed-rank. With this test we verify the hypothesis that the method with the largest number of wins, in terms of the evaluation measure in consideration, is statistically superior to the other one.

4.2 Evaluating Probability Estimation

We start by evaluating the results in terms of probability estimation, using the Brier score. The results are shown in Fig. 4 and summarized in Table 1.

Table 1. Number of wins for each method, in terms of the Brier score, along with the p-value resulting from a two-sided Wilcoxon signed-rank test and the average estimates

Clus-TPCE vs. Clus-Orig			Clus-TPCE vs. Clus-Ens			Average Brier score		
Clus-TPCE	Ties	Clus-Orig	Clus-TPCE	Ties	Clus-Ens	Clus-TPCE	Clus-Orig	Clus-Ens
36	0	12	21	0	27	0.087	0.109	0.096
p-value < 0.0001			p-value = 0.5686					

Clus-TPCE obtains 36/0/12 wins/ties/losses compared to Clus-Orig, and a smaller average Brier score. When compared to Clus-Ens, Clus-TPCE obtains

---

[2] As we use pruned trees in Clus-TPCE, the results we report for Clus-Orig are also based on pruned trees. Additional experiments revealed that the conclusions presented in this paper also hold for the case when we compare with Clus-Orig based on unpruned trees. Namely, Clus-TPCE outperforms Clus-Orig based on unpruned trees w.r.t. probability, ranking, and reliability estimations.

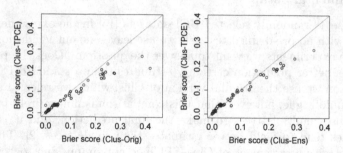

Fig. 4. Brier scores: Clus-TPCE vs. Clus-Orig (left); Clus-TPCE vs. Clus-Ens (right)

a smaller number of wins (21 wins against 27 wins for Clus-Ens), but has a smaller average Brier score. Interestingly, for the cases where both Clus-TPCE and Clus-Ens have a larger Brier score, the advantage is always in favor of Clus-TPCE (c.f. Fig. 4). For the cases with low scores, the results are in favor of Clus-Ens, but with a smaller difference. This explains why Clus-TPCE has a smaller average Brier score, even though it has a smaller number of wins.

4.3 Evaluating Ranking Estimation

We now compare the methods w.r.t. their ranking ability. We generate a ROC curve for each class against all the other ones and report the average AUC value. The results are shown in Fig. 5 and summarized in Table 2.

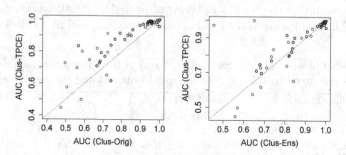

Fig. 5. AUC: Clus-TPCE vs. Clus-Orig (left); Clus-TPCE vs. Clus-Ens (right)

The results show that Clus-TPCE has a better ranking ability than the other methods. It obtains 37/3/8 wins/ties/losses compared to Clus-Orig and 29/1/18 wins/ties/losses compared to Clus-Ens. Note that these results are more in favor of Clus-TPCE than for the probability estimation evaluation. This is not unexpected. Clus-TPCE tends to shift the probability distribution output for some instances towards the uniform probability. This effect is stronger for the cases for which the method finds evidence that there is a high degree of uncertainty associated to their predictions. While this results in a better ranking estimation,

Table 2. Number of wins for each method, in terms of AUC, along with the p-value resulting from a two-sided Wilcoxon signed-rank test and the average estimates

Clus-TPCE vs. Clus-Orig			Clus-TPCE vs. Clus-Ens			Average AUC		
Clus-TPCE	Ties	Clus-Orig	Clus-TPCE	Ties	Clus-Ens	Clus-TPCE	Clus-Orig	Clus-Ens
37	3	8	29	1	18	0.868	0.814	0.849
p-value < 0.0001			p-value = 0.1706					

it might affect negatively the probability estimation, since the Brier score assumes that the method should ideally report a probability of 1 for the true class and 0 for the other classes.

4.4   Evaluating Reliability Estimation

Finally, we evaluate Clus-TPCE w.r.t. reliability estimation. To that aim, we use the probability output for the predicted class as the reliability that the prediction is correct. We apply the same procedure for Clus-Orig and Clus-Ens. For Clus-K&K, we use the procedure proposed by Kukar and Kononenko [5]. To evaluate the reliability estimations, we use AUC reliability, as discussed in Section 2.1. The results are shown in Fig. 6 and summarized in Table 3.

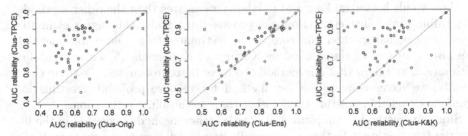

Fig. 6. AUC reliability: Clus-TPCE vs. Clus-Orig" (left); Clus-TPCE vs. Clus-Ens (center); Clus-TPCE vs. Clus-K&K (right)

Table 3. Number of wins for each method, in terms of AUC reliability, along with the p-value resulting from a two-sided Wilcoxon signed-rank test and the average estimates

Clus-TPCE vs. Clus-Orig			Clus-TPCE vs. Clus-Ens			Clus-TPCE vs. Clus-K&K			Average AUC reliability	
Clus-TPCE	Ties	Clus-Orig	Clus-TPCE	Ties	Clus-Ens	Clus-TPCE	Ties	Clus-K&K	Clus-TPCE	Clus-Orig
39	3	6	33	1	14	33	2	13	0.7969	0.6498
p-value < 0.0001			p-value = 0.0208			p-value < 0.0001			Clus-Ens	Clus-K&K
									0.7681	0.6836

Clus-TPCE outperforms the other three methods in terms of reliability estimation: it obtains 39/3/6 wins/ties/losses compared to Clus-Orig, 33/1/14 wins/ties/losses compared to Clus-Ens, and 33/2/13 wins/ties/losses compared to Clus-K&K. Furthermore, Clus-TPCE obtains the largest average AUC reliability, and the statistical tests indicate that the difference in the results is statistically significant. These results confirm the good performance of our method in ranking probabilities.

5   Conclusions

We proposed a method for estimating prediction certainty. The new method was implemented as an extension of the DT learner Clus, but the ideas investigated in this paper can also be applied to other machine learning methods, in particular those where the label of a single example can influence the learned model.

Our new method builds a DT for an input data consisting of the training data plus the instance to be classified, which is labeled with one of the possible class values. This procedure is repeated for every class, and in the end all induced trees are compared. This comparison allows us to identify instances that might present difficulties to be correctly classified, and to attribute some uncertainty to their predictions. We evaluated our method on 48 UCI datasets, and compared it to the original Clus and to standard bagging. The results showed that the new method yields better ranking and reliability estimates than the other methods. Regarding probability estimation, the proposed method yields better estimates than the original method and comparable estimates to the ensemble method. We also compared to the method by Kukar and Kononenko [5] w.r.t. reliability estimation, and show that our method produces better estimates. Based on these results, we recommend to use our method for relative probability estimation (ranking or reliability estimation) rather than for absolute estimation.

Since our method is complementary to the other methods suggested in the literature (see Section 2.2), they can be easily combined: simply use those methods instead of Clus to learn multiple trees for different versions of the test instances. Investigating these combinations is a very interesting avenue for future work.

Note that our method makes predictions for only one instance at a time, which increases its computational cost. This raises the question if the method can be extended in order to be able to perform the whole process once for a whole batch of unseen instances. A straightforward extension of the method consists of generating the same number of trees as the number of possible labels (once for all instances), where for every tree each instance to be classified receives a random label, with the constraint that the same instance will never receive the same label in two or more trees. This constraint assures that each instance receives each possible label once, allowing us to apply the same strategy to combine predictions used in this paper. However, as each generated tree is not only subject to the influence of the labeling of a single instance (but a batch of instances instead), it is not trivial to analyze if the prediction obtained for an instance is a result of how that instance was labeled, and/or how the other instances were labeled.

In fact, we have performed preliminary experiments with this extended method to test its ranking ability, and the results showed that it produces better results than a standard tree learner, but slightly worse results than standard bagging. Therefore, the extension of the proposed method remains for future research.

Acknowledgments. This research was supported by projects G.0413.09 "Learning from data originating from evolution" and G.0682.11 "Declarative experimentation", funded by the Research Foundation - Flanders (FWO-Vlaanderen).

References

1. Ferri, C., Flach, P.A., Hernández-Orallo, J.: Improving the AUC of probabilistic estimation trees. In: Lavrač, N., Gamberger, D., Todorovski, L., Blockeel, H. (eds.) ECML 2003. LNCS (LNAI), vol. 2837, pp. 121–132. Springer, Heidelberg (2003)
2. Provost, F., Domingos, P.: Tree induction for probability-based ranking. Machine Learning 52(3), 199–215 (2003)
3. Zadrozny, B., Elkan, C.: Obtaining calibrated probability estimates from decision trees and naive bayesian classifiers. In: Proceedings of the 18th International Conference on Machine Learning (ICML), pp. 609–616 (2001)
4. Brier, G.W.: Verification of forecasts expressed in terms of probability. Monthly Weather Review 78(1), 1–3 (1950)
5. Kukar, M., Kononenko, I.: Reliable classifications with machine learning. In: Elomaa, T., Mannila, H., Toivonen, H. (eds.) ECML 2002. LNCS (LNAI), vol. 2430, pp. 219–231. Springer, Heidelberg (2002)
6. Quinlan, J.R.: Induction of decision trees. Machine Learning 1(1), 81–106 (1986)
7. Hüllermeier, E., Vanderlooy, S.: Why fuzzy decision trees are good rankers. IEEE Transactions on Fuzzy Systems 17(6), 1233–1244 (2009)
8. Margineantu, D.D., Dietterich, T.G.: Improved class probability estimates from decision tree models. In: Denison, D.D., Hansen, M.H., Holmes, C.C., Mallick, B., Yu, B. (eds.) Nonlinear Estimation and Classification. Lecture Notes in Statistics, vol. 171, pp. 169–184. Springer (2001)
9. Liang, H., Yan, Y.: Improve decision trees for probability-based ranking by lazy learners. In: Proceedings of the 18th IEEE International Conference on Tools with Artificial Intelligence, pp. 427–435 (2006)
10. Ling, C.X., Yan, R.J.: Decision tree with better ranking. In: Proceedings of the 20th International Conference on Machine Learning, pp. 480–487 (2003)
11. Wang, B., Zhang, H.: Improving the ranking performance of decision trees. In: Fürnkranz, J., Scheffer, T., Spiliopoulou, M. (eds.) ECML 2006. LNCS (LNAI), vol. 4212, pp. 461–472. Springer, Heidelberg (2006)
12. Li, M., Vitanyi, P.: An Introduction to Kolmogorov Complexity and its Applications. Springer (1997)
13. Vovk, V., Gammerman, A., Saunders, C.: Machine-learning applications of algorithmic randomness. In: Proceedings of the 16th International Conference on Machine Learning, pp. 444–453 (1999)
14. Shafer, G., Vovk, V.: A tutorial on conformal prediction. J. Mach. Learn. Res. 9, 371–421 (2008)
15. Bache, K., Lichman, M.: UCI machine learning repository. University of California, Irvine, School of Information and Computer Sciences (2013), http://archive.ics.uci.edu/ml
16. Quinlan, J.R.: C4.5: Programs for Machine Learning. Morgan Kaufmann (1993)

Interactive Discovery
of Interesting Subgroup Sets

Vladimir Dzyuba and Matthijs van Leeuwen

Department of Computer Science, KU Leuven, Belgium
{vladimir.dzyuba,matthijs.vanleeuwen}@cs.kuleuven.be

Abstract. Although subgroup discovery aims to be a practical tool for exploratory data mining, its wider adoption is hampered by redundancy and the re-discovery of common knowledge. This can be remedied by parameter tuning and manual result filtering, but this requires considerable effort from the data analyst. In this paper we argue that it is essential to involve the user in the discovery process to solve these issues. To this end, we propose an interactive algorithm that allows a user to provide feedback *during search*, so that it is steered towards more interesting subgroups. Specifically, the algorithm exploits user feedback to guide a diverse beam search. The empirical evaluation and a case study demonstrate that uninteresting subgroups can be effectively eliminated from the results, and that the overall effort required to obtain interesting and diverse subgroup sets is reduced. This confirms that within-search interactivity can be useful for data analysis.

Keywords: Interactive data mining, pattern set mining.

1 Introduction

Informally, subgroup discovery [12,18] is concerned with finding subsets of a dataset that have a substantial deviation in a property of interest when compared to the entire dataset. It can be regarded as an exploratory data analysis task, with a strong emphasis on obtaining comprehensible patterns in the form of subgroup descriptions. In the context of a bank providing loans, for example, we could find that 16% of all loans with *purpose = used car* are not repaid, whereas for the entire population the proportion is only 5%. Subgroup discovery algorithms can cope with a wide range of data types, from simple binary data to numerical attributes and structured data. Various quality measures have been proposed to quantify subgroup interestingness, for which generally both the amount of deviation and the size of the subset are taken into account.

Subgroup discovery aims to be a *practical* tool for data exploration, and many case studies on real-world applications have been performed; see Herrera et al. [11] for a recent overview. Unfortunately, obtaining interesting results is usually a time-consuming job for which expertise on subgroup discovery is required. This is due to two main reasons: 1) large amounts of subgroups are found, of which many are redundant, and 2) background knowledge of the domain expert is not taken

A. Tucker et al. (Eds.): IDA 2013, LNCS 8207, pp. 150–161, 2013.

into account. To remedy these issues, careful tuning of the algorithm parameters and manual filtering of the results is a necessity. This requires considerable effort and expertise from the data analyst, and this clearly hampers the wider adoption of subgroup discovery as a tool for data exploration.

To address the *pattern explosion* in subgroup discovery, Diverse Subgroup Set Discovery (DSSD) [14] was recently proposed in an attempt to attain diverse rather than redundant subgroup sets. The main idea is to integrate pattern set mining into a levelwise search, so that diversity is maintained throughout search. Specifically, heuristic methods for selecting diverse subgroup sets are used to select a beam on each level, resulting in a diverse beam search.

Case Study: Sports Analytics. To illustrate the problems of existing methods and the potential of our proposed approach, we investigate the use of subgroup discovery in the context of sports analytics. There has recently been a significant interest in data mining in the professional sports community[1]. 'Black box' approaches that do not explain their outcomes would never be accepted, but subgroup discovery has the advantage that its results are *interpretable*.

The case study concerns a dataset containing information about games played by the Portland Trail Blazers in the 2011/12 season of the NBA[2]. Each tuple corresponds to a segment of a game played by the same group of 10 players (including 5 players on the opposing team). The attributes include 18 binary variables indicating presence of a particular player on the court, a nominal variable representing the opposing team, a numeric attribute pace[3], and 3 binary variables per team indicating whether offensive rating[4] and offensive/defensive rebound rates[5] of a team are higher than the season average.

We select *offensive rating* as the target property of interest, and the commonly used Weighted Relative Accuracy as the quality measure (see Section 5 for further details). This means that high-quality subgroups represent common situations in which the team is likely to have a high offensive rating, described in terms of players, opponents, and the course of the game.

To assess whether Diverse Subgroup Set Discovery gives satisfactory results, we ran the algorithm on the NBA data with default settings (cover-based heuristic with default quality-diversity trade-off [14]). We asked for the discovery of five subgroups, which are all shown in Table 1. The results suffer from two severe problems: 1) the results are clearly redundant, i.e. diversity could not be attained with the default parameter settings, and 2) none of the discovered subgroups is interesting to a domain expert, as the descriptions contain no surprising and/or actionable information. For example, it is a trivial fact for experts that poor defensive rebounding by an opponent ($opp\_def\_reb = F$) eases the task of scoring, whereas

---

[1] See for example http://www.sloansportsconference.com/
[2] Data source: http://basketballvalue.com/downloads.php
[3] *Pace* captures the speed of the game and is indicative of the team's playing style.
[4] *Offensive rating* is computed as the average number of points per shot.
[5] *Rebound rate* estimates how effective a team is at gaining possession of the ball after a missed shot, either by an opponent or by one of its own players.

Table 1. Five subgroups discovered by Diverse Subgroup Set Discovery [14]; cover-based approach with default quality-diversity trade-off. For each discovered subgroup its description, size and quality are given.

Description	Size	Quality
$opp\_def\_reb = F \wedge opponent \neq ATL \wedge thabeet = F$	219	0.0692
$opp\_def\_reb = F \wedge opponent \neq ATL$	222	0.0689
$opp\_def\_reb = F \wedge opponent \neq ATL \wedge ajohnson = F$	222	0.0689
$opp\_def\_reb = F \wedge thabeet = F \wedge opponent \neq PHI$	225	0.0685
$opp\_def\_reb = F \wedge opponent \neq PHI$	228	0.0682

absence of *reserve* players (*thabeet* and *ajohnson*) is not useful for decision making either.

Aims and Contributions. We argue that it is essential to *actively involve the user in the discovery process* to ensure diverse and interesting results. Even when diversity can be obtained through a fully automated discovery process, on itself this is not sufficient to guarantee interesting results. The main reason is that the user's background knowledge and goals are completely ignored. Some existing algorithms that try to leverage expert knowledge require specifying it in advance, but this is a hard task and may therefore be barely less time-consuming than post-processing humongous result sets.

We propose an interactive subgroup discovery algorithm that allows a user to provide feedback with respect to provisional results and steer the search away from regions that she finds uninteresting, towards more promising ones. The intuition behind our approach is that the 'best' subgroups often correspond to common knowledge, which is usually uninteresting. Users expect to obtain novel, unexpected insights, and therefore our system is designed to eliminate such uninteresting subgroups already during search.

The Interactive Diverse Subgroup Discovery (IDSD) framework that we propose builds upon DSSD by re-using the diverse beam search. However, we augment it by making the beam selection interactive: on each level of the search, users are allowed to influence the beam by *liking* and *disliking* subgroups. One of two subgroup similarity measures is then used to generalise this feedback to all subgroups for a specific level, by re-weighing qualities. The adjusted quality measure affects the (diverse) beam selection and hence the search can be guided.

Since it is hard to evaluate interactive data mining methods, we perform two types of evaluations. First, we perform an extensive quantitative evaluation in which user feedback is emulated. For this we treat a set of high-quality subgroups as 'background knowledge' in which the user is not interested, based on which we emulate the user feedback. The purpose of these experiments is to show that undesired results can be effectively avoided, which in return leaves space for novel, potentially more interesting results.

Second, we turn back to the case study that we introduced in this section. We asked a domain expert to use our interactive discovery system, and he successfully found more interesting patterns than with the standard diverse approach. This confirms that human-computer interaction makes it possible to discover interesting subgroups with much less effort than using standard algorithms.

2   Related Work

Subgroup discovery can be seen as an instance of *supervised descriptive rule discovery* [13], like contrast set [3] and emerging pattern mining [5]. Apart from DSSD, which was inspired by pattern set mining, several local approaches to redundancy elimination have been proposed: closed sets for labeled data [9] applies only to binary targets, a recent approach uses quadratic programming to do feature selection prior to the discovery process [15].

The importance of taking user knowledge and goals into account was first emphasised by Tuzhilin [17]. More recently De Bie argued that traditional objective quality measures are of limited practical use and proposed a general framework that models background knowledge as a *Maximum Entropy distribution* [4].

Applications of subgroup discovery in various domains often involve iterative refinement of results based on feedback of experts, e.g. in medicine [7,8]. A classification of background knowledge relevant to subgroup discovery was developed [1], and some of the insights were used in the VIKAMINE tool, which enables knowledge transfer between otherwise independent search sessions [2]. SVMs were applied to learn subgroup rankings from user feedback [16], but the feedback phase was not integrated into search.

Outside subgroup discovery, ideas regarding interactive search have been explored in Redescription Mining [6], but we go much further with the influence of users on beam selection. Finally, MIME is an interactive tool that allows a user to explore itemsets using traditional interestingness measures [10].

3   Preliminaries

We consider datasets that are bags of tuples. Let $A = \{A_1, \ldots, A_{l-1}, A_l\}$ denote a set of attributes, where each attribute A_j has a domain of possible values $\mathrm{Dom}(A_j)$. Then a dataset $\mathcal{D} = \{x_1, \ldots, x_n\} \subseteq \mathrm{Dom}(A_1) \times \ldots \times \mathrm{Dom}(A_l)$ is a bag of tuples over A. The attribute A_l is a *binary target attribute*, i.e. the property of interest. Attributes $D = \{A_1, \ldots, A_{l-1}\}$ are *description attributes*.

The central concept is the *subgroup*, which consists of a *description* and a corresponding *cover*. In this paper, a *subgroup description* d is a conjunction of boolean expressions over D, e.g. $D_1 = a \wedge D_2 > 0$. A *subgroup cover* G is a bag of tuples that satisfy the predicate defined by d: $G_d = \{\forall t \in \mathcal{D} : t \in G \Leftrightarrow d(t) = true\}$, the size of the cover $|G|$ is also called *subgroup coverage*.

Subgroup quality measures generally balance the degree of deviation and the size of a subgroup. We use *Weighted Relative accuracy*, given by $\varphi_{WRAcc}(G) = \dfrac{|G|}{|\mathcal{D}|} \times (1^G - 1^{\mathcal{D}})$, where 1^G (resp. $1^{\mathcal{D}}$) is the proportion of positive examples in G (resp. $\mathcal{D}$). The previous allows us to define *top-k subgroup discovery*:

Problem 1 (Top-k Subgroup Discovery). Given a dataset $\mathcal{D}$, a quality measure φ, and integer k, find the k top-ranking subgroups with respect to φ.

Bottom-up search is usually applied to solve this problem. The search space consists of all possible descriptions and is traversed from short to long descriptions. Common parameters to restrict the search space are a *minimum coverage* threshold (*mincov*), and a *maximum depth* (*maxdepth*). Either exhaustive search or *beam search* can be used, where the latter has the advantage that it is also feasible on larger problem instances. It explores the search space in a levelwise manner, and at each level only the w highest ranking candidates with respect to φ (the *beam*) are selected for further refinement, where *beam width* w is a user-supplied parameter. This makes it ideal for our current purposes.

Diverse Subgroup Set Discovery. We recently introduced the DSSD framework [14], which uses heuristic pattern set selection to select a more diverse beam on each level of beam search. The purpose of this approach is to achieve globally less redundant and therefore more interesting results.

The diverse beam selection strategies add a candidate subgroup to the beam only if it is sufficiently different from already selected subgroups. In this paper we use *description-based beam selection* because preliminary experiments showed that it works well for our purposes; our prototype discovery system primarily presents subgroup descriptions to the user. It first sorts all candidates descending by quality and initialises $Beam = \emptyset$, then iteratively considers each subgroup in order until $|Beam| = w$, and selects it only if there is no subgroup in the (partial) beam that has equal quality and the same conditions *except for one*.

We use *cover redundancy* (CR) to quantify redundancy of a subgroup set, i.e.

$$CR(\mathcal{G}) = \frac{1}{|\mathcal{D}|} \sum_{t \in \mathcal{D}} \frac{|c(t, \mathcal{G}) - \hat{c}|}{\hat{c}}, \text{ where } \mathcal{G} \text{ is a set of subgroups, } c(t, \mathcal{G}) \text{ is the } cover$$

count of a transaction, i.e. the number of subgroups that cover t, and $\hat{c}$ is the average cover count over all $t \in \mathcal{D}$. Essentially, it measures the deviation of the cover counts from the uniform distribution. Although absolute values are not very meaningful, CR is useful when comparing subgroup sets of similar size for the same dataset: a lower CR indicates that fewer tuples are covered by more subgroups than expected, therefore the subgroup set must be more diverse.

4   Interactive Diverse Subgroup Discovery

We now present the Interactive Diverse Subgroup Discovery (IDSD) algorithm, which employs user feedback to guide a beam search. Main design goals are to develop 1) a simple interaction mechanism that 2) requires little user effort. We rely on two observations to achieve these goals. First, it is easier for a user to assess patterns rather than individual transactions or attributes. Second, it is possible to generalise user feedback using similarities between subgroups.

To involve the user already during the discovery process, the central idea is to *alternate between mining and user interaction*: the algorithm mines a set of patterns, a user is given the opportunity to provide feedback, the feedback is used to steer the search, and the algorithm mines a new set of patterns.

Algorithm 1. Interactive Diverse Subgroup Discovery (IDSD)

Input: Dataset $\mathcal{D}$; beam selection S; subgroup similarity σ; *mincov, w, maxdepth*
Output: Set of k subgroups R

1: $beam \leftarrow \{\emptyset\}$, $I \leftarrow \emptyset$, $R \leftarrow \emptyset$, $depth \leftarrow 1$
2: **repeat**   ▷ Generate all candidates for this level
3:      $cands = \{c \in Extensions(beam) \mid Coverage(c, \mathcal{D}) \geq mincov \wedge$
              $\neg \exists n \in I_{neg} : IsExtension(c, n)\}$
4:      $beam \leftarrow \emptyset$
5:      **repeat**   ▷ Selection and interaction loop
6:          $beam \leftarrow SelectBeam(S, cands, \varphi', w, beam)$
7:          $I \leftarrow I \cup GetFeedback(beam)$
8:          $R \leftarrow R \cup I_{pos}$
9:          $beam \leftarrow beam \setminus I_{neg}$, $cands \leftarrow cands \setminus I_{neg}$
10:     **until** $|beam| = w$ ▷ No patterns were disliked
11:     **for all** $c \in cands$ **do**
12:         $UpdateTopK(R, k \times 100, c, \varphi'(c, I, \sigma))$
13:     $depth \leftarrow depth + 1$
14: **until** $depth > maxdepth$
15: **return** $SelectBeam(S, R, \varphi', k, \emptyset) \cup I_{pos}$ ▷ Selection from large overall top-k

---

As a levelwise search procedure that takes only a limited number of intermediate solutions to the next level, beam search provides an excellent framework to
implement this high-level procedure. That is, on each level we let the user influence the beam by *liking* and *disliking* subgroups. Patterns that are disliked are
immediately removed from the beam and replaced by others, effectively guiding
search away from those apparently uninteresting branches of the search space.

This approach has the advantage that it is relatively easy to evaluate subgroups with short descriptions at early levels, while this has a strong influence
on search. Providing feedback at later levels allows fine-tuning, and search parameters such as *maxdepth* and *w* allow a user to manage her efforts.

Algorithm Details. Algorithm 1 presents the method that we propose. In the
following we focus on how the DSSD diverse beam search, as briefly explained in
the Preliminaries, is modified to incorporate user feedback. Essentially it is still
a level-wise beam search, but with a modified strategy for selecting the beam
that both achieves diversity and allows for interactivity.

Feedback elicitation – Feedback elicitation is performed on line 7, after a beam
has been selected (line 6, see also below). All w selected subgroups are presented
to the user in a GUI, and she can provide feedback before continuing.

As feedback, the user can mark each subgroup in a beam as *interesting* ('like')
or *uninteresting* ('dislike'). Let I_{pos} (resp. I_{neg}) denote the set of all positively
(resp. negatively) evaluated subgroups. Additionally, let $I = I_{pos} \cup I_{neg}$ be the
set of all evaluated subgroups. Note that a user is not obliged to provide any
feedback, hence I might not include all subgroups that are in the current beam
and it might even be empty. In the latter case the resulting search is equal to
that of (non-interactive) DSSD.

If any subgroups are disliked in this phase, they are removed from the beam (line 9) and lines 5-9 are repeated until a complete beam consisting of w subgroups is obtained.

Candidate generation – On each level, initially all *direct extensions* of all subgroups in the current beam are generated as candidates (line 3). Here, a direct extension is a subgroup description augmented with one additional condition. Subgroups with too small coverage and all direct extensions of negatively evaluated subgroups in I_{neg} are removed. Note that this does not necessarily result in the complete pruning of the corresponding branch in the search tree. Consider the following example: $A \wedge B \wedge C$ may be generated as extension of $B \wedge C$, even if A was disliked and thus added to I_{neg} at *depth*-1. This preserves the capability to discover high-quality subgroups via other branches.

Feedback-driven subgroup selection – Since feedback only concerns individual subgroups, we need to generalise it to the complete candidate set. We achieve this through modification of the qualities of all subgroups in *Cands*: starting from the 'prior' given by φ, the qualities are updated according to how similar they are to the evaluated subgroups. This way, we obtain a quality measure φ' that takes user feedback into account and effectively re-ranks all possible subgroups. Subsequently, we use the regular diverse beam selection strategy on line 6, with the only difference that the modified qualities are used.

For this to work we need a notion of subgroup similarity: subgroups that are similar to interesting subgroups get a higher quality, whereas subgroups that are similar to uninteresting subgroups get a lower quality. Let $c \in Cands$ and $i \in I$, and let d_x resp. G_x denote the description resp. cover of a subgroup x. We propose the following two subgroup similarity measures:

$$\sigma_{description}(c, i) = \frac{|d_c \cap d_i|}{|d_c \cup d_i| - 1}, \ \sigma_{cover}(c, i) = \frac{|G_c \cap G_i|}{|G_c|} \qquad (1)$$

Description similarity is almost equal to Jaccard similarity; -1 is added to the denominator so that a subgroup and any of its direct extensions have maximal similarity of 1. Cover similarity is based on the overlap coefficient and has the same property for direct extensions.

Finally, given a subgroup similarity measure σ, the modified subgroup quality measure φ' that takes user feedback into account is defined as:

$$\varphi'(G) = \frac{1 + \sum_{i \in I_{pos}} \sigma(G, i)}{1 + \sum_{i \in I_{neg}} \sigma(G, i)} \times \varphi(G) \qquad (2)$$

It re-weighs the 'base quality' φ with a factor based on similarity to evaluated subgroups in I. Note that φ' is equivalent to φ when $I = \emptyset$. Also, values of φ' change immediately after each round of feedback elicitation, hence feedback has an immediate effect on (incremental) beam selection.

Overall results – During search a large overall 'top-k' is maintained using the re-weighed quality measure φ' (line 12). At the end of the algorithm (line 15), a set of k subgroups is selected from this overall large top-k using the subgroup selection procedure just described. Note that all positively evaluated subgroups are added to this final result set R regardless of their qualities.

5   Experiments

5.1   Quantitative Evaluation

In order to be able to perform a large series of experiments, we emulate user feedback. We select a set of high-quality subgroups BK that serves as *background knowledge*; subgroups in BK are already known and should therefore be avoided as much as possible. The intuition behind this approach echoes the example in Section 1: top subgroups usually correspond to common knowledge and are therefore uninteresting.

BK is selected from the output of a standard subgroup discovery algorithm. Selection depends on the subgroup similarity measure used: when using description similarity, we iteratively select the highest quality subgroup having a description with ≤ 3 conditions that does not overlap with the description of any previously selected subgroup; for cover similarity, we only select the highest quality subgroup. During search, BK is used to emulate evaluations: any subgroup s in the beam for which $\exists b \in BK : \sigma(s, b) > \beta$ is automatically 'disliked' by adding it to I_{neg}. Parameter β allows varying the amount of evaluated subgroups: larger values result in fewer negative judgements. Note that no positive feedback is emulated.

To evaluate the effectiveness of the algorithm in eliminating undesired conditions or tuples from the results, we compute the overlap of the discovered subgroups with elements of BK. Depending on the subgroup similarity measure this is either $overlap_{desc}(s, BK) = \max_{b \in BK} |d_s \cap d_b|$ or $overlap_{cov}(s, BK = \{b\}) = |G_s \cap G_b|$. We report the average overlap for all subgroups included in result set R, i.e. $overlap(R, BK) = \frac{1}{|R|} \times \sum_{s \in R} overlap(s, BK)$.

Dataset properties are listed in Table 2, together with size, average description length, and coverage of the generated background knowledge. Except for *nba*, which was introduced in Section 1, all were taken from the UCI repository[6]. The datasets were pre-processed as follows: transactions with missing values were removed, and all numeric attributes were discretised into 6 bins using equal-width binning.

Table 2. Datasets and background knowledge

	$|\mathcal{D}|$	$|A|$	Desc.BK		Cov.BK
			$|BK|$	$\overline{|d_{BK}|}$	$|G_{BK}|$
breast-w [bw]	683	11	2	2.0	137
credit-a [ca]	653	17	3	3.0	325
credit-g [cg]	1000	22	2	3.0	394
diabetes [db]	768	10	2	3.0	82
liver [lv]	345	8	2	3.0	107
nba	923	26	6	2.8	228

Search parameters were set to the following values in all experiments: $mincov = 0.1 \times |\mathcal{D}|$, $maxdepth = 5$, $w = 20$, $k = 100$. Note that small values of w and k are chosen in order to match the limited processing capabilities of a human user. We first focus on a single dataset, *credit-g*, and then discuss the results of experiments with multiple datasets.

---

[6] http://archive.ics.uci.edu/ml/datasets.html

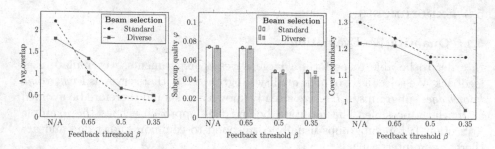

Fig. 1. Results for *credit-g*, description subgroup similarity. *N/A* corresponds to the non-interactive algorithm. The subgroup quality plot depicts averages, standard deviations and maxima of the individual (unweighed) qualities in the subgroup sets found.

A Characteristic Experiment in Detail. Figure 1 shows the results that we obtained on *credit-g* for various values of β, which controls the amount of emulated negative feedback. Also, we experimented with both standard and (description-based) diverse beam search. The left plot shows that the average overlap with BK decreases considerably when β decreases and thus more subgroups are evaluated negatively. This demonstrates that re-weighing subgroup quality using description-based similarity is effective at eliminating undesired conditions from the beam and final results. Only modest numbers of negative evaluations (8 to 12) were required to achieve this.

The middle plot shows that both maximum and average qualities of the subgroup sets decrease. This is as expected though: the user emulation scheme was designed to prune high-quality subgroups. Nevertheless, the algorithm succeeds in discovering (other) subgroups of quite high quality. Finally, the right plot shows that redundancy slightly decreases as I grows. Although the differences between the standard and diverse beam search appear to be small, the diverse results are clearly less redundant: in particular for lower β, cover redundancy is lower and standard deviation of the subgroup qualities is higher.

Overall Results. For the experiments in Table 3 we use diverse beam selection, with either description ($\beta = 0.35$) or cover ($\beta = 0.5$) subgroup similarity.

In general, both approaches adequately eliminate undesired subgroups from the result set, whether it is through negatively marked conditions or tuples. This is demonstrated by the consistently decreasing overlap with the background knowledge. Importantly, the number of evaluations required to achieve this is generally modest, i.e. $|I| \leq 25$ for all cases except two. This makes the approach practically useful and usable for a domain expert. As expected, average and maximum subgroup qualities decreases. Only the effect on cover redundancy varies depending on the dataset, but the difference is often small. We conclude that interaction and quality re-weighing work well together with the diverse beam selection.

Table 3. Overall results of feedback emulation

σ	$\mathcal{D}$	$|I|$	Standard diverse $\rightarrow$ Interactive			
			Avg.overlap(BK)	φ_{avg}	φ_{max}	CR
Desc	bw	8	1.37 → 0.80	.187 → .171	.200 → .184	1.20 → 1.26
	ca	24	1.65 → 0.89	.178 → .088	.181 → .131	1.00 → 0.90
	cg	8	1.80 → 0.66	.072 → .046	.074 → .049	1.22 → 1.15
	db	12	1.99 → 0.25	.077 → .048	.084 → .056	0.66 → 1.17
	lv	12	1.20 → 0.38	.042 → .041	.047 → .045	0.83 → 0.97
	nba	78	1.57 → 1.09	.067 → .058	.071 → .064	1.50 → 1.57
Cover	bw	17	125.37 → 120.09	.187 → .176	.200 → .200	1.20 → 1.12
	ca	157	318.95 → 232.68	.178 → .013	.181 → .025	1.00 → 0.18
	cg	7	353.49 → 168.50	.072 → .046	.074 → .049	1.22 → 1.15
	db	1	61.22 → 62.13	.077 → .077	.084 → .084	0.66 → 0.64
	lv	2	57.57 → 27.66	.042 → .041	.047 → .047	0.83 → 1.26
	nba	1	218.28 → 117.88	.067 → .036	.071 → .058	1.50 → 0.75

5.2   Case Study: Sports Analytics

As we have seen in Section 1, the subgroups discovered by DSSD were unsatisfactory. To demonstrate that the proposed interactive approach can be used to improve on this, we asked a basketball journalist to use IDSD and evaluate the results. In the following we set the search parameters to $mincov = 50$, $w = 10$, $maxdepth = 3$, and $k = 5$, and we use description similarity.

The domain expert evaluated 18 subgroups during an interactive search session, 13 of length-1 and 5 of length-2. Examples of *liked* subgroups include $crawford = F$, $pace < 88.977$, and $matthews = T \land hickson = T$ (7 subgroups in total). Examples of *disliked* subgroups are $opp\_def\_reb = F$, $thabeet = F$, and $pace < 88.977 \land opponent \neq MIA$ (11 subgroups in total).

Table 4. Top five subgroups discovered by IDSD with description-based similarity. For each discovered subgroup its description, size and quality are given.

Description	Size	Quality
$crawford = F \land matthews = T$	96	0.0382
$hickson = T$	186	0.0219
$crawford = F \land hickson = T$	328	0.0211
$matthews = T \land hickson = T$	290	0.0163
$matthews = T \land pace < 88.518$	303	0.0221

The discovered subgroups are presented in Table 4. Although the objective qualities of the subgroups are lower than the maximum, the results were considered more interesting as they provided novel insights about relevant players.

A user needs to consider subgroups one by one when processing results and providing feedback. Hence, we can estimate user effort E by counting the subgroups she had to consider. The effort induced by non-interactive diverse subgroup discovery is then equal to the lowest rank of an interesting subgroup in the result set sorted by quality. The effort induced by the interactive approach also includes the number of subgroups presented during the search:

$E_{IDSD} = maxdepth \times w + |I_{neg}|$. In this case we have $E_{DSSD} = 1049$ and $E_{IDSD} = 5 + (3 \times 10 + 11) = 46$, which confirms that within-search interaction substantially reduces the effort required to discover interesting results.

Discussion. Although this is a good example of a successful interactive session, in some other sessions the domain expert deemed the results unsatisfactory. In some cases the search space was pruned too eagerly, or positive and negative evaluations were not properly balanced. Also, the expert did not find the approach using cover similarity useful, as this resulted in descriptions that were were not interpretable. Training of the domain expert might solve this and results obviously also depend on the data, but this also shows that it is worth investigating more elaborate similarity measures to generalize user feedback.

Another crucial drawback is an unintuitive effect on beam selection, e.g. *disliking* a large subgroup based on its description (cf. *reserve_player = F*) steers the search away from promising regions. Another concern is the capacity to discover novel subgroups (as opposed to simply replicating the feedback). Multiple sessions might be required to explore unrelated regions. However, given the significantly lower effort, the cumulative effort is still reduced.

6   Conclusions and Future Work

We argued that it is essential to actively involve the user in the discovery process to obtain results that she finds interesting. To this end, we proposed the Interactive Diverse Subgroup Discovery (IDSD) algorithm that allows a user to provide feedback to provisional results already *during search*. It augments a diverse beam search by letting the user 'like' and 'dislike' subgroups in the beam. Although this interaction mechanism is conceptually simple and easy to use, it allows a user to guide the search effectively.

In the quantitative evaluation, we emulated the feedback of a user that wants to avoid the re-discovery of common knowledge. Experiments show that undesired results can be eliminated, whereas other, potentially more interesting subgroups are found. Furthermore, we conducted a case study in which a domain expert was able to find more interesting patterns when compared to the results of standard algorithms. This confirms that within-search human-computer interaction can contribute to a substantial reduction in the effort needed to discover interesting subgroups.

Future Work. This paper presents only a first step towards user-driven pattern discovery, but since the user is too often still neglected we believe it is an important step. In the future, one obvious line of research is to investigate what features other than inclusion/exclusion of individual conditions and tuples are relevant to the user, and are therefore useful to infer subjective interestingness.

A second direction that will be essential to research is pattern visualisation. In our prototype, we mainly focused on presenting subgroup descriptions, but in the future it will be important to visualise the different aspects of the patterns. Not only descriptions and covers should be visualised, but also other relevant features such as traditional interestingness and surprisingness measures. We deem this particularly important for larger datasets and/or beam widths. Only then will it be possible for the user to interactively explore the data in an intuitive way.

Acknowledgements. Matthijs van Leeuwen is supported by a Rubicon grant of the Netherlands Organisation for Scientific Research (NWO).

References

1. Atzmüller, M.: Exploiting background knowledge for knowledge-intensive subgroup discovery. In: Proceedings of IJCAI 2005, pp. 647–652 (2005)
2. Atzmüller, M., Puppe, F.: Semi-automatic visual subgroup mining using vikamine. Journal of Universal Computer Science 11(11), 1752–1765 (2005)
3. Bailey, J., Dong, G.: Contrast data mining: Methods and applications. Tutorial at ICDM 2007 (2007)
4. De Bie, T.: An information theoretic framework for data mining. In: Proceedings of KDD 2011, pp. 564–572 (2011)
5. Dong, G., Zhang, X., Wong, L., Li, J.: CAEP: Classification by aggregating emerging patterns. In: Arikawa, S., Nakata, I. (eds.) DS 1999. LNCS (LNAI), vol. 1721, pp. 30–42. Springer, Heidelberg (1999)
6. Galbrun, E., Miettinen, P.: A Case of Visual and Interactive Data Analysis: Geospatial Redescription Mining. In: Instant Interactive Data Mining Workshop at ECML-PKDD 2012 (2012)
7. Gamberger, D., Lavrac, N.: Expert-guided subgroup discovery: Methodology and application. Journal of Artificial Intelligence Research 17, 501–527 (2002)
8. Gamberger, D., Lavrac, N., Krstacic, G.: Active subgroup mining: a case study in coronary heart disease risk group detection. Artificial Intelligence in Medicine 28(1), 27–57 (2003)
9. Garriga, G.C., Kralj, P., Lavrac, N.: Closed sets for labeled data. Journal of Machine Learning Research 9, 559–580 (2008)
10. Goethals, B., Moens, S., Vreeken, J.: MIME: a framework for interactive visual pattern mining. In: Proceedings of KDD 2011, pp. 757–760 (2011)
11. Herrera, F., Carmona, C.J., González, P., Jesus, M.J.: An overview on subgroup discovery: foundations and applications. Knowledge and Information Systems 29(3), 495–525 (2011)
12. Klösgen, W.: Explora: A Multipattern and Multistrategy Discovery Assistant. In: Advances in Knowledge Discovery and Data Mining, pp. 249–271 (1996)
13. Kralj Novak, P., Lavrač, N., Webb, G.I.: Supervised descriptive rule discovery: A unifying survey of contrast set, emerging pattern and subgroup mining. Journal of Machine Learning Research 10, 377–403 (2009)
14. van Leeuwen, M., Knobbe, A.: Diverse subgroup set discovery. Data Mining and Knowledge Discovery 25, 208–242 (2012)
15. Li, R., Kramer, S.: Efficient redundancy reduced subgroup discovery via quadratic programming. In: Ganascia, J.-G., Lenca, P., Petit, J.-M. (eds.) DS 2012. LNCS, vol. 7569, pp. 125–138. Springer, Heidelberg (2012)
16. Rüping, S.: Ranking interesting subgroups. In: Proceedings of ICML 2009, pp. 913–920 (2009)
17. Tuzhilin, A.: On subjective measures of interestingness in knowledge discovery. In: Proceedings of KDD 1995, pp. 275–281 (1995)
18. Wrobel, S.: An algorithm for multi-relational discovery of subgroups. In: Komorowski, J., Żytkow, J.M. (eds.) PKDD 1997. LNCS, vol. 1263, pp. 78–87. Springer, Heidelberg (1997)

Gaussian Mixture Models for Time Series Modelling, Forecasting, and Interpolation

Emil Eirola[1] and Amaury Lendasse[1,2,3]

[1] Department of Information and Computer Science, Aalto University,
FI–00076 Aalto, Finland
emil.eirola@aalto.fi
[2] IKERBASQUE, Basque Foundation for Science, 48011 Bilbao, Spain
[3] Computational Intelligence Group, Computer Science Faculty,
University of the Basque Country, Paseo Manuel Lardizabal 1,
Donostia/San Sebastián, Spain

Abstract. Gaussian mixture models provide an appealing tool for time series modelling. By embedding the time series to a higher-dimensional space, the density of the points can be estimated by a mixture model. The model can directly be used for short-to-medium term forecasting and missing value imputation. The modelling setup introduces some restrictions on the mixture model, which when appropriately taken into account result in a more accurate model. Experiments on time series forecasting show that including the constraints in the training phase particularly reduces the risk of overfitting in challenging situations with missing values or a large number of Gaussian components.

Keywords: time series, missing data, Gaussian mixture model.

1 Introduction

A time series is one of the most common forms of data, and has been studied extensively from weather patterns spanning centuries to sensors and microcontrollers operating on nanosecond scales. The features and irregularities of time series can be modelled through various means, such as autocovariance analysis, trend fitting, or frequency-domain methods. From a machine learning perspective, the most relevant tasks tend to be prediction of one or several future data points, or interpolation for filling in gaps in the data. In this paper, we study a model for analysing time series, which is applicable to both tasks.

For uniformly sampled stationary processes, we propose a versatile methodology to model the features of the time series by embedding the data to a high-dimensional regressor space. The density of the points in this space can then be modelled with Gaussian mixture models [1]. Such an estimate of the probability density enables a direct way to interpolate missing values in the time series and conduct short-to-medium term prediction by finding the conditional expectation of the unknown values. Embedding the time series in a higher-dimensional space imposes some restrictions on the possible distribution of points, but these constraints can be accounted for when fitting the Gaussian mixture models.

A. Tucker et al. (Eds.): IDA 2013, LNCS 8207, pp. 162–173, 2013.
© Springer-Verlag Berlin Heidelberg 2013

The suggested framework can readily be extended to situations with several related time series, using exogenous time series to improve the predictions of a target series. Furthermore, any missing values can be handled by the Gaussian mixture model in a natural manner.

This paper is structured as follows. Section 2 presents the procedure for modelling time series by Gaussian mixture models, the constraints on the Gaussian mixture model due to time series data are discussed in Section 3, and some experiments showing the effect of selecting the number of components and introducing missing values are studied in Section 4.

2   Mixture Models for Time Series

Given a time series z of length n, corresponding to a stationary process:

$$z_0, z_1, z_2, \ldots, z_{n-2}, z_{n-1},$$

by choosing a regressor length d we can conduct a delay embedding [2] and form the design matrix $\mathbf{X}$,

$$\mathbf{X} = \begin{bmatrix} z_0 & z_1 & \cdots & z_{d-1} \\ z_1 & z_2 & \cdots & z_d \\ \vdots & \vdots & & \vdots \\ z_{n-d} & z_{n-d+1} & \cdots & z_{n-1} \end{bmatrix} = \begin{bmatrix} \boldsymbol{x}_0 \\ \boldsymbol{x}_1 \\ \vdots \\ \boldsymbol{x}_{n-d} \end{bmatrix}. \tag{1}$$

The rows of $\mathbf{X}$ can be interpreted as vectors in $\mathbb{R}^d$. We can model the density of these points by a Gaussian mixture model, with the probability density function

$$p(\boldsymbol{x}) = \sum_{k=1}^{K} \pi_k \mathcal{N}(\boldsymbol{x} \mid \boldsymbol{\mu}_k, \boldsymbol{\Sigma}_k) \tag{2}$$

where $\mathcal{N}(\boldsymbol{x} \mid \boldsymbol{\mu}_k, \boldsymbol{\Sigma}_k)$ is the probability density function of the multivariate normal distribution, $\boldsymbol{\mu}_k$ represents the means, $\boldsymbol{\Sigma}_k$ the covariance matrices, and π_k the mixing coefficients for each component k ($0 < \pi_k < 1$, $\sum_{k=1}^{K} \pi_k = 1$).

Given a set of data, the standard approach to training a Gaussian mixture model is the EM algorithm [3,4] for finding a maximum-likelihood fit. The log-likelihood of the N data points is given by

$$\log \mathcal{L}(\theta) = \log p(\mathbf{X} \mid \theta) = \sum_{i=1}^{N} \log \left(\sum_{k=1}^{K} \pi_k \mathcal{N}(\boldsymbol{x}_i \mid \boldsymbol{\mu}_k, \boldsymbol{\Sigma}_k) \right), \tag{3}$$

where $\theta = \{\pi_k, \boldsymbol{\mu}_k, \boldsymbol{\Sigma}_k\}_{k=1}^{K}$ is the set of parameters defining the model. The log-likelihood can be maximised by applying the EM algorithm. After some initialisation of parameters, the E-step is to find the expected value of the log likelihood function, with respect to the conditional distribution of latent variables $\mathbf{Z}$ given the data $\mathbf{X}$ under the current estimate of the parameters $\theta^{(t)}$:

$$Q(\theta \mid \theta^{(t)}) = \mathrm{E}_{\mathbf{Z} \mid \mathbf{X}, \theta^{(t)}} \left[\log \mathcal{L}(\theta; \mathbf{X}, \mathbf{Z}) \right] \tag{4}$$

This requires evaluating the probabilities t_{ik} that $\boldsymbol{x}_i$ is generated by the kth Gaussian using the current parameter values:

$$t_{ik}^{(t)} = \frac{\pi_k^{(t)} \mathcal{N}(\boldsymbol{x}_i \mid \boldsymbol{\mu}_k^{(t)}, \boldsymbol{\Sigma}_k^{(t)})}{\sum_{j=1}^{K} \pi_j^{(t)} \mathcal{N}(\boldsymbol{x}_i \mid \boldsymbol{\mu}_j^{(t)}, \boldsymbol{\Sigma}_j^{(t)})} . \tag{5}$$

In the M-step, the expected log-likelihood is maximised:

$$\theta^{(t+1)} = \arg \max_{\theta} Q(\theta \mid \theta^{(t)}), \tag{6}$$

which corresponds to re-estimating the parameters with the new probabilities:

$$\boldsymbol{\mu}_k^{(t+1)} = \frac{1}{N_k} \sum_{i=1}^{N} t_{ik}^{(t)} \boldsymbol{x}_i, \tag{7}$$

$$\boldsymbol{\Sigma}_k^{(t+1)} = \frac{1}{N_k} \sum_{i=1}^{N} t_{ik}^{(t)} (\boldsymbol{x}_i - \boldsymbol{\mu}_k^{(t+1)})(\boldsymbol{x}_i - \boldsymbol{\mu}_k^{(t+1)})^T, \tag{8}$$

$$\pi_k^{(t+1)} = \frac{1}{N} \sum_{i=1}^{N} t_{ik}^{(t)} . \tag{9}$$

Here $N_k = \sum_{i=1}^{N} t_{ik}^{(t)}$ is the effective number of samples covered by the kth component. The E and M-steps are alternated repeatedly until convergence. As the algorithm tends to occasionally converge to sub-optimal solutions, the procedure can be repeated to find the best fit.

2.1   Model Structure Selection

The selection of the number of components K is crucial, and has a significant effect on the resulting accuracy. Too few components are not able to model the distribution appropriately, while having too many components causes issues of overfitting.

The number of components can be selected according to the Akaike information criterion (AIC) [5] or the Bayesian information criterion (BIC) [6]. Both are expressed as a function of the log-likelihood of the converged mixture model:

$$\mathrm{AIC} = -2 \log \mathcal{L}(\theta) + 2P, \tag{10}$$

$$\mathrm{BIC} = -2 \log \mathcal{L}(\theta) + \log(N)P, \tag{11}$$

where $P = Kd + \frac{1}{2}Kd(d+1) + K - 1$ is the number of free parameters. The EM algorithm is run for several different values of K, and the model which minimises the chosen criterion is selected. As $\log(N) > 2$ in most cases, BIC more aggresively penalises an increase in P, generally resulting in a smaller choice for K than by AIC.

2.2 Forecasting

The model readily lends itself to being used for short-to-medium term time series prediction. For example, if a time series is measured monthly and displays some seasonal behaviour, a Gaussian model could be trained with a regressor size of 24 (two years). This allows us to take the last year's measurements as the 12 *first* months, and determine the conditional expectation of the following 12 months.

The mixture model provides a direct way to calculate the conditional expectation. Let the input dimensions be partitioned into past values P (known) and future values F (unknown). Then, given a sample $\boldsymbol{x}_i^P$ for which only the past values are known and a prediction is to be made, calculate the probabilities of it belonging to each component

$$t_{ik} = \frac{\pi_k \mathcal{N}(\boldsymbol{x}_i^P \mid \boldsymbol{\mu}_k, \boldsymbol{\Sigma}_k)}{\sum_{j=1}^{K} \pi_j \mathcal{N}(\boldsymbol{x}_i^P \mid \boldsymbol{\mu}_j, \boldsymbol{\Sigma}_j)}, \tag{12}$$

where $\mathcal{N}(\boldsymbol{x}_i^P \mid \boldsymbol{\mu}_k, \boldsymbol{\Sigma}_k)$ is the *marginal* multivariate normal distribution probability density of the observed (i.e., past) values of $\boldsymbol{x}_i$.

Let the means and covariances of each component also be partitioned according to past and future variables:

$$\boldsymbol{\mu}_k = \begin{bmatrix} \boldsymbol{\mu}_k^P \\ \boldsymbol{\mu}_k^F \end{bmatrix}, \quad \boldsymbol{\Sigma}_k = \begin{bmatrix} \boldsymbol{\Sigma}_k^{PP} & \boldsymbol{\Sigma}_k^{PF} \\ \boldsymbol{\Sigma}_k^{FP} & \boldsymbol{\Sigma}_k^{FF} \end{bmatrix}. \tag{13}$$

Then the conditional expectation of the future values with respect to the component k is given by

$$\tilde{\boldsymbol{y}}_{ik} = \boldsymbol{\mu}_k^F + \boldsymbol{\Sigma}_k^{FP}(\boldsymbol{\Sigma}_k^{PP})^{-1}(\boldsymbol{x}_i^P - \boldsymbol{\mu}_k^P) \tag{14}$$

in accordance with [7, Thm. 2.5.1]. The total conditional expectation can now be found as a weighted average of these predictions by the probabilities t_{ik}:

$$\hat{\boldsymbol{y}}_i = \sum_{k=1}^{K} t_{ik} \tilde{\boldsymbol{y}}_{ik}. \tag{15}$$

It should be noted that the method directly estimates the full vector of future values at once, in contrast with most other methods which would separately predict each required data point.

2.3 Missing Values and Imputation

The proposed method is directly applicable to time series with missing values. Missing data in the time series become diagonals of missing values in the design matrix. The EM-algorithm can in a natural way account for missing values in the samples [8,9].

An assumption here is that data are Missing-at-Random (MAR) [10]:

$$P(M \mid x_{\text{obs}}, x_{\text{mis}}) = P(M \mid x_{\text{obs}}),$$

i.e., the event M of a measurement being missing is independent from the value it would take (x_{mis}), conditional on the observed data (x_{obs}). The stronger assumption of Missing-Completely-at-Random (MCAR) is not necessary, as MAR is an ignorable missing-data mechanism in the sense that maximum likelihood estimation still provides a consistent estimator [10].

To conduct missing value imputation, the procedure is the same as for prediction in Section 2.2. The only difference is that in this case the index set P contains all known values for a sample (both before and after the target to be predicted), while F contains the missing values that will be imputed.

2.4   Missing-Data Padding

When using an implementation of the EM algorithm that is able to handle missing values, it is reasonable to consider that every value before and after the recorded time series consists is missing. This can be seen as "padding" the design matrix $\mathbf{X}$ with missing values (marked as '?'), effectively increasing the number of samples available for training from $n - d + 1$ to $n + d - 1$ (cf. Eq. (1)):

$$
\mathbf{X} =
\begin{bmatrix}
? & ? & \cdots & ? & z_0 \\
? & ? & \cdots & z_0 & z_1 \\
\vdots & \vdots & & \vdots & \vdots \\
? & z_0 & \cdots & z_{d-3} & z_{d-2} \\
z_0 & z_1 & \cdots & z_{d-2} & z_{d-1} \\
\vdots & \vdots & & \vdots & \vdots \\
z_{n-d} & z_{n-d+1} & \cdots & z_{n-2} & z_{n-1} \\
z_{n-d+1} & z_{n-d+2} & \cdots & z_{n-1} & ? \\
\vdots & \vdots & & \vdots & \vdots \\
z_{n-1} & ? & \cdots & ? & ?
\end{bmatrix}
=
\begin{bmatrix}
\boldsymbol{x}_0 \\
\boldsymbol{x}_1 \\
\vdots \\
\boldsymbol{x}_{d-2} \\
\boldsymbol{x}_{d-1} \\
\vdots \\
\boldsymbol{x}_{n-1} \\
\boldsymbol{x}_n \\
\vdots \\
\boldsymbol{x}_{n+d-2}
\end{bmatrix}
\tag{16}
$$

Fitting the mixture model using this padded design matrix has the added advantage that the sample mean and variance of (the observed values in) each column is guaranteed to be equal. The missing-data padding can thus be a useful trick even if the time series itself features no missing values, particularly if only a limited amount of data is available.

3   Constraining the Global Covariance

The Gaussian mixture model is ideal for modelling arbitrary continuous distributions. However, embedding a time series to a higher-dimensional space cannot lead to an arbitrary distribution. For instance, the mean and variance for each dimension should equal the mean and variance of the time series. In addition, all second-order statistics, such as covariances, should equal the respective autocovariances of the time series. These restrictions impose constraints on the mixture model, and accounting for them appropriately should lead to a more accurate model when fitting to data.

In the EM algorithm, we estimate means μ_k, covariances Σ_k, and mixing coefficients π_k for each component k, and then the global mean and covariance of the distribution defined by the model is

$$\mu = \sum_{k=1}^{K} \pi_k \mu_k, \qquad \Sigma = \sum_{k=1}^{K} \pi_k \left(\Sigma_k + \mu_k \mu_k^T \right) - \mu \mu^T. \tag{17}$$

However, the global mean and covariance correspond to the mean and autocovariance matrix of the time series. This implies that the global mean for each dimension should be equal. Furthermore, the global covariance matrix should be symmetric and Toeplitz ("diagonal-constant"):

$$\Sigma \approx \mathbf{R}_z = \begin{bmatrix} r_z(0) & r_z(1) & r_z(2) & \dots & r_z(d-1) \\ r_z(1) & r_z(0) & r_z(1) & \dots & r_z(d-2) \\ \vdots & \vdots & \vdots & & \vdots \\ r_z(d-1) & r_z(d-2) & r_z(d-3) & \dots & r_z(0) \end{bmatrix}$$

where $r_z(l)$ is the autocovariance of the time series z at lag l.

In practice, these statistics usually do not exactly correspond to each other, even when training the model on the missing-data padded design matrix discussed in Section 2.4. Unfortunately, the question of how to enforce this constraint in each M-step has no trivial solution. Forcing every component to have an equal mean and Toeplitz covariance structure by its own is one possibility, but this is far too restrictive.

Our suggestion is to calculate the M-step by Eqs. (7–9), and then modify the parameters as little as possible in order to achieve the appropriate structure. As $\theta = \{\mu_k, \Sigma_k, \pi_k\}_{k=1}^{K}$ contains the parameters for the mixture model, let Ω be the space of all possible values for θ, and $T \subset \Omega$ be the subset such that all parameter values $\theta \in T$ correspond to a global mean with equal elements, and Toeplitz covariance matrix by Eq. (17).

When maximising the expected log-likelihood with the constraints, the M-step should be

$$\theta^{(t+1)} = \arg\max_{\theta \in T} Q(\theta \mid \theta^{(t)}), \tag{18}$$

but this is not feasible to solve exactly. Instead, we solve the conventional M-step

$$\theta' = \arg\max_{\theta \in \Omega} Q(\theta \mid \theta^{(t)}), \tag{19}$$

and then project this θ' onto T to find the closest solution

$$\theta^{(t+1)} = \arg\min_{\theta \in T} d(\theta, \theta') \tag{20}$$

for some interpretation of the distance $d(\theta, \theta')$. If the difference is small, the expected log-likelihood $Q(\theta^{(t+1)} \mid \theta^{(t)})$ should not be too far from the optimal

$\max_{\theta \in T} Q(\theta \mid \theta^{(t)})$. As the quantity is not maximised, though it can be observed to increase, this becomes a Generalised EM (GEM) algorithm. As long as an increase is ensured in every iteration, the GEM algorithm is known to have similar convergence properties as the EM algorithm [3,4].

Define the distance function between sets of parameters as follows:

$$d(\theta, \theta') = \sum_k \|\boldsymbol{\mu}_k - \boldsymbol{\mu}'_k\|^2 + \sum_k \|\mathbf{S}_k - \mathbf{S}'_k\|_F^2 + \sum_k (\pi_k - \pi'_k)^2, \qquad (21)$$

where $\mathbf{S}_k = \boldsymbol{\Sigma}_k + \boldsymbol{\mu}_k \boldsymbol{\mu}_k^T$ are the second moments of the distributions of each component and $\|\cdot\|_F$ is the Frobenius norm. Using Lagrange multipliers, it can be shown that this distance function is minimised by the results presented below in Eqs. (22) and (23).

3.1   The Mean

After an iteration of the normal EM-algorithm by Eqs. (7–9), find the vector with equal components which is nearest to the global mean $\boldsymbol{\mu}$ as calculated by Eq. (17). This is done by finding the mean m of the components of $\boldsymbol{\mu}$, and calculating the discrepancy δ of how much the current mean is off from the equal mean:

$$m = \frac{1}{d} \sum_{j=1}^{d} \mu_j, \qquad \delta = \boldsymbol{\mu} - m\mathbf{1},$$

where $\mathbf{1}$ is a vector of ones. Shift the means of each component to compensate, as follows:

$$\boldsymbol{\mu}'_k = \boldsymbol{\mu}_k - \frac{\pi_k}{\sum_{j=1}^{K} \pi_j^2} \delta \qquad \forall k. \qquad (22)$$

As can be seen, components with larger π_k take on more of the "responsibility" of the discrepancy, as they contribute more to the global statistics. Any weights which sum to unity would fulfil the constraints, but choosing the weights to be directly proportional to π_k minimises the distance in Eq. (21).

3.2   The Covariance

After updating the means $\boldsymbol{\mu}_k$, recalculate the covariances around the updated values as

$$\boldsymbol{\Sigma}_k \leftarrow \boldsymbol{\Sigma}_k + \boldsymbol{\mu}_k \boldsymbol{\mu}_k^T - \boldsymbol{\mu}'_k \boldsymbol{\mu}'^T_k \qquad \forall k.$$

Then, find the nearest (in Frobenius norm) Toeplitz matrix $\mathbf{R}$ by calculating the mean of each diagonal of the global covariance matrix $\boldsymbol{\Sigma}$ (from Eq. (17)):

$$r(0) = \frac{1}{d} \sum_{j=1}^{d} \Sigma_{j,j}, \quad r(1) = \frac{1}{d-1} \sum_{j=1}^{d-1} \Sigma_{j,j+1}, \quad r(2) = \frac{1}{d-2} \sum_{j=1}^{d-2} \Sigma_{j,j+2}, \quad \text{etc.}$$

The discrepancy $\boldsymbol{\Delta}$ from this Toeplitz matrix is

$$
\boldsymbol{\Delta} = \boldsymbol{\Sigma} - \mathbf{R}, \quad \text{where } \mathbf{R} = \begin{bmatrix} r(0) & r(1) & r(2) & \dots & r(d-1) \\ r(1) & r(0) & r(1) & \dots & r(d-2) \\ \vdots & & \vdots & & \vdots \\ r(d-1) & r(d-2) & r(d-3) & \dots & r(0) \end{bmatrix}.
$$

In order to satisfy the constraint of a Toeplitz matrix for the global covariance, the component covariances are updated as

$$
\boldsymbol{\Sigma}'_k = \boldsymbol{\Sigma}_k - \frac{\pi_k}{\sum_{j=1}^{K} \pi_j^2} \boldsymbol{\Delta} \quad \forall k, \tag{23}
$$

the weights being the same as in Eq. (22). Eqs. (22) and (23), together with $\pi'_k = \pi_k$, minimise the distance in Eq. (21) subject to the constraints.

3.3 Heuristic Correction

Unfortunately the procedure described above seems to occasionally lead to matrices $\boldsymbol{\Sigma}'_k$ which are not positive definite. Hence an additional heuristic correction c_k is applied in such cases to force the matrix to remain positive definite:

$$
\boldsymbol{\Sigma}''_k = \boldsymbol{\Sigma}_k - \frac{\pi_k}{\sum_{k=1}^{K} \pi_k^2} \boldsymbol{\Delta} + c_k \mathbf{I} \quad \forall k. \tag{24}
$$

In the experiments, the value $c_k = 1.1|\lambda_{k0}|$ is used, where λ_{k0} is the most negative eigenvalue of $\boldsymbol{\Sigma}'_k$. The multiplier needs to be larger than unity to avoid making the matrix singular.

A more appealing correction would be to only increase the negative (or zero) eigenvalues to some acceptable, positive, value. However, this would break the constraint of a Toeplitz global covariance matrix, and hence the correction must be applied to all eigenvalues, as is done in Eq. (24) by adding to the diagonal.

3.4 Free Parameters

The constraints reduce the number of free parameters relevant to calculating the AIC and BIC. Without constraints, the number of free parameters is

$$
P = \underbrace{Kd}_{\text{means}} + \underbrace{\frac{1}{2}Kd(d+1)}_{\text{covariances}} + \underbrace{K-1}_{\text{mixing coeffs}},
$$

where K is the number of Gaussian components, and d is the regressor length. There are $d-1$ equality constraints for the mean, and $\frac{1}{2}d(d-1)$ constraints for the covariance, each reducing the number of free parameters by 1. With the constraints, the number of free parameters is then

$$
P' = \underbrace{(K-1)d+1}_{\text{means}} + \underbrace{\frac{1}{2}(K-1)d(d+1)+d}_{\text{covariances}} + \underbrace{K-1}_{\text{mixing coeffs}}.
$$

The leading term is reduced from $\frac{1}{2}Kd^2$ to $\frac{1}{2}(K-1)d^2$, in effect allowing one additional component for approximately the same number of free parameters.

3.5   Exogenous Time Series or Non-contiguous Lag

If the design matrix is formed in a different way than by taking consecutive values, the restrictions for the covariance matrix will change. Such cases are handled by forcing any affected elements in the matrix to equal the mean of the elements it should equal. This will also affect the number of free parameters.

As this sort of delay embedding may inherently have a low intrinsic dimension, optimising the selection of variables could considerably improve accuracy.

4   Experiments: Time Series Forecasting

To show the effects of the constraints and the number of components on the prediction accuracy, some experimental results are shown here. The studied time series is the Santa Fe time series competition data set A: Laser generated data [11]. The task is set at predicting the next 12 values, given the previous 12. This makes the regressor size $d = 24$, and the mixture model fitting is in a 24-dimensional space. The original 1000 points of the time series are used for training the model, and the continuation (9093 points) as a test set for estimating the accuracy of the prediction. Accuracy is determined by the mean squared error (MSE), averaging over the 12 future values for all samples. No variable selection is conducted, and all 12 variables in the input are used for the model. The missing-data padded design matrix of Section 2.4 is used for the training, even when the time series otherwise has no missing values.

4.1   The Number of Components

Gaussian mixture models were trained separately for 1 through 30 components, each time choosing out of 10 runs the best result in terms of log-likelihood. In order to provide a perspective on average behaviour, this procedure was repeated 20 times both with and without the constraints detailed in Section 3.

The first two plots in Fig. 1 show the MSE of the prediction on the training and test sets, as an average of the 20 repetitions. It is important to note that the model fitting and selection was conducted by maximising the log-likelihood, and not by attempting to minimise this prediction error. Nevertheless, it can be seen that the training error decreases when adding components, and is consistently lower than the test error, as expected. Notably, the difference between training and test errors is much smaller for the constrained mixture model than the unconstrained one. Also, the training error is consistently decreasing for both models when increasing the number of components, but for the test error this is true only for constrained model. It appears that the unconstrained model results in overfitting when used with more than 10 components. For 1 to 10 components,

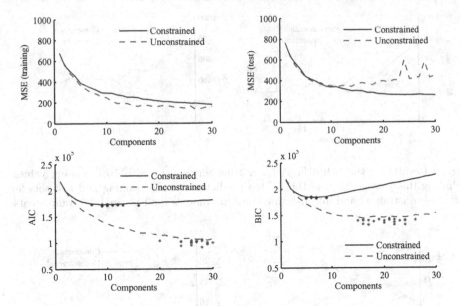

Fig. 1. Results on the Santa Fe A Laser time series data, including the average MSE of the 12-step prediction on the training and test sets, AIC and BIC values for both the constrained and unconstrained mixture models for 1 through 30 components.

there is no notable difference in the test error between the two models, presumably because around 10 components are required for a decent approximation of the density. However, for 10 or more components, the constraints provide a consistent improvement in the forecasting accuracy.

The third and fourth plots in Fig. 1 shows the evolution of the average AIC and BIC of the converged model. The line plots show the average value of the criterion, and the asterisks depict the minimum AIC (or BIC) value (i.e., the selected model) for each of the 20 runs. As results on the test set are not available in the model selection phase, the number of components should be chosen based on these criteria. As the log-likelihood grows much faster for the unconstrained model, this results in a consistently larger number of components as selected by both criteria. Comparing the AIC and BIC, it is clear that BIC tends to choose fewer components, as expected. However, the test MSE for the constrained model keeps decreasing even until 30 components, suggesting that both criteria may be exaggerating the penalisation in this case when increasing the model size.

4.2   Missing Data

To study the effect of missing values, the modelling of the Santa Fe Laser time series is repeated with various degrees of missing data (1% through 50%). In the training phase, missing data is removed at random from the time series before forming the padded design matrix. To calculate the testing MSE, missing values

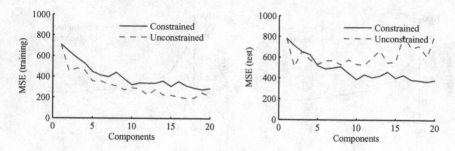

Fig. 2. Results on the Santa Fe A Laser time series data with 10% missing values, including the average MSE of the 12-step prediction on the training and test sets for both the constrained and unconstrained mixture models for 1 through 20 components.

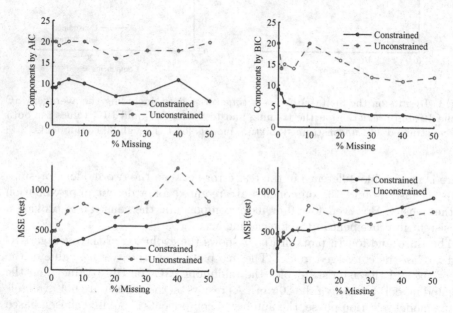

Fig. 3. Results on the Santa Fe A Laser time series data for various degrees of missing values, including the number of components selected by AIC (left) and BIC (right) and the resulting MSEs of the corresponding test set predictions.

are also removed from the inputs (i.e., the past values from which predictions are to be made) at the same probability. The MSE is then calculated as the error between the forecast and the actual time series (with no values removed).

The training and test MSE for 10% missing values are shown in Fig. 2. The behaviour is similar to the corresponding plots in Fig. 1, although the difference in the testing MSE appears more pronounced, and for a lower number of components. This supports the notion that the constraints help against overfitting.

Fig. 3 shows the number of components selected by AIC and BIC, and the corresponding test MSEs, for various degrees of missing values. As expected, the forecasting accuracy deteriorates with an increasing ratio of missing data. The number of components selected by the AIC remains largely constant, and the constrained model consistently performs better. The BIC, on the other hand, seems to selected far too few components for the constrained model (the MSE plots in Figs. 1 and 2 suggest five components are far from sufficient), resulting in a reduced forecasting accuracy.

Figs. 1 and 3 reveal largely similar results between using AIC and BIC for the unconstrained case. However, for the constrained model, BIC is clearly too restrictive, and using AIC leads to more accurate results.

5   Conclusions

Time series modelling through Gaussian mixture models is an appealing method, capable of accurate short-to-medium term prediction and missing value interpolation. Certain restrictions on the structure of the model arise naturally through the modelling setup, and appropriately including these constraints in the modelling procedure further increases its accuracy. The constraints are theoretically justified, and experiments support their utility. The effect is negligible when there are enough samples or few components such that fitting a mixture model is easy, but in more challenging situations with a large number of components or missing values they considerably reduce the risk of overfitting.

References

1. McLachlan, G., Peel, D.: Finite Mixture Models. Wiley Series in Probability and Statistics. John Wiley & Sons, New York (2000)
2. Kantz, H., Schreiber, T.: Nonlinear Time Series Analysis. Cambridge nonlinear science series. Cambridge University Press (2004)
3. Dempster, A.P., Laird, N.M., Rubin, D.B.: Maximum likelihood from incomplete data via the EM algorithm. Journal of the Royal Statistical Society 39(1), 1–38 (1977)
4. McLachlan, G., Krishnan, T.: The EM Algorithm and Extensions. Wiley Series in Probability and Statistics. John Wiley & Sons, New York (1997)
5. Akaike, H.: A new look at the statistical model identification. IEEE Transactions on Automatic Control 19(6), 716–723 (1974)
6. Schwarz, G.: Estimating the dimension of a model. The Annals of Statistics 6(2), 461–464 (1978)
7. Anderson, T.W.: An Introduction to Multivariate Statistical Analysis, 3rd edn. Wiley-Interscience, New York (2003)
8. Ghahramani, Z., Jordan, M.: Learning from incomplete data. Technical report, Lab Memo No. 1509, CBCL Paper No. 108, MIT AI Lab (1995)
9. Hunt, L., Jorgensen, M.: Mixture model clustering for mixed data with missing information. Computational Statistics & Data Analysis 41(3-4), 429–440 (2003)
10. Little, R.J.A., Rubin, D.B.: Statistical Analysis with Missing Data. 2nd edn. Wiley-Interscience (2002)
11. Gershenfeld, N., Weigend, A.: The Santa Fe time series competition data (1991), http://www-psych.stanford.edu/~andreas/Time-Series/SantaFe.html

When Does Active Learning Work?

Lewis P.G. Evans[1], Niall M. Adams[1,2], and Christoforos Anagnostopoulos[1]

[1] Department of Mathematics, Imperial College London
[2] Heilbronn Institute for Mathematical Research, University of Bristol

Abstract. Active Learning (AL) methods seek to improve classifier performance when labels are expensive or scarce. We consider two central questions: Where does AL work? How much does it help? To address these questions, a comprehensive experimental simulation study of Active Learning is presented. We consider a variety of tasks, classifiers and other AL factors, to present a broad exploration of AL performance in various settings. A precise way to quantify performance is needed in order to know when AL works. Thus we also present a detailed methodology for tackling the complexities of assessing AL performance in the context of this experimental study.

Keywords: classification, active learning, experimental evaluation, algorithms.

1 Introduction

Active Learning (AL) is an important sub-field of classification, where a learning system can intelligently select unlabelled examples for labelling, to improve classifier performance. The need for AL is often motivated by practical concerns: labelled data is often scarce or expensive compared to unlabelled data [9].

We consider two central questions: Where does AL work? How much does it help? These questions are as yet unresolved, and answers would enable researchers to tackle the subsequent questions of how and why AL works.

Several studies have shown that it is surprisingly difficult for AL to outperform the simple benchmark of random selection ([3,8]). Further, both AL methods and random selection often show high variability which makes comparisons difficult. There are many studies showing positive results, for example [9,5]. Notably there are several studies showing negative results, for example [2,8]. While valuable, such studies do not permit any overview of where and how much AL works. Moreover, this contradiction suggests there are still things to understand, which is the objective of this paper.

We take the view that a broader study should try to understand which factors might be expected to affect AL performance. Such factors include the classification task and the classifier; see Section 2.3. We present a comprehensive simulation study of AL, where many AL factors are systematically varied and subsequently subjected to statistical analysis.

Careful reasoning about the design of AL experiments raises a number of important methodological issues with the evaluation of AL performance. This

A. Tucker et al. (Eds.): IDA 2013, LNCS 8207, pp. 174–185, 2013.

paper contributes an assessment metholodology in the context of simulation studies to address those issues.

For practical applications of AL, there is usually no holdout test dataset with which to assess performance. That creates major unresolved difficulties, for example the inability to assess AL method performance, as discussed in [8]. Hence this study focusses on simulated data, so that AL performance can be assessed.

The structure of this paper is as follows: we present background on classification and AL in Section 2, then describe the experimental method and assessment methodology in Sections 3 and 3.1. Finally we present results in Section 4 and conclude in Section 5.

2 Background

This section presents the more detailed background on classification and AL.

2.1 Classification

Notationally, each classification example has features $\mathbf{x}_i$ and a corresponding label y_i. Thus each example is denoted by $\{\mathbf{x}_i, y_i\}$, where $\mathbf{x}_i$ is a p-dimensional feature vector, with a class label $y_i \in \{C_1, C_2, ..., C_k\}$.

A dataset consists of n examples, and is denoted $D = \{\mathbf{x}_i, y_i\}_1^n$. A classifier is an algorithm that predicts classes for unseen examples, with the objective of good generalisation on some performance measure. A good overview of classification is provided by [7, Chapter 1,2].

2.2 Active Learning

The context for AL is where labelled examples are scarce or expensive. For example in medical image diagnosis, it takes doctors' valuable time to label images with their correct diagnoses; but unlabelled examples are plentiful and cheap. Given the high cost of obtaining a label, systematic selection of unlabelled examples for labelling might improve performance. An AL method can guide selection of the unlabelled data, to choose the most useful or informative examples for labelling. In that way the AL method can choose unlabelled data to best improve the generalisation objective. A small set of unlabelled examples is first chosen, then presented to an expert (*oracle*) for labelling.

Here we focus on batch AL; for variations, see [9,4]. A typical scenario would be a small number of initially labelled examples, a large pool of unlabelled examples, and a small budget of label requests. An AL method spends the budget by choosing a small number of unlabelled examples, to receive labels from an oracle.

An example AL method is uncertainty sampling using Shannon Entropy (denoted SE). SE takes the entropy of the whole posterior probability vector for all classes. Informally, SE expresses a distance metric of unlabelled points from the classifier decision boundary.

$$Entropy = -\sum_{i=1}^{k} p(y_i|\mathbf{x}_j) \times \log(p(y_i|\mathbf{x}_j)).$$

Another example AL method is Query By Committee (denoted QBC), described in section 4.2.

The oracle then satisfies those label requests, by providing the labels for that set of unlabelled examples. The newly-labelled data is then combined with the initially-labelled data, to give a larger training dataset, to train an improved classifier.

The framework for AL described above is batch pool-based sampling; for variations see [4,9].

2.3 Active Learning Factors

Intuitively there are several factors that might have an important effect on AL performance. An experimental study can vary the values of those factors systematically to analyse their impact on AL performance.

One example of an AL factor is the nature of the classification task, including its difficulty and the complexity of the decision boundary. The classifier can be expected to make a major difference, for example whether it can express linear and non-linear decision boundaries, and whether it is parametric. The smoothness of the classification task input, for example continuous or discretised, might prove important since that smoothness affects the diversity of unlabelled examples in the pool. Intuitively we might expect a discretised task to be harder than a continuous one, since that diversity of pool examples would decrease. Other relevant factors include the number of initial labels ($N_{initial}$) and the size of the label budget (N_{budget}).

Some of these factors may be expected to materially determine AL performance. How the factors affect AL performance is an open question. This experimental study evaluates AL methods for different combinations of factor values, i.e. at many points in factor space. The goal here is to unravel how the factors affect AL performance. A statistical analysis of the simulation study reveals some answers to that question, see Section 4.1.

Below the factor values are described in detail.

Four different simulated classification tasks are used, to vary the nature and complexity of the classification problem. We restrict attention to binary classification problems. Figure 1 shows the classification tasks. These tasks are created from mixtures of Gaussian clusters. The clusters are placed to create decision boundaries, some of which are simple curves and others are more involved. In this way the complexity of the classification problem is varied across the tasks.

Still focussing on the classification task, task difficulty is varied via the Bayes Error Rate (BER). Input smoothness is also varied, having the values continuous,

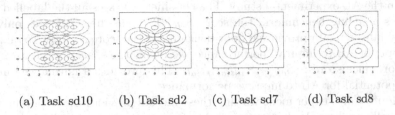

(a) Task sd10   (b) Task sd2   (c) Task sd7   (d) Task sd8

Fig. 1. Density contour plots to elucidate the classification problems

discretised, or a mixture of both. BER is varied by modifying the Gaussian clusters for the problems; input smoothness is varied by transforming the realised datasets.

Another factor to vary is the input dimension p, by optionally adding extra dimensions independent of the class. An interaction is expected between p and the initial amount of labelled data $N_{initial}$, since higher dimensional data should require more datapoints to classify successfully.

Four different classifiers were used: Logistic Regression (LogReg), Quadratic Discriminant Analysis (QDA), Random Forest (RF) and Support Vector Machines (SVM), to provide a variety of classifiers: linear and non-linear, parametric and non-parametric. These classifiers are described in [7]. The default parameters for RF are the defaults from R package RandomForest version 4.6-7; the default parameters for SVM are the defaults from R package RWeka version 0.4-14 (the complexity parameter C is chosen by cross-validation, the kernel is polynomial).

The amount of initial labelled data $N_{initial}$ is also varied. This factor is expected to be important, since too little data would give an AL method nothing to work with, and too much would often mean no possible scope for improvement.

The AL factors are summarised in Table 1.

Table 1. Active Learning Factors

Name	Values
Classification Task	sd10, sd2, sd7, sd8 (see Figure 1)
Task Input Type	Continuous, Discretised, Mixed
Task Input Dimension	2, 10
Classifier	LogReg, QDA, RF, SVM
$N_{initial}$	10, 25, 50, 100
Bayes Error Rate	0.1, 0.2, 0.35
Classifier Optimum Error Rate	[inferred]
Space for AL	[inferred]

The optimum error rate for the classifier on a specific task is evaluated experimentally, by averaging the results of several large train-test datasets, to provide a ceiling benchmark.

We also consider the potential space for AL to provide a performance gain. In the context of simulated data all labels are known, and some labels are hidden to perform the AL experimental study. The classifier that sees all the labelled data provides a ceiling benchmark, the score S_{all}. The classifier that sees only the initially labelled data provides a floor benchmark, the score $S_{initial}$. To quantify the scope for AL to improve performance, we define the space for AL as a ratio of performance scores: $(S_{all} - S_{initial})/S_{all}$. This provides a normalised metric of the potential for AL to improve performance.

A Monte Carlo experiment varying these factors provides the opportunity to statistically analyse the behaviour of AL. To get to this point, both a careful experiment and a refined methodology of performance assessment are required.

3   Experimental Method

AL is applied iteratively in these experiments: the amount of labelled data grows progressively, as the AL method spends a budget chunk at each time point. We may choose to spend our overall budget all in one go, or iteratively, in smaller batches.

In that sense the experimental setup resembles that of the AL challenge described in [5]. We use this iteration over budget because it is realistic for practical AL applications, and because it explores the behaviour of AL as the number of labelled examples grows. Experiments consider AL methods SE and QBC.

To motivate our experimental method, we present the summary plots of the relative performances of AL and RS over time, see Figures 2a and 2b.

The experimental setup is as follows. Firstly, sample a pair of datasets $[D_{train}, D_{test}]$ from the classification task. To simulate label scarcity, split the training dataset into initally labelled data $D_{initial}$ and an unlabelled pool D_{pool}.

The output of one experiment can be described in a single plot, for example Figure 2a. That figure shows the trajectory of performance scores obtained from progressive labelling, as follows. At each time point the AL method chooses a small set of examples for labelling, which is added to the existing dataset of labelled data. This selection happens repeatedly, creating a trajectory of selected datasets from the unlabelled pool. Each time point gives a performance score, for example error rate, though the framework extends to any performance metric. This gives the overall result of a trajectory of scores over time, denoted S_i: an empirical learning curve. Here i denotes the time point as we iteratively increase the amount of labelled data, with $i \in [0, 100]$.

Given several instances of RS, we form an empirical boxplot, called a sampling interval. Figure 2a shows the trajectory of scores for the AL method, and the vertical boxplots show the sampling intervals for the scores for RS.

Once this iterative process is done, we obtain a set of scores over the whole budget range, denoted S_i. Those scores are used to calculate various performance comparisons, specifically to see whether AL outperformed RS, see Section 3.1.

The AL method now has a score trajectory S_i: a set of scores over the whole budget range. All trajectories begin at the floor benchmark score $S_{initial}$ and terminate at the ceiling benchmark score S_{all}. From the score trajectory S_i a set of score differences δS_i is calculated via $\delta S_i = S_i - S_{i-1}$. The need for and

usage of the score differences is detailed in Section 3.1. The chosen AL method
is evaluated alongside several instances of RS, the latter providing a benchmark.
Experiments are repeated to generate several instances of RS, since RS shows
substantial variability.

To illustrate the trajectories of the performance scores S_i, Figure 2a shows
those scores for the AL method SE and comparison with RS.

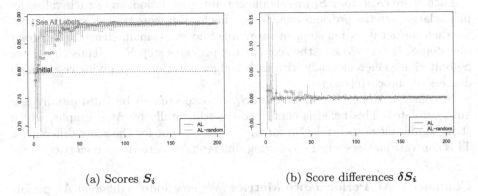

(a) Scores S_i (b) Score differences δS_i

Fig. 2. Scores S_i in subgraph (a) and score differences δS_i in subgraph (b), for AL
method Shannon Entropy vs Random Selection

3.1   Methodology to Evaluate AL Performance

This section elucidates the difficulties with existing AL performance metrics, and
contributes a novel assessment metholodology to address those complexities. The
primary goal of every AL performance metric is to quantify the AL performance
gain from a given experiment, as a single scalar summary.

Any AL performance methodology must first address two preliminary issues:
the benchmark for AL to outperform, and how to handle the variability of that
benchmark. The first issue is to decide which benchmark should be used to
compare AL methods against. One option is to compare AL performance to the
initial classifier. However, that ignores the fact that the labelled dataset is larger
in the case of AL: even random selection of further examples for labelling would
be expected to improve performance on average, since the classifier sees a larger
training dataset. Thus a better benchmark for AL is random selection (RS), that
sees exactly the same amount of labelled data as the AL method.

The second issue concerns the high variability of the benchmark, given that
experiments show RS to have high variability. The approach used here is to
evaluate multiple instances of RS, to get a reasonable estimate of both location
and dispersion of performance score. From those multiple instances we can form
a sampling interval of the RS score, and thus capture its high variability.

Score Trajectories under Experimental Budget Iteration. Having established the benchmark of RS, we consider the score trajectories in the exper-

imental context of budget iteration, to better understand how to compare AL against its benchmark.

We begin with the score trajectories S_i derived from the budget iteration process. The budget is iterated over the entire pool in 100 steps; during that iteration, the amount of labelled data grows from its minimum $N_{initial}$ to its maximum N_{train}. At each budget iteration step, the available budget is small compared to the total size of the pool. This is illustrated in Figure 2a.

Each score trajectory S_i has significant autocorrelation, since each value depends largely on the previous one. To see this for the score trajectory, recall that for each budget iteration step, the available budget is small. Hence the score at one step S_i is very close to the score at the previous step S_{i-1}. Thus the scores S_i only change incrementally with each budget iteration step, giving rise to high degrees of autocorrelation.

In contrast, the score differences δS_i are expected to be substantially less autocorrelated. This belief is confirmed experimentally by ACF graphs, which show significant autocorrelation for the scores but not for the score differences. This contrast matters when comparing different AL performance metrics.

Comparing AL Performance Metrics. We now address different AL performance metrics, each designed to measure the performance of AL methods. Two common AL performance metrics are direct comparisons of the score trajectories, and the Area Under the Active learning curve (AUA) (see [5]).

The autocorrelation of score trajectories S_i means that directly comparing two score trajectories is potentially misleading. For example, if an AL method does well against RS only for a small time at the start, and then does equally well, this would lead to the AL method's score trajectory dominating that of the RS over the whole budget range. This would present a false picture of where the AL method is outperforming RS. Much of the AL literature suggests that this early AL performance zone is precisely to be expected ([9]), and thus this comparison may often be partially flawed. Further, this same case shows that the AUA (see [5]) would overstate the AL performance gain; see Figure 2a which shows the score trajectories.

Here we resolve that problem by considering the score differences δS_i, not the scores themselves S_i. Those differences show much less autocorrelation than the scores (this is shown by ACF graphs).

An example of the score differences δS_i is shown in Figure 2b. Our new methodology is based on examining these score differences.

A New Methodology to Evaluate AL Performance. Our new methodology is based on comparing the score differences δS_i, as a way to compare AL against its benchmark RS. This is done in two stages.

The first stage is to seek a function that quantifies the result of the comparison between two score differences, δS_i^{SE} for AL method SE and δS_i^{RS} for RS. To ensure fair comparisons, ties need to be scored differently to both wins and losses. The approach adopted here is to use a simple comparison function f:

$$f(x,y) = \begin{cases} 1 & : x > y \\ 0.5 & : x = y \\ 0 & : x < y. \end{cases}$$

This comparison function f is applied to two score differences, e.g. $f(\delta S_i^{SE}, \delta S_i^{RS})$. The motivation here is to carefully distinguish wins, losses and ties, and to capture those three outcomes in one scalar summary. Applying that comparison function to compare all the score differences of SE and RS generates a set of comparison values, denoted C_i, each value $\in [0,1]$. Several instances of RS generate several such sets of values, one for each instance.

We use several instances of RS to capture its high variability, the number of RS instances being N_{RS}. Each instance j has its own set of comparison scores C_i^j. Those comparison values C_i^j are then averaged to form a single set of averaged comparison values, denoted $A_i = \frac{1}{N_{RS}} \sum_{j=1}^{N_{RS}} C_i^j$. Further, each value $A_i \in [0,1]$.

That single set of values A_i provides a summary of the overall performance comparison between the AL method and RS. That comparison is illustrated in Figure 3 which shows those average comparison values A_i over the whole budget range.

The final stage of the new method is interpreting the averaged comparison values A_i. The aim is to extract the relationship between A_i and budget, with a confidence interval band.

The lower 80% confidence interval is chosen to form a mildly pessimistic estimate of AL performance gain. We fitted a Generalised Additive Model (GAM) to this set of values (given the need for inference of confidence intervals). The GAM is chosen using a logit link function, with variable dispersion to get better confidence intervals under potential model mis-specification (see [6]). The GAM is implemented by R package mgcv version 1.7-22; the smoother function default is thin plate regression splines. The GAM relates the expected value of the distribution to the covariates thus:

$$g(\mathrm{E}(Y)) = \beta_0 + \beta_1 f_1(x_1).$$

The fitted GAM is shown in Figure 3. The estimated effect seems roughly linear. The baseline level of 0.5 is shown as a dotted line, which represents an AL method that ties with RS, i.e. does not outperform it.

Given the intricacies of evaluating AL performance, a primary goal for this methodology is to quantify AL performance from a given experiment as a single result. The GAM curve shows where the AL performance zone is, namely the initial region where the curve is significantly above 0.5. We consider the initial region, as much AL literature suggests that the AL performance gain occurs early in the learning curve, see [9]. Thus the length of the AL performance zone is the single result that summarises each experiment.

Overall, this methodology addresses some of the complexities of assessment of AL performance in simulation contexts. As such it provides a milestone on the road to more accurate and statistically significant measurements of AL performance. This is important given that many authors find that the AL performance effect can be elusive (e.g. [5,3]).

This methodology is illustrated with specific results in Figures 2a, 2b and 3.

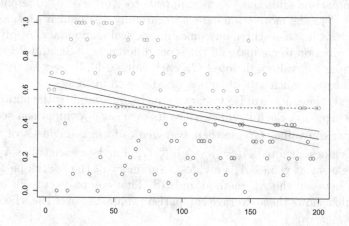

Fig. 3. Averaged Comparison Values A_i with Generalised Additive Model curve and pointwise 80% confidence interval

4   Results and Discussion

The dependent variable is the AL performance zone length, an integer count. That value is obtained via the methodology described above, which includes fitting a GAM to ensure statistical significance. The factors are given in Table 1.

4.1   Negative Binomial Regression Analysis

The experimental output includes the AL performance zone length (derived from the GAM) to the AL factors. Given the form of the aggregate experimental output, the appropriate initial analyses were Poisson and Negative-Binomial regression. A Poisson regression model was found to be over-dispersed. We fit a negative binomial regression generalised linear model, which fits reasonably well with significant under-dispersion:

$$Y_i \sim \text{NegBin}(\mu_i, \kappa)$$

with

$$log(\mu_i) = \mathbf{x}_i \cdot \boldsymbol{\beta}$$

where κ is a dispersion parameter.

The significant results of that model are summarised in Table 2.

Table 2. Negative Binomial Significant Results

Name	Coefficient	p-value
Intercept	−1.695	1.70e-12
Specific Classifier, Logistic Regression	1.142	<2e-16
Specific Task, labelled sd7	−0.481	0.000405
Input Type, Continuous	0.578	1.11e-09
Input Type, Discrete	−1.235	<2e-16

There are several results from the negative binomial regression which were not anticipated. For example, LogReg shows more improvement, all things being equal, than SVM, for AL method SE.

This may be due to classifier mis-match: one might conjecture that AL works better when the classifier is mis-matched to the task, because the range of example quality within the unlabelled pool might be much higher under mis-match.

Here classifier mis-match means the experimental metric of the classifier's sub-optimality on a given task. Classifier mis-match is the performance difference between this classifier and the optimal Bayes classifier on the task. Informally, mis-match measures how ill-suited a classifier is to a given task.

Under correct classifier match, most examples will improve a classifier's performance, whereas under mis-match, some examples may reduce the performance while others improve it, leading to a greater range of example quality under mis-match.

The choice of task is significant: the third task is worse than the fourth. The fourth task has a more complex decision boundary than the third, leading to expected greater model mis-match for this task. The fact that the third is worse for AL than the fourth is also consistent with the conjecture described above, that AL works better under mis-match.

There is a widespread belief in the AL literature that the AL performance zone is early in the budget range (see [9]). In other words, as we progressively increase the amount of the labelled data, AL provides its performance gain earlier more than later. AL methods are expected to select the more useful examples from the pool, and the greatest range of usefulness would exist early on. In practical applications, AL is usually required work earlier rather than later, since the essential context of AL is label scarcity. This belief is confirmed by the analysis: for the experiments that showed an AL performance gain, the mean and median lengths of the AL performance zone length were 38 and 32 respectively, out of a maximum of 200.

It is notable that input dimension turns out not to be significant.

It was quite rare for AL to show a performance gain at all, compared to RS, only in around 11% of experiments. This confirms existing studies that the AL performance gain is often elusive ([2,3,8]). It also emphasises the clear need for a precise reasoned methodology to analyse AL performance, hence the detailed methodology described in Section 3.1.

4.2 Results from QBC

To explore the importance of the AL method used, experiments evaluated a different AL method, QBC, using average KL-divergence as the disagreement measure. The two AL methods SE and QBC are very different in both algorithmic details and overall motivation (see [9,4]), making it worthwhile to compare their results.

The AL method QBC takes a committee of classifiers, and scores an unlabelled example $\mathbf{x}_j$ by how much disagreement there is within the committee. Disagreement measures include Vote Entropy and Average K-L Divergence; see [4,9]. For QBC the classifier committee was Logistic Regression, k-nearest-neighbour (with $k = 5$ and $k = 21$), Support Vector Machine and Random Forest.

The experimental setup was identical, and the results were analysed in the same way: by a negative binomial regression analysis. That model fits reasonably well. The results from QBC are somewhat different to those from the SE.

The QBC analysis confirms that input type is significant, with continuous input giving significantly greater AL performance than mixed; and mixed significantly greater than discrete. This confirms that a discretised task is harder than a continuous one, with discretisation reducing the the diversity of pool examples.

It is interesting that two very different AL methods lead to similar results for how AL performance depends on specific factors. We may explain this behaviour in part as follows. With Active Learning there are two distinct stages: firstly the selection of examples for labelling, and secondly the use of those examples in training a particular classifier. With SE the same classifier is used for both stages, whereas QBC uses a classifier committee for selection. The QBC results found that classifier was not significant, in contrast to the SE results which found that Logistic Regression is significantly better than SVM.

This suggests that QBC may be selecting examples which are useful independently of the classifier: good datapoints which benefit any classifier. That in itself is interesting, as it is a very plausible prior belief that the quality of datapoints would be strongly classifier dependent.

5 Conclusion

There are two central questions: Where does AL work? How much does it help? By examining a variety of experiments across a range of points in AL factor space, some conclusions can be drawn.

Overall AL failed to demonstrate a performance gain far more often than not (11% for SE, 6% for QBC). This is consistent with several other authors who reported largely negative results using AL ([1,8]). The analysis also confirmed the general belief in the literature that AL provides its performance gain early on in the budget range. Both AL methods, SE and QBC, showed that the smoothness of the input type makes a significant difference to AL performance.

In future we will extend this work, for example by including many more datasets, some from simulated data, other from real applications, e.g. [5]. Future results should enable recommendations of AL method for applications, by relating the type of classification task to the relative performances of different AL methods.

This experimental study has generated some unexpected results about the factors that determine where AL works. This study has shown many complexities with the assessment of AL performance. It has contributed a new methodology to assess AL performance.

Acknowledgement. The work of Lewis P. G Evans is supported by an EPSRC doctoral training award.

References

1. Bach, F.: Active learning for misspecified generalized linear models. Technical report, Centre de Morphologie Mathematique (2006)
2. Baldridge, J., Osborne, M.: Active learning and the total cost of annotation. In: Proceedings of Conference on Empirical Methods in Natural Language Processing, vol. 1, pp. 9–16 (2004)
3. Cawley, G.: Baseline methods for active learning. Journal of Machine Learning Research Workshop and Conference Proceedings 15, 47–57 (2011)
4. Fu, Y., Zhu, X., Li, B.: A survey on instance selection for active learning. Knowledge & Information Systems 35, 249–283 (2013)
5. Guyon, I., Cawley, G., Dror, G., Lemaire, V.: Results of the active learning challenge. Journal of Machine Learning Research 16, 19–45 (2011)
6. Hastie, T., Tibshirani, R.: Generalized Additive Models. Chapman & Hall / CRC (1990)
7. Hastie, T., Tibshirani, R., Friedman, J.: The Elements of Statistical Learning, 2nd edn. Springer (2009)
8. Provost, F., Attenberg, J.: Inactive learning difficulties employing active learning in practice. ACM SIGKDD 12, 36–41 (2010)
9. Settles, B.: Active learning literature survey. Computer Sciences Technical Report 1648, University of Wisconsin–Madison (2009)

OrderSpan: Mining Closed Partially Ordered Patterns

Mickaël Fabrègue[1,2], Agnès Braud[2], Sandra Bringay[3], Florence Le Ber[2],
and Maguelonne Teisseire[1]

[1] TETIS, IRSTEA, Montpellier, France
{mickael.fabregue,maguelonne.teisseire}@teledetection.fr
[2] ICube, Strasbourg University, CNRS, ENGEES, Illkirch France
florence.leber@engees.unistra.fr, agnes.braud@unistra.fr
[3] LIRMM, Montpellier 3 University, CNRS, France
bringay@lirmm.fr

Abstract. Due to the complexity of the task, partially ordered pattern mining of sequential data has not been subject to much study, despite its usefulness. This paper investigates this data mining challenge by describing OrderSpan, a new algorithm that extracts such patterns from sequential databases and overcomes some of the drawbacks of existing methods. Our work consists in providing a simple and flexible framework to directly mine complex sequences of itemsets, by combining well-known properties on prefixes and suffixes. Experiments were performed on different real datasets to show the benefit of partially ordered patterns.

Keywords: Data Mining, Sequential patterns, Partial orders.

1 Introduction

Sequential pattern mining is a very active research area, which is linked to the exponential growth of temporal and spatio-temporal databases. Many studies have demonstrated the usefulness of such patterns for analysis [1], classification [2][3] or prediction [4]. These patterns were introduced in [5] and are an extension of association rules [6]. Information is totally ordered according to a specific criterion, which is most often temporal. For instance, let us take an environmental database where the pattern ⟨(*Pollution*)(*Dead_animals*)⟩ is found. This means that the event *Pollution* is temporally followed by the event *Dead_animals*. Mining such related items according to temporal aspects is very useful for specialists in various domains such as software engineering [7], medicine [8] or marketing [9]. Despite their advantages, sequential patterns have often limited value since they only provide totally ordered information about data. For example, let us consider a second pattern, ⟨(*Pollution*)(*Dead_vegetation*)⟩, discovered in the same database. The coexistence of the two patterns is not taken into account using this method. However, they can be synthesized based on partial ordering.

Figure 1 presents a so-called partially ordered pattern that combines the two previous sequential patterns. This pattern means that the pollution event is followed

A. Tucker et al. (Eds.): IDA 2013, LNCS 8207, pp. 186–197, 2013.
© Springer-Verlag Berlin Heidelberg 2013

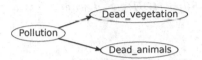

Fig. 1. Example of partially ordered pattern

by two other events *Dead_animals* and *Dead_vegetation* which themselves are not
ordered. We focus our study on mining closed partially ordered patterns since they
reduce information redundancy and limit result size. Furthermore, in the general
case, partially ordered patterns can be used in all kinds of sequential databases and
have many other advantages: (1) they provide more detailed information on order
among elements; (2) they are represented using a directed acyclic graph, which is
easy to understand; (3) they summarize sequential pattern sets. In this paper, we
present a new method designed to directly extract partially ordered patterns in a
general case of sequences of itemsets with item repetitions. We propose an algorithm called OrderSpan that explores the search space and mines the complete set
of closed partially ordered patterns. Relying on existing work on sequential pattern
mining, OrderSpan performs the extraction of partially ordered patterns based on
the prefix and suffix properties of sequences.

    This paper is organized as follows. Section 2 gives some preliminar definitions
on sequences and partially ordered patterns. Section 3 studies existing works
on partially ordered pattern mining. Section 4 defines the OrderSpan algorithm
and its different steps. The experimental results are presented in Section 5. We
conclude our study while providing some prospects in Section 6.

2   Problem Definition

Before presenting the notion of partially ordered pattern, we provide some important definitions relative to closed sequential pattern mining. As we will see
later, a partially ordered pattern is a more open structure composed of closed
sequential patterns.

Definition 1 *(Sequence)*
*Let $\mathcal{I} = \{I_1, I_2, \ldots, I_m\}$ be a set of **items**. An **itemset** IS is a non empty,
unordered, set of **items** denoted as $(I_{j_1} \ldots I_{j_k})$ where $I_{j_i} \in \mathcal{I}$. Let $\mathcal{IS}$ be the
set of all **itemsets** built from $\mathcal{I}$. A **sequence** S is a non empty ordered list of
itemsets denoted as $\langle IS_1 IS_2 \ldots IS_p \rangle$ where $IS_j \in \mathcal{IS}$.*

Definition 2 *(Sub-sequence)*
*A **sequence** $S_\alpha = \langle IS_1 IS_2 \ldots IS_p \rangle$ is a **sub-sequence** of another **sequence**
$S_\beta = \langle IS'_1 IS'_2 \ldots IS'_m \rangle$, denoted as $S_\alpha \preceq_s S_\beta$, if $p \leq m$ and if there are integers
$j_1 < j_2 < \ldots < j_k < \ldots < j_p$ such that $IS_1 \subseteq IS'_{j_1}, IS_2 \subseteq IS'_{j_2}, \ldots, IS_p \subseteq IS'_{j_p}$.*

Definition 3 *(Sequential pattern)*
*Let SP be a **sequence**. Let $\mathcal{S}'$ be the set of **sequences** such that $\forall S_i \in \mathcal{S}', SP \preceq_s
S_i$. $|\mathcal{S}'|$ is called the support of SP. SP is called a **sequential pattern**, denoted*

by **seq-pattern**, when $Support(SP) \geq \theta$ where θ is a given value (minimum support).

Definition 4 *(Closed sequential pattern)*
Let SP be a **sequential pattern**, SP is a **closed sequential pattern** if there is no other **sequential pattern** SP' such that $SP \preceq_s SP'$ and $Support(SP) = Support(SP')$.

Table 1. An example of a database of sequences

ID	Sequence
S_1	$\langle (af)(d)(e)(a) \rangle$
S_2	$\langle (e)(abf)(g)(bde) \rangle$
S_3	$\langle (e)(a)(b)(g) \rangle$

Table 2. Set of closed sequential patterns with the minimum support $\theta = 2$

ID	Sequence
$\{S_1, S_2, S_3\}$	$\langle (e)(a) \rangle$
$\{S_1, S_2\}$	$\langle (af)(d) \rangle$
$\{S_1, S_2\}$	$\langle (af)(e) \rangle$
$\{S_2, S_3\}$	$\langle (e)(a)(b) \rangle$
$\{S_2, S_3\}$	$\langle (e)(a)(g) \rangle$
$\{S_2, S_3\}$	$\langle (e)(b)(g) \rangle$

To illustrate these different aspects, we use the database in Table 1 as a reference example. This database contains three different sequences of itemsets based on the alphabet $\Sigma = \{a, b, d, e, f, g\}$. Given this database with a minimum support $\theta = 2$, the sub-sequences $\langle (e) \rangle$, $\langle (a) \rangle$ and $\langle (e)(a) \rangle$ are supported by sequences S_1, S_2 and S_3, and their support is equal to 3. Thus, these sub-sequences are seq-patterns because $3 \geq \theta$. But sequences $\langle (e) \rangle$ and $\langle (a) \rangle$ are not closed sequences since the sequence $\langle (e)(a) \rangle$ is such as $\langle (e) \rangle \preceq_s \langle (e)(a) \rangle$ and $\langle (a) \rangle \preceq_s \langle (e)(a) \rangle$ with an equivalent support. Therefore, these two sequences are redundant and can be skipped during pattern extraction to only retain closed sequences. Finally, the sequences $\langle (e)(a) \rangle$, $\langle (af)(d) \rangle$, $\langle (af)(e) \rangle$, $\langle (e)(a)(b) \rangle$, $\langle (e)(a)(g) \rangle$ and $\langle (e)(b)(g) \rangle$ in Table 2, give the complete set of closed seq-patterns. These closed sequential patterns can be extracted using an algorithm such that BIDE [10] or ClopSpan [11]. Below, we highlight the link between seq-patterns and partially ordered patterns, this permits us to convert this sequence mining problem into a directed acyclic graph mining problem.

Table 2 gives the complete set of closed seq-patterns with $\theta = 2$ and their associated set of supporting sequences. Some sets of closed seq-patterns are supported by the same set of sequences. For instance $\langle (e)(a)(b) \rangle$, $\langle (e)(a)(g) \rangle$ and $\langle (e)(b)(g) \rangle$ are supported by S_2 and S_3. Thus, a partial order can be defined to obtain a synthetic representation of these closed seq-patterns relative to the sequence set $\{S_1, S_2\}$. We can define as many partial orders as there are corresponding sets of sequences. In our example, these sets are $\{S_1, S_2, S_3\}$, $\{S_1, S_2\}$ and $\{S_2, S_3\}$ respectively. Figures 2, 3 and 4 give the three partial orders corresponding to each set of sequences. We use two vertices labeled "$\langle$" and "$\rangle$", representing the beginning and the end of each pattern.

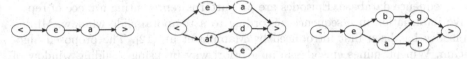

Fig. 2. Closed partial order G_1 on S_1, S_2 and S_3

Fig. 3. Closed partial order G_2 on S_1 and S_2

Fig. 4. Closed partial order G_3 on S_2 and S_3

In the following, these structures are called partially ordered patterns and are defined as follows:

Definition 5 (Partially ordered pattern)
*A **partial order** is a **directed acyclic graph** $G = (\mathcal{V}, \mathcal{E})$. $\mathcal{V}$ is the set of **vertices** and $\mathcal{E}$ is a set of directed **edges** such that $\mathcal{E} \subseteq \mathcal{V} \times \mathcal{V}$. With such a structure, we can determine a partial order on vertices, i.e. $u < v$ if there is a directed path from u to v. However, if there is not a path from u to v, these elements are not comparable. Each path in a partial order is a seq-pattern. Let θ be the minimum support and S' be the set of sequences that support all paths in G, such that $|S'|$ is the support of G. G is considered as a partially ordered pattern, denoted by **po-pattern**, if $|S'| \geq \theta$.*

Definition 6 (Sub-partially ordered pattern)
*A **partially ordered** G_α is a **sub-partially ordered pattern** of another **partially ordered pattern** G_β, denoted $G_\alpha \preceq_g G_\beta$, if for all paths $path_{\alpha i}$ in G_α there is a path $path_{\beta j}$ in G_β such that $path_{\alpha i} \preceq_s path_{\beta j}$.*

Definition 7 (Closed partially ordered pattern)
*Let G be a **partially ordered pattern**, G is a closed **partially ordered pattern** if there is no other **partially ordered pattern** G' such that $G \preceq_g G'$ and $Support(G) = Support(G')$.*

Let us consider the po-pattern G_3 given in Figure 4, which covers the sequences S_2 and S_3. There are two paths between the itemset (e) and the itemset (g) given by the sequences $\langle (e)(b)(g) \rangle$ and $\langle (e)(a)(g) \rangle$, thus $(e) < (g)$ and $Support(G_3) = 2$. Given the po-pattern G_1, we observe that $G_1 \preceq_g G_3$ since the path $\langle (e)(a) \rangle$ in G_1 is included in the paths $\langle (e)(a)(g) \rangle$ and $\langle (e)(a)(b) \rangle$. Po-patterns are separated from each other according to the set of sequences they are included in.

3 Related Work

In the literature, po-patterns mining has been studied in two main contexts. The first involves mining po-patterns as frequent episodes occurring within a single sequence of events. In the second one, po-patterns are mined over a sequence database.

Mining episodes in a single sequence is very different from mining po-patterns in a sequence database. Episodes are po-patterns representing a piece of repetitive information in a sequence, according to a temporal slide window. Mining episodes has been first introduced by Mannila et al. [12]. The proposed algorithm, Winepi, mines episodes in an Apriori way by using a sliding window of fixed width. This method has drawbacks since mining huge databases leads to a significant overhead. In addition, the algorithm does not merely extract closed patterns but the complete set of po-patterns, therefore the number of patterns can be quite large. Some authors [13,14] propose new algorithms to only mine closed episodes. Both use a pattern-growth paradigm in order to explore the search space. They have to keep in computer memory the complete set of extracted patterns to verify if each episode is closed or not, during the process [14] or alternatively during a post-processing step [13].

In [15], Pei et al. studied the problem of mining po-patterns in string databases. Despite the good performance of the Frecpo algorithm, they are able to only extract patterns on simple sequences that have non-repetitive items and no itemsets. This considerably reduces the potential applications of the algorithm as nowadays, temporal databases are composed of multiple types of information for the same timestamp, and the same piece of information can appear several times in a sequence. Alternatively, Garriga [16] presents another algorithm to extract closed po-patterns. It first extracts closed seq-patterns using an algorithm such as CloSpan [11] or BIDE [10], and then performs a postprocessing operation to convert a set of closed seq-patterns into po-patterns. Thus, the algorithm does not directly extract patterns, but uses an existing closed seq-pattern algorithm.

The proposal in this paper shows how to directly mine closed po-patterns in a sequence database using a pattern-growth approach. Our method manages repetitive items and sequences composed of itemsets. In addition, closeness checking is directly performed during the process without having to consider already extracted po-patterns. Furthermore, mining only closed po-patterns gives a smaller set of results without any information loss. The information obtained is also more semantically relevant compared to results obtained with seq-pattern mining.

4   The OrderSpan Algorithm

We now present the OrderSpan algorithm which is designed to meet the previously enumerated challenges inherent to po-pattern mining: (1) mining directly po-patterns from a sequential database; (2) focusing extraction on closed po-patterns, in order to reduce the result size; (3) considering sequences of itemsets with repetitive items. This algorithm relies on a two-phase approach, based on **prefix** and **suffix** properties of sequences. The following subsection presents the Pattern-Growth paradigm which is well-known in seq-pattern mining and is used as a basis for our method.

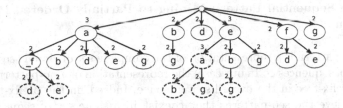

Fig. 5. The prefix tree representing the search space of the database in Table 1 with a minimum support $\theta = 2$

4.1   The Sequential Pattern Mining Paradigm

Several useful methods had been proposed to tackle the problem of mining sequences. We choose to base our algorithm on the Pattern-Growth paradigm which uses a divide-and-conquer approach. This paradigm was first implemented in the PrefixSpan algorithm [17], which is currently one of the most efficient algorithm for extracting seq-patterns, both in terms of computation time and memory consumption.

To illustrate this paradigm, let us consider the tree in Figure 5, which gives us the complete seq-pattern mining search space from the database of Table 1, with $\theta = 2$. Each node represents the last item of a sequence and, starting from the root, it is recursively possible to retrieve the complete set of seq-patterns. The number above each vertex is the support of the corresponding seq-pattern. Thereby, given the sub-tree starting from the labeled vertex 'a' under the root, we obtain the following seq-patterns: $\langle(a)\rangle$, $\langle(af)\rangle$, $\langle(af)(d)\rangle$, $\langle(af)(e)\rangle$, $\langle(a)(b)\rangle$, $\langle(a)(d)\rangle$, $\langle(a)(e)\rangle$, $\langle(a)(g)\rangle$. Two operations are available in this tree, the I-Extension and the S-Extension. Conceptually, given a sequence $S = \langle IS_1 IS_2...IS_p\rangle$ and an item α, the operation $S \diamond \alpha$ concatenates the item α to S. The I-Extension, noted $S \diamond_i \alpha$, concatenates α in the last itemset IS_p of S, e.g. $\langle(a)\rangle \diamond_i f$ gives the sequence $\langle(af)\rangle$. The S-Extension concatenates α in a new itemset following IS_p, e.g. $\langle(af)(e)\rangle$ is a S-Extension of $\langle(af)\rangle$ with e. In Figure 5, a node starting with symbol '_' represents a I-Extension, otherwise it represents a S-Extension.

Based on a depth-first search approach, the Pattern-Growth paradigm recursively divides the database by using database projections. Such projections are made according to a seq-pattern, called a **prefix** [17]. Given a sequence **prefix** p, a projected sequence S from p is noted $S|_p$, that is a **suffix** of S according to the first appearance of the **prefix** p in S. For instance, $\langle(ab)(bc)(ac)(b)\rangle|_{\langle(a)(c)\rangle} = \langle(ac)(b)\rangle$ and $\langle(ab)(bc)(ac)(b)\rangle|_{\langle(a)(b)\rangle} = \langle(\_c)(ac)(b)\rangle$. For each node in the tree, it is possible to build a projection $\mathcal{DB}|_p$ of the complete database according to the pattern p given by the node. Given $\mathcal{DB}$ the sample database in Table 1, $\mathcal{DB}|_{\langle(e)(a)\rangle} = \{\langle\rangle, \langle(\_bf)(g)(bde)\rangle, \langle(b)(g)\rangle\}$. By scanning this projected database, we find the frequent items b and g with a support equal to 2. Thus, $\mathcal{DB}$ will be recursively projected by **prefixes** $\langle(e)(a)(b)\rangle$ and $\langle(e)(a)(g)\rangle$.

4.2 From Sequential Pattern Mining to Partially Ordered Pattern Mining

A closed po-pattern is a representation of the complete set of closed seq-patterns given a set of sequences S. In our case, the representation of a po-pattern is given by a sub-prefix-tree in the overall search space. Indeed, in the prefix-tree, it is easy to retrieve the seq-patterns that coexist in a same set of sequences. Let us consider the po-pattern G_3 (Figure 4). Its sub-prefix-tree representation is given Figure 6. This sub-prefix-tree represents the complete set of seq-patterns appearing in sequences S_2 and S_3 of the sample database given in Table 1. For all seq-patterns SP in the tree, there is at least one path p in G_3 such that $SP \preceq_s p$. Therefore, for each sequence subset in a sequential database, there is a sub-prefix-tree representing all the seq-patterns covering them.

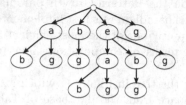

Fig. 6. The sub-prefix-tree representing the po-pattern G_3

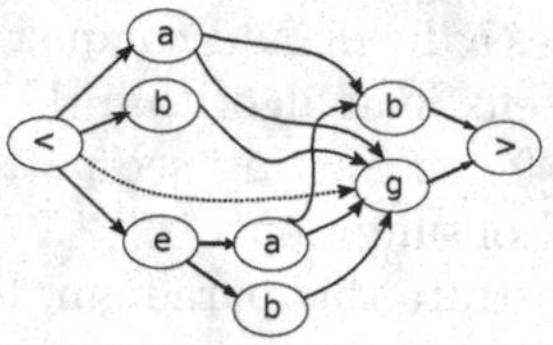

Fig. 7. Merging operation on the sub-prefix-tree corresponding to Figure 6

The first algorithm step is based on this property. Algorithm 1 extracts the complete set of sub-prefix-trees. Let us take a set of sequences S, the algorithm first initializes a sub-prefix-tree which covers all sequences in S. Then, all frequent sub-prefix-trees on sub-sets in S are extracted recursively. This is based on the following assumption: for a sequence sub-set in the database, there is only one closed po-pattern, i.e. a sub-prefix-tree, describing its sequences [16]. To address this issue and prevent redundancy in extracted patterns, we use a data structure called *ListSet*. It contains the list of sequence sub-sets covered by a previously extracted po-pattern.

Lines [1-3]: a directed acyclic graph is initialized. To fit definitions given in Section 2, directed acyclic graphs contain two nodes representing the beginning and the end of the po-pattern, labeled "⟨" and "⟩" respectively. Each node contains: (1) a label representing the extracted information, i.e. an itemset; (2) the projected database containing the suffixes of sequences in S, according to the prefix, i.e seq-pattern, given from the root in the sub-prefix-tree until the node. A queue called *NodeQueue* is initialized with the "begin" node of the po-pattern. **Lines [4-20]:** *NodeQueue* is empty when there are no more nodes to be extended, i.e. the sub-prefix-tree is extracted. **Lines [7-12]:** this part extends the current po-pattern, i.e. sub-prefix-tree, when an occurrence has a support equal to $|S|$ to cover sequences in S. The I-Extension and the S-Extension are performed in line 10. **Lines [13-19]:** discovering occurrences that have a support lower than the sequence set cardinality $|S|$, which means that there is a more

---

Algorithm 1: ForwardTreeMining

    input : S a sequence set, θ a minimum support, *ListSet* the set of sequence
                database sub-sets already explored

1 *PartialOder* $\leftarrow$ new DAG;
2 *PartialOder.begin.database* = new ProjectedDatabase(S);
3 *NodeQueue* $= \emptyset \cup$ *PartialOder.begin*;
4 **while** *NodeQueue is not empty* **do**
5 | *Node* = *NodeQueue.pop_back*;
6 | *ListOcc* $\leftarrow$ getListOccurences(*Node.databse*, θ);
7 | **foreach** *Occ in ListOcc such that Occ.support* $= |S|$ **do**
8 | | N' = *Node.new*;
9 | | N'.*database* = *N.database.projectOn(Occ.item)*;
10 | | Extends the current node N with N';
11 | | *NodeQueue* = *NodeQueue* $\cup N'$;
12 | **end**
13 | **foreach** *Occ in ListOcc such that Occ.support* $< |S|$ **do**
14 | | **if** *ListSet does not contain the set of sequences covered by the*
                    occurrence Occ **then**
15 | | | S' = the set of sequences covered by the occurrence *Occ*;
16 | | | *ListSet* = *ListSet* $\cup S'$;
17 | | | ForwardTreeMining(S', θ, *ListSet*);
18
19 | **end**
20 **end**
21 MergingSuffixTree(PartialOder.end);
22 print(PartialOder);

---

specific po-pattern covering a subset S' such that $S' \subset S$. Therefore, if there is not a po-pattern covering this subset S', a new one is extracted by calling recursively *ForwardTreeMining* on S'. **Line [21]**: after having extracted the complete sub-prefix-tree, the method $MergingSuffixTree$ is applied on the po-pattern in order to prune and merge the redundant vertices. This operation is presented in the next section.

4.3 Pruning and Merging a Sub-suffix-Tree

Since we use the prefix property to mine the complete set of sub-prefix-trees, we obtain a lot of redundancy. Let us consider the sub-prefix-tree given by Figure 6: (1) It represents the complete set of seq-patterns on sequences S_2 and S_3. Some of these seq-patterns are not closed and generate redundant information. For instance the sequences $\langle (a)(b) \rangle$ and $\langle (a)(g) \rangle$ are included in sequences $\langle (e)(a)(b) \rangle$ and $\langle (e)(a)(g) \rangle$, respectively; (2) By starting from the sub-prefix-tree leaves, vertices with the same label can be merged by using the **suffix** property on sequences. Indeed, merging these vertices maintains element order. For example, sequences $\langle (e)(a)(g) \rangle$ and $\langle (e)(b)(g) \rangle$ are both suffixed by the **suffix** $\langle (g) \rangle$, and in the sub-prefix-tree, vertices with the label $\{g\}$ can be merged into only one vertex $\{g\}$.

Then, by starting from the end vertex $\{\rangle\}$ and by using the suffix property on sequences, it is possible to recursively merge all the redundant vertices within a po-pattern, i.e. a sub-prefix-tree.

Figure 7 provides an example of such an operation on the sub-prefix-tree extracted from sequences S_2 and S_3. To illustrate the method, the operation is only performed on the parent nodes of the ending vertex. First, the vertices representing the **suffix** $\langle (b) \rangle$ and labeled $\{b\}$, are merged together by retaining edges from their respective parent nodes. Next, the same operation is performed on the vertices labeled $\{g\}$ for the **suffix** $\langle (g) \rangle$. Once a set of vertices related to a same **suffix** are merged together into one vertex V, this operation is recursively carried out on each parent nodes. Therefore, the operation executed on parent nodes of the merged vertex $\{g\}$ will merge all vertices labeled $\{a\}$ and subsequently all vertices labeled $\{b\}$ which corresponds to **suffixes** $\langle (a)(g) \rangle$ and $\langle (b)(g) \rangle$, respectively.

During this process an issue can emerge. Retaining edges from all parent nodes can generate transitive redundancy in the po-pattern. Indeed in Figure 7, the dotted edge from the beginning to the merged vertex $\{(g)\}$ is redundant since the order information is already given by paths represented by sequences $\langle (b)(g) \rangle$ and $\langle (e)(a)(g) \rangle$. Thus, after having applied the $MergingSuffixTree$ operation on a vertex V, we check for each parent node V' of V, if there is another parent node V'' such that V' is a parent of V''. If this is the case, the edge (V', V) is removed from the po-pattern. Thus, by removing transitive redundancy during the process, this operation ensures that all non closed paths, i.e. non closed seq-patterns, are removed. Using prefix and suffix properties ensures that all shared middle parts of sequences are considered in po-patterns. The $MergingSuffixTree$ operation removes transitive redundancy and, at the same time, recursively merges the commonly shared information.

5   Experiments

In this section, some tests conducted on different kinds of sequential databases are presented. The experiments were performed on a laptop computer with an Intel Core i7 and 8 Gb of main memory, running on Debian stable 7.0. We implemented the OrderSpan algorithm in C++. To illustrate the usefulness of directly mining closed po-patterns in the case of complex sequences, we compared the number of total closed po-patterns with the number of closed seq-patterns. Two real, but very different datasets are chosen. The *BreastCancer* dataset [18] contains 677 gene sequences having a constant length of 128 non-repetitive items. This dataset is very dense as each gene appears in each sequence, therefore, each sequence is composed of 128 different items. A huge set of po-patterns can be extracted at very high support levels such as 90%. The *Fresqueau* dataset[1] is composed of 2505 sequences with a length variability between 1 and 277. It can have repetitive items and itemsets for a maximal length of 856. This dataset is very sparse with 4000 different items and a very heterogeneous distribution.

---

[1] http://engees-fresqueau.unistra.fr/

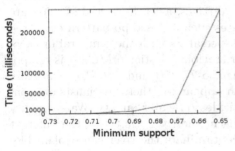
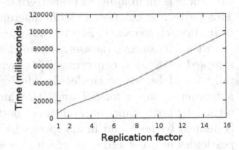

Fig. 8. Performances

Fig. 9. Scalability test

We conducted a performance assessment of OrderSpan on the *Fresqueau* dataset. Figure 8 shows the computation time (in milliseconds) of the algorithm according to the relative minimum support (in percents). Computation cost grows exponentially as the minimum support decreases. For example, given a relative minimum support of 69% on the *Fresqueau* dataset, runtime is equal to 5.892 seconds and increases to 24.964 seconds given a support of 67%. Then, we tested the performance scalability of the algorithm on this dataset. We fixed the minimum support at 69% and we replicated the sequences from 2 to 16 times. Figure 9 shows a linear scalability of OrderSpan given a minimum support equal to 69%. It increases from 5.892 seconds to 97.545 seconds when the database is replicated 16 times, with 97.545/5.892=16.555.

Table 3. Comparing the number of patterns on the two datasets

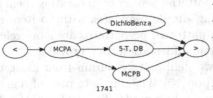

Fig. 10. A po-pattern extracted from the *Fresqueau* dataset

BreastCancer dataset		
θ	Closed po pat.	Closed seq pat.
100%	1	382
95%	234	1678
90%	1297	8733
Fresqueau dataset		
θ	Closed po pat.	Closed seq pat.
72%	14	19
69%	84	133
67%	347	758

Closed po-patterns and closed seq-patterns have been mined over a minimum support θ. Table 3 compares the number of extracted closed po-patterns and closed seq-patterns in the two datasets. When the minimum support θ is varied the number of closed po-patterns is always smaller than the number of closed seq-patterns. This is because closed po-patterns synthesize sets of closed seq-patterns appearing in the same sub-sets of sequences in a database. The experiments show that there is an ubiquitous po-pattern that covers all the *BreastCancer* dataset. This pattern is composed of 128 different vertices, as each of the 128 genes appears in each sequence. Figure 10 illustrates such a closed po-pattern extracted from the *Fresqueau* dataset. It provides specialists with the temporal order of sampled pesticides, concerning 1741 river stations. Pesticide *MCPA* is temporally found before pesticides *DichloBenza*, 5-*T*, *DB* and *MCPB*, which are not ordered except for 5-*T* and *DB* which appear together. Specialists are concerned by environmental impacts of pesticides on water quality. Water quality is defined by flora and fauna species living in the river. Detecting the different pesticides and the order in which they contaminate the river can explain biodiversity changes. Thus, extracting such patterns instead of closed seq-patterns is of great interest for analysts. It provides a better information about the order between elements covering the data, with a smaller result set. Furthermore, with the ability to mine sequences of itemsets, we can extract knowledge from temporal databases such as the *Fresqueau* dataset.

6   Conclusion

Po-pattern mining requires the development of new techniques to extract such patterns in the general context of large temporal databases. The work of discovering partial orders instead of total orders implies greater complexity and a much vaster search space.

This paper details a two-phase approach, called OrderSpan, which can be used to mine the complete set of closed po-patterns. Our method uses both the prefix and suffix properties of seq-patterns, based on *ForwardTreeMining* and *Merging-SuffixTree* operations, respectively. It is able to directly mine all kind of sequential databases with sequences of itemsets. It overcomes previous work limitations, by directly mining po-patterns neither by generating them from seq-pattern sets, nor by only considering simple sequences with no item repetitions. Moreover, in comparison to seq-patterns, po-patterns are more relevant for specialist analysis due to the synthetic information they provide.

During sub-prefix-trees extraction, we deliberately refrain from using optimizations such as those proposed in [11] or [10]. The aim is to present a general and flexible framework, designed to be easily complexified and optimized using existing works in closed seq-pattern mining. Concerning future works, we aim at optimizing the OrderSpan method and generalizing the process to others patterns.

Acknowledgment. This work was funded by the French National Research Agency (ANR), as part of the ANR11_MONU14 Fresqueau project.

References

1. Geng, L., Hamilton, H.J.: Interestingness measures for data mining: A survey. ACM Computing Survey (2006)
2. Cheng, H., Yan, X., Han, J., Hsu, C.: Discriminative frequent pattern analysis for effective classification. In: Conference on Data Engineering, ICDE (2007)
3. Cheng, H., Yan, X., Han, J., Yu, P.S.: Direct discriminative pattern mining for effective classification. In: Conference on Data Engineering, ICDE (2008)
4. Wang, M., Shang, X.-q., Li, Z.-h.: Sequential pattern mining for protein function prediction. In: Tang, C., Ling, C.X., Zhou, X., Cercone, N.J., Li, X. (eds.) ADMA 2008. LNCS (LNAI), vol. 5139, pp. 652–658. Springer, Heidelberg (2008)
5. Agrawal, R., Srikant, R.: Mining sequential patterns. In: International Conference on Data Engineering, ICDE (1995)
6. Agrawal, R., Srikant, R.: Fast algorithms for mining association rules in large databases. In: Int. Conference on Very Large Data Bases, VLDB (1994)
7. Ren, J., Wang, L., Dong, J., Hu, C., Wang, K.: A novel sequential pattern mining algorithm for the feature discovery of software fault. In: Int. Conference on Computational Intelligence and Software Engineering, CiSE (2009)
8. Sallaberry, A., Pecheur, N., Bringay, S., Roche, M., Teisseire, M.: Sequential patterns mining and gene sequence visualization to discover novelty from microarray data. Journal of Biomedical Informatics (2011)
9. George, A., Binu, D.: Drl-prefixspan: A novel pattern growth algorithm for discovering downturn, revision and launch (drl) sequential patterns. Central European Journal of Computer Science (2012)
10. Wang, J., Han, J.: Bide: efficient mining of frequent closed sequences. In: International Conference on Data Engineering, ICDE (2004)
11. Yan, X., Han, J., Afshar, R.: CloSpan: Mining Closed Sequential Patterns in Large Datasets. In: SIAM International Conference on Data Mining, SDM (2003)
12. Mannila, H., Toivonen, H., Verkamo, A.I.: Discovery of Frequent Episodes in Event Sequences. In: Data Mining and Knowledge Discovery (1997)
13. Tatti, N., Cule, B.: Mining closed strict episodes. In: Data Mining and Knowledge Discovery (2012)
14. Zhou, W., Liu, H., Cheng, H.: Mining closed episodes from event sequences efficiently. In: Zaki, M.J., Yu, J.X., Ravindran, B., Pudi, V. (eds.) PAKDD 2010, Part I. LNCS, vol. 6118, pp. 310–318. Springer, Heidelberg (2010)
15. Pei, J., Wang, H., Liu, J., Wang, K., Wang, J., Yu, P.S.: Discovering frequent closed partial orders from strings. IEEE Transactions on Knowledge and Data Engineering (2006)
16. Casas-Garriga, G.: Summarizing sequential data with closed partial orders. In: SIAM International Conference on Data Mining, SDM (2005)
17. Pei, J., Han, J., Mortazavi-Asl, B., Wang, J., Pinto, H., Chen, Q., Dayal, U., Hsu, M.C.: Mining sequential patterns by pattern-growth: The prefixspan approach. IEEE Transactions on Knowledge and Data Engineering (2004)
18. Fabregue, M., Bringay, S., Poncelet, P., Teisseire, M., Orsetti, B.: Mining microarray data to predict the histological grade of a Breast Cancer. Journal of Biomedical Informatics (2011)

Learning Multiple Temporal Matching
for Time Series Classification

Cedric Frambourg, Ahlame Douzal-Chouakria, and Eric Gaussier

Université Joseph Fourier, Grenoble 1 / CNRS / LIG
{Cedric.Frambourg,Ahlame.Douzal,Eric.Gaussier}@imag.fr

Abstract. In real applications, time series are generally of complex
structure, exhibiting different global behaviors within classes. To discriminate such challenging time series, we propose a multiple temporal matching approach that reveals the commonly shared features within classes,
and the most differential ones across classes. For this, we rely on a new
framework based on the variance/covariance criterion to strengthen or
weaken matched observations according to the induced variability within
and between classes. The experiments performed on real and synthetic
datasets demonstrate the ability of the multiple temporal matching approach to capture fine-grained distinctions between time series.

1 Introduction

The problem of exploring, classifying or clustering multivariate time series arises
in a natural way in a lot of domains, inducing a notable increase activity in this
area of research these last years. The Dynamic Time Warping (DTW) [1] is frequently and successfully used in many domains to classify time series that share
similar global behaviors within classes subject to some delays. However it fails on
complex time series, namely, that present different global shapes within classes,
or similar ones between classes. In fact, the applied DTW alignment yields a local view, as it is performed in light of a single pair of time series, ignoring all
other time series dynamics within and between clusters; furthermore, the alignment process used is achieved regardless of the analysis process (as clustering
or classification), weakening its efficiency on complex data. Several variants of
DTW have been proposed to improve performance in classification or clustering.
They mostly aim to more finely estimate the DTW parameters, namely, warping
constraints, the time weighting, or the underlying divergence function between
mapped values. Without being exhaustive, the first part of these works mainly
rely on the Sakoe-Chiba, Itakura or Rabiner [2] approches to constrain globally
or locally the DTW warping space [3]. The second propositions concentrate on the
estimation of time weighting functions [4], whereas the last works pay particular
attention to the definition of adaptive divergence functions involving both values
and behaviors components of time series [5]. Although these approaches yield
more accurate temporal alignments, time series of the same class are assumed
to share a single global behavior. In real application time series are generally of

A. Tucker et al. (Eds.): IDA 2013, LNCS 8207, pp. 198–209, 2013.
© Springer-Verlag Berlin Heidelberg 2013

more complex structure. In particular, time series may exhibit different global behaviors within classes, or similar ones between classes. Consequently, for classification purpose, it appears important that the temporal alignment relies on the commonly shared features within the classes and the most differential ones between classes. Such challenging problem is addressed in some recent works. In particular, Ye [6] proposed shapelets extraction for time series classification. A pool of candidate shapelets is first generated by using a sliding window to extract all of the possible subsequences of all possible lengths from the time series dataset. Then, similar to the split criterion defined in [7], an euclidean distance and a maximization information gain criterion are used to both learn distance thresholds and to select the discriminative shapelets. For the same aim, a top-down and one-pass extraction of the discriminative sub-sequences is proposed in [5], while the involved metric is adaptive within a classification tree induction. In [8], [9] a learning pattern graphs from sequential data is proposed. For time series, such linkages are hardly reachable by conventional alignments strategies that are mainly limited to monotone warping functions preserving temporal order constraints [1].

We propose in this paper a new approach for multiple temporal alignment that highlights class-specific characteristics and differences. The main idea rely on a discriminant criterion based on variance/covariance to strengthen or weaken links according to their contributions to the variances within and between classes. The variance/covariance measure is used in many approaches, including exploratory analysis, discriminant analysis, clustering and classification [10]. However, to the best of our knowledge, it has never been investigated to define temporal alignment for time series classification. To this end, we propose a new formalization of the classical variance/covariance for a set of time series, as well as for a partition of time series (Section 2). In Section 3, we present a method for training the intra and inter class time series matching, driven by within-class variance minimization and between-class variance maximization. Subsequently, the learned discriminative matching is used to define a locally weighted time series metric that restricts the time series comparison to discriminative features (Section 4). In Section 5, the experiments carried out on both simulated and real datasets reveal the proposed approach able to capture fine-grained distinctions between time series, all the more so that time series of a same class exhibit dissimilar behaviors.

2   Variance/Covariance for Time Series

We first recall the definition of the conventional variance/covariance matrix, prior to introducing its formalization for time series data. Let X be the $(n \times p)$ data matrix containing n observations of p numerical variables. The conventional $(p \times p)$ variance/covariance matrix expression is:

$$V = X^t(I - UP)^t P(I - UP)X \qquad (1)$$

where, I is the diagonal identity matrix, U the matrix of ones, and P a diagonal weight matrix of general term $p_i = \frac{1}{n}$ for equally weighted observations.

In the following, we provide a generalization of the variance/covariance expression Eq.(1) to multivariate time series observations.

Variance Induced by a Set of Time Series. For a set of time series, let X be the $(nT \times p)$ matrix providing the description of n multivariate time series $S_1, ..., S_n$ by p numerical variables at T time stamps. The matching between n time series can be described by a matrix M of positive terms composed of n^2 block matrices $M^{ll'}$ $(l = 1, ..., n; l' = 1, ..., n)$. A block $M^{ll'}$ is a $(T \times T)$ matrix that specifies the matching between S_l and $S_{l'}$, of general term $m_{ii'}^{ll'} \in [0, 1]$ giving the weight of the link between the observation i of S_l and i' of $S_{l'}$. Then, the $(p \times p)$ variance/covariance matrix V_M induced by a set of time series $S_1, ..., S_n$ connected to one another according to the matching matrix M can be defined on the basis of Eq.(1), as:

$$V_M = X^t(I - M)^t P(I - M)X \qquad (2)$$

where P is a $(nT \times nT)$ diagonal matrix of weights, with $p_i = \frac{1}{nT}$ for equally weighted observations. Note that for a complete linkage matching, M is equal to UP and V_M leads to a conventional variance covariance V Eq.(1). For clarity and to simplify notation, we focus for the theoretical developments on univariate time series. The extension to the multivariate case is direct and will be used in the experiments. Thus, let x_i^l be the value of the variable X taken by S_l $(l = 1, ..., n)$ at the ith time stamp $(i = 1, ..., T)$.

Definition 1. *The variance V_M of the variable X is given by:*

$$V_M = \sum_{l=1}^{n} \sum_{i=1}^{T} p_i(x_i^l - \sum_{l'=1}^{n} \sum_{i'=1}^{T} m_{ii'}^{ll'} x_{i'}^{l'})^2 \qquad (3)$$

Note that each value x_i^l is centered relative to the term $\sum_{l'=1}^{n} \sum_{i'=1}^{T} m_{ii'}^{ll'} x_{i'}^{l'}$ estimating the average of X in the neighborhood of the time i of S_l. The neighborhood of i is the set of instants i' of $S_{l'}$ $(l' = 1...n)$ connected to i with $m_{ii'}^{ll'} \neq 0$. We now proceed to define the variance within and between classes when the set of time series is partitioned into classes.

Variance Induced by a Partition of Time Series. Let us now consider a set of time series $S_1, ..., S_n$ partitioned into K classes, with $y_i \in \{1, ..., K\}$ the class label of S_i and n_k the number of time series belonging to class C_k. The definition of the *within variance* (i.e. the variance within classes) and the *between variance* (i.e. the variance between classes) induced by K classes is obtained by using the expression given in Eq.(2) based on a matching M specified below.

Definition 2. *The **within variance** with an intra-class matching matrix M is given by:*

$$WV_M = \frac{1}{nT} \sum_{k=1}^{K} \sum_{l=1}^{n_k} \sum_{i=1}^{T} (x_i^l - \sum_{l'=1}^{n_{k^{\cdot}}} \sum_{i'=1}^{T} m_{ii'}^{ll'} x_{i'}^{l'})^2$$

with

$$M^{ll'} = \begin{cases} \mathbf{I} & if\ l = l' \\ \neq \mathbf{0} & if\ y_l = y_{l'}\ and\ l \neq l' \\ \mathbf{0} & if\ y_l \neq y_{l'} \end{cases} \quad (4)$$

where $\mathbf{I}$ *and* $\mathbf{0}$ *are the* $(T \times T)$ *identity and zero matrices, respectively.*

The general setting for the blocks $M^{ll'}$ of the intra-class matching M is based on three considerations: (a) the Euclidean alignment $(M^{ll} = \mathbf{I})$ linking each time series to itself ensures a variance of zero when comparing a time series with itself, (b) time series within the same class should be connected, while (c) time series of different classes are not connected, as they do not contribute to the within variance. Similarly, we have:

Definition 3. *The* **between variance** *with an inter-class matching matrix* M *is given by:*

$$BV_M = \tfrac{1}{nT} \sum_{k=1}^{K} \sum_{l=1}^{n_k} \sum_{i=1}^{T} (x_i^l$$
$$-(m_{ii}^{ll} x_i^l + \sum_{k' \neq k} \sum_{l'=1}^{n_{k'}} \sum_{i'=1}^{T} m_{ii'}^{ll'} x_{i'}^{l'}))^2$$

with

$$M^{ll'} = \begin{cases} \mathbf{I} & if\ l = l' \\ \mathbf{0} & if\ y_l = y_{l'}\ and\ l \neq l' \\ \neq \mathbf{0} & if\ y_l \neq y_{l'} \end{cases} \quad (5)$$

where $\mathbf{I}$ *and* $\mathbf{0}$ *are the* $(T \times T)$ *identity and zero matrices, respectively.*

The setting of the inter-class matching M is symmetric with respect to the preceding one, matching between time series of the same class being forbidden, while matching between time series of different classes is taken into account.

As one can note, the matching matrix M plays a crucial role in the definition of the within and between variances. The main issue for time series classification is therefore to learn a discriminative matching that highlights shared features within classes and distinctive ones between classes. To do so, we look for the matching matrix M, under the general settings given in Eqs. (4) and (5), that minimizes the within variance and maximizes the between variance. We present an efficient way to do this in the following section.

3   Learning Discriminative Matchings

We present here an efficient method to learn the matching matrix M, so as to connect time series based on their discriminative features. The proposed approach consists of two successive phases. In the first phase, the intra-class matching is learned to minimize the within variance. The learned intra-class matching reveals time series connections based on class-specific characteristics. In the second phase, the learned intra-class matching is refined to maximize the between variance.

Learning the Intra-class Matching. We are interested in inferring commonly shared structure within classes, that is in identifying the set of time stamps i' connected to each time stamp i regardless of their weights. Thus, the problem of learning the intra-class matching matrix M to minimize the within variance. We introduce here an efficient approach that iteratively evaluates the contribution of each linked observation (i, i') to the within variance; the weights $m_{ii'}^{ll'}$ are then penalized for all links (i, i') that significantly increase the within variance. For a given class C_k, this process, called *TrainIntraMatch*, is described in Algorithm 1 and involves the following steps.

Algorithm 1 *TrainIntraMatch*(X, α, k)

M = complete intra-class matching   Step 1
for all (l, l') *with* $y_l = y_{l'} = k$ and $l \neq l'$ **do**
  for all $(i, i') \in [1, T] \times [1, T]$ **do**
    $C_{ii'}^{ll'}$ evaluation with Eq. (7)   Step 2
  end for
end for
repeat
  $LinkRemoved = false$
  for all $(i, l) \in [1, T] \times [1, n]$ **do**
    $Link = \arg\max_{i', l'}(C_{ii'}^{ll'})$ satisfying Eq. (9)   Step 3
    if $Link \neq \emptyset$ **then**
      Remove $Link$ ($m_{i,i'}^{l,l'} = 0$) and
      Update weights with Eq. (8)
      Update contributions
      $LinkRemoved = true$
    end if
  end for
until $\neg LinkRemoved$ Step 4
$return(M_{Intra} = M)$

---

1. **Initialization (Step 1).** A complete linkage is used to initialize the intra-class matching matrix M, to ensure that all possible matchings are considered and that no *a priori* constraints on the type of matching one should look for are introduced.

$$M^{ll'} = \begin{cases} \mathbf{I} & \text{if } l = l' \\ \frac{1}{T}\mathbf{U} & \text{if } y_l = y_{l'} \text{ and } l \neq l' \\ \mathbf{0} & \text{if } y_l \neq y_{l'} \end{cases} \quad (6)$$

2. **Computing link contributions (Step 2).** We define the contribution $C_{i_1 i_2}^{l_1 l_2}$ of the link (i_1, i_2) between S_{l_1} and S_{l_2} ($y_{l_1} = y_{l_2}$) as the induced variation on the within variance after the link (i_1, i_2) has been removed:

$$C_{i_1 i_2}^{l_1 l_2} = WV_M - WV_{M \backslash (i_1, i_2, l_1, l_2)} \quad (7)$$

where $M \backslash (i_1, i_2, l_1, l_2)$ denotes the matrix obtained from M by setting $m_{i_1 i_2}^{l_1 l_2}$ to 0 and re-normalizing its i_1^{th} row:

$$m_{i_1 i'}^{l_1 l'} \leftarrow \frac{m_{i_1 i'}^{l_1 l'}}{1 - m_{i_1 i_2}^{l_1 l_2}} \quad (8)$$

The evaluated contributions reveal two types of links: the links of positive contribution $C_{ii'}^{ll'} > 0$ that decrease the within variance if removed, and the links of negative contribution $C_{ii'}^{ll'} < 0$ that increase the within variance if removed.

3. **Link deletion (Step 3).** The deletion of a link with positive contribution ensures that the within variance will decrease. In addition, if all links within a row have a negligible contribution to the variance, one can dispense with removing them in order to (a) avoid overtraining and (b) speed up the process. Thus, a link (i, i') between S_l and $S_{l'}$ is deleted if it satisfies:

$$C_{ii'}^{ll'} > \alpha.WV_{M_1} \quad \text{and} \quad \sum_{i''=1,(i''\neq i')}^{T} m_{ii''}^{ll'} > 0 \tag{9}$$

where $\alpha \in [0,1]$ and WV_{M_1} is the initial within variance.
Because the normalization in Eq.(8) performed after the deletion of (i_1, i_2) impacts only the weights of the i_1^{th} row, deleting a single link per row at each iteration of the process guarantees that the global within variance will decrease. Thus, at each iteration one can simply delete the link on each row of maximal contribution compliant with Eq.(9).

4. **Stopping the learning process (Step 4).** The algorithm iterates steps 2, 3 and 4 until there are no more links satisfying the conditions specified in Eq.(9).

From the learned intra-class matching obtained at step 4, noted M_{Intra}, one may induce for each time series S_l one intra-block M_{Intra}^l to indicate the characteristic linkage between S_l and time series of the same class. This intra-block is obtained by summing the block matrices learned for S_l, as follows:

$$M_{Intra}^l = \sum_{l' \in 1,...,n_k} M_{Intra}^{ll'} \tag{10}$$

Learning the Inter-class Matching. The goal of this second phase is to refine the highlighted connections in M_{Intra} (i.e., that connects shared features within classes) to capture the links that are additionally differentiating classes. For this, we refer to a similar algorithm called *TrainInterMatch*, where the inter-class matching is initialized with M_{Intra}, then trained to maximize the between variance BV_M of Definition 3. As for the within variance minimization problem, we adopt the same approach, which consists in iteratively evaluating the contribution of each linked observations (i, i') to the between variance; the weights $m_{ii'}^{ll'}$ are then penalized for all links (i, i') significantly decreasing the between variance. We now turn to the application of the learned matching matrix to time series classification.

4   Time Series Classification Based on the Learned Matching

Our aim here is to present a way of using learned discriminative matching to locally weight time series for k-nearest neighbor classification. The purpose of the proposed weighting is to restrict the time series comparison to the discriminant (characteristic and differential) features. Let M_* be the discriminative matching learned by the *TrainIntraMatch* and *TrainInterMatch* algorithms, where discriminant linkages are highly weighted. For each S_l of the training sample, we note $M_*^{l\cdot}$ the average of the learned matrices $M_*^{ll'}$ ($y_{l'} \neq y_l = k$):

$$M_*^{l\cdot} = \frac{1}{(n-n_k)T} \sum_{l'} M_*^{ll'}$$

It defines the linkage schema of S_l to a given time series (of the same or of different class) according to S_l own discriminative features. To damp the effect of outliers, the geometric mean could be used for $M_*^{l\cdot}$ as well.

In k-nearest neighbor classification, one can compare a new time series S_{test} to a sample series S_l of C_k based on its learned discriminative matching $M_*^{l\cdot}$. This can be achieved by looking for the delay r that leads to the minimal distance between S_{test} and S_l:

$$D_l(S_l, S_{test}) = \min_{r \in \{0,..,T-1\}} \left(\sum_{|i-i'| \leq r;\, (i,i') \in [1,T]^2} \frac{m_{ii'}^{l\cdot}}{\sum_{|i-i'| \leq r} m_{ii'}^{l\cdot}} (x_i^l - x_{i'}^{test})^2 \right) \quad (11)$$

where r corresponds to the Sakoe-Chiba band width [2]. Note that for $r = 0$, D_l defines a locally weighted Euclidean distance involving the diagonal weights $m_{ii}^{l\cdot}$.

5   Experiments

Synthetic Datasets. The first objective of these experiments is to show through challenging synthetic datasets that the proposed approach successes to recover the *a priori* known discriminative features. For this, two synthetic datasets BME and UMD are considered, where a given class may be composed of time series of different global behaviors and including amplitude and delay variations. BME is composed of three classes *Begin*, *Middle*, and *End* of time series of length 128. Figure 1 illustrates the time series variability within each class, it shows the profile of one time series (in black) compared to the remaining time series (in grey) of the class. In the *Begin* (respectively the *End*) class, time series are characterized by a small bell arising at the initial (respectively final) period. The overall behavior may be different within a same class depending on whether the large bell is up or down positioned. Furthermore, time series of the *Begin* and the *End* classes composed of an up-positioned large bell are quite similar to the *Middle* class time series. The second dataset UMD, composed of three classes *Up*, *Middle*, and *Down* (time series length of 150), introduces an additional complexity with the *Up* and *Down* classes characterized by a small bell that may occur at different time stamps, as illustrated in Figure 1.

Electric Power Consumption Classification. The proposed approach is motivated by a classification problem of a real electrical power consumption of customers, to adequately meet consumer demands. To classify such challenging data, we refer to the proposed approach to: a) localize the periods that characterize the

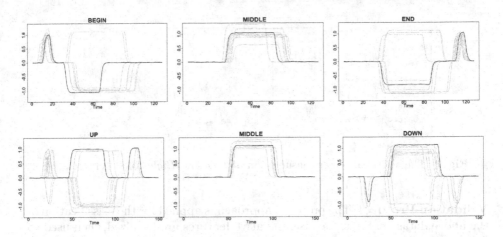

Fig. 1. BME (top three classes) and UMD (bottom three classes) datasets

daily power consumption of each class, b) highlight periods that differentiate the power consumption of different classes, c) and classify new power consumption based on the learned discriminative features.

The application relies on two public datasets[1] CONSLEVEL and CONSSEASON providing the electric power consumption recorded in a personal home over almost one year (349 days). Each time series consists of 144 measurements that give the power consumption of one day with a 10 minute sampling rate. CONSLEVEL is composed of 349 time series distributed in two classes (*Low* and *High*) depending on whether the average electric power during the peak demand period [6:00pm-8:00pm] is lower or greater than the annual average consumption of that period. Figure 2 shows the electric consumption profiles within the CONSLEVEL classes; the red frames delineate the time interval [108,120], corresponding to the peak period [6:00pm-8:00pm]. On the other hand, CONSSEASON is composed of 349 time series distributed in two season classes (*Warm* and *Cold*) depending on whether the power consumption is recorded during the warm (from April to September) or cold (from October to March) seasons (Figure 2). Note that the electric power consumption profiles differ markedly within classes in both datasets.

Character Trajectories Classification. The objective of this latter dataset is to verify whether the proposed approach can recover standard time series structures within classes, namely, when the classes are mainly composed of time series of similar global behaviors. For this, we have considered a standard dataset on character trajectories TRAJ [12], where time series share a quite similar global behavior within classes (20 classes of 50 time series each).

---

[1] These data are available at http://bilab.enst.fr/wakka.php?wiki=HomeLoadCurve, and analyzed in [11]

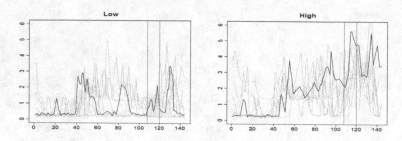

Fig. 2. The electrical power consumption of *Low* and *High* CONSLEVEL classes

Validation Protocol. The proposed approach is applied for the classification of the above datasets. First, the discriminative features are localized, then used to define a locally weighted time series metric D as given in Eq.(11). The relevance of the learned discriminative features and of the induced metric is then studied through a k-nearest neighbor classification for several neighborhood sizes $k = 1, 3, 5, 7$. For BME and UMD datasets a training and test sets of 360 and 1440 time series, respectively, are considered. For the real datasets the performances are evaluated based on 10-fold cross-validation protocol. Finally, the results obtained are compared to two baselines: the Euclidean DE and dynamic time warping DTW distances (Table 1).

Results and Discussion. The algorithms *TrainIntraMatch* and *TrainInter-Match* are applied to the above datasets with $\alpha = 0.5\%$. As an example, let us first illustrate, for the BME dataset, the progression of the within and between variances during the learning processes (Figure 3). The clearly monotonically decreasing (respectively increasing) behavior of the within (respectively between) class variance, which ends at a plateau, assesses: a) the pertinence of the conducted links penalization to minimize the within variance and maximize the between variance, b) the convergence of the proposed algorithms.

For CONSLEVEL, Figure 4 shows the learned intra-class (left) and inter-class (right) blocks for a given time series of the *Low* class. The intra-class block reveals a checkerboard structure, indicating that the electric power consump-

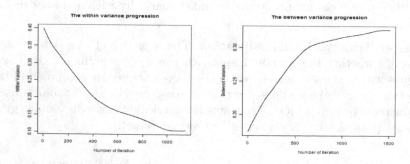

Fig. 3. The within and between variance progression for BME dataset

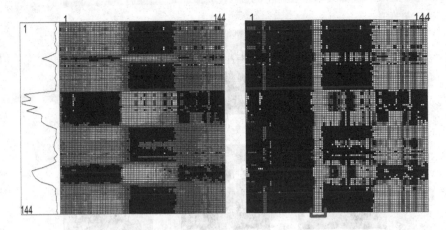

Fig. 4. The intra (left) and inter (right) class matching learned for *Low* class

tion within the *Low* class alternates, in a daily period, between a low and a moderately high consumption. The corresponding inter-class block shows the discriminative matching between the considered *Low* class time series and time series of the *High* class (on column). This block displays many discriminative regions; for example, it shows that the power consumption within the *High* class within the period underlined in red (prior to 6:00pm-8:00pm) is especially important in predicting the consumption during the peak period. For each above described dataset, a locally weighted time series metric D is defined based the learned discriminative matching, as given in Eq.(11), then used for the time series classification. The relevance of the proposed approach and of the induced metric are studied according to the validation process described above. The results obtained are compared to two baselines: the Euclidean DE and dynamic time warping DTW distances.

The misclassification error rates obtained in Table 1 show the efficiency of the proposed locally weighted metric D in discriminating between complex time series classes, compared to standard metrics for time series. In particular, one can note that for all datasets but TRAJ, the best results (in bold) are obtained

Table 1. *k*-Nearest Neighbor classification error rates on synthetic data

	k	D	DE	DTW
	1	**0.032**	0.165	0.130
BME	3	0.034	0.208	0.132
	5	0.062	0.234	0.136
	7	0.079	0.297	0.191
	1	**0.055**	0.173	0.121
UMD	3	0.111	0.333	0.177
	5	0.173	0.343	0.225
	7	0.222	0.378	0.274

Table 2. *k*-Nearest Neighbor classification error rates on real data

	k	D	DE	DTW
	1	0.056	0.306	0.289
CONSLEVEL	3	0.044	0.267	0.261
	5	0.028	0.233	0.239
	7	**0.017**	0.233	0.233
	1	**0.094**	0.239	0.283
CONSSEASON	3	0.128	0.228	0.311
	5	0.205	0.200	0.300
	7	0.111	0.222	0.306
	1	0.014	**0.012**	0.019
TRAJ	3	0.018	0.017	0.022
	5	0.022	0.021	0.028
	7	0.019	0.021	0.026

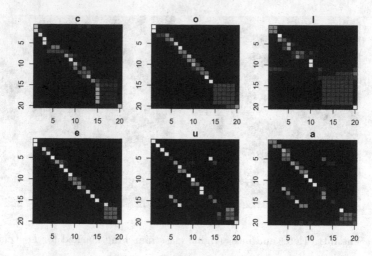

Fig. 5. The learned discriminative matching for the characters "c", "o", "i","e","u", and "a" of TRAJ dataset

with D. For TRAJ, the three metrics lead to comparable results suggesting that the Euclidean alignment is an appropriate matching for this dataset. In Figure 5, we can see that the learned discriminative matching, for example, for "c", "o", "i","e","u", and "a" characters is close to the Euclidean one, which shows the ability of the proposed approach to recover standard time series alignments. In addition, one can see that for nearly all datasets the best performances are obtained for $k = 1$. For CONSLEVEL, a slight improvement is reached for $k = 7$, indicating a great clusters overlap for this dataset.

Conclusion and Future Works

The DTW is frequently and successfully used in many domains to classify time series that share similar global behaviors subject to some delays within classes. All the interest of the proposed approach rely on complex time series for which DTW fails ; namely, that describe different global shapes within classes, or similar ones between classes. The achieved results support the promising abilities of the learned temporal matching to discriminate finely such challenging time series. However, the current learning algorithms are not scalable owing to the induced complexity. Thus, our future work will mainly focus on calculus complexity reduction to ensure the proposed method be useable for large scale data. The main idea consists to sparse the initial intra-class matching matrix M. For this, we first segment each time series into a set of salient sub-segments, subsequently, each bloc $M_{ll'}$ is defined as a set of diagonal alignments linking all pairs of sub-segments of S_l and S'_l. Performances of the scalable variant of the approach will then be compared to alternative methods on large scale and complex time series data. On the other hand, we aim to study new ways to define weighted metrics based on the discriminative masks $M_*^{l\cdot}$, for instance, by generalizing conventional DTW to achieve alignments limited to the discriminative regions of $M_*^{l\cdot}$.

References

1. Kruskall, J., Liberman, M.: The symmetric time warping algorithm: From continuous to discrete. In: Time Warps, String Edits and Macromolecules. Addison-Wesley (1983)
2. Sakoe, H., Chiba, S.: Dynamic programming algorithm optimization for spoken word recognition. IEEE Transactions on Acoustics, Speech, and Signal Processing 26(1), 43–49 (1978)
3. Yu, D., Yu, X., Hu, Q., Liu, J., Wu, A.: Dynamic time warping constraint learning for large margin nearest neighbor classification. Information Sciences 181, 2787–2796 (2011)
4. Jeong, Y., Jeong, M., Omitaomu, O.: Weighted dynamic time warping for time series classification. Pattern Recognition 44, 2231–2240 (2011)
5. Douzal-Chouakria, A., Amblard, C.: Classification trees for time series. Pattern Recognition 45(3), 1076–1091 (2012)
6. Ye, L., Keogh, E.: Time series shapelets: A new primitive for data mining. Data Min. Knowl. Disc. 22, 149–182 (2011)
7. Yamada, Y., Suzuki, E., Yokoi, H., Takabayashi, K.: Decision-tree induction from time-series data based on standard-example split test. In: Proceedings of the 20th International Conference on Machine Learning, pp. 840–847. Morgan Kaufmann (2003)
8. Peter, S., Höppner, F., Berthold, M.: Pattern graphs: A knowledge-based tool for multivariate temporal pattern retrieval. In: IEEE Conf. Intelligent Systems (2012)
9. Peter, S., Höppner, F., Berthold, M.R.: Learning pattern graphs for multivariate temporal pattern retrieval. In: Hollmén, J., Klawonn, F., Tucker, A. (eds.) IDA 2012. LNCS, vol. 7619, pp. 264–275. Springer, Heidelberg (2012)
10. Fisher, R.: The use of multiple measures in taxonomic problems. Annals of Eugenics 7, 179–188 (1936)
11. Hebrail, G., Hugueney, B., Lechevallier, Y., Rossi, F.: Exploratory analysis of functional data via clustering and optimal segmentation. Neurocomputing 73, 1125–1141 (2010)
12. Asuncion, A., Newman, D.: UCI Machine learning repository (2007)

On the Importance of Nonlinear Modeling
in Computer Performance Prediction

Joshua Garland and Elizabeth Bradley

University of Colorado, Boulder CO 80309-0430, USA

Abstract. Computers are nonlinear dynamical systems that exhibit complex and sometimes even chaotic behavior. The low-level performance models used in the computer systems community, however, are linear. This paper is an exploration of that disconnect: when linear models are adequate for predicting computer performance and when they are not. Specifically, we build linear and nonlinear models of the processor load of an Intel i7-based computer as it executes a range of different programs. We then use those models to predict the processor loads forward in time and compare those forecasts to the true continuations of the time series.

1   Introduction

Accurate prediction is important in any number of applications. In a modern multi-core computer, for instance, an accurate forecast of processor load could be used by the operating system to balance the workload across the cores in real time. The traditional methods that are used in the computer systems community to model low-level performance metrics like processor and memory loads are based on linear, time-invariant (and often stochastic) techniques, e.g., autoregressive moving average (ARMA), multiple linear regression, etc. [13]. While these models are widely accepted—and for the most part easy to construct—they cannot capture the nonlinear interactions that have recently been shown to play critical roles in computer performance [23]. As computers become more and more complex, these interactions are beginning to cause problems—e.g., hardware design "improvements" that do not work as expected. Awareness about this issue is growing in the computer systems community [22], but the modeling strategies used in that field have not yet caught up with those concerns.

An alternative approach is to model a computer as a nonlinear dynamical system [23,24]—or as a *collection* of nonlinear dynamical systems, i.e., an iterated function system [1]. In this view, the register and memory contents are treated as state variables of these dynamical systems. The logic hardwired into the computer, combined with the code that is executing on that hardware, defines the system's dynamics—that is, how its state variables change after each processor cycle. As described in previous IDA papers [2,10], this framework lets us bring to bear the powerful machinery of nonlinear time-series analysis on the problem of modeling and predicting those dynamics. In particular, the technique called *delay-coordinate embedding* lets one reconstruct the state-space dynamics of the system from a time-series measurement of a single state variable[1].

---

[1] Technically, the measurement need only be a smooth function of at least one state variable.

A. Tucker et al. (Eds.): IDA 2013, LNCS 8207, pp. 210–222, 2013.

One can then build effective prediction models in this embedded space. One of the first uses of this approach was to predict the future path of a ball on a roulette wheel, as chronicled in [3]. Nonlinear modeling and forecasting methods that rely on delay-coordinate embedding have since been used to predict signals ranging from currency exchange rates to Bach fugues; see [5,28] for good reviews.

This paper is a comparison of how well those two modeling approaches—linear and nonlinear—perform in a classic computer performance application: forecasting the processor loads on a CPU. We ran a variety of programs on an Intel i7-based machine, ranging from simple matrix operation loops to SPEC cpu2006 benchmarks. We measured various performance metrics during those runs: cache misses, processor loads, branch-prediction success, and so on. The experimental setup used to gather these data is described in Section 2. From each of the resulting time-series data sets, we built two models: a garden-variety linear one (multiple linear regression) and a basic nonlinear one: the "Lorenz method of analogues," which is essentially nearest-neighbor prediction in the embedded state space [18]. Details on these modeling procedures are covered in Section 3. We evaluated each model by comparing its forecast to the true continuation of the associated time series; results of these experiments are covered in Section 4, along with some explanation about when and why these different models are differently effective. In Section 5, we discuss some future directions and conclude.

Modeling low-level hardware metrics like processor and memory loads is a very different, and arguably harder, problem than modeling the actions of the software. Nonlinear machine-learning techniques have been used quite successfully to model computer dynamics at the software level—e.g., what compiler flag settings work well for a given program [6] or what code characteristics cause massive slowdowns [12]. These models can be very useful in *post facto* design and optimization. The application tackled here is fundamentally different: we are interested in *real-time* prediction of *low-level* performance metrics in real machines—something to which machine-learning techniques have not, to our knowledge, been applied. (One group has done some preliminary work on neural-net based runtime algorithms for allocating the workload of several programs across the resources of a given computer [20], but this has only been tested in *simulated* computers, whose low-level performance dynamics have been shown to differ greatly from those of real machines [7,23].)

2 Experimental Methods

The testbed for these experiments was an HP Pavilion Elite computer with an Intel Core® i7-2600 CPU running the 2.6.38-8 Linux kernel. This so-called "Nehalem" chip is representative of modern CPUs; it has eight cores running at 3.40Ghz and an 8192 kB cache. Its kernel software allows the user to monitor events on the chip, as well as to control which core executes each thread of computation. This provides a variety of interesting opportunities for model-based control. An effective prediction of the cache-miss rate of individual threads, for instance, could be used to preemptively migrate threads that are bogged down waiting for main memory to a lower-speed core, where they can spin their wheels without burning up a lot of power[2].

To build models of this system, we instrumented the kernel software to capture performance traces of various important internal events on the chip. These traces are recorded from the hardware performance monitors (HPMs), specialty registers that are built into most modern CPUs in order to log hardware event information. We used the libpfm4 library, via PAPI [4], to interrupt the executables periodically and read the contents of the HPMs. At the end of the run, this measurement infrastructure outputs the results in the form of a time series. In any experiment, of course, one must be attentive to the possibility that the act of measurement perturbs the dynamics under study. For that reason, we varied the rate at which we interrupted the executables, compared the results, and used that comparison to establish a sample rate that produced a smooth measurement of the underlying system dynamics. A detailed explanation of the mechanics of this measurement process can be found in [2,23,24].

The dynamics of a running computer depend on both hardware and software. We ran experiments with four different C programs: two benchmarks from the SPEC cpu2006 benchmark suite (the 403.gcc compiler and the 482.sphinx speech recognition system) and two four-line programs (col_major and row_major) that repeatedly initialize a matrix—in column-major and row-major order, respectively. These choices were intended to explore the range of current applications. The two SPEC benchmarks are complex pieces of code, while the simple loops are representative of repetitive numerical applications. 403.gcc works primarily with integers, while 482.sphinx is a floating-point benchmark. Row-major matrix initialization works naturally with modern cache design, whereas the memory accesses in the col_major loop are a serious challenge to that design, so we expected some major differences in the behavior of these two simple loops. Figure 1 shows traces of the instructions executed per cycle, as a function of time, during the execution of the two SPEC benchmarks on the computer described in the first paragraph of this section. There are clear patterns in the processor load during the operation of 482.sphinx. During the first 250 million instructions of this program's execution, roughly two instructions are being carried out every cycle, on the average, by the Nehalem's eight cores. Following that period, the processor loads oscillate, then stabilize at an average of one instruction per cycle for the period from 400-800 million instructions. Through the rest of the trace, the dynamics move between different regimes, each with characteristics that reflect how well the different code segments can be effectively executed across the cores. The processor load during the execution of 403.gcc , on the other hand, appears to be largely stochastic. This benchmark takes in and compiles a large number of small C files, which involves repeating a similar process, so the lack of clear regimes makes sense. The dynamics of row_major and col_major (not shown due to space constraints) were largely as expected. The computer cannot execute as many instructions during col_major because of the mismatch between its memory-access pattern and the design of the cache, so the baseline level of the col_major trace is much lower than

---

[2] Kernels and operating systems do some of this kind of reallocation, of course, but they do so using current observations (e.g., if a thread is halting "a lot") and/or using simple heuristics that are based on computer systems knowledge (e.g., locality of reference).

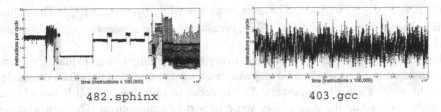

482.sphinx   403.gcc

Fig. 1. Processor load traces of the programs studied here

row_major . Temporally, the row_major trace looks very much like 403.gcc : largely stochastic. col_major , on the other hand, has a square-wave pattern because of the periodic stalls that occur when it requests data that are not in the cache.

The following section describes the techniques that we use to build models of these time-series data sets.

3   Modeling Computer Performance Data

3.1   Overview

The goal of this paper is to explore the effectiveness of linear and nonlinear models of computer performance. Many types of models, of both varieties, have been developed by the various communities that are interested in data analysis. We have chosen multiple linear regression models as our *linear* exemplar because that is a standard in the computer systems literature [22]. In order to keep the comparison as fair as possible, we chose the Lorenz method of analogues, which is the simplest of the many models used in the nonlinear time-series analysis community, as our *nonlinear* exemplar. In the remainder of this section, we give short overviews of each of these methods; Sections 3.2 and 3.3 present the details of the model-building processes.

Multiple Linear Regression Models.   Suppose we have a set of n observations from a system, where the ith observation includes $m + 1$ measurements: a scalar response variable r_i and a vector of m explanatory variables $[e_{i1}, \ldots, e_{im}]$. Working from those n observations $\{([e_{i1}, \ldots, e_{im}], r_i)\}_{i=1}^{n}$, multiple linear regression (MLR) models the future value of the response variable via a linear combination of the current explanatory variables—that is:

$$r_{i+1} = [1, e_{i1}, ..., e_{im}]\beta \tag{1}$$

where $\beta = [\beta_1, \ldots, \beta_{m+1}]^T$ is a vector of $m + 1$ fit parameters [13]. We estimated β by using ordinary least squares to minimize the sum of squared residuals (SSR) between the set of observations and the set of hyperplanes defined in equation (1):

$$SSR(\beta) = \sum_{i=1}^{n-1}(r_{i+1} - [1, e_{i1}, ..., e_{im}]\beta)^2 = (r - E\beta)^T(r - E\beta)$$

Here, E is a $(n-1)$ by $(m+1)$ matrix, with rows $\{[1, e_{i1}, ..., e_{im}]\}_{i=1}^{n-1}$. The β that minimizes this sum, $\hat{\beta} = (E^T E)^{-1} E^T r$, is called the *ordinary-least-squares estimator* of β.

In practice, it is customary to measure everything that might be an explanatory variable and then use a method called stepwise backward elimination to reduce the model down to what is actually important. And in order to truly trust an MLR model, one must also verify that the data satisfy some important assumptions. These procedures are discussed in more detail in Section 3.2.

> *Advantages:* MLR models are simple and easy to construct and use, as they only require a small vector-vector multiplication at each prediction step.

> *Disadvantages:* strictly speaking, MLR can only be used to model a linear deterministic system that is measured without any error. MLR models involve multiple explanatory variables and cannot predict more than one step ahead.

Nonlinear Models. Delay-coordinate embedding [25,26,27] allows one to reconstruct a system's full state-space dynamics from time-series data like traces in Figure 1. There are only a few theoretical requirements for this to work. The data, $x_i(t)$, must be evenly sampled in time (t) and both the underlying dynamics and the measurement function— the mapping from the unknown d-dimensional state vector Y to the scalar value x_i that one is measuring—must be smooth and generic. When these conditions hold, the delay-coordinate map

$$F(\tau, d_{embed})(x_i) = ([x_i(t),\ x_i(t+\tau),\ \ldots,\ x_i(t+d_{embed}\tau)]) \tag{2}$$

from a d-dimensional smooth compact manifold M to $\mathbb{R}^{2d+1}$ is a diffeomorphism on M [26,27]: in other words, that the reconstructed dynamics and the true (hidden) dynamics have the same topology. This method has two free parameters, the delay τ and the embedding dimension d_{embed}, which must also meet some conditions for the theorems to hold, as described in Section 3.3. Informally, delay-coordinate embedding works because of the internal coupling in the system—e.g., the fact that the CPU cannot perform a computation until the values of its operands have been fetched from some level of the computer's memory. This coupling causes changes in one state variable to percolate across other state variables in the system. Delay-coordinate embedding brings out those indirect effects explicitly and geometrically.

The mathematical similarity of the true and reconstructed dynamics is an extremely powerful result because it guarantees that F is a good model of the system. As described in Section 1, the nonlinear dynamics community has recognized and exploited the predictive potential of these models for some time. Lorenz's method of analogues, for instance, is essentially nearest-neighbor prediction in the embedded space: given a point, one looks for its nearest neighbor and then uses *that* point's future path as the forecast [18]. Since computers are deterministic dynamical systems [23], these methods are an effective way to predict their performance. That claim, which was first made in [10], was the catalyst for this paper—and the motivation for comparison of linear and nonlinear models that appears in the following sections.

Advantages: models based on delay-coordinate embeddings capture nonlinear dynamics and interactions, which the linear models ignore, and they can be used to predict forward in time to arbitrary horizons. They only require measurement of a single variable.

Disadvantages: these models are more difficult to construct, as estimating good values for their two free parameters can be quite challenging in the face of noise and sampling issues in the data. The prediction process involves near-neighbor calculations, which are computationally expensive.

3.2 Building MLR Forecast Models for Computer Performance Traces

In the experiments reported here, the response variable is the number of instructions per cycle (IPC) executed by the CPU. Following [22], we chose the following candidate explanatory variables *(i)* instructions retired *(ii)* total L2 cache[3] misses *(iii)* number of branches taken *(iv)* total L2 instruction cache misses *(v)* total L2 instruction cache hits and *(vi)* total missed branch predictions. The first step in building an MLR model is to "reduce" this list: that is, to identify any explanatory variables—aka *factors*—that are meaningless or redundant. This is important because unnecessary factors can add noise, obscure important effects, and increase the runtime and memory demands of the modeling algorithm.

We employed the stepwise backward elimination method [8], with the threshold value (0.05) suggested in [13], to select meaningful factors. This technique starts with a "full model"—one that incorporates every possible factor—and then iterates the following steps:

1. If the p-value of any factor is higher than the threshold, remove the factor with the largest p-value
2. Refit the MLR model
3. If all p-values are less than the threshold, stop; otherwise go back to step 1

For all four of the traces studied here, this reduction algorithm converged to a model with three factors: L2 total cache misses, number of branches taken, and L2 instruction cache misses.

To predict IPC (the response variable) using the reduced MLR model, one takes a measurement of each of the factors that appear in the reduced model (say, $[e_{n1}, \ldots, e_{nm}]$) and then simply evaluates the function $[1, e_{n1}, \ldots, e_{nm}]\hat{\beta}$ and assigns that outcome to the *next* time-step, i.e, r_{n+1}. That is how the predictions in the next section were constructed.

Like any model, MLR is technically only valid if the data meet certain conditions. Two of those conditions are *not* true for computer-performance traces: linear relationship between explanatory and response variables (which was disproved in [23]) and normal distribution of errors, which is clearly not the case in our data, given the nonlinear trends in residual quantile-quantile plots of our data (not shown). Despite this, MLR

---

[3] Modern CPUs have many levels of data and instruction caches: small, fast memories that are easy for the processor to access. A key element of computer design is anticipating what to "fetch" into those caches.

models are used routinely in the computer systems community [22]. And they actually work surprisingly well, as indicated by the results in Section 4.

3.3 Building Nonlinear Forecast Models for Computer Performance Traces

The first step in constructing a nonlinear forecast model of a time-series data set like the ones in Figure 1 is to perform a delay-coordinate embedding using equation (2). We followed standard procedures [14] to choose appropriate values for the embedding parameters: the first minimum of the mutual information curve [9] as an estimate of the delay τ and the false-nearest neighbors technique [16], with a threshold of 10-20%, to estimate the embedding dimension d_{embed}. For both traces in Figure 1, $\tau = 100000$ instructions and $d_{embed} = 12$. A plot of the reconstructed dynamics of these two traces appears in Figure 2. The coordinates of each point on these plots are differently delayed elements of the IPC time series: that is, IPC at time t on the first axis, IPC at time $t + \tau$ on the second, IPC at time $t + 2\tau$ on the third, and so on. An equivalent embedding (not shown here) of the row_major trace looks very like 403.gcc: a blob of points. The embedded col_major dynamics, on the other hand, looks like *two* blobs of points because of its square-wave pattern. Recall from Section 3.3 that these trajectories are guaranteed to have the same topology as the true underlying dynamics, provided that τ and d_{embed} are chosen properly. And structure in these kinds of plots is an indication of determinism in that dynamics.

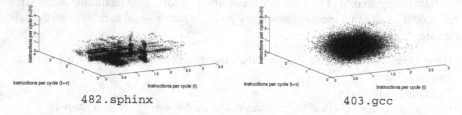

482.sphinx   403.gcc

Fig. 2. 3D projections of delay-coordinate embeddings of the traces from Figure 1

The nonlinear dynamics community has developed dozens of methods that use the structure of these embeddings to create forecasts of the dynamics; see [5,28] for overviews. The Lorenz method of analogues (LMA) is one of the earliest and simplest of these strategies [18]. LMA creates a prediction of the future path of a point x_o through the embedded space by simply finding its nearest neighbor and then using *that* point's future path as the forecast[4]. The nearest neighbor step obviously makes this algorithm very sensitive to noise, especially in a nonlinear system. One way to mitigate that sensitivity is to find the l nearest neighbors of x_o and average their future paths. These comparatively simplistic methods work surprisingly well for computer-performance prediction, as reported at IDA 2010 [10]. In the following section, we compare the prediction accuracy of LMA models with the MLR models of Section 3.2.

---

[4] The original version of this method requires that one have the true state-space trajectory, but others (e.g., [15]) have validated the theory and method for the kinds of embedded trajectories used here.

4 When and Why Are Nonlinear Models Better at Predicting Computer Performance?

4.1 Procedure

Using the methods described in Sections 3.2 and 3.3, respectively, we built and evaluated linear and nonlinear models of performance traces from the four programs described on page 212 (403.gcc, 482.sphinx, col_major and row_major), running on the computer described in Section 2. The procedure was as follows. We held back the last k points of each time series (referred to as "the test signal," c_i). We then constructed the model with the remaining portion of the time series ("the learning signal") and used the model to build a prediction $\hat{p}_i$. We computed the root mean squared prediction error between that prediction and the test signal in the usual way:

$$RMSE = \sqrt{\frac{\sum_{i=1}^{k}(c_i - \hat{p}_i)^2}{k}}$$

To compare the results across signals with different units, we normalized the RMSE as follows:

$$nRMSE = \frac{RMSE}{max\{c_i\} - min\{c_i\}}$$

The smaller the nRMSE, obviously, the more accurate the prediction.

4.2 Results and Discussion

First, we compared linear and nonlinear models of the two SPEC benchmark programs: the traces in the top row of Figure 1. For 403.gcc, the nonlinear LMA model was better than the linear MLR model (0.128 nRMSE versus 0.153). For 482.sphinx, the situation was reversed: 0.137 nRMSE for LMA and 0.116 for MLR. This was contrary to our expectations; we had anticipated that the LMA models would work better because their ability to capture both the gross and detailed structure of the trace would allow them to more effectively track the regimes in the 482.sphinx signal. Upon closer examination, however, it appears that those regimes overlap in the IPC range, which could negate that effectiveness. Moreover, this head-to-head comparison is not really fair. Recall that MLR models use *multiple* measurements of the system—in this case, L2 total cache misses, number of branches taken, and L2 instruction cache misses—while LMA models are constructed from a *single* measurement (here, IPC). In view of this, the fact that LMA beats MLR for 403.gcc and is not too far behind it for 482.sphinx is impressive, particularly given the complexity of these programs. Finally, we compared the linear and nonlinear model results to a simple "predict the mean" strategy, which produces a 0.140 and 0.250 nRMSE for 403.gcc and 482.sphinx, respectively—higher than either MLR or LMA.

In order to explore the relationship between code complexity and model performance, we then built and tested linear and nonlinear models of the row_major and col_major traces. The resulting nRMSE values for these programs, shown in the third and fourth row of Table 1, were lower than for 403.gcc and 482.sphinx,

Table 1. Normalized root-mean-squared error between true and predicted signals for linear (MLR), nonlinear (LMA), and "predict the mean" forecast strategies

Program	Interrupt Rate (cycles)	LMA nRMSE	MLR nRMSE	naive nRMSE
403.gcc	100,000	0.128	0.153	0.140
482.sphinx	100,000	0.137	0.116	0.250
row_major	100,000	0.063	0.091	0.078
col_major	100,000	0.020	0.032	0.045
403.gcc	1,000,000	0.196	0.208	0.199
482.sphinx	1,000,000	0.137	0.187	0.462
row_major	1,000,000	0.057	0.129	0.103
col_major	1,000,000	0.028	0.305	0.312

supporting the intuition that simpler code has easier-to-predict dynamics. Note that the nonlinear modeling strategy was more accurate than MLR for *both* of these simple four-line matrix initialization loops. The repetitive nature of these loops leaves its signature in their dynamics: structure that is exposed by the geometric unfolding of the embedding process. LMA captures and uses that global structure—in effect, "learning" it—while MLR does not. Again, LMA's success here is even more impressive in view of the fact that the linear models require more information to construct. Finally, note that the LMA models beat the naive strategy for both row_major and col_major, but the linear MLR model did not.

Another important issue in modeling is sample rate. We explored this by changing the sampling rate of the traces while keeping the overall length the same: i.e., by sampling the same runs of the same programs at 1,000,000 instruction intervals, rather than every 100,000 instructions. This affected the accuracy of the different models in different ways, depending on the trace involved. For 403.gcc, MLR was still better than LMA, but not by as much. For 482.sphinx, the previous result (MLR better than LMA) was reversed. For row_major and col_major, the previous relationship not only persisted, but strengthened. In both of these traces, predictions made from MLR models were less accurate than simply predicting the mean; LMA predictions were *better* than this naive strategy. See the bottom four rows of Table 1 for a side-by-side comparison of these results to the more sparsely sampled results described in the previous paragraphs.

To explore which model worked better as the prediction horizon was extended, we changed that value—the k in Section 4.1—and plotted nRMSE. (Note that this does *not* involve an iterative reimplementation of the prediction strategy, but rather a series of disjoint experiments designed to test whether the accuracy degrades with increasing horizon. The results of the shorter-horizon runs are not used to inform the longer-horizon runs.) In three of the four traces—all but 482.sphinx—the nonlinear model held and even extended its advantage as the prediction horizon lengthened; see Figure 3 for some representative plots. The initial variability in these plots is an artifact of normalizing over short signal lengths and should be disregarded. The vertical discontinuities (e.g., at the 1300 and 2100 marks on the horizontal axis of the col_major plot, as well as the 1600 and 3000 marks of 482.sphinx) are also normalization artifacts[5].

---

[5] When the signal moves into a heretofore unvisited regime, that causes the $max - min$ term in the nRMSE denominator to jump.

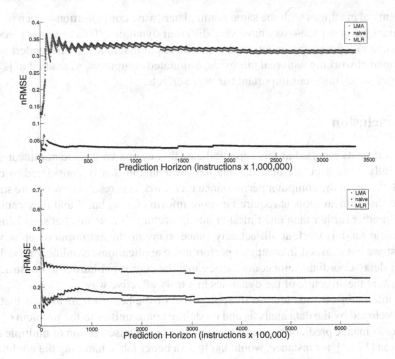

Fig. 3. nRMSE versus prediction horizon.Top: col_major Bottom: 482.sphinx

The sawtooth pattern in the top two traces in col_major nRMSE are due to the cyclical nature of that loop's dynamics. LMA captures this structure, while MLR and the naive strategy do not, and thus produces a far better prediction.

Experimental computer science involves a number of unexpected challenges, including the variability and nonstationarity of hardware and software. Since software plays such an important role in the dynamics, and since modern computers run so much housekeeping software, this is a real issue. The routine software updates that most computers receive periodically via the web, for instance, can completely destroy experimental repeatability. To control for these effects, the computer used in the experiments reported here was completely isolated from any network and its routine software update facility was disabled. Even so, experimental repeatability was an issue. Several months after the results above were gathered, we went back and re-ran the same 482.sphinx and 403.gcc tests repeatedly in order to do some statistical testing on the results. One of these tests caused the machine to halt; after that, the dynamics were completely different. We are currently working to understand this effect, which we suspect is due to an old version of the PAPI software that reads the values of the hardware performance registers.

The issues raised in the previous paragraph are a large part of why it is so important to build predictive models in real time—as is done here. *Post facto* machine-learning strategies that learn patterns from a large collection of examples cannot track the quickly changing behavior of modern computer systems. Variability is another important issue. Identical machines with the same nominal hardware configuration—even from the same manufacturing run—can have very different dynamics. This obviously makes it a challenge to repeat experiments across ensembles of machines. It also underlines the importance of working with *real* machines. Simulated computers, as studied in [20], do not model these subtle but important hardware effects.

5 Conclusion

The experiments reported in this paper indicate that even a very basic nonlinear model is generally more accurate than the simple linear model that is considered to be the state-of-the-art in the computer performance literature. This result is even more striking because those linear models require far more information to build and they cannot be used to predict further than one timestep into the future. It is somewhat surprising that these linear models work at all, actually, since many of the assumptions upon which they rest are not satisfied in computer performance applications. Nonlinear models that work in delay-coordinate embedding space may be somewhat harder to construct, but they capture the structure of the dynamics in a truly effective way.

It would be of obvious interest to apply some of the better linear models that have been developed by the data analysis and modeling communities to the problem of computer performance prediction. Even a segmented or piecewise version of multiple linear regression [17,21], for instance, would likely do a better job at handling the nonlinearity of the underlying system. This idea, which we are currently exploring, involves a major challenge: how to choose the breakpoints between the segments. Because of the complex nonlinear behavior, simple change-point detection techniques are not up to this task; because of the need for *real-time* detection of the regime shifts, sophisticated techniques that employ geometric or topological techniques (e.g., [1]) do not apply. And MLR is not really designed to be a temporal predictor anyway; linear predictors like the ones presented in [19] might be much more effective. Better *nonlinear* models are another obvious piece of our research agenda. There are regression-based nonlinear models that we could use, for instance, such as [11], as well as the many other more-sophisticated models in the nonlinear dynamics literature [5,28]. It might be useful to develop nonlinear models that use sliding windows in the data in order to adapt to regime shifts, but the choice of window size is an obvious issue. Nonlinear models that use multiple probes—like MLR does—could be extremely useful, but the underlying mathematics for that has not yet been developed. All of this work could be usefully informed by a broader set of experiments that brought out the connections between performance dynamics patterns and model accuracy. This, too, is underway in our lab.

Acknowledgements. This material is based upon work sponsored by the NSF under Grants CMMI 1162440 and SMA-0720692. Any opinions, findings, and conclusions or recommendations expressed in this material are those of the author(s) and do not necessarily reflect the views of the National Science Foundation.

References

1. Alexander, Z., Bradley, E., Garland, J., Meiss, J.: Iterated function system models in data analysis: Detection and separation. CHAOS 22(2) (April 2012)
2. Alexander, Z., Mytkowicz, T., Diwan, A., Bradley, E.: Measurement and dynamical analysis of computer performance data. In: Cohen, P.R., Adams, N.M., Berthold, M.R. (eds.) IDA 2010. LNCS, vol. 6065, pp. 18–29. Springer, Heidelberg (2010)
3. Bass, T.: The Eudaemonic Pie. Penguin, New York (1992)
4. Browne, S., Deane, C., Ho, G., Mucci, P.: PAPI: A portable interface to hardware performance counters. In: Proceedings of Department of the Defense HPCMP Users Group Conference (1999)
5. Casdagli, M., Eubank, S. (eds.): Nonlinear Modeling and Forecasting. Addison Wesley (1992)
6. Cavazos, J., Fursin, G., Agakov, F., Bonilla, E., O'Boyle, M.F.P., Temam, O.: Rapidly selecting good compiler optimizations using performance counters. In: Proceedings of the International Symposium on Code Generation and Optimization (2007)
7. Desikan, R., Burger, D., Keckler, S.W.: Measuring experimental error in microprocessor simulation. In: Proceedings of the International Symposium on Computer Architecture (2001)
8. Faraway, J.J.: Practical regression and ANOVA in R (2002),
   http://cran.r-project.org/doc/contrib/Faraway-PRA.pdf
9. Fraser, A., Swinney, H.: Independent coordinates for strange attractors from mutual information. Physical Review A 33(2), 1134–1140 (1986)
10. Garland, J., Bradley, E.: Predicting computer performance dynamics. In: Gama, J., Bradley, E., Hollmén, J. (eds.) IDA 2011. LNCS, vol. 7014, pp. 173–184. Springer, Heidelberg (2011)
11. Grassberger, P., Hegger, R., Kantz, H., Schaffrath, C., Schreiber, T.: On noise reduction methods for chaotic data. Chaos 3, 127 (1993)
12. Grechanik, M., Fu, C., Xie, Q.: Automatically finding performance problems with feedback-directed learning software testing. In: Proceedings of the International Conference on Software Engineering (2012)
13. Jain, R.: The Art of Computer Systems Performance Analysis: Techniques for Experimental Design, Measurement, Simulation, and Modeling, 2nd edn. John Wiley & Sons (1991)
14. Kantz, H., Schreiber, T.: Nonlinear Time Series Analysis. Cambridge University Press (1997)
15. Kennel, M., Isabelle, S.: Method to distinguish possible chaos from colored noise and to determine embedding parameters. Phys. Rev. A 46, 3111 (1992)
16. Kennel, M.B., Brown, R., Abarbanel, H.D.I.: Determining minimum embedding dimension using a geometrical construction. Physical Review A 45, 3403–3411 (1992)
17. Liu, J., Wu, S., Zidek, J.: On segmented multivariate regression. Statistica Sinica 7, 497–525 (1997)
18. Lorenz, E.N.: Atmospheric predictability as revealed by naturally occurring analogues. Journal of the Atmospheric Sciences 26, 636–646 (1969)
19. Makhoul, J.: Linear prediction: A tutorial review. Proceedings of the IEEE 63(4), 561–580 (1975)
20. Martinez, J.F., Ipek, E.: Dynamic multicore resource management: A machine learning approach. IEEE Micro 29(5), 8–17 (2009)
21. McGee, V.E., Carleton, W.T.: Piecewise regression. Journal of the American Statistical Association 65(331), 1109–1124 (1970)
22. Moseley, T., Kihm, J.L., Connors, D.A., Grunwald, D.: Methods for modeling resource contention on simultaneous multithreading processors. In: Proceedings of the International Conference on Computer Design (2005)

23. Myktowicz, T., Diwan, A., Bradley, E.: Computers are dynamical systems. Chaos 19, 033124 (2009), doi:10.1063/1.3187791
24. Mytkowicz, T.: Supporting experiments in computer systems research. Ph.D. thesis, University of Colorado (November 2010)
25. Packard, N., Crutchfield, J., Farmer, J., Shaw, R.: Geometry from a time series. Physical Review Letters 45, 712 (1980)
26. Sauer, T., Yorke, J., Casdagli, M.: Embedology. Journal of Statistical Physics 65, 579–616 (1991)
27. Takens, F.: Detecting strange attractors in fluid turbulence. In: Rand, D., Young, L.S. (eds.) Dynamical Systems and Turbulence, Springer, Berlin (1981)
28. Weigend, A., Gershenfeld, N. (eds.): Time Series Prediction: Forecasting the Future and Understanding the Past. Santa Fe Institute (1993)

Diversity-Driven Widening

Violeta N. Ivanova and Michael R. Berthold

Dept. of CIS and Graduate School Chemical Biology (KoRS-CB)
University of Konstanz, 78457 Konstanz, Germany
`firstname.lastname@uni-konstanz.de`

Abstract. This paper follows our earlier publication [1], where we introduced the idea of tuned data mining which draws on parallel resources to improve model accuracy rather than the usual focus on speed-up. In this paper we present a more in-depth analysis of the concept of Widened Data Mining, which aims at reducing the impact of greedy heuristics by exploring more than just one suitable solution at each step. In particular we focus on how diversity considerations can substantially improve results. We again use the greedy algorithm for the set cover problem to demonstrate these effects in practice.

1 Introduction

In [1], we claim that utilizing parallel compute resources to improve the accuracy of data mining algorithms and to obtain better models is of merit and is an important, emerging area of research in the the field of (parallel) data mining. The main reason for data mining algorithms not finding optimal solutions, is the usually enormous solution space, which requires the use of a – often greedy – heuristic. While this helps in finding a solution in reasonable time, it limits the exploration of the solution space and often leads to suboptimal solutions. In [1] we presented two generic strategies for using parallel resources to improve a greedy data mining heuristic, namely *Deepening* and *Widening*. Deepening focuses on smarter strategies to pick temporary solutions by looking several steps ahead and selecting a temporary solution, which has shown potential to perform better further down the search. The goal of Widening, in contrast, lies on achieving better accuracy by exploring more solutions simultaneously, not just the locally optimal one. We then demonstrated that both of these tuning methods offer potential for improvement using a widened versions of the greedy base algorithm for the set cover problem and a deepened decision tree learner.

In this paper we again focus on our key goal: the development of algorithm and architecture-independent generic strategies, which can be applied to a broad spectrum of data mining heuristics in order to improve their accuracy, while keeping the runtime constant. At the same time we wish to abstract away from implementation details, such as parallelization models. We will focus primarily on Widening in this paper and mechanisms to further improve the search.

The ideal goal of Widening is, given a sufficient (and hence usually enormous) number of parallel resources, to enable full exploration of the search space and

A. Tucker et al. (Eds.): IDA 2013, LNCS 8207, pp. 223–236, 2013.

guarantee the discovery of the optimal solution. In practice this is, of course, usually not feasible. Instead we need to make sure we make best use of every parallel resource available towards the goal of improvement of solution quality, while still keeping the runtime of the widened heuristic the same as the runtime of the original heuristic. The goal of cost-effective investment of parallel resources is closely related to the concept of *diverse exploration* of the search space, which is the main focus of this paper. The main goal of *Diversity-driven Widening* is to force different workers to investigate substantially different models, hence resulting in a diverse set of final solutions. Ideally this increases our chances to find an even better model than the standard heuristic. We describe simple ways to achieve diversity and illustrate how enforcing diversity in widening techniques helps to further improve the accuracy of the data mining heuristic. However, ensuring diversity often adds computational overhead which contradicts the second goal of Widening: keeping the runtime constant. We therefore investigate and compare Widening techniques with and without communication between the parallel workers. We demonstrate these different practical techniques on the greedy algorithm for the set cover problem.

2   Related Work

Trading quantity (of computational resources) for quality (of discovered solutions) has already been published before. In [2] the authors focus on a broad range of applications, ranging from cryptography to game playing. We, instead, focus on data mining algorithms which allows us to formalize a number of constraints based on the underlying model search space.

Plenty of related work exists in others areas, e.g. parallel data mining. We do not have the space to discuss all of this in detail and only briefly summarize the main trends and mainly focus on improvements of search heuristics through the use of diversity, as this is most relevant for the focus of this paper.

Speed-Up through Parallelization. For the vast majority of parallelizations of data mining algorithms, the aim is to improve efficiency. Comprehensive surveys are found in [3,4,5,6]. A large amount of work focuses on the parallelization of decision tree learning. One of the earliest distributed decision tree algorithms, SPRINT [7], has served as the basis for many subsequent parallel decision tree approaches. Some noteworthy examples include [8] (employing data parallelism), [9] (using task parallelism), and [10,11] (presenting hybrid approaches). Probably the second most researched area is parallel association rule mining algorithms. Extensive surveys exist in this area as well [12]. Parallelism in clustering algorithms has been used for both efficient cluster discovery and more efficient distance computations. Partitioning clustering algorithms are parallelized mostly using message-passing models, examples are presented in [13,14]. Examples for hierarchical clustering, which is more costly, include [15,16]. However, as discussed above, speed-up is not the primary goal of Widening.

In recent years specialized frameworks have also emerged, allowing data mining algorithms to be implemented in a distributed fashion and/or operating on distributed data. MapReduce [17] is the most prominent paradigm for processing parallel tasks across very large datasets and allows for massive scalability across thousands of servers in a cluster or a grid. Due to its inherent structure, MapReduce requires specialized versions of the data mining algorithms to be developed. Chu et al. [18] present a general approach by using a summation representation of the algorithms. While offering amazing scalability, MapReduce has inherent flaws – it is not designed to deal well with moderate-size data with complex dependencies; it is not suitable for algorithms that require communication between the parallel workers or impose other dependencies across iterations (see [19] for more details). So whereas MapReduce offers the potential for creating better models based on processing more data, Widening focuses on improving model quality using normal amounts of data. This does not exclude Widened MapReduce style implementations, of course, but it is not at the core of this paper.

Model Quality Improvement. A number of papers also concentrate on improving the accuracy of the models. Some approaches learn more models to be used in concert (ensembles) or in a randomized fashion (meta heuristics).

Ensembles use multiple models to obtain better predictive performance than could be obtained from any of the constituent models. The most notable examples are bootstrap aggregating or bagging [20], boosting [21], and random forests [22]. However, a high degree of accuracy comes at the price of interpretability, as these methods do not result in a single interpretable model, which is contrary to the goal of Widened Data Mining.

Learners based on stochastic learning algorithms, such as genetic algorithms are naturally parallelizable. Parallelization can be achieved by way of independent parallel execution of independent copies of a genetic algorithm, followed by selecting the best of the obtained results. This results in improved accuracy [23]. This is similar to Widening, however, Widening aims at exploring the search space in a structured way as opposed to the randomized nature of these other methods.

Using Diversity to Improve Search Heuristics. There is a wealth of literature focusing on the improvement of (greedy) search algorithms in general. In [24], an approach is presented for incorporating diversity within the cost function, which is used to select the intermediate solution. In [25], the authors use the observation that, in most cases, failing to find the optimal solution can be explained by a small number of erroneous decisions along the current path. Therefore, their improved search first explores the left-most child at a given depth as suggested by the current heuristic function. If no solution is found, all leaf nodes are tried that differ in just one decision point from the heuristic function choice. The algorithm iteratively increases the deviation from the greedy heuristic. The Widening proposed here performs a similar search for alternatives, but in parallel. In [26] the idea of adding diversity by a simple K-best first search was explored and shown empirically to be superior to the greedy (best-first) search heuristic.

Improving Set Covering. And finally, we should not forget to mention that a vast amount of literature addresses improving the greedy algorithm for the set cover problem [27,28]. Since we use this to illustrate the effects of Widening, we hasten to add that we do not intend to compete with these approaches! However, the greedy base set covering algorithm allows the benefits of Widening to be demonstrated well and intuitively.

3   General Widening of a Greedy Heuristic

We can view many of the data mining algorithms as a *greedy search* through a space of potential solutions, the model *search space*. This search space consists of model candidates, from which the greedy algorithm chooses a locally optimal solution at each step, until a sufficiently good solution is found, based on some stopping criteria. The greedy search can, therefore, be schematically presented as an iterative application of two operators: *refinement r* and *selection s*.

During the refinement operation, a temporary model m is made more specific to generate new, potentially better, models (which we refer to as *refinements*). The *selection* operator chooses the locally best model from all possible refinements.

For the purpose of this paper it is sufficient to assume the existence of a family of models $\mathcal{M}$, that constitutes the domain of the two operators. The refinement operator is model and algorithm specific and the selection operator is usually driven by the training data. We will investigate the selection operator in more detail, as it will be the tool we use to widen a greedy heuristic. It is usually based on a given quality measure ψ, which evaluates the quality of a model m from a family of models $\mathcal{M}$ (and therefore also its refinements):

$$\psi : \mathcal{M} \to \mathbb{R}.$$

Employing this notation, we can present one iterative step of the greedy search as follows:

$$m' = s_{\text{best}}(r(m)),$$

where

$$s_{\text{best}}(M) = \arg\max_{m'' \in M} \{\psi(m'')\},$$

that is, the model from the subset $M \subseteq \mathcal{M}$ which is ranked highest by the quality measure is chosen at each step. Figure 1 depicts this view of a greedy model searching algorithm.

We can now also specify how we got to a certain model and define the concept of *selection path*, which defines how a specific model is reached:

$$\mathcal{P}_s(m) = \{m^{(1)}, m^{(2)}, \dots, m^{(n)}\},$$

where the order is specified via the refinement/selection steps, that is

$$\forall i = 1, \dots, n-1 : m^{(i+1)} = s(r(m^{(i)})).$$

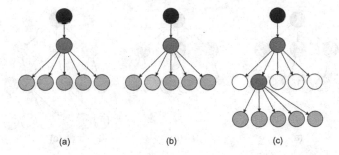

Fig. 1. The classic heuristic (often greedy) search algorithm. On the left (a), the current model m is depicted in green, the refinement options $r(m)$ are shown gray. The selection operator s picks the yellow refinement (b) and the next level then continues the search based on this choice.

and $m^{(n)} = m$ and $m^{(1)}$ is a root model for which no other model exists that it is a refinement of. Note that the selection path depends heavily on the chosen selection operator s, which will come in handy later.

3.1 Widening of a Greedy Heuristic

In order to improve the accuracy of the greedy algorithm one has to deal with its inherent flaw – the fact that a locally optimal choice may in fact not lead us towards the globally optimal solution. To address this issue, we can explore several options in parallel – which is precisely what Widening is all about. How those parallel solution candidates are picked is the interesting question, which we will address later, but let us first look into widening itself in a bit more detail. Using the notation introduced above, one iteration of Widening can be represented as follows:

$$M' = \{m'_1, \ldots, m'_k\} = s_{\text{widened}} \left(\bigcup_{m \in M} r(m) \right).$$

That is, at each step, the *widened* selection operator s_{widened} considers the refinements of a set M of original models and returns a new set M' of k refined models for further investigation. We will refer to parameter k as the *width* of the widened search. Intuitively, it is clear that the larger the width, i.e. the more models (and hence selection paths) are explored in the solution space, the higher our chances are of finding a better model in comparison to the normal greedy search. Figure 2 illustrates this process.

An easy implementation of the above (what we will later refer to as top-k Widening) is a beam search. Instead of following one greedy path, the path of k best solution candidates is explored. However, this does not guarantee that we are indeed exploring alternative models – on the contrary, it is highly likely that we are exploring only closely related variations of the locally best model.

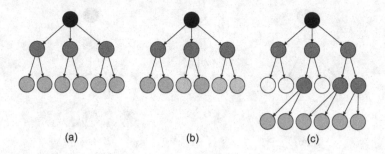

Fig. 2. Widening. From a set of models M (green circles), the refinement operator creates several sets of models (gray), shown on the left (a). The selection now picks a subset of the refined models (yellow circles in (b)) and the search continues from these on the right (c).

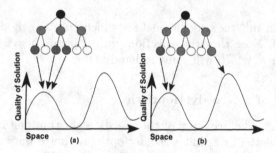

Fig. 3. Normal Widening may lead to local exploitation only (a). Adding diversity constraints encourages broader exploration of the model space (b).

In the area of genetic algorithms this effect is known as *exploitation*, that is, we are essentially fine tuning a model in the vicinity of an (often local) optimum.

3.2   Diversity-Driven Widening

In order to avoid the local exploitation, as discussed above, we can add diversity constraints which enforces the search to more broadly investigate our search space. In genetic algorithms this is often called *exploration*. Figure 3 illustrates the difference between local exploitation and global exploration. This effect has also been shown to improve results quite considerably for other beam search type problems, as we briefly discussed in Section 2. By forcing the parallel workers to consider not only multiple selection paths, but also *diverse* ones simultaneously, we aim to obtain better exploration of the search space and escape entrapment by local optima. Techniques for this type of Diversity-driven Widening are the main focus of this paper.

4   Techniques for (Diversity-Driven) Widening

In this section we will describe several specific techniques for Widening. We start by establishing the base *top-k Widening*, describe the diversity-driven version, and then focus on approaches that require less or no communication effort.

Top-k Widening. In [1] we already described this approach to Widening. It is the most obvious approach and identical to a classic beam search. In each iteration of top-k Widening each parallel worker selects the top k choices for the refinements of its model and from the resulting k^2 choices, the top k are chosen:

$$\{m'_1, \ldots, m'_k\} = s_{\text{top-}k}\left(\bigcup_{i=1,\ldots,k} s_{\text{top-}k}\left(r(m_i)\right)\right)$$

where $s_{\text{top-}k}$ selects the top k models from a set of models according to the given quality measure ψ. Obviously, $s_{\text{top-}1} = s_{\text{best}}$.

In [1] we demonstrate that top-k Widening leads to an improved quality, with larger *width* k leading to better accuracy. However, two main flaws exist. The first problem, as mentioned above already, is that possibly only a small neighborhood of the best solutions is explored. Secondly, continuous communication is required between threads which contradicts our goal of wanting to keep the time constant.

Diverse Top-k Widening. As discussed above we can tackle the first flaw of Top-k Widening by enforcing diversity. One simple way to add diversity can be achieved by using a fixed diversity threshold θ, a distance function δ, and by modifying the selection operator $s_{\text{top-}k,\delta}$ to iteratively pick the best k refinements, that satisfy the given diversity threshold. This can be summarized as follows:

```
1:  M_all = ∪_{i=1,...,k} r(m_i)              create set of all possible refinements
2:  m_1 = arg max_{m∈M_all}{ψ(m)}            pick the locally optimal model as first model
3:  M_1 = {m_1}                              add as initial model to solution
4:  for i = 2,...,k:                         iteratively pick next, sufficiently diverse model:
5:    m_i = arg max_{m∈M_all}{ψ(m) | ¬∃m' ∈ M_{i-1} : δ(m,m') < θ}
6:    M_i = M_{i-1} ∪ {m_i}
7:  endfor
8:  return M_k
```

This is a known approach for diverse subset picking, however, our second problem persists: we still require frequent communication among our parallel workers to make sure we pick a diverse solution subset among all intermediate solutions at each iteration.

Communication-Free Widening. In order to achieve communication-free Widening, we must force each worker to focus on its own subset of models without continuously synchronizing this with the other workers. Ideally, communication-free Widening is achieved via partitioning the model search space between the parallel workers in such a way, that the full search space is covered by the partitioning,

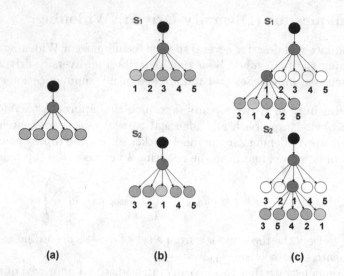

Fig. 4. Communication-free Widening: two different selection paths generated by two different selection operators s_1 and s_2.

every model is reachable in at least one partition, and there is no overlap between different partitions. Even more ideally, those partitions contain sufficiently diverse solutions. However, those objectives are difficult to meet in practice. Our current goal is more modest: we aim to enforce diverse selection paths to be explored by the different parallel workers, which will hopefully lead to diverse final solutions. We approach this goal indirectly, by assigning a modified quality measure ψ_i to each selection operator s_i (of worker i), which, when given a choice, has a personalized preference for a (different) subset of models. If this individualized assignment is properly implemented, each parallel worker i will explore a different selection path $\mathcal{P}_{s_i}$ if refinements with sufficiently similarly high quality exist.

Figure 4 illustrates two different selection paths generated by two different selection operators s_1 and s_2. Our goal that each individualized selection operator explores a different and diverse path through the search space can be achieved by a modification of the selection operators which we describe below.

Diverse Communication-Free Widening. As described above, we need to ensure that the explored selection paths by the parallel workers $i = 1, \ldots, k$ are *sufficiently diverse*. To increase the chances for different parallel workers to explore diverse paths, we modify the k quality measures ψ_i of operator s_i so that each ψ_i assigns different preferences for the models in the search space. In the most subtle case, this will only break ties between models differently when we have more than one refinement with the (locally) optimal quality. However, to achieve real exploration, we will also want to lift slightly worse models above better ones *for some* of our workers but we need to ensure that at least one worker still investigates the locally optimal choice. It is important to note that while we want to explore different and diverse selection paths, we also do not want to focus solely on diversity. Random diverse exploration may lead to investing resources in the discovery of many degenerate solutions. We will need to balance between the two notions "diversity" and "quality". This trade-off was already visible in the case of the diverse top-k approach discussed above, where the threshold θ determines how much the selection operator $s_{\text{top-}k,\delta}$ is influenced by diversity and how much by the quality of the remaining solutions.

Enforcing diversity by modifying the underlying quality measures tends to be fairly algorithmic specific. Generally, the quality measures can assign diverse preferences directly to models (or parts of models), which we call *model-driven diversity* or by assigning preferences based on data points, which we term *data-driven diversity*. We will show examples for these approaches in the following section.

5 Diversity-Driven Widening of Set Covering

Various data mining problems employ strategies that are similar to the set cover problem. We have already used this algorithm in [1] to illustrate the benefits of Widening and will use it here as well.

5.1 The Set Cover Problem

We consider the standard (unweighted) set cover problem. Given a universe X of n items and a collection $\mathcal{S}$ of m subsets of $X : \mathcal{S} = \{S_1, S_2, \ldots, S_m\}$. We assume that the union of all of the sets in $\mathcal{S}$ is X: $\bigcup_{S_i \in \mathcal{S}} S_i = X$. The aim is to find a sub-collection of sets in $\mathcal{S}$, of minimum size, that contain ("cover") all elements of X.

The standard iterative algorithm [29] follows a greedy strategy, which, at each step, selects the subset with the largest number of remaining uncovered elements. Using the formalizations introduced above, a single iterative step of the algorithm operates as follows: if m is the temporary cover, a refinement generated by $r_{\text{greedySCP}}(m)$ represents the addition of a single subset, not yet part of m, to m. From all of the possible refinements, generated by $r_{\text{greedySCP}}(m)$, $s_{\text{greedySCP}}$ picks the one with the largest number of covered elements as the new intermediate cover. The quality measure ψ, used by the selection operator, $s_{greedySCP}$, therefore simply ranks the models based on the number of elements they cover.

5.2 Diversity-Driven Widening of Set Covering

In the following we will discuss how we can use the widening strategies described above for the greedy algorithm for the set cover problem. Note that our goal here is not to outperform other algorithmic improvements of the standard greedy set covering algorithm but instead use this to illustrate the benefits of Widening itself.

Top-k Widening and Diversity. Instead of selecting one locally best intermediate cover, the top-k Widening of the greedy SCP algorithm selects k best
covers at each given step. To implement diversity, we can use a a simple threshold based on the Jaccard distance and enforce that the chosen k intermediate
covers chosen by the selection operator $s_{\text{top-}k,\delta}$ at each step have a minimum
distance:

$$\delta(m_i, m_j) = 1 - \frac{|m_i \cup m_j|}{|m_i \cap m_j|}.$$

(Each model m covers a set of elements, so we are interested in picking intermediate models that are sufficiently different.)

Communication-Free Widening: Model-Driven Diversity. Enforcing diversity without continuously comparing intermediate models is more difficult.
We can define individual quality measures ψ_i, by enforcing different preferences
for different subsets. Given an intermediate cover m, ψ_i evaluates the refinement
$m' = m \cup S_j$ for an additional subset S_j based on the original quality measure
and an individual preference weight $w_i \in (0, 1)$ for the subset S_j:

$$\psi_i(m \cup S_j) = \psi(m \cup S_j) + t * w_i(S_j),$$

The set of weights $w_i(\cdot)$ for a given ψ_i defines an order π_i on the set of subsets
S for a particular parallel worker i:

$$w_i(S_{\pi_i(1)}) > \cdots > w_i(S_{\pi_i(|S|)})$$

Our goal is to have k diverse orders $\pi_1, \ldots, \pi_k$ of the subsets by ensuring that
the *inversion distances* between different orders are maximized. The inversion
distance between two ordered sets calculates how many pairs of elements are
present in a different order in the two orders π_p and π_q:

$$d_{\text{inv}}(\pi_p, \pi_q) = \sum_{k \neq l} \begin{cases} 1 & \text{if } (\pi_p(k) - \pi_p(l)) \cdot (\pi_q(k) - \pi_q(l)) < 0 \\ 0 & \text{else} \end{cases}.$$

Assigning preferences in this fashion will steer the selection operators based on
characteristics of the models (or model fragments), hence the term *model-driven
diversity*.

By varying parameter t, we can control how much the selection paths of the
parallel workers deviate from the selection paths explored by the greedy SCP
algorithm. The parameter t controls the relative importance of the factors quality
and diversity. For parameter $t \leq 1$, the parallel workers explore different selection
paths of the greedy algorithm, only considering different paths that have equally
good, local quality. Here the different orders π_i only serve to differentiate the tie
breaking. For parameter values $t > 1$, the selection paths of the parallel workers
also include locally sub-optimal solutions. Large values of t ($\gg 1$) will lead to
random exploration of the search space.

Communication-Free Widening: Data-Driven Diversity. In contrast to the model (fragment) driven diversity described above, we can also ensure diversity by weighting data elements. To accomplish this we enforce diverse preferences for the elements from X for the different selection operators s_i:

$$\psi_i(m \cup S_j) = \psi(m \cup S_j) + t \cdot \frac{1}{|\{e \in S_j \wedge e \notin m\}|} \sum_{e \in S_j \wedge e \notin m} w_i(e),$$

the preference for different elements is again defined via weights $w_i(e)$ and the weights define an ordering on the elements where we again aim for k different orderings via sufficient inversion distance. Note that this approach bears some similarities to boosting because we weight the impact of data elements on the model quality measure differently.

It must be noted, that, while using diverse quality measures can help steer the parallel workers into diverse selection paths, it by no means guarantees it. Choosing different models at each step can still lead to having the same final solution, just generated along a different paths. Implementing selection in such a way that diversity of the obtained final solutions is guaranteed is an interesting focus of future work. In the following section, however, we will demonstrate that regardless of the lack of theoretical guarantees, our simple approaches to Diversity-driven Widening are beneficial.

6   Experimental Evaluation and Discussion

In this section we demonstrate the impact of the Widening techniques discussed above using three benchmark data sets: $rail507$, $rail516$, and $rail582$ [30]. We aim to demonstrate how different widths of the search affect the quality of the solution and the additional benefit of enforcing diversity on the widened searches. Each experiment was repeated 50 times. For diverse top-k Widening, a Jaccard distance threshold of 0.01 was used in all experiments.

Figure 5 shows the results for top-k Widening with and without diversity. Figure 6 shows the results for communication-free Widening without diversity

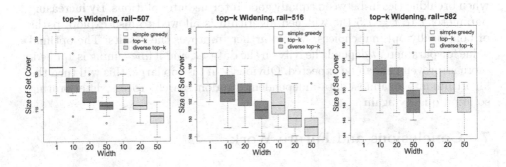

Fig. 5. Results from the evaluation top-k Widening with and without diversity

234 V.N. Ivanova and M.R. Berthold

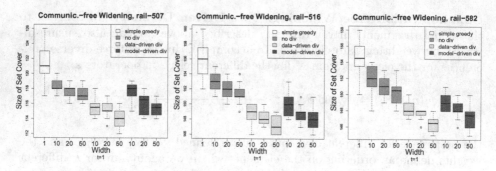

Fig. 6. Results from communication-free Widening without diversity, with data-driven and model-driven diversity, for parameter $t = 1$

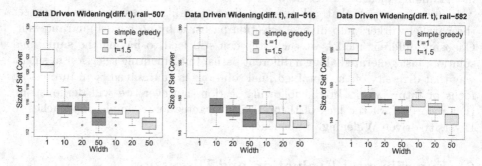

Fig. 7. Results from communication-free Widening using data-driven diversity for different values of parameter t

and with data- resp. model-driven diversity enforcement using a fixed trade-off parameter $t = 1$, while Figure 7 demonstrates the impact of trade-off parameter t for data-driven diversity.

From the above results two main trends become clear. As expected, a larger width of the search does improve the quality of the solution. Enforcing diversity improves the results even further. For communication-free Widening, the first set of tests simply enhances the greedy algorithm by exploring different options when breaking ties in-between equally good intermediate solutions. By increasing parameter t to $t = 1.5$ the widened algorithm is allowed to also explore paths of non-locally optimal choices, which further improves the results. The optimal value for parameter t depends heavily on the dataset, and if fine-tuning is applied, more improvement can be expected. Obviously, if t is too large, this will turn the algorithm into an almost data-independent, random search process, deteriorating solution quality again.

7 Conclusions and Future Work

We continued earlier work on the impact of Widening on data mining algorithms. A number of practical techniques to implement Widening focusing on reduction of communication and enhancing the exploration of the solution space were presented. The latter shows promise to further increase the accuracy of Widening as we have demonstrated using the base greedy set covering algorithm. Focusing on better ways to enforce diversity without the need for extensive communications is an area of future work, as is the application of the presented techniques to other data mining algorithms.

Acknowledgements. We thank Arno Siebes and Oliver Sampson for many interesting discussions and constructive feedback. V. Ivanova is supported by the DFG under grant GRK 1042 (Research Training Group "Explorative Analysis and Visualization of Large Information Spaces").

References

1. Akbar, Z., Ivanova, V.N., Berthold, M.R.: Parallel data mining revisited. Better, not faster. In: Hollmén, J., Klawonn, F., Tucker, A. (eds.) IDA 2012. LNCS, vol. 7619, pp. 23–34. Springer, Heidelberg (2012)
2. Akl, S.G.: Parallel real-time computation: Sometimes quantity means quality. Computing and Informatics 21, 455–487 (2002)
3. Kumar, V.: Special Issue on High-performance Data Mining. Academic Press (2001)
4. Kargupta, H., Chan, P.: Advances in Distributed and Parallel Knowledge Discovery. AAAI/MIT Press (2000)
5. Zaki, M.J., Ho, C.-T. (eds.): KDD 1999. LNCS (LNAI), vol. 1759. Springer, Heidelberg (2000)
6. Zaki, M.J., Pan, Y.: Introduction: Recent developments in parallel and distributed data mining. DPD 11(2), 123–127 (2002)
7. Shafer, J., Agrawal, R., Mehta, M.: Sprint: A scalable parallel classifier for data mining. In: VLDB, pp. 544–555 (1996)
8. Zaki, M.J., Ho, C.-T., Agrawal, R.: Parallel classification for data mining on shared-memory multiprocessors. In: ICDE, pp. 198–205 (1999)
9. Darlington, J., Guo, Y.-K., Sutiwaraphun, J., To, H.W.: Parallel induction algorithms for data mining. In: Liu, X., Cohen, P., Berthold, M. (eds.) IDA 1997. LNCS, vol. 1280, pp. 437–445. Springer, Heidelberg (1997)
10. Srivastava, A., Han, E.-H., Kumar, V., Singh, V.: Parallel formulations of decision-tree classification algorithms. DMKD 3(3), 237–261 (1999)
11. Kufrin, R.: Decision trees on parallel processors. In: PPAI, pp. 279–306 (1995)
12. Zaki, M.J.: Parallel and distributed association mining: a survey. IEEE Concurrency 7(4), 14–25 (1999)
13. Judd, D., McKinley, P.K., Jain, A.K.: Large-scale parallel data clustering. TPAMI 20(8), 871–876 (1998)
14. Dhillon, I., Modha, D.: A data-clustering algorithm on distributed memory multiprocessors. In: Large-scale Parallel KDD Systems Workshop, ACM SIGKDD, pp. 245–260 (2000)
15. Olson, C.F.: Parallel algorithms for hierarchical clustering. JPC 21, 1313–1325 (1995)
16. Garg, A., Mangla, A., Gupta, N., Bhatnagar, V.: PBIRCH: A scalable parallel clustering algorithm for incremental data. In: IDEAS, pp. 315–316 (2006)

17. Dean, J., Ghemawat, S.: MapReduce: simplified data processing on large clusters. Commun. ACM 51(1), 107–113 (2008)
18. Chu, C.-T., Kim, S.K., Lin, Y.-A., Yu, Y.Y., Bradski, G.R., Ng, A.Y., Olukotun, K.: Map-reduce for machine learning on multicore. In: NIPS, pp. 281–288 (2006)
19. Ma, Z., Gu, L.: The limitation of MapReduce: A probing case and a lightweight solution. In: Intl. Conf. on Cloud Computing, GRIDs, and Virtualization, pp. 68–73 (2010)
20. Breiman, L.: Bagging predictors. JML 24(2), 123–140 (1996)
21. Schapire, R.E.: The strength of weak learnability. JML 5, 28–33 (1990)
22. Breiman, L.: Random forests. JML 45(1), 5–32 (2001)
23. Talia, D.: Parallelism in knowledge discovery techniques. In: Fagerholm, J., Haataja, J., Järvinen, J., Lyly, M., Råback, P., Savolainen, V. (eds.) PARA 2002. LNCS, vol. 2367, pp. 127–136. Springer, Heidelberg (2002)
24. Shell, P., Rubio, J.A.H., Barro, G.Q.: Improving search through diversity. In: AAAI (1994)
25. Harvey, W.D., Ginsberg, M.L.: Limited discrepancy search. IJCAI, 607–615 (1995)
26. Felner, A., Kraus, S., Korf, R.E.: KBFS: K-best-first search. AMAI 39(1-2), 19–39 (2003)
27. Berger, B., Rompel, J., Shor, P.W.: Efficient nc algorithms for set cover with applications to learning and geometry. JCSS 49(3), 454–477 (1994)
28. Blelloch, G.E., Peng, R., Tangwongsan, K.: Linear-work greedy parallel approximate set cover and variants. In: SPAA, pp. 23–32 (2011)
29. Johnson, D.S.: Approximation algorithms for combinatorial problems. In: STOC, pp. 38–49 (1973)
30. Beasley, J.E.: Or-library: Distributing test problems by electronic mail. The Journal of the Operational Research Society 41(11), 1069–1072 (1990)

Towards Indexing of Web3D Signing Avatars

Kabil Jaballah and Mohamed Jemni

LaTICE Research Laboratory, ENSIT, University of Tunis,
5 Avenue Taha Houssein, 1008 Tunis, Tunisia
Kabil.jaballah@utic.rnu.tn, Mohamed.jemni@fst.rnu.tn

Abstract. Signing avatars are becoming common and being published on the World Wide Web at an incredible rate. They are actually used in education to help deaf children and their parents to learn sign language thanks to many 3D signing avatars systems. In Tunisia, signing avatars have been used since several years as part of the WebSign project. During the last few years, thousands of 3D signing avatars have been recorded using WebSign and few other systems. One of the major challenges that we was facing is how to index and retrieve efficiently this huge quantity of 3D signed scenes. Indexing and cataloging these signed scenes is beyond the capabilities of current text-based search engines. In this paper, we describe a system that collects 3D signed scenes, processes and recognizes the signed words inside them. The processed scenes are then indexed for later retrieval. We use a novel approach for sign language recognition from 3D scenes based on the Longest Common Subsequence algorithm. Our system is able to recognize signs inside 3D scenes at a rate of 96.5 % using a 3D model dictionary. We present also a novel approach for search and results ranking based on the similarity rates between 3D models. Our method is more efficient than Hidden Markov Models in term of recognition time and in the case of co-articulated signs.

Keywords: Sign language, virtual reality, signing avatars, websign, content based, search engine, 3D retrieval.

1   Introduction

Sign language is a visual/spacial language used by deaf individuals. No one of sign language is universal; different sign languages are used in different countries or regions [1] [1]. In the USA, the language used by the deaf community is ASL (American Sign language). In Tunisia, different sign languages are used in different regions. Unlike spoken language, sign languages are based on iconic signs [2, 3, 4] [2, 3, 4] which make this means of communication more complex. Unfortunately deaf youngsters lack access to many sources of information and are not exposed to media (radio, television, conversations around the dinner table etc.) where concepts related to science may be discussed. Therefore, some concepts that hearing children learn incidentally in everyday life may have to be explicitly taught to deaf pupils in school.

Signing avatars have the potential to overcome the barriers faced by deaf people to access to sources of information. These 3D animated characters are able to interpret and provide sign language translation to any type of media including educational

A. Tucker et al. (Eds.): IDA 2013, LNCS 8207, pp. 237–248, 2013.

resources. Even though videos of real signers are the common way for providing signed translations to educational material as well as full dictionaries[1] for many sign languages[2], signing avatars are progressively getting interest and seem to have promising future despite the fact that understandability issues of 3D sign language have been reported [5]. In this context, in Tunisia, we have been using signing avatars to teach deaf children and their parents since several years. More than one thousands of 3D signed "phrases" have been created as part of the WebSign project [6] [6] for teaching purposes [7] as well as translation purposes. H-animator [8] and Vsigns [9] have been used to create teaching material[3] as well. Unfortunately, the huge amount of 3D signed scenes collected thus far is not indexed and retrieved efficiently. In fact, to search for a sign inside the data repository, the user has two choices: either explore and play all the scenes until the wanted sign is found or insert a query in order to match scenes according to their filenames. Both of two choices might generate irrelevant results because the scenes are not indexed according to what is being signed.

The main contribution of this study is that we propose effective and efficient indexing and matching algorithm for content-based signing avatars retrieval. The proposed framework's anatomy is similar to a search engine where we proposed new approaches for:

- Motion extraction and normalization from Web3D signing avatars
- 3D Signs automatic recognition using dynamic programming
- 3D scenes indexing using SML (Sign Markup Language)
- User query interpretation and results sorting according to similarity measures

The proposed system can be used by anyone who uses signing avatars which are compliant with Web3D standards (H-anim and X3D/VRML). A pre-treatment is necessary and done automatically before a scene is indexed. In fact, the system first checks the processed scene structure and description and verifies its compliance with Web3D standards. The remainder of this paper is organized as follows. In section 2 we make review on previous work on sign language recognition and 3D motion indexing and retrieval. In section 3, we give details of the proposed system's architecture and the main approaches used to recognize and index 3D sign language scenes. Before concluding, experimental results are discussed in section 4.

2 Related Work

Since 3D signing avatars contain rich spatial-temporal data, our work actually belongs to both sign language recognition (classic approaches applied to videos of real

---

[1] http://www.aslpro.com
http://www.signingsavvy.com
http://www.sematos.eu/lsf.html
http://www.lsfplus.fr/
http://www.pisourd.ch/index.php?theme=dicocomplet
http://www.wikisign.org
[2] Sign languages for different countries and different regions.
[3] Websites, interactive learning environments , etc.

signers) and 3D motion retrieval (gestures and motion recognition from 3D virtual humans). In this section, we make a brief review on previous work on 3D motion analysis and synthesis as well as a detailed review on classical approaches used in the context of sign language recognition from video data.

2.1   3D Motion Retrieval

Automatic retrieval of actions/gestures in 3D space is a challenge but useful technique. Motion tracking and recognition of human interactions by a multi-layer finite state machine is presented in [10]. By using body pose vectors, human gestures recognition is presented in [11][11]. Retrieval of actors in a 3D scene becomes a useful technology [12] [13] [13], if animated actors in a scene database are to be reused [14]. Instead of using user pre-defined metadata [14], 3D animations should be retrieved based on the existing animation models. We are actually involving in one particular domain which is VRML/X3D animated characters that "speak" sign language. These animated characters are recorded and reused in the context of virtual learning environments (VLE). The appropriate technique for 3D motion recognition and retrieval depends on the input data. We distinguish 2 main methods: model based methods and appearance based methods. In our case, knowing that we deal with VRML/X3D scenes, where important parameters like palm position and joints angles are available, we use skeletal based methods. In fact, skeletal algorithms make use of segments mapping of the body. With the skeletal representation of the human body[4], it is possible to use pattern matching against a model database.

2.2   Sign Language Recognition

Researches on sign language automatic recognition began in 1990s. At the beginning, the recognition concerned mainly finger spelling from isolated signs. In this context, Takahashi and Kishino [15] in 1991 proposed an approach that relied on wired gloves the user had to wear. The system was capable of recognizing handshapes, but ignored any other features and was limited to finger spelling. Their experiments showed that 30 out of 46 pre-defined gestures of the Japanese kana manual alphabet could be recognized. For specific gestures, additional information such as "fingertips touching" were required, but not supplied by the gloves. A later approach in 1995 by Starner and Pentland featured real-time recognition of American Sign language from video and made use of Hidden Markov Models (HMMs). It required the user to wear solid colored gloves for better stability. Several restrictions were imposed, for instance some features of ASL (referencing objects by pointing at them, as well as all facial features) were ignored, and only sentences of a specific type (personal pronoun,verb, noun, adjective, same personal pronoun) were regarded. The system had a word accuracy of 99.5% of 395 sentences on training with grammar restrictions enabled, and 92% with grammar restrictions disabled. A problem with disabled restrictions was multiple insertions of the same sign, even if it was only signed once.

---

[4] H-anim description of the skeleton.

Vogler and Metaxas[16] used 3D data for their system in 1998 by setting up a system of three orthogonally placed webcams. After recovering the body parts from video, the data were used as input for HMMs for continuous (rather than isolated) signer-dependent sign language recognition. They experimented with 2D data input and the results showed that by using 3D data, higher word accuracy could be achieved.

In 2008, Dreuw et al [17] [17] proposed a signer-independent system for sign language recognition based on speech recognition techniques, using a single webcam, without the need for any gloves. After several optimizations, the word error rate could be decreased to 17.9% on a vocabulary of 104 signs signed by three speakers. Unlike previous works that concentrated on manual features only, a more recent approach in 2009 by Kelly et al [18] also incorporates a non-manual feature, namely head movement. The system relies on a single webcam and the user wearing colored gloves for continuous sign recognition. Testing of the framework consisted of 160 video clips of unsegmented sign language sentences and a small vocabulary of eight manual signs and three head movement gestures. A detection ratio of 95.7% could be achieved.

3   Framework Overview

As we mentioned previously, the target scenes are compliant with H-anim. The Humanoid Animation standard (H-anim) [5] was developed by the Humanoid Animation Working Group of Web3D consortium[6]. H-anim specifies a hierarchy of 89 joint with 89 segments, and in its brand new version, a node called *End Effector* that supports animation using *Inverse-Kinematics* has been added. This standard specifies how to define the form and the behavior of 3D humanoids with standard Virtual Reality Modeling Language (VRML) [19] or its successor Extensible 3D Graphics (X3D) [20]. Knowing that X3D is XML-based; it is actually widely used for rendering 3D web applications. The user only needs an X3D/VRML compliant player to view the rendered animations[7]. More recently, X3D has been integrated with HTML5[8] thanks to its X3DOM node and does not subsequently need any plugin since it uses only WebGL[9] and JavaScript[10] to render 3D contents.

3.1   Motion Extraction and Representation

The extraction of key-frames from Web3D scenes comes after validating those scenes to be sure that they are compliant with H-anim. During the validation phase, we use an XSD (XML schema Description) provided by the H-anim working group. We surely treat only the joints where some motion is applied in the scene graph. Even though the processed scenes are compliant with H-anim, we still need to normalize the rotation values that have been grabbed from the animated joints. In this context, the majority of signing avatars systems use the Quaternion representation of 3D

---

[5] http://www.h-anim.org/

[6] http://www.web3d.org/

[7] http://www.web3d.org/x3d/content/examples/X3dResources.html

[8] http://www.w3.org/TR/html5/

[9] http://www.khronos.org/webgl/

[10] http://www.web3d.org/x3d/wiki/index.php/
    X3D_and_HTML5#X3DOM_proposal

rotations, some other scenes uses the Euler representation. During the motion extraction process, a conversion to Quaternion is applied.

The table below shows a portion of a returned array by our algorithm. For every time slot, we obtain all rotation values of the avatar's animated joints. To represent 3D rotations, we adopt the representation based on quaternion (x,y,z,w). The conversion from any representation to quaternions is presented *Euclidean space*[11]

Table 1 is a portion of extracted rotation values from an avatar signing the word "bonjour". We selected the rotation values concerning two joints "r_elbow" and "r_index". This motion was detected at t = 0.5 seconds.

Table 1. Extracted motion sequence using Fetch_Rotation Algorithm

	JointName	r_elbow			
	RotationValue	*X*	*Y*	*Z*	*W*
		-0.498537	1.092814	-0.108278	1.955234
0.5	JointName	r_index1			
	RotationValue	*X*	*Y*	*Z*	*W*
		-12.37785	0.000000	0.000000	0.161756

Once the motion sequence is extracted and normalized, we have to represent the extracted data through an adequate model. For this reason, we use SML (Sign language Markup language) [6], which based on a phonologic representation. The SML tree is used to index motion in an efficient way that allows us to retrieve and reuse this motion. Below is the figure 1 that represents the index structure used to store 3D sign language scenes in the dictionary. The tags "wordn" are only inserted if a signed word from the dictionnary has been matched.

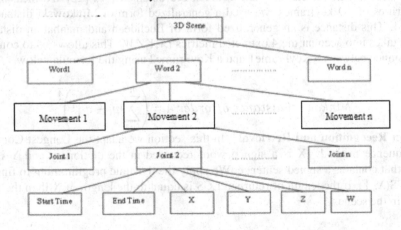

Fig. 1. SML tree for data representation and storage

---

[11] http://www.euclideanspace.com

3.2   Motion Retrieval

The extracted data (see table1) represents a sequence of 3D key-frames that have to be retrieved against a database (Dictionary). Known that scenes to be processed are issued from different systems, the number of keyframes and their rotation values could be different from those recorded in our dictionary. Thus, our problem is similar to the Longest Common Subsequence (LCS) problem [21] where:

-   The scene to be processed is the Text: T[Key-Frame₁,..,Key-Frameₙ], where key-frames are the returned 3D rotation values by our *Fetch_Rotations* algorithm.

-   The Signed word (recorded in the dictionary) to search in the scene is the pattern: P[KeyFrame₁,..,KeyFrameₘ].

-   Every Key-Frame belongs to $\sum$ alphabet which is composed by joints names as described by H-anim.

    o $\Sigma = \{r_{elbow}, l_{elbow}, r_{wrist}, l_{wrist}, ...\}$

3D Key-Frames Comparison: Key-frames are 3D postures extracted from the scenes with quaternion rotation values. We are in the context of a 4-dimension space. In this context, the similarity of two postures is computed by a metric distance in the 4-dimensional space of feature vectors. In the literature, the Minkowski family of distances (L_s) is commonly used in the case of 3D objects matching. Examples of these functions are (L_1), which is the Manhattan distance, (L_2), which is called the Euclidean distance. Minkowski distance is used in our context. This feature-based approach has several advantages compared to other approaches for implementing similarity search. The extraction of features from multimedia data like 3D scenes is usually fast and easily parametrizable. Metrics for FVs (Feature vectors) such as the Minkowski distances can also be efficiently computed. Spatial access methods or metric access methods can be used to index the obtained FVs [22]. To measure the similarities of 3D keyframes, we used a generalized form of Minkowski distance of order 4. This distance is an generalized form of Euclidean and manhattan distances which takes into account the 4 extracted metrics *(X,Y,Z,W)* This allows us to compute the distance between a KeyFrame1 and a KeyFrame2 using the formula below.

$$Minkowski\ distance\ of\ order\ 4 = \left(\sum_{i=1}^{n} |p_i - q_i|^4 \right)^{\frac{1}{4}}$$

Motion Recognition and Retrieval : In this section we adapt the Longest Common Subsequence method. X is a signed word recorded in the dictionary and Y is the scene that contains a signed sentence. We apply the dynamic programming to find out the LCS(*X, Y*). If the length of returned LCS is equal to the length of X then the word exists in the scene.

- *X*=[r_wrist(0 0 0 1), r_elbow(0.5 0 0 0.5), r_index(1 1 1 1)]
- *Y*= [r_wrist(0 0 0 1), r_elbow(0.5 0 0 0.5), r_elbow(0 0 1 0), r_index(1 1 1 1), r_elbow(0.1 0 0 -1.3), r_wrist(0 0 0 0) , l_wrist(1.5 0 0 1)].

To solve the LCS problem we proceed in two steps as follows:

1. Distance matrix computing using the following adapted formula :

$$T[i,j] = \begin{cases} 0 & if\ i = 0\ or\ j = 0 \\ T[i-1,j-1]+1 & if\ i > 0\ and\ j > 0\ and\ Similarity\ (x_i, y_j) \geq threshold \\ \max\ \{T[i-1,j], T[i,j-1]\} & if\ i > 0\ and\ j > 0\ and\ Similarity\ (x_i, y_j) < threshold \end{cases}$$

2. The second step consists on performing a back-trace to find the LCS(X,Y). We adapted the backtrace algorithm so that it can use similarity as well. The algorithm takes the scene Y of length n, the pattern X of length m and the distance matrix T computed in the previous step. The table 2 shows the way our algorithm works on the example given previously where the pattern r_wrist, r_elbow, r_index was indeed found.

Table 2. Illustration of the Back-Trace algorithm

			-1	0	1	2	3	4	5	6
				r_wrist	r_elbow	r_elbow	r_index	r_elbow	r_wrist	l_wrist
				0 0 0 1	0.5 0 0 0.5	0 0 1 0	1 1 1 1	0.1 0 0 1.3	0 0 0 0	1.5 0 0 1
-1			0	0	0	0	0	0	0	0
0	r_wrist	0 0 0 1	0	1	1	1	1	1	1	1
1	r_elbow	0.5 0 0 0.5	0	1	2	2	2	2	2	2
2	r_index	1 1 1 1	0	1	2	2	3	3	3	3

In the above case, the length of the LCS of the pattern X and the scene Y is equal to 3 which is the length of the pattern X. this means that the pattern was totally matched in the scene. The min length of the LCS is a parameter that we can set as a threshold to decide whether there is a match or not. In the experimentations, several thresholds have been tested and the value of 0.9 was noticed as the best value.

4   Evaluation of the System

The system described above has been implemented and tested. The data used during the testing phased was composed of:

- A database (dictionary) of patterns which are 3D signed words (FSL) where every word is referenced with its adequate SML code as mentioned in section 2.1. This database is used in order to process text-based queries where the user is expected to search for 3D signed translation of a given word. The dictionary contains over 800 3D signed words that have been recorded using H-animator, and WebSign. Both of the two used systems are compliant with H-anim knowing that all the records have been preprocessed with the indexing algorithm in order to extract the key-frames and generate the SML tree for all of them.

- The processed data is composed of more than 600 scenes grabbed from the internet and/or recorded using Vsigns[12], H-animator and WebSign. We should notice that the scenes recorded using WebSign are continuous scenes[13] and have not been generated automatically, meaning that the dictionary of signs has been used. The test data is composed of 3 sets: 3D scenes containing 1 word, 3D continuous scenes where 2 words are combined and a set of 3 signed words scenes.

The system we designed was implemented; a user interface has been developed as well. The user interface allows users to insert two types of queries. Text based queries and 3D signed scenes queries. The first type of queries consists of submitting a word glosses[14] and launch a search in the index. The processed scenes are preprocessed in order to extract key-frames and generate the SML tree. In the second step, the dictionary of patterns (FSL 3D signed words) is used to retrieve eventual words contained in those scenes.

4.1   Queries Description

A user interface has been developed aiming to allow us to insert two types of queries:

Text-Based Queries: a collection of keywords to search for in the index. This type of query is used when we want to search for a sequence of words which are recorded in the dictionary and recognized (or not) in the processed scenes. Once a word is recognized from a 3D scene, it is stored in the index using the path of the scene, and the similarity rate. Then, we compute the signature of the word according to MD5 hashing functions[15] and put it in the index precisely in a field called *WordID*. Concretely, after splitting the user's query, a signature for every token is generated using the same hashing function as in the indexing stage. This enables as to have faster matching and relevant results. Next, we use the following ranking algorithm to sort the matched scenes:

```
For each word from query list
  For each scene containing the word
    Add the similarity of the word to the scene's
current score
Sort the results according to their score values by
descending
Return results
```

3D –Based Queries: the system also processes 3D scenes as queries. The inserted scenes are first checked to validate their conformance with Web3D and H-anim

---

[12] http://vsigns.iti.gr:8080/VSigns/index.html?query=car
[13] Continuous scenes contain signed phrases : combining signed words.
[14] Keywords which don't necessary constitue a meaningful sentence.
[15] http://en.wikipedia.org/wiki/MD5

standards. Then, the key-frames are extracted and the SML tree is generated. The last step is the matching against the constructed index. The 3D queries processing is logically slower than the text-based queries due to the preprocessing of the scenes. The results are ranked using the same algorithm as for text based queries.

4.2   Retrieval Performance

To evaluate the performance of our matching method, two parameters are necessary to apply our adaptation of the LCS algorithm for retrieving motion: The Minkowski similarity that measures the distance between two key-frames and the accepted length of a potential alignment of two motions. During the tests, several values of these two parameters were used in order to decide which value is the best. The figure below shows the results of the performance tests. We have to notice that after several tests, we decided to set the length of acceptable alignments to 0.9 of the shortest sequence. Each of the 3 sets of data was matched against the dictionary of signed words (French Sign language). The average precision graph shows the fraction of relevant 3D signs that have been retrieved. The average recall shows the fraction of the retrieved 3D signs which are relevant. It is clear that the best performance rates have been noted for the WebSign 3D scenes which is normal since those scenes are generated using the same dictionary as for recognition process. The few mismatches concerning WebSign scenes are due to the concatenation of two or more signed words where the 3D rendering engine applies automatically interpolation between two signs. The result is a co-articulation meaning that there is no reset to original position and subsequently, similarity could not reach 100%. Concerning the H-animator, the set of 3D scenes were generated manually in conformance with FSL dictionary. The best precision rate reached 80% for a Minkowski value of 0.94. Concerning Vsigns, the average precision rate reached 75%. This relatively low performance is due to the different (from WebSign and H-animator) key-framing method used by Vsigns. Moreover, the collected signs looked like similar to what we have recorded with WebSign and H-animator. However, the gestures were far from being similar.

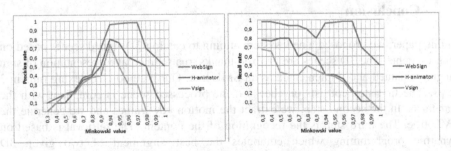

Fig. 2. Retrieval precision and recall

In order to compare our approach of dynamic programming with the most commonly used approach for both visual and model-based sign language; we implemented Hidden Markov Model recognizer. For this purpose, all the processed data for each posture are sent to the HMM toolkit called HTK[16] which handles all training and recognition steps. Regarding the training and processed data, we collected 20 isolated signs from each system describe above, 15 continuous sequences containing 2 co-articulated signs and 15 sequences containing 3 co-articulated signs. All of the scenes include only right-handed signs. HMM assumes a probabilistic model for the input data. The model of each posture P produces an observation Y with probability $P(P,Y)$. The aim is to decode the posture data with observation so that the posture reaches the maxim a posteriori probability.

To estimate the probability of an extracted key-frame from our 3D scenes, we used the distribution of 8 joints rotation values for the hand in a specific posture. Since the joints values are continuous, we used a discrete function to obtain intervals of values that are used as membership thresholds. During the training phases, We built 30 left-right Hidden Markov Models for the 30 basic postures with all states transition probabilities in order to calculate their $P(Y|P)$. For recognition, the HTK's Viterbi recognizer was used. The recognition precision rate reached 97.61% (table 4) which is higher than our method, however, when it is comes to analyze co-articulated signs, the rate falls dramatically. This can be explained by the fact that signs are inflected[17] which affects the probabilities. These rates could be improved by the use of grammars [16]. Moreover, the execution time in our method is better than in HMM based method. In fact, building of large matrices of probabilities combined with dynamic programming the Viterbi's back-trace is more time consuming than the LCS method.

Table 3. Comparison of the HMM and LCS performances

	Recognition Rate average (Precision)		Recognition Time average	
	HMM	LCS	HMM	LCS
Isolated Signs	97.61%	96.2%	1.2s	0.97s
2 Coarticulated Signs	78.23%	92.11%	6.9s	6.31s
3 coarticulated Signs	66.3%	92.10%	10.02s	9.33s

5   Conclusion

In this paper, we proposed a new system aiming to catalog 3D signed scenes based on their contents. The 3D scenes are indexed thanks to a segmental representation of sign language named and implemented as an SML tree. Our method of indexing runs in 3 steps. The first step consists on checks if the processed scene is compliant with the standards. In the second step, we extract the motion key-frames and we generate the SML tree. The third step is the recognition of the content. The retrieval is based on dynamic programming which computes the best alignment of two given 3D sequences. We used Minkowski similarity to compare each of the key-frames extracted from the scenes. Moreover, the set of answers are ranked thanks a new

---

[16] http://htk.eng.cam.ac.uk/
[17] The the beginning of one sign is influenced by the end of the previous sign.

method of ranking based on the cumulated Minkowski similarity values. To test the performance of our approaches, we implemented the algorithm in a framework and used 3 sets of data issued from WebSign, H-animator and Vsigns. The precision average of the retrieval method reached up to 96.5% of relevant matches. The system is able to process 3D scenes queries as well as text-based queries since we used an FSL Word/sign dictionary. Our method is more efficient than HMM based in term of time consuming. It also produced better recognition rates when the processed signs are co-articulated.

During the tests, several distance measures have been tested and the Minkowski similarity showed the best performances. However, other distances and method exist and will be studied in the future. At time of writing, we work on the possibility to generalize our method to different 3D signed scenes with different geometries and which are not compliant to the standards. To eliminate the dependence to dictionaries, a method based on sign language parameters (Hand shape, orientation, movement, location) classification is ongoing.

References

1. Stokoe, W.C.: Sign Language Structure: An Outline of the Visual Communication Systems of the American Deaf. Studies in Linguistics: Occasional Papers (8) (1960)
2. Cuxac, C.: La langue des signes française, les voies de l'iconicité, Orphys (2000)
3. Klima, E.S., Bellugi, U.: The Signs of Language. Harvard University Press (1988)
4. Frishberg, N.: Arbitrariness and Iconicity: Historical Change in American Sign Language. Language 51(23) (1975)
5. Huenerfauth, M., Pengfei, L., Rosenberg, A.: Evaluating American Sign Language generation through the participation of native ASL signers. In: ACM SIGACCESS Conference on Computers and Accessibility, New York (2007)
6. Jemni, M., Elghoul, O.: A system to make signs using collaborative approach. In: Miesenberger, K., Klaus, J., Zagler, W.L., Karshmer, A.I. (eds.) ICCHP 2008. LNCS, vol. 5105, pp. 670–677. Springer, Heidelberg (2008)
7. El Ghoul, O., Jemni, M.: Multimedia courses generator for deaf children. The International Arab Journal for Information Technology 6(5) (2009)
8. Buttussi, F., Chittaro, L., Nadalutti, D.: H-Animator: A Visual Tool for Modeling, Reuse and Sharing of X3D Humanoid Animations. In: International Conference on 3D Web Technology, New York (2006)
9. Papadogiorgaki, M., Grammalidis, N., Sarris, N., Strintzis, M.G.: Synthesis of Virtual Reality Animations from SWML using MPEG-4 Body Animation Parameters. In: Sign Processing Workshop, Lisbon (2004)
10. Sangho, P., Jihun, P., Jake, K.A.: Video retrieval of human interactions using model-based motion tracking and multi-layer finite state automata. In: CIVR (2003)
11. Ben-Arie, J., Wang, Z., Pandit, P., Rajaram, S.: Human activity recognition using multidimensional indexing. IEEE Transactions on Pattern Analysis and Machine Intelligence 24(8), 1091–1104 (2002)
12. Dejan, V., Saupe, D.: 3D model retrieval. In: Spring Conference on Computer Graphics, pp. 89–93 (May 2000)
13. Zhang, C., Chen, T.: Indexing and retrieval of 3D models aided by active learning. In: ACM International Conference on Multimedia, Ottawa (2001)

14. Akanksha, H.Z., Prabhakaran, B., Ruiz, J.: Reusing motions and models in animations. In: EGMM (2001)
15. Takahashi, T., Kishino, F.: Hand Gesture Coding Based On Experiments Using A Hand Gesture Interface Device. ACM SIGCHI Bulletin 23(2), 67–73 (1991)
16. Vogler, C., Metaxas, D.: ASL recognition based on a coupling between HMMs and 3D motion analysis. In: Sixth International Conference on Computer Vision (1998)
17. Dreuw, P., Neidle, C., Athitsos, V., Sclaroff, S., Ney, H.: Benchmark Databases for Video-Based Automatic Sign Language Recognition. In: International Conference on Language Resources and Evaluation, Morocco (2008)
18. Kelly, D., Delannoy, J.R., McDonald, J., Markham, C.: A Framework for Continuous Multimodal Sign Language Recognition. In: International Conference on Multimodal Interfaces (2009)
19. Brutzman, D., Daly, L.: X3D: extensible 3D graphics for Web authors. Morgan Kaufmann Publishers (2006)
20. Daly, L., Brutzman, D.: X3D: Extensible 3D Graphics for Web Authors. Morgan Kaufmann Publishers (2006)
21. Jaballah, K., Jemni, M.: Toward automatic Sign Language recognition from Web3D based scenes. In: ICCHP, Vienna (2010)
22. Benjamin, B., Daniel, A., Dietmar, S., Tobias, S., Dejan, V.: Feature-Based Similarity Search in 3D Object Databases. ACM Computing Surveys 27(4), 345–387 (2005)

Variational Bayesian PCA versus k-NN on a Very Sparse Reddit Voting Dataset

Jussa Klapuri, Ilari Nieminen, Tapani Raiko, and Krista Lagus

Department of Information and Computer Science,
Aalto University, FI-00076 AALTO, Finland
{jussa.klapuri,ilari.nieminen,tapani.raiko,krista.lagus}@aalto.fi

Abstract. We present vote estimation results on the largely unexplored Reddit voting dataset that contains 23M votes from 43k users on 3.4M links. This problem is approached using Variational Bayesian Principal Component Analysis (VBPCA) and a novel algorithm for k-Nearest Neighbors (k-NN) optimized for high dimensional sparse datasets without using any approximations. We also explore the scalability of the algorithms for extremely sparse problems with up to 99.99% missing values. Our experiments show that k-NN works well for the standard Reddit vote prediction. The performance of VBPCA with the full preprocessed Reddit data was not as good as k-NN's, but it was more resilient to further sparsification of the problem.

Keywords: Collaborative filtering, nearest neighbours, principal component analysis, Reddit, missing values, scalability, sparse data.

1 Introduction

Recommender systems (RS) are software tools and techniques providing suggestions for items or objects that are assumed to be useful to the user [11]. These suggestions can relate to different decision-making processes, such as which books might be interesting, which songs you might like, or people you may know in a social network.

In this paper, we focus on an RS in the context of reddit.com [10], which is a type of online community where *users* can vote *links* either up or down, i.e. *upvote* or *downvote*. Reddit currently has a global Alexa rank of 120 and 44 in the US [2]. Our objective is to study whether the user will like or dislike some content based on a priori knowledge of the user's voting history.

This kind of problem can generally be approached efficiently by using *collaborative filtering* (CF) methods. In short, collaborative filtering methods produce user specific recommendations of links based on voting patterns without any need of exogenous information about either users or links [11]. CF systems need two different entities in order to establish recommendations: *items* and *users*. In this paper items are referred to as *links*. With users and links, conventional techniques model the data as a sparse user-link matrix, which has a row for each user and a column for each link. The nonzero elements in this matrix are the votes.

A. Tucker et al. (Eds.): IDA 2013, LNCS 8207, pp. 249–260, 2013.

The two main techniques of CF to relate users and links are *the neighborhood approach* and *latent factor models*. The neighborhood methods are based on finding similar neighbors to either links or users and computing the prediction based on these neighbors' votes, for example, finding k nearest neighbors and choosing the majority vote. Latent factor models approach this problem in a different way by attempting to explain the observed ratings by uncovering some latent features from the data. These models include neural networks, Latent Dirichlet Allocation, probabilistic Latent Semantic Analysis and SVD-based models [11].

1.1   Related Work

Probably the most known dataset for vote prediction is the Netflix dataset [8]. This paper is heavily based on a master's thesis [7], and uses the same preprocessed datasets as introduced in the thesis. The approach of estimating votes from the Reddit dataset in this paper is similar to [9], although they preprocessed the dataset in a different way and used root-mean-square error as the error measure while this paper uses classification error and class average accuracy. There has also been some research on comparing a k-NN classifier to SVM with several datasets of different sparsity [3], but the datasets they used were lower-dimensional, less sparse and not all users were evaluated in order to speed up the process. Still, their conclusion that k-NN starts failing at a certain sparsity level compared to a non-neighborhood model concurs with the results of this paper.

2   Dataset

Reddit dataset originated from a topic in social news website Reddit [6]. It was posted by a software developer working for Reddit in early 2011 in hopes of improving their own recommendation system. The users in the dataset represent almost 17% of the total votes on the site in 2008.

The original dataset consists of $N = 23\,091\,688$ votes from $n = 43\,976$ users over $d = 3\,436\,063$ links in 11 675 subreddits. Subreddits are similar to subforums, containing links that are similar in some aspect, but this data was not used in the experiments. The dataset does not contain any additional knowledge on the users or the content of the links, only user-link pairs with a given vote that is either -1 (a downvote) or 1 (an upvote).

Compared to the Netflix Prize dataset [8], Reddit dataset is around 70 times sparser and has only two classes. For the missing votes, there is no information on whether a user has seen a link and decided not to vote or simply not having seen the link at all. In this paper, the missing votes are assumed to be missing at random (MAR) [12]. This is a reasonable assumption due to high dimensionality of the data and low median number of votes per user.

2.1   Preprocessing

The dataset can be visualized as a bipartite undirected graph (Figure 1). Even though no graph-theoretic approaches were used in solving the classification problem, this visualization is particularly useful in explaining the preprocessing and the concept of core of the data. In graph theory, the degree of a vertex v equals to the number of edges incident to v. If the degree of any vertex is very low, there may not be enough data about the vertex to infer the value of an edge, i.e. the vote. An example of this can be found in Figure 1 between *link_4* and *user_7*, where the degree of *user_7* is 1 and the degree of *link_4* is 3. This is a manifestation of the *new user problem* [1], meaning that the user would have to vote more links and the link would have to get more votes in order to accurately estimate the edge.

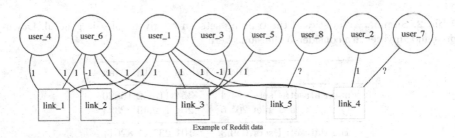

Example of Reddit data

Fig. 1. Bipartite graph representation of Reddit data. Subreddits are visualized as colors/shades on the links (square vertices). The vote values {-1,1} are represented as numbers on the arcs, where "?" means no vote has been given.

These kinds of new users and unpopular links make the estimation task very difficult and should be pruned out of the data. This can be done by using *cutoff* values such that all the corresponding vertices having a degree below the cutoff value are pruned out of the data. With higher cutoff values, only a subset of the data, i.e. *core of the data* remains, which is similar to the idea of k-cores [13]. This subset contains the most active users who have voted a large part of all the remaining links, and the most popular links which are voted by almost every user. The full Reddit dataset was pruned into two smaller datasets, namely, the big dataset using 4 as the cutoff value for all vertices (users and links), and the small dataset with a stricter cutoff value of 135. See [7] for more details on these parameters.

The resulting $n \times d$ data matrices M were then split randomly into a training set M_{train} and a test set M_{test}, such that the training set contained 90% of the votes and the test set the remaining 10%, respectively. The splitting algorithm worked user-wise, i.e., it randomly divided a user's votes between the training set and test set for all users such that at least one vote was put into the test set, even if the user had given less than 10 votes.

Training set properties are described in Table 1 which shows that the ratio of upvotes gets higher as the pruning process gets closer to the core of the data. In general the estimation of downvotes is a lot more difficult than upvotes. This is partly due to the fact that downvotes are rarer and thus the prior probability for upvotes is around six to nine times higher than for the downvotes. Summary statistics of the nonzero values in the training sets are given in Table 2. Small test set contained 661,903 votes and big test set 1,915,124 votes.

Table 1. Properties of votes given by users and votes for links for the preprocessed datasets

Dataset	Users (n)	Links (d)	Votes (N)	Density $(\frac{N}{nd})$	Ratio of upvotes
Small	7,973	22,907	5,990,745	0.0328	0.9015
Big	26,315	853,009	17,349,026	0.000773	0.8688
Full	43,976	3,436,063	23,079,454	0.000153	0.8426

Table 2. Summary statistics of the training sets on all given votes in general. Downvotes and upvotes are considered the same here.

	Mean	Median	Std	Max
Small dataset: Users	751.4	354	1156.2	15820
Links	261.5	204	159.49	2334
Big dataset: Users	659.3	101	2335.7	82011
Links	20.3	7	55.6	3229

3 Methods

This section describes the two methods used for the vote classification problem: the k-nearest neighbors (k-NN) classifier and Variational Bayesian Principal Component Analysis (VBPCA). The novel k-NN algorithm is introduced in Section 3.2.

3.1 k-Nearest Neighbors

Nearest neighbors approach estimates the behavior of the active user u based on users that are the most similar to u, or likewise, find links that are the most similar to the voted links. For the Reddit dataset in general, link-wise k-NN seems to perform better than user-wise k-NN [7], so every k-NN experiment was run link-wise.

For the k-NN model, *vector cosine-based similarity* and *weighted sum of others' ratings* were used [14]. More formally, when $\mathcal{I}$ denotes the set of all links, $i, j \in \mathcal{I}$ and i, j are the corresponding vectors containing votes for the particular links from all users, the similarity between links i and j is defined as

$$w_{ij} = \cos(i, j) = \frac{i \cdot j}{\|i\|\|j\|}. \tag{1}$$

Now, let $\mathcal{N}_u(i)$ denote the set of closest neighbors to link i that have been rated by user u. The classification of $\hat{r}_{ui}$ can then be performed link-wise by choosing an odd number of k neighboring users from the set $\mathcal{N}_u(i)$ and classifying $\hat{r}_{ui}$ to the class that contains more votes. Because the classification depends on how many neighbors k are chosen in total, all the experiments were run for 4 different values of k. For better results, the value of parameter k can be estimated through K-fold cross validation, for example.

However, this kind of simple neighborhood model does not take into account that users have different kinds of voting behavior. Some users might give down-votes often while some other users might give only upvotes. For this reason it is good to introduce rating normalization into the final k-NN method [14]:

$$\hat{r}_{ui} = \bar{r}_i + \frac{\sum\limits_{j \in \mathcal{N}_u(i)} w_{ij}(r_{uj} - \bar{r}_j)}{\sum\limits_{j \in \mathcal{N}_u(i)} |w_{ij}|}. \tag{2}$$

Here, the term $\bar{r}_i$ denotes the average rating by all user to the link in i and is called the mean-centering term. Mean-centering is also included in the nominator for the neighbors of i. The denominator is simply a normalizing term for the similarity weights w_{ij}.

The algorithm used for computing link-wise k-NN using the similarity matrix can generally be described as:

1. Compute the full $d \times d$ similarity matrix S from M_{test} using cosine similarity. For a $n \times d$ matrix M with normalized columns, $S = M^T M$.
2. To estimate $\hat{r}_{ui}$ using Eq. (2), find k related links j for which r_{uj} is observed, with highest weights in the column vector S_i.

For the experiments, this algorithm is hereafter referred to as "k-NN full".

3.2   Fast Sparse k-NN Implementation

For high-dimensional datasets, computing the similarity matrix S may be infeasible. For example, the big Reddit dataset would require computing half of a $853,009 \times 853,009$ similarity matrix, which would require some hundreds of gigabytes of memory. The algorithm used for computing fast sparse k-NN is as follows:

1. Normalize link feature vectors (columns) of M_{train} and M_{test}.
2. FOR each user u (row) of M_{test} DO
   (a) Find all the links j for which r_{uj} is observed and collect the corresponding columns of M_{train} to a matrix A_u.
   (b) Find all the links j for which $\hat{r}_{uj}$ is to be estimated and collect the corresponding columns of M_{test} to a matrix B_u.
   (c) Compute the pseudosimilarity matrix $S_u = A_u^T B_u$ which corresponds to measuring cosine similarity for each link between the training and test sets for user u.
   (d) Find the k highest values (weights) for each column of S_u and use Eq. (2) for classification.
3. END

This algorithm is referred to as "k-NN sparse" in Sec. 4. Matrix S_u is called a pseudosimilarity matrix since it is not a symmetric matrix and most likely not even a square matrix in general. This is because the rows correspond to the training set feature vectors and the columns to the test set feature vectors. The reason this algorithm is fast is due to the high sparsity (smaller N) and differentiating between only 2 classes. Also the fact that only the necessary feature vectors from training and test sets are multiplied, as well as parallelizing this in matrix operation. On the contrary, for less sparse datasets this is clearly slower than computing the full similarity matrix S from the start.

If the sparsifying process happens to remove all votes from a user from the training set, a naive classifier is used which always gives an upvote. This model could probably be improved by attempting to use user-wise k-NN in such a case. With low mean votes per user, the matrices S_u stay generally small, but for the few extremely active users, S_u can still be large enough not to fit into memory (over 16 GB). In these cases, S_u can easily be computed in parts without having to compute the same part twice in order to classify the votes.

It is very important to note the difference to a similar algorithm that would classify the votes in M_{test} one by one by computing the similarity only between the relevant links voted in M_{train}. While the number of similarity computations would stay the same, the neighbors $\mathcal{N}_u(i)$ would have to be retrieved from M_{train} again for each user-link pair (u, i), which may take a significant amount of time for a large sparse matrix though the memory usage would be much lower.

3.3   Variational Bayesian Principal Component Analysis

Principal Component Analysis (PCA) is a technique that can be used to compress high dimensional vectors into lower dimensional ones and has been extensively covered in literature, e.g. [5].

Assume we have n data vectors of dimension d represented by $r_1, r_2, \ldots, r_n$ that are modeled as

$$r_u \approx W x_u + m, \tag{3}$$

where W is a $d \times c$ matrix, x_j are $c \times 1$ vectors of principal components and m is a $d \times 1$ bias vector.

PCA estimates W, x_u and m iteratively based the observed data r_u. The solution for PCA is the unique principal subspace such that the column vectors of W are mutually orthonormal and, furthermore, for each $k = 1, \ldots, c$, the first k vectors form the k-dimensional principal subspace. The principal components can be determined in many ways, including singular value decomposition, least-square technique, gradient descent algorithm and alternating W-X algorithm. All of these can also be modified to work with missing values. More details are discussed in [4].

Variational Bayesian Principal Component Analysis (VBPCA) is based on PCA but includes several advanced features, including regularization, adding the noise term into the model in order to use Bayesian inference methods and introducing prior distributions over the model parameters. VBPCA also includes *automatic relevance determination* (ARD), resulting in less relevant components tending to zero when the evidence for the corresponding principal component for reliable data modeling is weak. In practise this means that the decision of the number of components to choose is not so critical for the overall performance of the model, unless the number of maximum components is too low. The computational time complexity of VBPCA using the gradient descent algorithm per one iteration is $O((N + n + d)c)$. More details on the actual model and implementation can be found in [4].

In the context of the link voting prediction, data vectors r_u contain the votes of user u and the bias vector m corresponds to the average ratings $\bar{r}_i$ in Eq. (2). The c principal components can be interpreted as features of the links and people. A feature might describe how technical or funny a particular link i is (numbers in the $1 \times c$ row vector of W_i), and how much a person enjoys technical or funny links (the corresponding numbers in the $c \times 1$ vector x_u). Note that we do not analyse or label different features here, but they are automatically learned from the actual votes in an unsupervised fashion.

4   Experiments

The experiments were performed on the small and big Reddit datasets as described in Section 2. Also, we further sparsified the data such that for each step, 10% of the remaining non-zero values in the training set were removed completely at random. Then the models were taught with the remaining training set and error measures were calculated for the original test set. In total there were 45 steps such that during the last step the training set is around 100 times sparser than the original. Two error measures were utilized, namely classification error and class average accuracy. The number of components for VBPCA was chosen to be the same as in [7], meaning 14 for the small dataset and 4 for the big dataset. In short, the heuristic behind choosing these values was based on the best number of components for SVD after 5-fold cross validation and doubling it, since VBPCA uses ARD. Too few components would make VBPCA underperform and too many would make it overfit the training set, leading to poorer performance.

The k-NN experiments were run for all 4 different k values simultaneously, which means that the running times would be slightly lower if using only a single value for k during the whole run.

Classification error was used as the error measure, namely $\#\{\hat{r}_{ui} \neq r_{ui}\}/N$. Class average accuracy is defined as the average between the proportions of correctly estimated downvotes and correctly estimated upvotes. For classification error, lower is better while for the value of class average accuracy higher is better.

VBPCA was given 1000 steps to converge while also using the *rmsstop* criterion, which stops the algorithm if either the absolute difference $|\text{RMSE}_{t-50} - \text{RMSE}_t| < 0.0001$ or relative difference $|\text{RMSE}_{t-50} - \text{RMSE}_t|/\text{RMSE}_t < 0.001$. Since VBPCA parameters are initialized randomly, the results and running times fluctuate, so VBPCA was ran 5 times per each step and the mean of all these runs was visualized.

Table 3. Metrics for different methods on the small Reddit dataset

Method	Accuracy	Classification error	Class average accuracy	Downvotes	Upvotes
Naive	0.9018	0.0982	0.5000	0.0000	**1.0000**
Random	0.8225	0.1775	0.4990	0.0965	0.9016
k-NN_User	0.9176	0.0824	0.6256	0.2621	0.9890
k-NN_Link	**0.9237**	**0.0763**	0.6668	0.3470	0.9865
VBPCA	0.9222	0.0778	**0.6837**	**0.3870**	0.9805

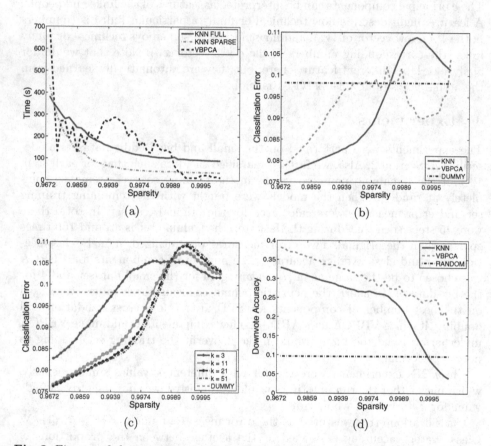

(a)   (b)

(c)   (d)

Fig. 2. Figures of the experiments on small dataset: Time versus sparsity (a). Classification error versus sparsity (b). Classification error versus sparsity for various k (c). Downvote estimation accuracy versus sparsity (d).

In addition to k-NN and VBPCA, naive upvote and naive random models were also implemented. Naive upvote model estimates all $\hat{r}_{ui}$ as upvotes and naive random model gives an estimate of upvote with the corresponding probability of

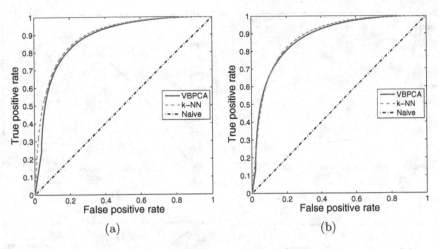

Fig. 3. ROC curves on small dataset (a). ROC curves on big dataset (b).

Table 4. Metrics for different methods on the big Reddit dataset

Method	Accuracy	Classification error	Class average accuracy	Downvotes	Upvotes
Naive	0.8687	0.1313	0.5000	0.0000	**1.0000**
Random	0.7720	0.2280	0.5002	0.1316	0.8689
k-NN_User	0.8930	0.1070	0.6738	0.3766	0.9711
k-NN_Link	**0.9048**	**0.0952**	**0.7091**	**0.4438**	0.9745
VBPCA	0.8991	0.1009	0.6929	0.4132	0.9726

upvotes from the training set, e.g. $p = 0.9015$ for small dataset, and a downvote with probability $1 - p$.

All of the algorithms and experiments were implemented in Matlab running on Linux using an 8-core 3.30 GHz Intel Xeon and 16 GB of memory. However, there were no explicit multicore optimizations implemented and thus k-NN algorithms were practically running on one core. While VBPCA toolbox is able to utilize multiple cores, the experiments were run using only one core for comparison.

4.1 Results for Small Reddit Dataset

The results for the methods on the original small Reddit dataset are seen in Table 3, which includes the user-wise k-NN before sparsifying the dataset further. The naive model gives a classification error of 0.0982, the dummy baseline for the small dataset. The running time of k-NN full is slightly lower than k-NN sparse in Figure 2a during the first step, but after sparsifying the training set, k-NN sparse performs much faster. It can be seen from Figure 2b that the k-NN classifier

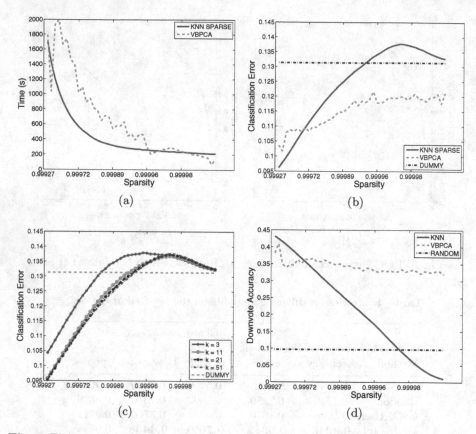

Fig. 4. Figures of the experiments on big dataset: Time versus sparsity (a). Classification error versus sparsity (b). Classification error versus sparsity for various k (c). Downvote estimation accuracy versus sparsity (d).

performs better to a certain point of sparsity, after which VBPCA is still able to perform below dummy baseline. This behavior may partly be explained by the increasing number of naive classifications by the k-NN algorithm caused by the increasing number of users in M_{train} with zero votes given. Figure 2c indicates that the higher the number of neighbors k, the better. Downvote estimation is consistently higher with VBPCA than with k-NN (Figure 2d). ROC curves are displayed in Figure 3a.

4.2   Results for Big Reddit Dataset

Classification error for the dummy baseline is 0.1313 for the big dataset. Figure 4b indicates that the fast k-NN classifier loses its edge against VBPCA quite early on, around the same sparsity level as for the small dataset. However, VBPCA seems to take more time on converging (Figure 4a). Higher k values lead to better performance, as indicated in Figure 4c. Downvote estimation with k-NN seems to suffer a lot from sparsifying the training set (Figure 4d), while VBPCA is only slightly affected. ROC curves for the big dataset are shown in Figure 3b.

5 Conclusions

In our experimental setting we preprocessed the original Reddit dataset into two smaller subsets containing some of the original structure and artificially sparsified the datasets even further. It may be problematic or even infeasible to use standard implementations of k-NN classifier for the high-dimensional very sparse 2-class dataset such as the Reddit dataset, but this problem was avoided using the fast sparse k-NN presented in Sec. 3. The results on the small dataset indicate that when the density of the dataset gets below around 3%, the fast sparse k-NN becomes faster than standard k-NN. This result most likely can be extended to similar datasets, since fast sparse k-NN gains advantages from sparsity. While initially k-NN classifier seems to perform better, VBPCA starts performing better when the sparsity of the datasets grow beyond approximately 0.99990. VBPCA is especially more unaffected by the increasing sparsity for the downvote estimation, which is generally more difficult than upvote estimation.

There are many ways to further improve the accuracy of k-NN predictions. The fact that the best results were obtained with the highest number of neighbors (k=51) hints that the cosine-based similarity weighting is more important to the accuracy than a limited number of the neighbors. One could, for instance, define a tunable distance metric such as $\cos(i, j)^\alpha$, and find the best α by cross-validation. The number of effective neighbors (sum of weights compared to the weight of the nearest neighbor) could then be adjusted by changing α while keeping k fixed to a large value such as 51.

Acknowledgements. We gratefully acknowledge the support from the Finnish Funding Agency for Technology and Innovation (VirtualCoach, 40043/11) and The Academy of Finland (Finnish Centre of Excellence in Computational Inference Research COIN, 251170).

References

1. Adomavicius, G., Tuzhilin, A.: Toward the next generation of recommender systems: A survey of the state-of-the-art and possible extensions. IEEE Transactions on Knowledge and Data Engineering 17(6), 734–749 (2005)
2. Alexa: Alexa - reddit.com site info, http://www.alexa.com/siteinfo/reddit.com (accessed August 5, 2013)
3. Grčar, M., Fortuna, B., Mladenič, D., Grobelnik, M.: knn versus svm in the collaborative filtering framework. In: Data Science and Classification, pp. 251–260. Springer (2006)

4. Ilin, A., Raiko, T.: Practical approaches to principal component analysis in the presence of missing values. Journal of Machine Learning Research (JMLR) 11, 1957–2000 (2010)
5. Jolliffe, I.T.: Principal Component Analysis, 2nd edn. Springer (2002)
6. King, D.: Want to help reddit build a recommender? – a public dump of voting data that our users have donated for research, http://redd.it/dtg4j (2010), (accessed August 5, 2013)
7. Klapuri, J.: Collaborative Filtering Methods on a Very Sparse Reddit Recommendation Dataset. Master's thesis, Aalto University School of Science (2013)
8. Netflix: Netflix prize webpage, http://www.netflixprize.com/ (2009) (accessed August 5, 2013)
9. Poon, D., Wu, Y., Zhang, D.Q.: Reddit recommendation system (2011), http://cs229.stanford.edu/proj2011/PoonWuZhang-RedditRecommendationSystem.pdf (accessed August 5, 2013)
10. Reddit: reddit: the front page of the internet, http://www.reddit.com/about/ (accessed August 5, 2013)
11. Ricci, F., Rokach, L., Shapira, B., Kantor, P.B. (eds.): Recommender Systems Handbook. Springer (2011)
12. Rubin, D.B.: Multiple Imputation for Nonresponse in Surveys. Wiley (1987)
13. Seidman, S.: Network structure and minimum degree. Social Networks 5(3), 269–287 (1983)
14. Su, X., Khoshgoftaar, T.M.: A survey of collaborative filtering techniques. Adv. in Artif. Intell. 2009, 4:2 (January 2009), http://dx.doi.org/10.1155/2009/421425

Analysis of Cluster Structure in Large-Scale English Wikipedia Category Networks

Thidawan Klaysri, Trevor Fenner, Oded Lachish,
Mark Levene, and Panagiotis Papapetrou

Department of Computer Science and Information Systems,
Birkbeck, University of London, UK

Abstract. In this paper we propose a framework for analysing the structure of a large-scale social media network, a topic of significant recent interest. Our study is focused on the Wikipedia category network, where nodes correspond to Wikipedia categories and edges connect two nodes if the nodes share at least one common page within the Wikipedia network. Moreover, each edge is given a weight that corresponds to the number of pages shared between the two categories that it connects. We study the structure of category clusters within the three complete English Wikipedia category networks from 2010 to 2012. We observe that category clusters appear in the form of well-connected components that are naturally clustered together. For each dataset we obtain a graph, which we call the *t-filtered* category graph, by retaining just a single edge linking each pair of categories for which the weight of the edge exceeds some specified threshold t. Our framework exploits this graph structure and identifies connected components within the *t-filtered* category graph. We studied the large-scale structural properties of the three Wikipedia category networks using the proposed approach. We found that the number of categories, the number of clusters of size two, and the size of the largest cluster within the graph all appear to follow power laws in the threshold t. Furthermore, for each network we found the value of the threshold t for which increasing the threshold to $t+1$ caused the "giant" largest cluster to diffuse into two or more smaller clusters of significant size and studied the semantics behind this diffusion.

Keywords: graph structure analysis, large-scale social network analysis, Wikipedia category network, connected component.

1 Introduction

Wikipedia is one of the most popular large social media networks and has experienced exponential growth in its first few years of existence in terms of articles, page edits, and users [4]. Moreover, this large network has been studied extensively; for example, analysis of the social networks emanating from the Wiki-talk page or discussions page reveals rich social interactions between editors [1,8,11]. On the other hand, the Wikipedia network of category links, which indirectly implies social relations when authors assign their articles into specific categories,

A. Tucker et al. (Eds.): IDA 2013, LNCS 8207, pp. 261–272, 2013.

has received much less attention from the research community, in particular in terms of large-scale structural social network analysis. Current research on the Wikipedia category network has mainly concentrated on content-based analysis.

The Wikipedia category network mainly consists of categories, where two categories are connected by an edge if they have some "similarity". In our setting, similarity is expressed by the number of pages shared between two categories. In other words, the weight of an edge is equal to the number of common pages between the categories, and hence expresses the similarity between them: the higher the weight, the higher the similarity.

Wikipedia categorisation refers to assigning an article to at least one category to which it logically belongs. The Wikipedia categorisation system is likely to be improved in the long run, as category policies are still being refined[1]. There is no limit to the size of the categories, but when a category becomes very large, it may be diffused (or broken down) into smaller categories or subcategories. This phenomenon is called *large category diffusion*.

Our objective in this paper is to examine the structural properties of the category clusters within the Wikipedia category network by identifying well-connected components in the graph. These components can be used for comparison with the Wikipedia category tree, based on the expectation that categories falling into same cluster should have a high degree of proximity within the Wikipedia tree.

Our key contributions are summarized as follows:

- We present *t-component*, a framework for identifying natural category clusters in the form of well-connected components in a category-links network that employs an edge-weight threshold t regulating the "strength" of the components.
- Using the proposed framework, we study several structural properties of the Wikipedia network, such as the number of non-trivial category clusters, the size of the largest category cluster, and the number of the smallest category clusters, and how they evolve as the edge-weight threshold increases.
- We observe the diffusion of the largest category cluster as a "giant"-cluster splitting into smaller sub-clusters and examine their contents.
- We find that the largest connected component shrinks at a power-law rate as the edge-weight threshold t increases. This is consistent with similar observations for various properties of social networks, such as the Barabasi-Albert model, which is considered a reasonable generative model of the Web.

2   Related Work

Analysis of web social networks has become a popular research area, especially in the context of online social networking applications. Some large-scale networks have been analysed recently. For example, social interactions have been analysed in Twitter [10], Wattenhofer et al. [18] analysed the nature of the YouTube

---

[1] http://en.wikipedia.org/wiki/Wikipedia:FAQ/Categorization

network, Sadilek et al. [12] modelled the spread of diseases by analysing health messages from Twitter, Volkovich et al. [17] analysed structural properties and spatial distances of the Spanish social network Tuenti, and finally Goel et al. [3] studied user browsing behaviour changes.

Wikipedia, which is one of the most popular social media networks has been studied extensively. For example, Hu et al. [4] analysed and predicted user collaborations, Leskovec et al. [9] investigated the promotion process from the point of view of the voters engaged in group decision-making, and Jurgens et al. [5] investigated trends of editor behaviour. The page links structure has also been studied. For instance, Buriol et al. [2] examined the page links structure and its evolution over time and Kamps et al. [6] compared the Wikipedia link structure to other similar web sites. Also, a survey on graph clustering methods by Schaeffer [13] provides a thorough review of different graph cluster definitions and measures for evaluating the quality of clusters.

There were a lot of studies of the Wikipedia user talk pages in the context of its induced social network, which contains rich social interactions in the "talk" domain. Examples include analysing the policy governance discussed on user talk pages [12] and detecting structural patterns forming a tree structure [8].

In general, category links in the Wikipedia category network have been studied using text analysis. For example, Schonhofen [14] attempted to identify document topics, while Kittur et al. [7] represented topic distribution mapping with category structure, and Jiali et al. [19] studied document topic extraction. While Zesch and Gurevych [20] analysed the Wikipedia category graph from a natural language processing perspective, the large-scale Wikipedia network structure has been studied much less. For example, Suchecki et al. [16] investigated the evolution of the English Wikipedia category structure from 2004 to 2008, but focusing merely on the structure of the documentation of knowledge.

3   Preliminaries

In this paper, we focus on the Wikipedia category network, which we describe next. Then we provide the necessary background definitions to be used in the remainder of this paper.

3.1   The Wikipedia Category Network

Wikipedia contains knowledge in the form of Wiki pages and is edited collaboratively by millions of volunteer authors in 285 different languages, among which the English Wikipedia contains the largest number of articles. In Wikipedia, each article is assigned to at least one category, while a categorised article should be assigned to all of the most specific subcategories to which it logically belongs.

Assigning pages to the categories induces a social network of pages and categories established by the editors. This network can be considered as a graph, representing a set of relationships between pages and categories, or only between categories.

3.2   Problem Setting

Let $P = \{p_1, \ldots, p_n\}$ be the set of n Wikipedia pages and C be the set of m Wikipedia categories. Each page $p_i \in P$ belongs to at least one category $c_j \in C$. A graph $G = (V, E)$ is defined as a set of vertices V and a set of edges E, such that each edge $e_k \in E$ connects two vertices $v_i, v_j \in V$, which is denoted as $v_i \to^{e_k} v_j$.

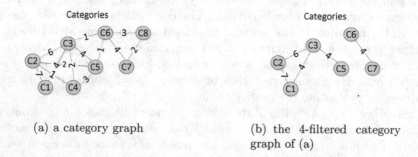

(a) a category graph

(b) the 4-filtered category graph of (a)

Fig. 1. Examples of two category graphs

Definition 1. (page-category graph)
A page-category graph is a bipartite graph G^{PC} that represents the network of connectivity between Wikipedia pages and Wikipedia categories. The set of vertices is $P \cup C$ and there is an edge $p \to c$ whenever page $p \in P$ belongs to category $c \in C$.

Each page in P belongs to at least one category in C. A page that belongs to a single category is called an *isolated page*.

Definition 2. (edge-weighted category graph)
An edge-weighted category graph G^{EW} is a graph where the set of vertices corresponds to the Wikipedia categories C. Each edge e_k between two vertices v_i and v_j (corresponding to categories c_i and c_j, respectively) is assigned with a weight $w_k \in \mathbb{N}$ equal to the number of common pages in both c_i and c_j, or equivalently

$$w_k = |\{p \in P \mid p \in c_i \text{ and } p \in c_j\}|$$

Note that an edge e_k with weight $w_k = 1$ is called *a feeble edge* and a category that is not sharing any page with any other category is called an *isolated* or *trivial category*. It follows that pages connected to an isolated category are necessarily isolated.

Definition 3. (t-filtered category graph)
A t-filtered category graph G_t^{EW} is obtained from an edge-weighted category graph G^{EW} by the removal of every edge e_k with weight less than $t \in \mathbb{N}$, i.e., e_k is in G_t^{EW} if and only if

$$w_k \geq t \, , \, \forall e_k \in E.$$

In Figure 1 we see two examples of category graphs. The first one (on the left) is a category graph where no filtering has been applied, while the second one (on the right) is the corresponding 4-filtered category graph of the one on the left. We note that the t-filtered graph is closely related to the m-core of a graph [15, pp. 110-114].

Definition 4. (category cluster)
A category cluster $C(G^{EW}, t)$ is a well-connected component of an edge-weighted category graph G^{EW}, and is obtained as a connected component of the corresponding t-filtered category graph G_t^{EW} for a specified threshold $t \geq 2$.

Using the above definitions we can now formulate the problem studied in this paper as follows.

Problem 1. *Given a Wikipedia page-category graph G^{PC} and a threshold $t \in \mathbb{N}$, identify the largest category cluster in the corresponding t-filtered category graph G_t^{EW}.*

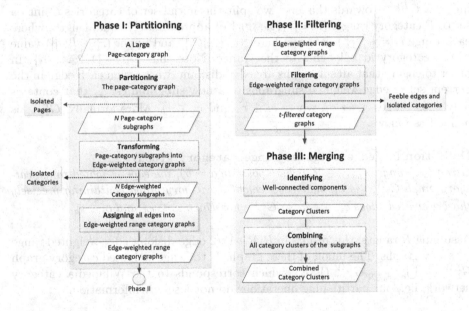

Fig. 2. An overview of the six steps of the t-component framework

4   The t-Component Framework

The *t-component* framework consists of the three main phases as shown in Figure 2: (I) partitioning the edge-weight category graph, (II) filtering the subgraphs, and (III) merging the subgraphs. Next we describe each phase in more detail.

4.1   Phase I: Partitioning

Due to the difficulty of manipulating the entire page-category graph, which is very large, in this phase, the input bipartite page-category graph G^{PC} is partitioned into a set of N page-category subgraphs $\{G_1^{PC}, \ldots, G_N^{PC}\}$. The split is performed ensuring that: (1) each page $p \in P$ appears in only one subgraph and (2) all subgraphs have approximately the same number of pages. In addition, all isolated pages are eliminated.

Next, each page-category subgraph G_i^{PC} is transformed into its corresponding edge-weighted category subgraph G_i^{EW}, following Definition 2. Note that all isolated categories are eliminated. An important observation here is that, during this process, the same edges may appear in more than one edge-weighted category subgraph. In other words, we could have different sets of pages shared between the same two categories c_i, c_j within different subgraphs. We note that G^{EW} is the union of all edge-weighted category subgraphs G_i^{EW}, for $i = 1, 2, \ldots, N$. Due to its size we cannot explicitly construct G^{EW}.

Therefore, we construct a collection of edge-disjoint category graphs whose union is G^{EW}. Towards this end, we split the initial set of categories C into a set of R category ranges $\mathcal{R} = \{r_1, \ldots, r_R\}$ of approximately equal length, where each range $r_t = [r_t^l, r_t^u)$ defines a lower ($r_t^l \in [|C|]$) and upper ($r_t^u \in [|C|]$) value of the category id belonging to that range. Note that $[n] = \{1, 2, \ldots, n\}$. In order to ensure that all subgraphs are edge-disjoint, we reassign each edge in the current set of edge-weighted subgraphs to a new subgraph G_{r_a,r_b}^{EW} that contains only those edges $c_i \to c_j$, where $c_i \in r_a$ and $c_j \in r_b$. More formally, G_{r_a,r_b}^{EW} is defined as follows.

Definition 5. (edge-weighted range category graph)
Given two ranges $r_a, r_b \in \mathcal{R}$ (where possibly $a = b$), the edge-weighted range category graph G_{r_a,r_b}^{EW} is the new edge-weighted category graph containing precisely those weighted edges $c_i \to^{e_k} c_j$ in G^{EW} for which $c_i \in r_a$ and $c_j \in r_b$.

Assuming R ranges, this results in $R(R+1)/2$ edge-disjoint edge-weighted range category graphs. The union of these graphs is the edge-weighted category graph $G^{EW} = \bigcup_{r_a,r_b \in \mathcal{R}} G_{r_a,r_b}^{EW}(V, E)$, which corresponds to the Wikipedia category network; i.e., our partitioning operations do not lose any information.

4.2   Phase II: Filtering

In the second phase, we introduce a threshold parameter $t \in \mathbb{N}$ that will be used to obtain the filtered category graph G_t^{EW}. To do this, for each $G_{r_a,r_b}^{EW} \subset G^{EW}$, all edges with weight less than t are removed. Hence, each $G_{r_a,r_b}^{EW_t}$ is converted to its corresponding t-filtered category graph $G_{r_a,r_b,t}^{EW}$. It is easy to see that $G_t^{EW} = \bigcup_{r_a,r_b \in \mathcal{R}} G_{r_a,r_b,t}^{EW}(V, E)$. Note that during this phase all isolated categories are removed.

4.3   Phase III: Merging

In the third phase, we first identify the connected components within each $G_{r_a,r_b}^{EW} \subset G^{EW}$ using Breadth First Search (BFS). Each connected component corresponds to a category cluster $\mathcal{C}(G_{r_a,r_b}^{EW}, t)$, by Definition 4. Merging all connected components of all these subgraphs of G_t^{EW} by combining components that share at least one category, we obtain the complete set of category clusters for threshold t as the connected components of G_t^{EW}.

5   Experiments

5.1   Setup

We used the English Wikipedia category network for evaluating the performance of the proposed framework. We studied three years: 2010, 2011, and 2012. Each year was studied separately as an individual dataset. The data is freely available online [2].

For each year we used the same number of partitions of the initial page-category graph, i.e., $N = 2,000$. During this process we eliminated all isolated pages and then transformed each page-category graph to its equivalent edge-weighted category graph by assigning the edge weights accordingly and eliminating all isolated categories. Next, we eliminated duplicate instances of category pairs within different partitions by splitting the categories into ranges (as indicated by Phase I of the framework). We used 70 ranges, i.e., $R = 70$, resulting in a total of 2,485 edge-disjoint edge-weighted range category graphs.

In addition, we studied different values for the t threshold, ranging from 2 to 4096. Note that all feeble edges (having weight equal to 1) were removed from the network as required by the framework.

The framework was implemented in Java on an Intel i5 processor. The execution time depends critically on the number of edges and the size of the clusters. For example, for $t = 2$ it took over a week, while for $t = 4096$ it took less than a minute to perform all the computations.

5.2   Results

Our experimental findings on the three English Wikipedia category networks are presented next. We present the structural properties of the networks and investigate the structural behaviour of the clusters with respect to the threshold t.

Structural Properties
The structural properties of the three Wikipedia category networks are summarised in Table 1. It can be observed that, from 2010 to 2012, the number of

---

[2] http://dumps.wikimedia.org/index.html

pages and categories increased by around 40% and 50%, respectively. However, it is interesting to note that, although the number of isolated pages increased by around 60%, the number of isolated categories was almost unchanged. A possible explanation for this is that, when new categories are added to the network, they are likely to be linked to existing pages as well as new pages. They will therefore be related to existing categories. We also note that the number of page-category links (i.e. edges in the page-category graph) increased by around 50%. A consequence of this, which can be checked using Table 1, is that the average degrees, both pages per category and categories per page, were substantially unchanged.

Table 1. Structural Properties of English Wikipedia Category Link Networks

Network Properties	English 2010	English 2011	English 2012
Number of pages	8,989,264	12,182,689	12,453,596
Number of categories	567,939	801,902	858,869
Number of page-category links	39,484,287	56,969,309	60,386,600
Number of isolated pages	1,083,655	1,735,857	1,755,160
Number of isolated categories	7,443	7,858	7,375
% Isolated pages	12.05%	14.25%	14.09%
% Isolated categories	1.31%	0.98%	0.86%

Structural Behaviour of Category Clusters

We studied how the category clusters depend on the weight threshold t. Specifically, we studied all values of t from 2 to 4096. Some of our most important findings are shown in Figure 3. The charts show the number of categories (excluding isolated categories) in the complete category networks in (a), the number of the category clusters in (b), the number of clusters of size 2 in (c), and the size of the largest clusters in (d). A very significant finding here is that all four log-log plots appear to exhibit power-law behaviour with respect to t. It also seems that the power law exponent is not significantly changing over the time period studied in this paper.

In addition, we note in Figure 3(d) that there is a threshold value for each of the three datasets where the size of the largest category cluster drops sharply. This suggests that each large category (for each of the three years) has diffused into smaller categories or a subcategories. Taking a closer look at these diffusion points, we observed that, in all three cases, the largest cluster was split into two large subclusters. Hence, we plot those diffusion points and display them individually for the 2010 (Figure 4) and 2012 (Figure 5) networks. Due to space limitations, we omit year 2011. In both figures we can see the significant diffusion points (t threshold values) and the corresponding sizes of the first and second largest category clusters.

Semantics of the Cluster Diffusion

We studied the semantics of the category cluster diffusion. Specifically, we compared the categories that appeared in the original large cluster and then those

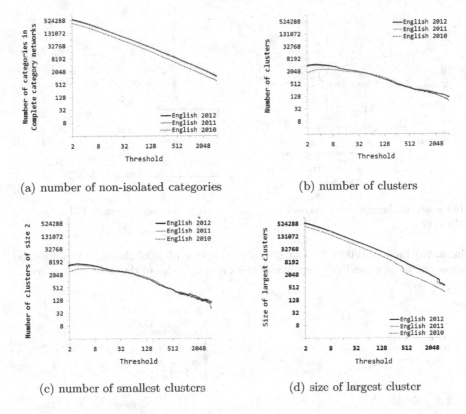

(a) number of non-isolated categories

(b) number of clusters

(c) number of smallest clusters

(d) size of largest cluster

Fig. 3. Log-log plots of (a) the number of non-isolated categories, (b) the number of category clusters, (c) the number of clusters of size 2, and (d) the size of the largest cluster, for different weight threshold values for the English Wikipedia Category Network 2010 - 2012.

that appeared in the two largest clusters right after the diffusion. Almost all categories were preserved after the diffusion, but were split between the two clusters so very few categories diffused into smaller components.

In addition, we note that, after the diffusion, a small fraction of the categories present in the initial cluster were not part of any of the new diffused clusters. In the case of year 2010, there were twelve categories missing, for 2011 there were none, while for 2012 there was only one. Hence, based on the previous observation, we investigated whether there exists any semantic connection or relation between the categories within the two diffused clusters. Specifically, we observed that the frequent categories in the two clusters were substantially different. One cluster would typically contain more general category types, such as "start-class", "stub-class", "people", and "articles", while the second cluster would contain more specific category types, such as "players", "american articles", and "footballers". Some examples of the dominant category titles can be seen in Figure 4(b) for the 2010 network and in Figure 5(b) for the 2012 network.

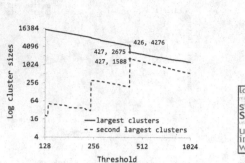

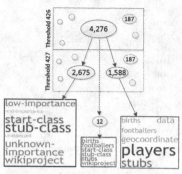

(a) sizes of largest and second largest clusters

(b) diffused category titles

Fig. 4. (a) English Wikipedia Category 2010 Log-log plots of the largest and second largest cluster sizes and (b) examples of the category titles of the diffused clusters

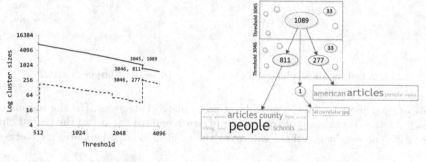

(a) sizes of largest-second largest clusters

(b) diffused category titles

Fig. 5. (a) English Wikipedia Category 2012 Log-log plots of the largest and second largest cluster sizes and (b) examples of the category titles of the diffused clusters

6   Summary and Conclusions

In this paper we presented a framework for manipulating a large Wikipedia page-category network. The proposed framework was used to analyze the structure of the network. We obtained, in the Wikipedia category network, global category clusters in the form of well-connected components.

In our experiments, we demonstrated the applicability of the proposed framework to several instances of the English Wikipedia category network and observed that, over the years 2010 to 2012, the number of pages, categories, page-category links and isolated pages all increased by 40-60%, but the number of isolated categories was fairly constant. The most significant finding was that the number of non-isolated categories, the number of clusters, the number of clusters of size two, and the size of the largest cluster all appear to follow power laws with respect to the threshold t. This behaviour is observed for each of the three years of the English Wikipedia category network studied in the paper. Furthermore, for each network we found the value of the threshold t for which increasing the threshold caused the largest cluster to diffuse into two smaller category clusters of significant size. We also observed that this diffusion is typically the result of a "giant" cluster splitting into smaller sub-clusters.

Future work includes the study of our framework using other language Wikipedia category networks. Based on our current investigations, the Wikipedia category graphs for several other languages appear to show a similar cluster structure. In addition, other possible graph clustering techniques are being considered, in particular, the k-core of the category graph and how it relates to the components of the t-filtered category graph.

References

1. Beschastnikh, I., Kriplean, T., McDonald, D.W., ICWSM: Wikipedian self-governance in action: Motivating the policy lens. In: Proceedings of ICWSM (2008)
2. Buriol, L.S., Castillo, C., Donato, D., Leonardi, S., Millozzi, S.: Temporal analysis of the wikigraph. In: IEEE/WIC/ACM International Conference on Web Intelligence (WI 2006), pp. 45–51. IEEE (2006)
3. Goel, S., Hofman, J.M., Sirer, M.I.: Who does what on the web: A large-scale study of browsing behavior. In: Proceedings of ICWSM (2012)
4. Hu, M., Lim, E.-P., Krishnan, R.: Predicting outcome for collaborative featured article nomination in wikipedia. In: Proceedings of ICWSM (2009)
5. Jurgens, D., Lu, T.-C.: Temporal motifs reveal the dynamics of editor interactions in wikipedia. In: Proceedings of ICWSM (2012)
6. Kamps, J., Koolen, M.: Is wikipedia link structure different? In: Proceedings of the Second ACM International Conference on Web Search and Data Mining, pp. 232–241. ACM (2009)
7. Kittur, A., Chi, E.H., Suh, B.: What's in wikipedia?: mapping topics and conflict using socially annotated category structure. In: Proceedings of the SIGCHI Conference on Human Factors in Computing Systems, pp. 1509–1512. ACM (2009)
8. Laniado, D., Tasso, R., Volkovich, Y., Kaltenbrunner, A.: When the wikipedians talk: Network and tree structure of wikipedia discussion pages. In: Proceedings of ICWSM (2011)
9. Leskovec, J., Huttenlocher, D.P., Kleinberg, J.M.: Governance in social media: A case study of the wikipedia promotion process. In: Proceedings of ICWSM (2010)
10. Macskassy, S.A.: On the study of social interactions in twitter. In: Proceedings of ICWSM (2012)
11. Massa, P.: Social networks of wikipedia. In: Proceedings of the 22nd ACM Conference on Hypertext and Hypermedia, pp. 221–230. ACM (2011)
12. Sadilek, A., Kautz, H.A., Silenzio, V.: Modeling spread of disease from social interactions. In: Proceedings of ICWSM (2012)
13. Schaeffer, S.E.: Graph clustering. Computer Science Review 1(1), 27–64 (2007)

14. Schönhofen, P.: Identifying document topics using the wikipedia category network. Web Intelligence and Agent Systems 7(2), 195–207 (2009)
15. Scott, J.: Social Network Analysis: a handbook. SAGE Publications, London (2011)
16. Suchecki, K., Salah, A.A.A., Gao, C., Scharnhorst, A.: Evolution of wikipedia's category structure. Advances in Complex Systems 15(supp. 1) (2012)
17. Volkovich, Y., Scellato, S., Laniado, D., Mascolo, C., Kaltenbrunner, A.: The length of bridge ties: Structural and geographic properties of online social interactions. In: Proceedings of ICWSM (2012)
18. Wattenhofer, M., Wattenhofer, R., Zhu, Z.: The youtube social network. In: Proceedings of ICWSM (2012)
19. Yun, J., Jing, L., Yu, J., Huang, H., Zhang, Y.: Document topic extraction based on wikipedia category. In: 2011 Fourth International Joint Conference on Computational Sciences and Optimization (CSO), pp. 852–856. IEEE (2011)
20. Zesch, T., Gurevych, I.: Analysis of the wikipedia category graph for nlp applications. In: Proceedings of the TextGraphs-2 Workshop (NAACL-HLT 2007), pp. 1–8 (2007)

1d-SAX: A Novel Symbolic Representation for Time Series

Simon Malinowski[1], Thomas Guyet[1], René Quiniou[2], and Romain Tavenard[3]

[1] AGROCAMPUS-OUEST/ IRISA-UMR 6074, F-35042 Rennes, France
first.last@agrocampus-ouest.fr
[2] Centre de Recherche INRIA Rennes Bretagne Atlantique, France
[3] IDIAP Research Institute, Martigny, Switzerland

Abstract. SAX (Symbolic Aggregate approXimation) is one of the main symbolization techniques for time series. A well-known limitation of SAX is that trends are not taken into account in the symbolization. This paper proposes 1d-SAX a method to represent a time series as a sequence of symbols that each contain information about the average and the trend of the series on a segment. We compare the efficiency of SAX and 1d-SAX in terms of goodness-of-fit, retrieval and classification performance for querying a time series database with an asymmetric scheme. The results show that 1d-SAX improves performance using equal quantity of information, especially when the compression rate increases.

1 Introduction

Time series data mining (TSDM) has recently attracted the attention of researchers in data mining due to the increase availability of data with temporal dependency. TSDM algorithms such as classification/clustering of time series, pattern extraction or similarity search require a distance measure between time series. The computation of these distances is mainly done using the classical Euclidean distance or the Dynamic Time Warping distance. These computations may lead to untractable costs for long series and/or huge databases. Hence, many approximate representations of time series have been developed over the last decade. Symbolic representation is one technique to approximate time series. The most used symbolization technique is called SAX (Symbolic Aggregate approXimation) [7]. It is a very simple technique to symbolize time series without the need for any a priori information. It has been shown to provide good performance for TSDM purposes. Some extensions to the SAX representation have been proposed to take into account the slope information in the time series segments [2,8,11]. In this paper we propose a novel symbolic representation for time series, based on the quantization of the linear regression of segments of the time series. We first show that this novel symbolic representation fits the original data more accurately than the SAX representation for the same number of symbols available. Then, this symbolic representation is used to make efficient similarity search in a time series database. We propose an asymmetric querying scheme based on our symbolic representation and compare its performance with the one based on SAX representation.

A. Tucker et al. (Eds.): IDA 2013, LNCS 8207, pp. 273–284, 2013.

2   Background and Related Work

In the domain of time series data mining, approximated representations of time series are needed to enable efficient processing. Many methods have been proposed for dimensionality reduction, most of them being numerical *e.g.* discrete Fourier transform (DFT), discrete wavelet transform (DWT), singular value decomposition (SVD), principal component analysis (PCA), adaptive piecewise constant approximation (APCA), piecewise aggregate approximation (PAA), *etc* (see [1] for a survey). Symbolic methods have also been widely used because, beyond simplicity, readability and efficiency for time series representation, algorithms from other domains such as text processing and information retrieval, or bioinformatics can be used. Among symbolic representation methods, SAX proposed by Lin *et al.* [7] earned a large success. SAX is based on PAA and assumes that PAA values follow a Gaussian distribution. SAX discretizes PAA values according to equal-sized areas under the Gaussian curve yielding so-called breakpoints. Using lower bounds that are cheap to compute helps focusing on a small subset of the database sequences for which exact distance can later be computed [10].

The quality of the SAX representation of a time series depends on i) the number of PAA coefficients, *i.e.* the number of segments the time- series is divided in, ii) the number of symbols used for quantization (the alphabet size), iii) the gaussianity assumption. Several works have addressed these problems. In [9], Pham *et al.* alleviate the gaussianity assumption by introducing adaptive breakpoint vectors acting on segment size and alphabet size. However, the simplicity of SAX is lost by introducing a preprocessing phase using a clustering method. Other approaches attempt to enrich the PAA representation and, further, the SAX symbols. Extended SAX (ESAX)[8] associates the symbolic minimum and maximum of the PAA segment to the related SAX symbol as well as the order of their occurrences. This defines an abstract shape that gives finer results in time series retrieval. However, the size of the ESAX representation is 3 times the size of the SAX representation. From an efficiency point of view, authors did not compare their method with a SAX representation of the same size. In [3], authors make use of piecewise linear approximation (PLA), but only in a post-processing step and without quantizing PLA values. Several very recent works attempt to introduce a symbolic representation of the segment trend into the SAX representation. In [2], Esmael *et al.* associate one of the trend values U (up), D (down) and S (straight) to each SAX symbols computed from the segment trend. This yields a symbolic representation alternating SAX and trend values, having twice the size of the SAX representation. Trend approximations are obtained by linear regression and quantizing the line slope. No details are given about slope quantization and about the justification of using only three values. In [11], Zalewski *et al.* represent the slope information by quantizing the differences between the PAA values of two successive segments in the time series. A quantization algorithm separates difference values into k classes and determines the related centroids. The symbolic value affected to some difference value is the symbol associated to the closest centroid. Several quantization methods are evaluated. The main drawback of this method is the loss of simplicity and readability by switching to a first order derivative representation. In [6], authors introduce TSX, a Trend-based Symbolic approXimation. TSX divides a segment into 3 sub-segments determined by the most peak (MP) point, the most dip (MD) point and the bounds of the PAA segment. Then, TSX associates to each SAX symbol the trend information (symbolic slope) of its related sub-segments. This yields a 4-tuple symbolic series representation of time series. This representation is close to the ESAX representation but it is finer. A static lookup table is given for selecting the slope breakpoints in the TSX representation. However, the authors did not take into account the fact that slope breakpoints are dependent on the selected segment size.

In this paper we propose a symbolic representation that quantizes both the average and slope values of the time series on each segment. It hence produces one symbol per segment, each symbol can be interpreted in terms of an average and a slope value related to the linear regression on the segments.

3   The 1d-SAX Symbolic Representation

Our novel symbolic representation for time series is detailed in this section. We first review the main principles of the SAX method, before explaining how we propose to extend it.

3.1   SAX Representation

SAX transforms a numerical time series into a sequence of symbols taking their values in a finite alphabet. This method is very simple and does not require any *a priori* information about the input time series (apart from the distribution should be Gaussian with zero mean and unit variance). SAX representation is based on three steps:

1. Divide a time series into segments of length L
2. Compute the average of the time series on each segment
3. Quantize the average values into a symbol from an alphabet of size N

SAX makes the assumption that time series values follow a Gaussian distribution. The quantization step makes use of $(N-1)$ breakpoints that divide the area under the Gaussian distribution into N equiprobable areas. These breakpoints can be found in lookup tables. Hence, the average values computed for each segment of the time series (step 2 above) are then quantized according to the breakpoints of the Gaussian distribution. Fig. 1 shows an example of the SAX representation of a time series with $N = 4$.

3.2   1d-SAX Representation for Time Series

In this section we detail a novel symbolic representation of time series. The rationale behind this representation is the following. The SAX representation explained above relies only on the average value of the time series on each segment.

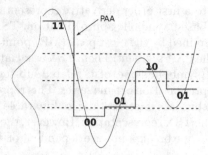

Fig. 1. Example of the SAX representation of a time series with $N = 4$. The dotted lines on the figure represent the three breakpoints of the Gaussian distribution $\mathcal{N}(0,1)$. SAX symbols are represented by their binary values.

Hence, two segments having different behaviors but with close averages will be quantized into the same symbol. For instance, a time series with an increasing trend can be mapped into the same bin as a time series with a decreasing trend if their respective means are close.

We propose here to integrate into the SAX representation an additional information about the trend of the time series on each segment. This new representation is denoted 1d-SAX in the following. It is based on three main steps, similarly to SAX:

1. Divide of the time series into segments of length L
2. Compute the linear regression of the time series on each segment
3. Quantize these regressions into a symbol from an alphabet of size N.

For each segment generated by step 1, the linear regression is computed and then quantized into a finite alphabet. The linear regression is computed using the least square estimation : let $V_1, \ldots, V_L$, be the values of a time series V on the time segment $T = [t_1, \ldots, t_L]$. The linear regression of V on T is the linear function $l(x) = sx + b$ that minimizes the distance between l and V on T. It is entirely described by the two values s and b. s represents the slope of l and b the value taken by l for $x = 0$. The least square optimization leads to:

$$s = \frac{\sum_{i=1}^{L}(t_i - \overline{T})(V_i - \overline{V})}{\sum_{i=1}^{L}(t_i - \overline{T})^2} \text{ , and } b = \overline{V} - s \times \overline{T}, \tag{1}$$

where $\overline{T}$ and $\overline{V}$ represent respectively the average values of V and T.

In the following, we choose to describe a linear regression of V on a time segment T by its slope value (s above), and the average value a of l on the segment. a is defined by $a = s \times (t_1 + t_L)/2 + b$. After this step (second step above), the original time series is represented by a pair (s, a) on each segment it has been divided in. We then need to quantize these pairs into an alphabet of N symbols. For that purpose, the two values are quantized separately and later combined into a symbol. Statistical properties of the linear regression ensure that

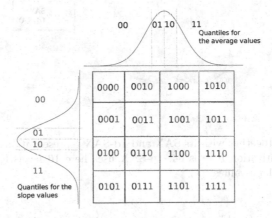

Fig. 2. Obtaining 1d-SAX binary symbols from the quantization of both the average and the slope values of the linear regression. The number of symbols is equal to 16, 4 levels are given to average values and 4 to the slope.

both the distribution of the average values and the slope values are Gaussian of mean 0. The variance of the average values is equal to 1, while the one of the slope values σ_L^2 is a decreasing function of L. According to these properties, quantization of the average and slope values can be done as in the SAX representation. The average values are quantized on N_a levels ($N_a < N$) according to the N_a quantiles of the Gaussian distribution $\mathcal{N}(0,1)$, while the slope values are quantized on N_s levels ($N_s < N$) according to the N_s quantiles of the Gaussian distribution $\mathcal{N}(0,\sigma_L^2)$. The choice of this parameter σ_L^2 is important. From the analysis of the impact of σ_L on many time series with Gaussian distribution, $\sigma_L^2 = 0.03/L$ appears to be a good compromise. The value of σ_L will be **fixed** for the experiments presented in Section 5. We assume here for clarity purposes that N_a and N_s are powers of two, *i.e.* $N_a = 2^{n_a}$, and $N_s = 2^{n_s}$. The quantization of the average value leads to a n_a-bit symbol, while the quantization of the slope value leads to a n_s-bit symbol. n_a is then interleaved with n_s to give a $(n_a + n_s)$-bit symbol, that represents the quantized value of the linear regression on $N = 2^{(n_a+n_s)}$ levels. Fig. 2 shows how to obtain $N = 16$ symbols with $N_a = N_s = 4$. $N_a - 1$ breakpoints are computed to define symbols for the average values, N_s breakpoints are computed to define the symbols for the slope values and these symbols are interleaved to get the final symbols on N levels.

The 1d-SAX representation allows for different configurations (for a same number N of levels) depending on the number of levels given to the average values and to the slope values. For instance, a symbolic representation on 64 levels can be obtained with 32 levels for the average and 2 levels for the slope, or 16 for the average and 4 for the slope, *etc.* The impact of these configurations will be discussed in Section 5.

This technique defines, for a given set of parameters (L, N_a, N_s), $N = N_a \times N_s$ binary symbols that represent linear functions on the segment $[1, \ldots L]$. A symbol obtained from a segment of a time series is a binary word ω of N bits. From ω, we

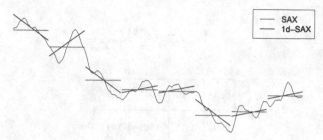

Fig. 3. A time series together with its SAX and 1d-SAX representation. The numbers of levels for the quantization here is 64. 1d-SAX uses here 16 levels for the average values and 4 for the slope values.

can extract ω_a and ω_s, the binary words representing respectively the quantized average and slope values of the linear regression of the time series on the segment. The value of ω_a indicates that the average value of the linear regression on the segment lies between two breakpoints of the $\mathcal{N}(0, 1)$ distribution : β_k^a and β_{k+1}^a for instance. Similarly, the slope values of the linear regression on the segment lies between two breakpoints of the $\mathcal{N}(0, \sigma_L^2)$ distribution : β_l^s and β_{l+1}^s. We can get from these intervals an approximation of the average and slope values of the linear regression by taking the median values on each interval. These median values are also given by the quantiles of the Gaussian distribution. Following that procedure, we can obtain a numerical approximation of a time series from its quantized version (with SAX or 1d-SAX).

Fig. 3 shows an example of a time series together with its SAX and 1d-SAX representations. On this example, 64 levels have been used for the quantization, 1d-SAX uses here 16 levels for the average values and 4 for the slope values. We can see on this example that the 1d-SAX representation fits the time series more accurately than the SAX representation. This result will be highlighted in the experimental results section.

4   Asymmetric Querying for Time Series Database Search

We have applied this novel time series representation to the 1-nearest neighbour search (1-NNS) problem. The aim of this application is the following. Let us consider that we have a database D containing $\#D$ time series. Given a query q, we want to find the time series in D that is most similar to q. We assume in the rest of this paper that all the time series in the database have the same length, also equal to the length of the queries. The brute force method consists in calculating the distances between the query and all the series of the database and return the one that is most similar to q. The number of distances to compute is hence equal to $\#D$. Taking advantage of the approximate representation to speed-up the search in big database of time series is interesting in that case. SAX representation has been for instance used to index and mine terabytes of time series [10].

We define in this section an asymmetric querying method for approximate search in time series database. The term asymmetric means that the queries are not quantized to avoid having a double quantization error (when both queries and series of the database are quantized). Performing asymmetric querying has been shown to improve the accuracy of the distance approximation for vector searches [4]. We propose a method based on this idea to perform approximate search in time series database.

Let us assume that D contains time series, as well as their symbolic representation (1d-SAX) for a given set of parameters (L, N_a, N_s). This set of parameters completely defines the $N_a \times N_s$ symbols $s_1, \ldots, s_N$ that are used to quantize the time series. The numerical approximation of these symbols can be computed as explained at the end of Section 3. The algorithm to search the 1-NN of a query q is :

1. Split q into segments of length L : $q = q_1, \ldots, q_w$
2. Compute the Euclidean distances between every segment of q and the symbols $s_j, 1 \leq j \leq N$. These distances are put in a lookup table $A = (a_{i,j})$ of dimension $w \times N$, where $a_{i,j} = ED(q_i, s_j)^2$. ED represents the Euclidean distance.
3. For every time series d in D, the quantized version of d, $\hat{d} = \hat{d}_1, \ldots, \hat{d}_w$ is available. An approximate distance $Dist_{asym}$ between q and $\hat{d}$ is obtained by

$$Dist_{asym}(q, \hat{d}) = \sum_{i=1}^{w} ED(q_i, \hat{d}_j)^2 = \sum_{i=1}^{w} a_{i,s_{\hat{d}_j}}, \qquad (2)$$

which is just obtained by accessing the lookup table and summing over w elements.

After these steps, the approximate distances between q and all the time series in D are available. These distances can be used to select the approximate nearest neighbours of q. The number of elementary arithmetical operations ν_q to compute for a query search using this method is

$$\nu_q = (3L - 1) \times w \times N + (w - 1) \times \#D, \qquad (3)$$

where the left part represents the cost of step 2 above and the right part the one of step 3. The number of elementary operations in the case of the brute force method is $(3Lw - 1) \times \#D$. The computation cost is lower with this approximate retrieval scheme for large databases where $N \leq \#D$.

5   Experimental Evaluations

In this section we evaluate the performance of our symbolic representation of time series in terms i) of goodness-of-fit to the input data, ii) of retrieval performance when used to query a database of time series using our asymmetric scheme and iii) quality of classification using a k-nearest-neighbour classification

Table 1. Average approximation error (in terms of Euclidean distance) between a time series and its symbolic representation (SAX, 1d-SAX with 4 levels for the slopes and 1d-SAX with 8 levels for the slopes) for $N = 256$. Results are evaluated for two different values for $L : L_1$ and L_2.

Dataset	$w \times L$	L_1	SAX	1d-SAX $Ns = 4$	1d-SAX $Ns = 8$	L_2	SAX	1d-SAX $Ns = 4$	1d-SAX $Ns = 8$
Beef	450	10	3.733	**2.545**	2.611	50	9.602	6.851	**6.286**
CBF	120	10	5.511	4.963	**4.862**	20	6.936	5.785	**5.570**
Coffee	250	10	4.288	2.538	**2.309**	50	10.907	8.061	**7.588**
FaceFour	350	10	10.825	8.186	**7.592**	50	17.826	17.081	**16.961**
Fish	450	10	2.027	**0. 957**	1.206	50	9.183	4.203	**2.792**
Gun-Point	150	10	2.250	1.208	**0.969**	25	4.559	2.529	**2.153**
Lighting2	500	10	10.730	**10.269**	10.394	50	14.564	12.983	**12.478**
Lighting7	250	10	8.371	**7.960**	8.043	50	11.538	10.316	**10.040**
OSULeaf	420	10	4.493	2.034	**1.935**	35	11.985	6.862	**6.271**
Random walks	500	10	3.924	2.939	**2.934**	50	8.558	6.365	**5.876**
Swedish Leaf	120	10	4.302	2.324	**2.063**	30	9.803	5.718	**4.886**
Wafer	150	10	6.597	5.2789	**4.799**	25	11.787	10.987	**10.789**
50Words	270	10	4.661	**3.091**	3.127	30	9.999	6.667	**5.878**
Yoga	420	10	2.371	**1.298**	1.319	35	7.475	3.424	**2.767**

scheme. We have used 13 datasets provided by the UCR Time Series Data Mining archive [5] and one dataset of random walks. Other data sets from the UCR archive have not been considered here due to small lengths of the time series particularly. Each of these dataset is decomposed into a training set and a test set. The query used for time series retrieval are taken in the test sets while the train sets represent the different databases. For all the results presented in this section, we fixed $\sigma_L^2 = 0.03/L$, which turned out to be a good trade-off for all the datasets.

5.1 Quality of Representation

We first evaluate the proposed symbolic representation of time series in terms of the approximation error induced by the quantization of a time series, that we define here as the Euclidean distance between an original time series and its numerical approximation obtained with the 1d-SAX method (2). The lowest this distance the better the fit to the original time series. We have plotted in Fig. 4 the average approximation error versus N for the 50Words dataset and two different values of L ($L = 10$ and $L = 30$) and compared the error obtained with SAX and 1d-SAX for the same numbers of symbols used for the symbolization (*i.e.* same quantity of information). We can see on the left part of this figure that when L is small, the gain brought by the slope information begins to be

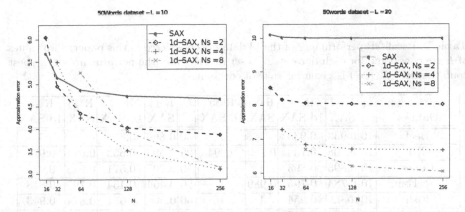

Fig. 4. Average approximation error versus N (50Words dataset) for two different values of L

significant for $N \geq 64$. The best configuration from $N = 64$ is the one with $N_s = 4$ (4 levels to quantify the slope values), while the one with $N_s = 8$ gets closer at $N = 256$. When L is higher (right part of Fig. 4), we can see that this phenomenon is amplified: the gain brought by 1d-SAX over SAX is much bigger, even for small values of N. In addition, we can see that for this value of L, the best configuration tends to be the one with $N_s = 8$. Similar results from all the datasets are given in Table 1. In this table, N is set to 256 and two values of L have been tried for each dataset. The same conclusion can first be drawn: the gain in terms of approximation error increases with L. This result makes sense: representing a time series on a small segment by its average value is less restrictive than on a long segment. We can also draw another conclusion: the number of levels N_a given to the average values should be higher than the one given to the slope values N_s. This means that a balance between N_a and N_s (assuming that their product is fixed and equal to N) has to be found to optimize the performance of the 1d-SAX representation. Most of the time, for all the datasets that we use, the best configuration for small L (less than 25) was obtained with $N_s = 2$ or $N_s = 4$, while it was obtained with $N_s = 4$ or $N_s = 8$ for longer values of L (more than 25).

5.2   Retrieval Performance

We exploited the property of having a symbolic representation closer to the original time series in a time series retrieval scheme. In this section we present some experimental results of the retrieval scheme presented in Section 4. The results are given in terms of the recall@R measure. This measure reflects the probability of finding the true nearest neighbor in the R sequences of the dataset that are closer to the query q in terms of the approximate distance. It is hence equal to the probability of retrieving the correct 1-NN if computing exact distance only for the best R candidates. Fig. 5 shows the recall@R performance of our scheme

Table 2. Recall@R performance of the 14 datasets considered in this paper. Two values of L are considered for each dataset. N is set to 256. Only the performance of the best configuration (N_a, N_s) is given for sake of conciseness.

Dataset	L_1	R@1 SAX	R@1 1d-SAX	R@5 SAX	R@5 1d-SAX	L_2	R@1 SAX	R@1 1d-SAX	R@5 SAX	R@5 1d-SAX
Beef	10	0.933	0.933	1	1	50	0.866	0.833	1	1
CBF	10	0.708	0.734	0.969	0.994	20	0.596	0.653	0.941	0.971
Coffee	10	0.893	0.893	1	1	50	0.607	0.571	0.928	0.928
FaceFour	10	0.784	0.795	0.989	1	50	0.454	0.454	0.795	0.818
Fish	10	0.92	0.954	1	1	50	0.508	0.611	0.874	0.943
Gun-Point	10	0.920	0.907	1	1	25	0.713	0.733	0.987	0.993
Lighting2	10	0.557	0. 574	0.918	0.934	50	0.328	0.377	0.738	0.836
Lighting7	10	0.397	0.466	0.849	0.877	50	0.205	0.288	0.589	0.726
OSULeaf	10	0.950	0.942	1	1	35	0.479	0.739	0.884	0.996
Random walks	10	0.903	0.902	1	1	50	0.479	0.657	0.943	0.992
Swedish Leaf	10	0.646	0.691	0.923	0.955	30	0.146	0.169	0.395	0.459
Wafer	10	0.653	0.627	0.920	0.919	25	0.472	0.440	0.830	0.824
50Words	10	0.863	0.859	0.998	0.998	30	0.488	0.618	0.800	0.954
Yoga	10	0.963	0.964	1	1	35	0.825	0.834	0.995	0.997

Table 3. Classification performance (percentage of correct classification) of SAX and 1d-SAX. Two values of L are considered for each dataset. N is set to 256.

Dataset	L_1	Correct classif. SAX	Correct classif. 1d-SAX	L_2	Correct classif. SAX	Correct classif. 1d-SAX
Beef	10	0.7	0.7	50	0.7	0.667
CBF	10	0.891	0.896	20	0.863	0.898
Coffee	10	0.964	0.964	50	0.679	0.857
FaceFour	10	0.636	0.648	50	0.477	0.716
Fish	10	0.760	0.766	50	0.606	0.749
Gun-Point	10	0.773	0.793	25	0.840	0.900
Lighting2	10	0.803	0. 803	50	0.705	0.738
Lighting7	10	0.630	0.658	50	0.644	0.658
OSULeaf	10	0.483	0.492	35	0.455	0.504
Swedish Leaf	10	0.758	0.769	30	0.448	0.491
Wafer	10	0.818	0.818	25	0.817	0.819
50Words	10	0.657	0.664	30	0.622	0.675
Yoga	10	0.963	0.964	35	0.825	0.834

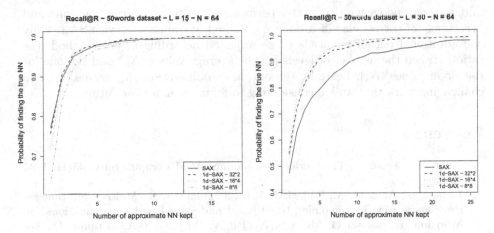

Fig. 5. Recall@R measure (50words dataset) for $N = 64$ and two values of L

using 1d-SAX and SAX for the 50Words dataset and two different values of L. We can see that for small values of L, the gain brought by our symbolic representation is not significant, while this gain increases with the length L of segments. As in Section 5.1, configurations with $N_s = 2$ or $N_s = 4$ are better for small L and configurations with $N_s = 4$ or $N_s = 8$ are better for long L. Recall@1 and Recall@5 values for all the datasets considered in this paper are given in Table 2. These values are obtained for $N = 256$, and only the best configuration for 1d-SAX is given in the table for sake of conciseness.

Finally, we also give some results in terms of quality of classification. In the considered data sets, every time series (in both training and test sets) is labeled by a class. We used these labels to evaluate the performance of 1d-SAX in terms of classification. For that purpose, we used the classical k-nearest-neighbour scheme. For each time series of the test set, we estimate its label by looking at the labels of the k nearest time series in the corresponding training set. Results are given in Table 3 for $k = 5$.

6   Conclusion and Discussion

In this paper we propose a novel symbolic representation for time series. This representation is based on the quantization of the linear regression of the time series on subsegments. Symbols take into account information about the average values and the slope values of the time series. One of the main advantage of the proposed method over other representations is that the quantity of information needed to represent a time series is the same as the one needed by SAX, for a same number of symbols N. We have shown that our 1d-SAX method allows for a better fitting of the original time series: the approximation error induced by the symbolization is reduced compared to SAX. Furthermore, we have used this representation to the application of time series retrieval and classification in databases and shown that better performance in terms of recall measure and percentage of correct classification are obtained in comparison with SAX. Only one additional parameter needs to be adjusted according to our method: the ratio between the number of levels for the average values (N_a) and the one for the slope values (N_s). Learning the optimal configuration $(N_a$ versus $N_s)$ is a challenging work that we are considering to focus on in a close future.

References

1. Esling, P., Agon, C.: Time-series data mining. ACM Comput. Surv. 45(1), 1–34 (2012)
2. Esmael, B., Arnaout, A., Fruhwirth, R.K., Thonhauser, G.: Multivariate time series classification by combining trend-based and value-based approximations. In: Murgante, B., Gervasi, O., Misra, S., Nedjah, N., Rocha, A.M.A.C., Taniar, D., Apduhan, B.O. (eds.) ICCSA 2012, Part IV. LNCS, vol. 7336, pp. 392–403. Springer, Heidelberg (2012)
3. Hung, N.Q.V., Anh, D.T.: Combining SAX and Piecewise Linear Approximation to improve similarity search on financial time series. In: Proc. of the Int. Symp. on Information Technology Convergence (ISITC), pp. 58–62 (2007)
4. Jégou, H., Douze, M., Schmid, C.: Product quantization for nearest neighbor search. IEEE Transactions on Pattern Analysis and Machine Intelligence 33(1), 117–128 (2011)
5. Keogh, E., Zhu, Q., Hu, B., Hao, Y., Xi, X., Wei, L., Ratanamahatana, C.A.: The UCR times series classification/clustering homepage (2011)
6. Li, G., Zhang, L., Yang, L.: TSX: A novel symbolic representation for financial time series. In: Anthony, P., Ishizuka, M., Lukose, D. (eds.) PRICAI 2012. LNCS, vol. 7458, pp. 262–273. Springer, Heidelberg (2012)
7. Lin, J., Keogh, E.J., Lonardi, S., Chiu, B.Y.: A symbolic representation of time series, with implications for streaming algorithms. In: Proc. of the 8th ACM SIGMOD Workshop on Research Issues in Data Mining and Knowledge Discovery, pp. 2–11 (2003)
8. Lkhagva, B., Suzuki, Y., Kawagoe, K.: New time series data representation esax for financial applications. In: Proc. of the 22nd Int. Conf. on Data Engineering Workshops, pp. 17–22 (2006)
9. Pham, N.D., Le, Q.L., Dang, T.K.: Two novel adaptive symbolic representations for similarity search in time series databases. In: Proc. of the 12th Asia-Pacific Web Conference (APWeb), pp. 181–187 (2010)
10. Shieh, J., Keogh, E.: iSAX: Indexing and mining terabyte sized time series. In: Proc. of the ACM SIGKDD Int. Conf. on Knowledge Discovery and Data Mining (2008)
11. Zalewski, W., Silva, F., Lee, H.D., Maletzke, A.G., Wu, F.C.: Time series discretization based on the approximation of the local slope information. In: Pavón, J., Duque-Méndez, N.D., Fuentes-Fernández, R. (eds.) IBERAMIA 2012. LNCS, vol. 7637, pp. 91–100. Springer, Heidelberg (2012)

Learning Models of Activities Involving Interacting Objects

Cristina Manfredotti[1], Kim Steenstrup Pedersen[2],
Howard J. Hamilton[3], and Sandra Zilles[3]

[1] LIP6, Pierre and Marie Curie University (UPMC), Paris, France
[2] DIKU, University of Copenhagen, Copenhagen, Denmark
[3] Department of Computer Science, University of Regina, Regina, Canada

Abstract. We propose the LEMAIO multi-layer framework, which makes use of hierarchical abstraction to learn models for activities involving multiple interacting objects from time sequences of data concerning the individual objects. Experiments in the sea navigation domain yielded learned models that were then successfully applied to activity recognition, activity simulation and multi-target tracking. Our method compares favourably with respect to previously reported results using Hidden Markov Models and Relational Particle Filtering.

1 Introduction

Many practical problems including activity recognition, multi-target tracking and detection of activity, require reasoning about the interactions of multiple related objects. A complex activity (such as exchanging goods between two ships) is a type of interaction usually realized as a sequence of lower-level actions that may involve multiple objects. Given a model of how an activity is decomposed, it is possible to effectively recognize an ongoing activity by observing low-level attributes (such as position, speed and color) [3,17]. However, the model for such activities is usually unknown. We investigate one possible solution: learning such a model directly from sensor data. This paper introduces a framework, called LEMAIO (LEarning Models of Activities involving Interacting Objects), for learning probabilistic models of complex activities involving multiple interacting objects from sensor data. This framework is capable of inferring the interactions between the objects, while also inferring how complex activities decompose into lower level actions.

An *activity* is usually recognized from the sequence of the attribute values of the interacting objects. An example of an activity in the soccer domain is "passing the ball", which can be recognized observing the sequence of positions of the players and the ball over time. The same activity can be undertaken in many ways, represented as different sequences of attribute values, e.g., all the ways in which the ball can be passed from one player to another. To avoid listing all possible realizations of an activity, we need an abstract representation (a *model*) of it. This model should be such that an automatic system can use

A. Tucker et al. (Eds.): IDA 2013, LNCS 8207, pp. 285–297, 2013.

it efficiently to recognize an activity from (noisy) data (*activity recognition*), simulate it (*activity generation*), or track it (*multi-target tracking*).

Since activities often involve multiple objects, modeling the relations between them is crucial for capturing their behaviour. Consider the difference between the activities of "passing" and "intercepting" a ball: both activities result in a new player having control of the ball but the former requires the two players to be on the same team (to be in the relation of having the same value for the team attribute), while the latter requires them to be on different teams. We distinguish between *atomic activities* (called simply "activities" in [1]), involving coordinated actions among multiple agents at one time, and *complex activities*, which are sequences of atomic activities. We do not assume we know the relations between objects: we know that objects might interact, but we do not know how. We assume all relations are pairwise.

We are given a training set where each instance is a sequence of attribute values describing the complete state of the world, along with a label identifying the complex activity represented and we adopt a probabilistic viewpoint: the problem is to learn from this training set a probabilistic model able to identify complex activities in new data, track individual objects while complex activities are occurring, and generate sequences of synthetic data simulating complex activities. The LEMAIO framework addresses this problem by learning a three-layer hierarchical model from the bottom up that can be mapped into a Dynamic Bayesian Network.

The main contributions of our work are: (1) a general top-supervised learning framework to learn a hierarchical probabilistic model for complex activities from low-level data; (2) the decomposition of complex activities into lower level actions and interactions between objects and the explicit modelling of objects' interactions; (3) an implementation of the framework based on Expectation Maximization and clustering; (4) empirical evidence of the effectiveness of our approach in learning models able to recognize, track, and generate complex activities.

2   The LEMAIO Framework

To describe the LEMAIO framework, we first explain the four levels of abstraction. Then we describe how a three-layered model is learned that allows values at any level to be generalized to the next higher level. Finally, we show how the model is used to generate synthetic data corresponding to specified activities.

Our learning approach is *top-supervised*: labels are available in the training data only at the top (complex activity) level. The learned model is able to assign a label to a complex activity represented by an unseen sequence of low-level data and is also able to recognize the lower-level constituents of the activity.

2.1   Levels of Abstraction in LEMAIO

The LEMAIO framework uses four levels of abstraction: (0) attribute values for objects (raw data), (1) single object activities (activities involving only one object), relations between pairs of objects at a single time, and changes in relations over time, (2) atomic activities and (3) complex activities. We assume that during any time interval an object can be involved in at most one single object activity and a set of related objects can only be involved in at most one atomic activity and at most one complex activity. In this section, we consider each level in turn. These levels of abstraction are general and the decomposition of a complex activity into atomic activities, relations, changes of relations and single object activities can be applied to a variety of domains.

Level 0: We collect all attribute values of the objects in the world at consecutive time steps while some complex activity is occurring. We assume the attribute values correspond to noiseless observations from sensors that coincide with the actual state of the world (in presence of noisy observation we could filter the data before learning). Let $s_t^{(i)}$ be the *state of the object* $o^{(i)}$ at time t. The *state of the world* s_t can be represented as the vector of the states of all individual objects in the world at time t. The training data consists of pairs of sequences of consecutive states of the world from time 1 to time T and labels of the complex activity represented by the sequence $(s_{1:T}, \gamma)$. Assuming that we have labels for complex activities is restrictive but fair: it is easier for a person to classify a complex activity (providing a labeled instance) than to describe such an activity.

Level 1: At level 1 we represent how objects behave individually, how they interact and how their interactions change over time.

- *single object activities* are associated with changes over time in the attribute values of single objects. $c_{(t-1,t)}^{(i)}$ represents the single object activity that $o^{(i)}$ performs in time interval $(t-1, t)$ and assumes values in $\mathcal{E} = \{\epsilon_1, \epsilon_2, \cdots, \epsilon_{n_E}\}$.
- *relations* represent degrees of similarity between attribute values of objects at the same time. We assume pairwise relations. $r_t^{(i,j)}$ represents the relation between $o^{(i)}$ and $o^{(j)}$ and assumes values in $\mathcal{R} = \{\rho_1, \rho_2, \cdots, \rho_{n_R}\}$.
- *changes in relation* represent changes in the degree of similarity over time. $d_{(t-1,t)}^{(i,j)}$ represents the change of relation between $o^{(i)}$ and $o^{(j)}$ during the time interval $(t-1, t)$ and assumes values in $\mathcal{D} = \{\delta_1, \delta_2, \cdots, \delta_{n_D}\}$.

From data classified into single object activities, relations and change in the relations we can learn distributions for atomic activities.

Level 2: Atomic activities describe how one object behaves with respect to another during the time interval between consecutive time points. $a_t^{(i,j)}$ represents an atomic activity involving the related objects $o^{(i)}$ and $o^{(j)}$ in the interval $(t-1, t)$ and it is learned from vectors of the form $a_t^{(i,j)} = [e_{(t-1,t)}^{(i)}, e_{(t-1,t)}^{(j)}, r_t^{(i,j)}, d_{(t-1,t)}^{(i,j)}]$. We define the set of all possible atomic activities as $\mathcal{A} = \{\alpha_1, \alpha_2, \cdots, \alpha_{n_A}\}$.

Level 3: A *complex activity* $c^{(i,j)}$ is represented by a sequence of atomic activities that involve $o^{(i)}$ and $o^{(j)}$; $c^{(i,j)}$ can assume values in $\mathcal{C} = \{\gamma_1, \gamma_2, \cdots, \gamma_{n_C}\}$.

Example: Let us focus on two possible complex activities (*Rendezvous* and *Avoidance*) that can occur at sea. They consist of two vessels approaching each

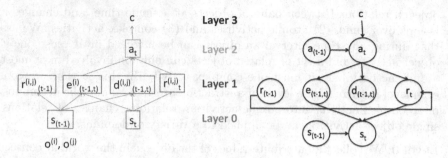

Fig. 1. left: The LEMAIO hierarchy with layers. **right:** The Dynamic Bayesian Network learned by LEMAIO.

other and subsequently going apart. In the latter only one of the two vessels changes its speed to avoid the other ship, but in the former, when the vessels are close to each other, they stay close with speed near zero to illegally exchange goods. The state of the world is indicated by the position, name and class of each vessel. The training set consists of pairs of a sequence of states and the label (R or A) of the complex activity. A single object activity can encode the movement of a ship (e.g., moving fast towards north); a relation can give the separation distance between two ships or if they are of different/equal type; a change in relation can tell whether the distance between two ships is increasing or decreasing during a time interval; and an atomic activity can describe the idea of "approaching" (e.g., two ships have decreasing distance over time). A complex activity modeling a *Rendezvous* could be composed as a sequence of atomic activities such as "approaching", "staying together" and "going apart" (each possibly repeated).

2.2   Learning with LEMAIO

In order to model the uncertainty about which activities are currently being performed and how a complex activity decomposes into lower-level constituents, we introduce a number of probability distributions. These are models that, given an observed pattern and using the Bayes theorem, (1) assign probabilities to the events that associate the pattern with any (single object, atomic, or complex) activity and (2) assign probabilities to the future. In this way, the learned model can be used for classification and generative purposes. We separate our presentation into levels based on the abstraction hierarchy shown in Fig. 1 left.

Single Object Activities and Relations: To learn models for these quantities we preprocess the data into three sets of differences.

- $\Delta^{(i)}_{(t-1,t)}$ denotes the difference between the states of $o^{(i)}$ at two consecutive time points: $\Delta^{(i)}_{(t-1,t)} = s^{(i)}_t - s^{(i)}_{t-1}$, where $t > 0$;

- $\Delta_t^{(i,j)}$ denotes the distance between the states of $o^{(i)}$ and $o^{(j)}$ at the same time step: $\Delta_t^{(i,j)} = dist(s_t^{(i)}, s_t^{(j)})$, where $t \geq 0$, $i \neq j$;
- $\Delta_{(t-1,t)}^{(i,j)}$ is short for: $\Delta_{(t-1,t)}^{(i,j)} = \Delta_t^{(i,j)} - \Delta_{t-1}^{(i,j)}$, where $t > 0$, $i \neq j$.

Since single object activities involve only one object, they can be seen as the change in the attribute values for one object. Given a sequence of states (our training data), for all the attribute values of every object of every pair of consecutive time steps we compute $\Delta_{(t-1,t)}^{(i)}$. From these data, we learn a model for single object activities. The model consists of the prior of the single object activity class, $p(e_{(t-1,t)}^{(i)} = \epsilon_k)$, and the probability density function $p(\Delta_{(t-1,t)}^{(i)} | e_{(t-1,t)}^{(i)} = \epsilon_k)$.

A relation is a difference between the attribute values of two objects at a single time point. We learn a probability distribution for relations from data of the form $\Delta_t^{(i,j)}$ obtained from the states of every pair of objects in the training data at the same time step. The model consists of the prior of the relation class, $p(r_t^{(i,j)} = \rho_k)$, and the probability density function $p(\Delta_t^{(i,j)} | r_t^{(i,j)} = \rho_k)$.

A change in relation is a difference between the relation of two objects over time. From $\Delta_{(t-1,t)}^{(i,j)}$, we learn the prior of the classes of changes in relations, $p(d_{(t-1,t)}^{(i,j)} = \delta_k)$, and the probability density function $p(\Delta_{(t-1,t)}^{(i,j)} | d_{(t-1,t)}^{(i,j)} = \delta_k)$.

According to the Bayes formula the probability of a single object activity class ϵ_k, given the observed data $\Delta_{(t-1,t)}^{(i)}$, is the posterior

$$p(e_{(t-1,t)}^{(i)} = \epsilon_k | \Delta_{(t-1,t)}^{(i)}) = \frac{p(e_{(t-1,t)}^{(i)} = \epsilon_k) p(\Delta_{(t-1,t)}^{(i)} | e_{(t-1,t)}^{(i)} = \epsilon_k)}{p(\Delta_{(t-1,t)}^{(i)})}. \tag{1}$$

Such a posterior can be used for classification purposes. Similar posteriors can be derived for relations and changes in relations.

Atomic Activities: To learn a model for atomic activities we first apply the probabilistic models learned at layer 1 (Eq. 1) to classify data into single object activities, relations and changes of relations. Secondly, by considering every time interval in every input sequence, we collect vectors $v^{(i,j)}$ of single object activities, relations and changes in relations for every pair of related objects:

$$v^{(i,j)} = [e_{(t-1,t)}^{(i)}, e_{(t-1,t)}^{(j)}, r_t^{(i,j)}, d_{(t-1,t)}^{(i,j)}]. \tag{2}$$

We then cluster these vectors for all (i, j) pairs according to a distance measure f. Next, for each cluster (α_k) we select one vector (v_{α_k}) to represent all vectors in the cluster. Finally we map each cluster into the set of labels in $\mathcal{A}$. We assume the probability that a vector $v^{(i,j)}$ is in cluster α_k, $p(v^{(i,j)} | a^{(i,j)} = \alpha_k)$, is given by one minus its normalized distance from v_{α_k}: $p(v^{(i,j)} | a^{(i,j)} = \alpha_k) = 1 - \bar{f}(v^{(i,j)}, v_{\alpha_k})$. We learn the prior $p(a^{(i,j)} = \alpha_k)$ proportional to the number of the data points in the training set that fall in cluster α_k. To classify vectors of the form of Eq. 2 into atomic activities we can use the posterior

$$p(a^{(i,j)} = \alpha_k | v^{(i,j)}) \propto p(a^{(i,j)} = \alpha_k) p(v^{(i,j)} | a^{(i,j)} = \alpha_k).$$

Complex Activities: Complex activities are defined as sequences of atomic activities. We group the data in the training set according to their complex activity label and, using the distributions learned at the previous layers, we map them into sequences of atomic activities. From these sequences we learn the probability that an atomic activity α_k follows a sequence of atomic activities $a_{1:t-1}^{(i,j)}$ given a particular complex activity γ_k:

$$p(a_t^{(i,j)} = \alpha_k | c = \gamma_k, a_{1:t-1}^{(i,j)}). \tag{3}$$

The probability $p(a_1 = \alpha_k)$ is proportional to the number of times it occurs at time $t = 1$ in the training set. The prior $p(c = \gamma_k)$ is proportional to the number of occurrences of γ_k in the training set. We classify a sequence of atomic activities $a_{1:t}^{(i,j)} = \{a_1^{(i,j)}, a_2^{(i,j)}, \cdots, a_t^{(i,j)}\}$ as the complex activity $c = \gamma_k$ that is associated with the highest value of $p(c = \gamma_k | a_{1:t}^{(i,j)})$, where

$$p(c = \gamma_k | a_{1:t}^{(i,j)}) = \frac{p(a_t^{(i,j)} | c = \gamma_k, a_{1:t-1}^{(i,j)}) p(a_{1:t-1}^{(i,j)}, c = \gamma_k)}{p(a_{1:t}^{(i,j)})}. \tag{4}$$

The overall learned model is depicted in Fig. 1 right.

2.3   Activity Generation with LEMAIO

Activity generation aims at generating sequences of states that match a given complex activity $c = \gamma_k$. Assume we are given a sequence of states $s_{0:t-1}$ that matches complex activity γ_k. Given the probability distributions learned so far, $s_{0:t-1}$ can be classified into sequences of atomic activities $a_{1:t-1}^{(i,j)}$. We want to generate the next atomic activity $a_t^{(i,j)}$ such that the sequence $a_{1:t}^{(i,j)}$ is constrained to be associated with complex activity $c = \gamma_k$. To do so, we sample from the probability distribution $p(a_t^{(i,j)} | c = \gamma_k, a_{1:t-1}^{(i,j)})$ learned at layer 3. Suppose $a_t^{(i,j)} = \alpha_k$. Next, we sample a vector $v^{(i,j)}$ from the probability distribution $p(v^{(i,j)} | a_t^{(i,j)} = \alpha_k)$ learned at layer 2. Suppose $v^{(i,j)} = [\epsilon_i, \epsilon_j, \rho_l, \delta_m]$, telling us the generated single object activities, relations and changes of relations.

Knowing the current state s_{t-1}, single object activities (ϵ_i and ϵ_j), relations (ρ_l) and change in relations (δ_m) we can generate the next state s_t by sampling from the probability distributions learned at layer 1. To model the change in relation, let us introduce a random variable D_m: $D_m \sim p(\Delta_{(t,t-1)}^{(i,j)} | d_{(t,t-1)}^{(i,j)} = \delta_m)$. Let q be the distribution of the random variable $\Delta_{t-1}^{(i,j)} + D_m$. We can sample a value for $\Delta_t^{(i,j)}$ from the probability

$$p(\Delta_t^{(i,j)} | d_{(t-1,t)}^{(i,j)} = \delta_m, r_t^{(i,j)} = \rho_l, \Delta_{t-1}^{(i,j)}) = q(\Delta_{t-1}^{(i,j)} + D_m) p(\Delta_t^{(i,j)} | r_t^{(i,j)} = \rho_l),$$

estimated from the distributions of the relations and their changes.

To simplify the following explanation, we assume the only objects in the world are $o^{(i)}$ and $o^{(j)}$. Given the sampled value for $\Delta_t^{(i,j)}$, we can sample $s_t = [s_t^{(i)}, s_t^{(j)}]$

from $p(s_t|\epsilon_i, \epsilon_j, s_{t-1}, \Delta_t^{(i,j)})$. That, assuming $p(s_t^{(i)}|s_{t-1}^{(i)}, \Delta_t^{(i,j)}) = p(s_t^{(i)}|s_{t-1}^{(i)})$, can be factored as:

$$p(s_t|\epsilon_i, \epsilon_j, s_{t-1}, \Delta_t^{(i,j)}) = \frac{p(s_t^{(i)}|\epsilon_i, s_{t-1}^{(i)})p(s_t^{(j)}|\epsilon_j, s_{t-1}^{(j)})p(s_t^{(j)}|s_{t-1}^{(j)}, \Delta_t^{(i,j)}, s_t^{(i)})}{p(s_t^{(j)}|s_{t-1}^{(j)})}. \quad (5)$$

With this assumption, we can first sample the state of $o^{(i)}$ and then sample the state of $o^{(j)}$ taking into account the state of $o^{(i)}$ already sampled and their relations. This assumption is equivalent to assuming one of the two objects ($o^{(i)}$ in this case) is the "leader" and can be loosened in practice by exchanging the order in which the objects are processed at each time step.

2.4   An Implementation

In our LEMAIO-1 implementation of the LEMAIO framework, we use mixtures of Gaussians, for the distributions at layer 1, a mixture of categorical distributions at layer 2 and a mixture of Markov chains at layer 3.

LEMAIO-1 Layer 1: Since the same approach is used for the three kinds of entities at layer 1, here we present only the procedure for learning single object activities. We use Expectation Maximization (EM) to learn the prior of the classes of single object activities and the probability density function of the data given the class that maximize the likelihood of the data [6]. Assuming we have observed a particular change in one object's attribute values ($\Delta_{(t-1,t)}^{(i)}$), the likelihood of the observed data given the parameters Θ of the distributions is calculated as: $p(\Delta_{(t-1,t)}^{(i)}|\Theta) = \sum_{k=1}^{K} p(\epsilon_k)p(\Delta_{(t-1,t)}^{(i)}|\epsilon_k, \Theta)$ where K is the number of classes represented in the data, chosen to minimize the Bayes Information Criterion (BIC) [7], and Θ is the vector of the means and variances of the chosen Gaussian distributions.

LEMAIO-1 Layer 2: To cluster vectors $v^{(i,j)}$ (cf. Eq. 2) into atomic activity classes we use the K-medoids clustering algorithm [14]. We compute the distance measure $f(v_1, v_2)$ on which the clustering is based in the following way: first we map each element in v_1 and in v_2 into the mean of the Gaussian distribution that best fits the class represented by the element; then we compute the Euclidean distance between these elements and average over the elements of the vectors. The number of clusters K is chosen such that it maximizes the intracluster similarity of our data. The number of distributions and the number of clusters K is chosen given a maximum number of sets $\mathcal{K}$.

LEMAIO-1 Layer 3: Given sequences of atomic activities labeled with the same complex activity ($c = \gamma_k$), we learn a Markov chain. In a Markov chain, the probability of an atomic activity at time step t depends only on the value of the atomic activity at time step $t-1$ and on the current complex activity $c = \gamma_k$. We thus have (Eq. 3): $p(a_t^{(i,j)} = \alpha_k|c = \gamma_k, a_{1:t-1}^{(i,j)}) = p(a_t^{(i,j)} = \alpha_k|c = \gamma_k, a_{t-1}^{(i,j)})$. Modeling the transitions between atomic activities with a Markov chain allows us to simplify the classification of complex activities writing Eq. 4 as:

$$p(c = \gamma_k | a_{1:t}^{(i,j)}) = p(c = \gamma_k) \prod_{t=2}^{T} p(a_t^{(i,j)} = \alpha_k | c = \gamma_k, a_{t-1}^{(i,j)}). \qquad (6)$$

3 Experiments

We experimented with our implementation on the sea navigation data set provided in [4]. This data set is composed of 37 sequences called *encounters*. An encounter is a sequence of 96 time steps recording the 2D positions of two ships involved in either a rendezvous or an avoidance; there are 19 rendezvous encounters. In the following, we describe how we applied LEMAIO-1 to learn models from this dataset and how we tested the resulting models.

In our data set the state $s_t^{(i)}$ is the vector $[x_t^i, y_t^i]$ of the position of $o^{(i)}$. As distance between $s_t^{(i)}$ and $s_t^{(j)}$ we use the Euclidean distance. In this way we learn the following probabilistic models: i) for single object activities from vectors representing the *movement* of individual objects, ii) for relations from *distances* between objects and iii) for changes in relations from *differences of distances during time*. Given the small number of encounters in the data set, we adopted the *leave-one-out* cross-validation technique [10].

At the first level, to choose K, we set $\mathcal{K}1$ to 10 and apply BIC. We restrict the EM algorithm to iterate for a maximum of 1000 times and add a small regularization factor $(1e-5)$ to the diagonal of the covariance matrices to ensure they are positive-definite. For the K-medoid algorithm, we fix $\mathcal{K}2$ to 20 and the maximum number of iterations to 500. On average, the number of Gaussians learned for single object activities is 9, for relations 7 and for changes of relations 9. On average, the number of clusters the K-medoid algorithm finds is 19. Similar results were obtained with different values of $\mathcal{K}$. We learned two Markov chains, one for each type of encounters. To avoid having transitions of probability 0 for unobserved patterns, we used *Laplace's succession rule*.

We tested the models learned by LEMAIO-1 on activity classification, activity generation, and multi-object tracking. Moreover, we tested these models on online activity recognition and compared the results to [17].

Experiment 1: Encounter Classification

Rendezvous (Positive)	19
Avoidance (Negative)	18
True Negative	11
True Positive	19
False Positive	7
False Negative	0
Accuracy	0.81
True Negative Rate	0.61
Recall (True Positive Rate)	1
Precision	0.73
F-measure	0.84

To test an unseen encounter, our classification method assigns it the label (rendezvous or avoidance) associated with the Markov chain with the highest likelihood. Since the data includes 37 encounters, we trained and tested 37 different models: each model was trained on 36 of the 37 encounters and tested on the remaining encounter. The results are reported in the table above. Our method had an F-measure of 0.84. A lower F-measure of 0.72 was previously reported [17] on the same data set using hidden Markov models, obtained with a supervised approach, whereas our system is (only) top-supervised.

Experiment 2: Encounter Generation

To evaluate the suitability of the models learned by LEMAIO-1 for generating encounters probabilistically, we ran two experiments of increasing complexity.

In *Experiment 2a* (generation given a sequence of atomic activities), for each learned model, we took the encounter part of the test set, and classified it to give a sequence of atomic activities. From this sequence we generated the low-level data representing an encounter, i.e., we generated the positions of the two ships. One of these encounters is shown in Fig. 2 left, where the original rendezvous from the test set is shown at the top and a rendezvous generated by the learned model is shown at the bottom. Notice that the generated tracks follow the paths of the original ones, as dictated by the recognized atomic activities.

In *Experiment 2b* (generation given a complex activity; by first generating a sequence of atomic activities and then generating encounters from them), for each model we generated a sequence of atomic activities $a_{1:T}$ for the rendezvous complex activity and another sequence for the avoidance activity. We sampled the atomic activity sequence according to the Markov chain learned, by first sampling the first atomic activity a_1 according to the vector of priors in the Markov chain (associated with the relevant encounter) and then sampling the

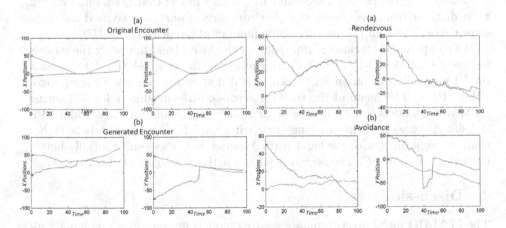

Fig. 2. left: An example encounter generated given the atomic activities: (a) the original encounter from the test set and (b) an encounter generated from the atomic activities recognized from the original encounter. **right:** Two examples of encounters generated given complex activities: (a) a rendezvous and (b) an avoidance.

atomic activity a_{t+1} according to the probability of transition from the atomic activity a_t. Fig. 2 on the right represents a rendezvous (top) and an avoidance (bottom) generated from one model learned by LEMAIO-1. In both cases the ships are approaching at the beginning and going apart at the end. In the rendezvous, there is a distinctive behaviour localized in the center where the two ships stay close together for a while; this does not happen in the avoidance.

For both Experiment 2a and 2b the generation of a sequence of positions given a sequence of atomic activities is done with sampling. From each atomic activity in the sequence we sample a particular vector of probabilistic models that gives us the single object activities of $o^{(i)}$ ($e_t^{(i)}$), and of $o^{(j)}$ ($e_t^{(j)}$), their relation $r_t^{(i,j)}$ and the change of their relations $d_t^{(i,j)}$ at time t. For each atomic activity we generate $M_2 = 100$ vectors and from each of these $M_1 = 100$ positions following Eq. 5. We sample M_1 positions of $o^{(i)}$ and $o^{(j)}$ independently from $p(s_t^i|\epsilon_i, s_{t-1}^i)$ and $p(s_t^j|\epsilon_j, s_{t-1}^j)$, resp. We sample M_1 distances $\Delta_t^{(i,j)}$ from Eq. 5 on the line (s_t^i, s_t^j) and, for each sample, we fix one of the sampled $s_t^{(i)}$ or $s_t^{(j)}$ and pick the other at the opposite side at distance $\Delta_t^{(i,j)}$. To avoid preferential treatment, we exchange the order in which $s_t^{(i)}$ or $s_t^{(j)}$ are chosen at each time step.

Experiment 3: Tracking

We evaluate the tracking ability of the models learned by LEMAIO-1. This experiment makes use of the 3PF algorithm presented in [17] coupled for the prediction step with the same transition model used for the generation experiments and learned by our LEMAIO-1 implementation. Each particle first samples the distribution of complex activities, then samples the atomic activities using the appropriate Markov chain, and then predicts the next position of each object based on the atomic activities. Thus, while tracking, the algorithm is also able to recognize the activity online. When an observation arrives the tracker filters it by weighting the particles according to a sensor model that takes into account their distance from the observation. For comparison purposes we used the same sensor model used in [17] and the same number of particles ($M = 100$).

We compare the tracking performance of the 3PF algorithm using the models learned by LEMAIO-1, with the performance of the original 3PF (that uses a model manually optimised for tracking) and a standard particle filtering algorithm (PF)[1]. The mean of the tracking errors on 37 encounters for 3PF using LEMAIO-1 models is 0.27, for the original 3PF is 0.15 and for the standard PF is 1.68. As expected, the tracking error with the LEMAIO-1 models is higher than that obtained with the hand-crafted model, but it was substantially better than the standard PF. The accuracy for the activity recognition task is 0.95.

4   Discussion

The LEMAIO multi-layer framework learns models for activities involving multiple interacting objects from sequences of attribute values for the individual

---

[1] The tracking error for an encounter is computed as the mean distance between the filtered and actual positions of the ships across all time points.

objects. Our experiments show the validity of the models learned on a publicly available data set. In particular, our results are better than previously reported results using Hidden Markov Models (for activity recognition) and Relational Particle Filtering (for tracking).

Numerous researchers have dealt with the problem of modeling and recognizing the actions of a single agent [2,19]. Single object activities can be used to recognize the action of an agent in a time interval. In practice, many activities of interest involve several agents, which interact with each other and with the environment. LEMAIO is better suited to such problems than the single agent approaches because it learns a model for the relations between interacting objects and the way these relations change over time, in contrast to other approaches that consider relations between objects by either limiting the interactions to particular types [9] or constraining the objects and their interactions to be fixed over time [8]. Many works have dealt with the problem of representing and recognizing complex activities from data [20,21,22]. These approaches typically rely on a model of the activity being provided by a domain expert. Such models are rarely available for real life systems featuring many variables with complex interdependencies. As well, many of these models are inflexible and can be used for recognition but not for generation or tracking. In contrast, LEMAIO learns its own model, which can be used for recognition, generation and tracking. Some existing approaches use a hierarchy of actions specified by a stochastic context free grammar [13,15], making them less flexible than LEMAIO.

Several approaches have dealt with learning concepts similar to the ones learned at the various layers of the LEMAIO framework. For example, the equivalent of single object activities has been learned in various computer vision systems [16]. Atomic activities have been used, for example, to recognize robot actions by various RoboCup competitors [18] taking as given the interpretation of low level attribute values. Bobick [1] distinguished the concepts of *actions*, characterized by simple motion patterns typically executed by a single agent, and *activities*, which are more complex and involve coordinated actions among multiple agents. To the best of our knowledge LEMAIO is one of the first approaches to put these concepts together and is the first approach to learn models for relations.

Other previous approaches studied the behaviour of several objects moving together by considering them as a single entity [5,11]. Our aim is to model the behaviour of interacting objects pursuing an activity that is permitted to be something other than moving together. For this reason, we chose to learn a model of the relations that is separate from the model of the activities of the single objects. The relations studied in this paper were based on distance or similarity. We hypothesize that the LEMAIO framework can learn models for relations that are not distance or similarity relations, such as "passing an obstacle on the left" or "being on the same team".

In future work, we will investigate the use of different probability distributions. We think that, especially at the second layer, the use of time windows may improve the accuracy of the model. Therefor, we will investigate the use of Temporal Nodes Bayesian Networks [12]. The assumption that relations are pairwise is certainly a limitation, but investigating all possible combinations of related objects while learning is computationally intractable for a large number of objects. We are investigating the use of non-parametric methods to discover which objects are related while performing a particular activity.

References

1. Bobick, A.F.: Movement, activity and action: the role of knowledge in the perception of motion. Phil. Trans. Lond. B 352, 1257–1265 (1997)
2. Bobick, A.F., Davis, J.W.: The recognition of human movement using temporal templates. IEEE Trans. Pattern Anal. Mach. Intell. 23(3), 257–267 (2001)
3. Borg, M., Thirde, D., Ferryman, J.M., Fusier, F., Valentin, V., Brémond, F., Thonnat, M.: Video surveillance for aircraft activity monitoring. In: AVSS, pp. 16–21 (2005)
4. CAIAC: The CAIAC intelligent systems challenge (2009),
   http://www.intelligent-systems-challenge.ca/challenge2009/
   problemDescriptionAndDataset/index.html
5. Cattelani, L., Manfredotti, C.E., Messina, E.: Multiple object tracking with relations. In: ICPRAM (1), pp.459–466 (2012)
6. Dempster, A.P., Laird, N.M., Rubin, D.B.: Maximum likelihood from incomplete data via the EM algorithm. J. Roy. Stat. Soc. B 39(1), 1–38 (1977)
7. Fraley, C., Raftery, A.E.: How many clusters? Which clustering method? Answers via model-based cluster analysis. The Computer Journal 41(8), 578–588 (1998)
8. Friedman, N., Getoor, L., Koller, D., Pfeffer, A.: Learning probabilistic relational models. IJCAI, 1300–1309 (1999)
9. Galata, A., Cohn, A.G., Magee, D.R., Hogg, D.: Modeling interaction using learnt qualitative spatio-temporal relations and variable length Markov models. In: ECAI, pp. 741–745 (2002)
10. Geisser, S.: Predictive Inference. Taylor & Francis (1993)
11. Gning, A., Mihaylova, L., Maskell, S., Pang, S., Godsill, S.: Group object structure and state estimation with evolving networks and Monte Carlo methods. IEEE Trans. Signal Processing 59(4), 1383–1396 (2011)
12. Hernandez-Leal, P., Gonzalez, J.A., Morales, E.F., Sucar, L.E.: Learning temporal nodes bayesian networks. Int. J. Approx. Reasoning 54(8), 956–977 (2013)
13. Ivanov, Y.A., Bobick, A.F.: Recognition of visual activities and interactions by stochastic parsing. IEEE Trans. Pattern Anal. Mach. Intell. 22(8), 852–872 (2000)
14. Kaufman, L., Rousseeuw, P.J.: Finding Groups in Data: An Introduction to Cluster Analysis. John Wiley (1990)
15. Lee, K., Kim, T.K., Demiris, Y.: Learning action symbols for hierarchical grammar induction. In: ICPR, pp. 3778-3782 (2012)
16. Li, K., Hu, J., Fu, Y.: Modeling complex temporal composition of actionlets for activity prediction. In: Fitzgibbon, A., Lazebnik, S., Perona, P., Sato, Y., Schmid, C. (eds.) ECCV 2012, Part I. LNCS, vol. 7572, pp. 286–299. Springer, Heidelberg (2012)
17. Manfredotti, C.E., Fleet, D.J., Hamilton, H.J., Zilles, S.: Simultaneous tracking and activity recognition. In: ICTAI, pp. 189–196 (2011)

18. Nguyen, N.T., Phung, D.Q., Venkatesh, S., Bui, H.H.: Learning and detecting activities from movement trajectories using the hierarchical hidden Markov models. In: CVPR, pp. 955–960 (2005)
19. Niebles, J.C., Wang, H., Fei-Fei, L.: Unsupervised learning of human action categories using spatial-temporal words. Int. J. Comput. Vision 79(3), 299–318 (2008)
20. Oh, S.M., Rehg, J.M., Balch, T.R., Dellaert, F.: Data-driven MCMC for learning and inference in switching linear dynamic systems. In: AAAI, pp. 944–949 (2005)
21. Ryoo, M.S., Aggarwal, J.K.: Stochastic representation and recognition of high-level group activities. Int. J. Comput. Vision 93(2), 183–200 (2011)
22. Ryoo, M.S., Aggarwal, J.K.: Semantic representation and recognition of continued and recursive human activities. Int. J. Comput. Vision 82(1), 1–24 (2009)

Correcting the Usage of the Hoeffding Inequality in Stream Mining*

Pawel Matuszyk, Georg Krempl, and Myra Spiliopoulou

Otto-von-Guericke University Magdeburg, Germany
{pawel.matuszyk,georg.krempl,myra}@iti.cs.uni-magdeburg.de

Abstract. Many stream classification algorithms use the Hoeffding Inequality to identify the best split attribute during tree induction.

We show that the prerequisites of the Inequality are violated by these algorithms, and we propose corrective steps. The new stream classification core, `correctedVFDT`, satisfies the prerequisites of the Hoeffding Inequality and thus provides the expected performance guarantees.

The goal of our work is not to improve accuracy, but to guarantee a reliable and interpretable error bound. Nonetheless, we show that our solution achieves lower error rates regarding split attributes and sooner split decisions while maintaining a similar level of accuracy.

1 Introduction

After the seminal work of Domingos and Hulten on a very fast decision tree for stream classification [1], several decision tree stream classifiers have been proposed, including CVFDT [7], Hoeffding Option Tree [9], CFDTu [11], VFDTc [3], as well as stream classification rules (e.g. [8,4]). All of them apply the Hoeffding Bound [6] to decide whether a tree node should be split and how. We show that the Hoeffding Inequality has been applied erroneously in numerous stream classification algorithms, to the effect that the expected guarantees are not given.

We propose `correctedVFDT`, which invokes the Inequality with correct parameter settings and uses a new split criterion that satisfies the prerequisites. Thus, `correctedVFDT` provides the expected performance guarantees. We stress that our aim is not a more accurate method, but a more *reliable* one, the performance of which can be properly interpreted.

The paper is organised as follows. In the next section, we present studies where problems with the usage of the Hoeffding Bound have been reported and alternatives have been proposed. In section 3, we explain why the usage of the Hoeffding Inequality in stream classification is inherently erroneous. In section 4 we propose a new method that alleviates these errors, and in section 5, we prove that it satisfies the prerequisites of the Hoeffding Inequality and thus delivers the expected performance guarantees. In section 6, we show that our approach

---

* Part of this work was funded by the German Research Foundation project SP 572/11-1 IMPRINT: Incremental Mining for Perennial Objects.

A. Tucker et al. (Eds.): IDA 2013, LNCS 8207, pp. 298–309, 2013.

has competitive performance on synthetic and real data. Section 7 summarizes the findings and discusses remaining open issues.

2 Related Work

Concerns on the reliability of stream classifiers using the Hoeffding Bound have been raised in [9]: Pfahringer et al. point out that "Despite this guarantee, decisions are still subject to limited lookahead and stability issues." In Section 6, we show that the instability detected in [9] is quantifiable.

Rutkowski et al. [10] claim that the Hoeffding Inequality [6] is too restrictive, since (A) it only operates on numerical variables and since (B) it demands an input that can be expressed as a sum of the independent variables; this is not the case for Information Gain and Gini Index. They recommend McDiarmid's Incquality instead, and design a 'McDiarmid's Bound' for Information Gain and Gini Index [10]. However, as we explain in Section 3, the most grave violation of the Hoeffding Inequality in stream classification concerns the independence of the variables (prerequisite B). This violation of the Inequality's assumptions is not peculiar to the Hoeffding Inequality, it also holds for the way the McDiarmid Bound uses the McDiarmid Inequality. Replacing one Inequality with another does not imply that the prerequsite is satisfied. Hence, we rather replace the split criterion with one that satisfies its prerequisites. We concentrate on the Hoeffding Inequality in this work. The McDiarmid Inequality is more general indeed and we can study it in future work. For the purposes of stream classification, though, the Hoeffding Inequality seems sufficient, because restriction (A) is irrelevant: the split functions return real numbers anyway.

3 Hoeffding Bound – Prerequisites and Pitfalls

The Hoeffding Inequality proposed by Wassily Hoeffding [6] states that for a random variable Z with range R, the true average of Z, $\overline{Z}$, deviates from the observed average $\widehat{Z}$ not more than ε, subject to an error-likelihood δ:

$$|\overline{Z} - \widehat{Z}| < \varepsilon \,, where \; \varepsilon = \sqrt{\frac{R^2 \cdot \ln(1/\delta)}{2n}} \tag{1}$$

where n is the number of instances. Inequality 1 poses following PREREQUISITES:

1. The random variables must be identically distributed and almost surely bounded; the variable ranges are used when computing the bound.
2. Random observations of the variables must be independent of each other.

In many stream classification algorithms, Z is the value returned by the function computing the 'goodness' of a split attribute. Given a significance level δ, the Hoeffding Inequality states whether the instances n seen thus far are enough for choosing the best split attribute. This is mission-critical, since wrong splits (especially for nodes close to the root) affect the performance of the classifier negatively. In the presence of drift, this may also lead to uninformed decisions about discarding or replacing a subtree. We show that stream classification methods violate the prerequisites of the Hoeffding Inequality (subsection 3.1) and that the decision bound is wrongly set (3.2).

3.1   Violation of Prerequisites

Domingos and Hulten [1] proposed Information Gain (IG) and Gini Index (GI) as exemplary split functions appropriate for the Hoeffding Bound: at each time point, the data instances in the tree node to be split are considered as the observations input to the Hoeffding Inequality, and the values computed upon them by IG/GI are assumed to be averages.

Violation 1: The Hoeffding Inequality applies to *arithmetical* averages only [6]. IG and GI "can not be expressed as a sum S of elements" (i.e. "of the independent variables") [10]. We do not elaborate further on this issue, since it is obvious.

Violation 2: The variables, i.e. the observations used for the computation of the split criterion must be independent (PREREQ. 2). However, consider a sliding window of length 4 and assume the window contents $w_1 = [x_1, x_2, x_3, x_4]$ and then $w_2 = [x_3, x_4, x_5, x_6]$, after the window has moved by two positions. Obviously, the window contents overlap. When a function like IG computes a value over the contents of each window, it considers some instances more than once. Thus, the computed values are not independent.

3.2   A Decision Bound That Cannot Separate between Attributes

Domingos and Hulten specify that the Hoeffding Bound should be applied as follows, quoting from [1], second page, where G is the split function:

> "Assume G is to be maximized, and let X_a be the attribute with highest observed $\widehat{G}$ after seeing n examples, and X_b be the second-best attribute. Let $\Delta\overline{G} = \overline{G}(X_a) - \overline{G}(X_b) \geq 0$ be the difference between their observed heuristic values. Then, given a desired δ, the Hoeffding bound guarantees that X_a is the correct choice with probability $1 - \delta$ if n examples have been seen at this node and $\Delta\overline{G} > \epsilon$." [1],[2]

Claim. The Hoeffding Bound does *not* provide the guarantee expected in [1].

Proof. Assume that the split candidates are X, Y with IG values G_X and G_Y, observed averages $\widehat{G_X}, \widehat{G_Y}$ and real averages $\overline{G_X}, \overline{G_Y}$ (cf. Figure 1). Considering

---

[1] In [1], this text is followed by a footnote on the "third-best and lower attributes" and on applying Bonferroni correction to δ if the attributes in the node are independent.

[2] Note: we use ε instead of ϵ, $\overline{Z}$ for the true average and $\widehat{Z}$ for the observed one.

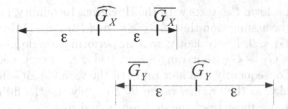

Fig. 1. Observed vs real averages of two random variables: the observed averages differ by more than ε, but the Hoeffding Bound does not guarantee that G_Y is superior

n observations in range R (of the split test), the probability that the real average $\overline{Z}$ deviates from the observed one $\widehat{Z}$ by more than ε is bounded by Ineq. 1 [6]:

$$Pr(\widehat{Z} - \overline{Z} \geq \varepsilon) \leq exp(\frac{-2n\varepsilon^2}{R^2}) \qquad (2)$$

In Figure 1, we see that $\widehat{G_Y}$ is greater then $\widehat{G_X}$ by more than ε, but this does not hold for the real averages $\overline{G_Y}$ and $\overline{G_X}$. Hence, a span of one ε is not sufficient to guarantee separation of the gain values.

This claim holds also when we consider $G_X - G_Y$ as a single random variable ΔG (as done in [1]): the range of ΔG is the sum of ranges of G_X and G_Y, again requiring a change of the decision bound. We give the correct bound in 4.1.

4   New Method for Correct Usage of the Hoeffding Bound

Our new core correctedVFDT encompasses a correction on the decision bound, and a new split function that satisfies the prerequisites of [6] (cf. section 3).

4.1   Specifying a Proper Decision Bound

Domingos and Hulten define $\Delta G = G_Y - G_X$ as a random variable with range $R = logc$ (for Information Gain IG, c is the number of classes) and check whether $\widehat{\Delta G} - \overline{\Delta G}$ exceeds ε [1], where ε is a positive number. However, this definition of ΔG assumes that it is already non-negative, i.e. there exists some non-negative constant k, so that $|G_Y - G_X| \geq k$ holds.

Assume that there exists a $k > 0$ so that the true average [3] $E(|G_Y - G_X|)$ is $\geq k$. The absolute value is a convex function and $|G_Y - G_X|$ does not follow a degenerate distribution, so Jensen's inequality holds in its strict form, i.e.:

$$E(|G_Y - G_X|) > |E(G_Y - G_X)| \equiv |E(G_Y) - E(G_X)| \qquad (3)$$

So, we cannot conclude that $|\overline{G_Y} - \overline{G_X}| \geq k$, i.e. even if the true average of $|G_Y - G_X|$ exceeds some positive value, we cannot say that Y is superior to X.

---

[3] We temporarily change the notation from $\overline{Z}$ to $E(Z)$ for better readability.

We must thus perform *two* tests with the Hoeffding Inequality, (1) for $\Delta G_1 := G_Y - G_X$ under the assumption that $\Delta G_1 \geq 0$, *and* (2) for $-\Delta G_1 := G_X - G_Y$, assuming that $\Delta G_1 < 0$. Equivalently, we can perform a single *modified test* on a variable $\Delta G := G_Y - G_X$ that ranges over $[-\log c; +\log c]$, i.e. it may take negative values! Consequently, the new range of the variable ΔG that we denote as R' is twice as high as the original range R. To apply the Hoeffding Inequality on such a variable, we must reset the decision bound to:

$$\varepsilon' = \sqrt{\frac{R'^2 \cdot \ln(1/\delta)}{2n}} = \sqrt{4\frac{R^2 \cdot \ln(1/\delta)}{2n}} = 2 \cdot \sqrt{\frac{R^2 \cdot \ln(1/\delta)}{2n}} \tag{4}$$

i.e. to twice the bound dictated by Ineq. 1. Then, the correctness of the split decision is guaranteed given δ. Alternatively, we can keep the original decision bound and adjust the error-likelihood to δ^4. Further, a larger number of instances is required to take a split decision. We study both effects in Section 6.

4.2 Specifying a Proper Split Function

Functions like Information Gain cannot be used in combination with the Hoeffding Inequality, because they are not arithmetic averages [10]. We term a split function that is an arithmetic average and satisfies the two prerequisites of the Hoeffding Inequality (cf. Section 3) as *proper*.

For a proper split function, we need to perform the computation of the expected quality of a node split on each element of the node independently. We propose *Quality Gain*, which we define as the improvement on predicting the target variable at a given node v in comparison to its parent $Parent(v)$, i.e.

$$QGain(v) = Q(v) - Q(Parent(v)) \tag{5}$$

where the quality function $Q()$ is the normalized sum:

$$Q(v) = \frac{1}{|v|} \sum_{o \in v} oq(o) \tag{6}$$

and $oq()$ is a function that can be computed for each instance o in v. Two possible implementations of $oq()$ are: *isCorrect()* (Eq. 7), whereas $Q()$ corresponds to the conventional accuracy, and *lossReduction()* (Eq. 8) that can capture the cost of misclassification in skewed distributions:

$$isCorrect(o) = \begin{cases} 1, & \text{if } o \text{ is classified correctly} \\ 0, & \text{is misclassified} \end{cases} \tag{7}$$

$$lossReduction(o) = 1 - misclassificationCost(o) \tag{8}$$

We use *isCorrect()* to implement $oq()$ hereafter, and term the so implemented $QGain()$ function as *AccuracyGain*. However, the validation in the next Section holds for all implementations of $oq()$. In the research regarding split measures the misclassification error has been indicated as a weaker metric than e.g. information gain [5]. Our goal is, however, not to propose a metric that yields higher accuracy of a model, but one that can be used together with the Hoeffding Bound without violating its prerequisites and thus allowing for interpretation of the performance guarantees given by this bound. In Section 6.2 we show that this metric is competitive to information gain in terms of accuracy and it reveals further positive features important for a streaming scenario.

5   Validation

We first show that our new split function satisfies the prerequisites of the Hoeffding Inequality. Next, we show that no correction for multiple testing is needed.

5.1   Satisfying the Assumptions of the Hoeffding Bound

Quality Gain, as defined in Eq. 5 using a quality function as in Eq. 6, satisfies the PREREQUISITES of the Hoeffding Inequality. PREREQ 1 (cf. Section 3) says that the random variable has to be almost surely bounded. The implementations of $oq()$ in Eq. 7, range in $[0, 1]$ and the same holds for the quality function $Q()$ in Eq. 6 by definition. Hence PREREQ 1 is satisfied.

PREREQ 2 (cf. Section 3) demands independent observations. In stream mining, the arriving data instances are always assumed to be independent observations of an unknown distribution. However, as we have shown in subsection 3.1, when Information Gain is computed over a sliding window, the content overlap and the combination of the instances for the computation of entropy lead to a violation of PREREQ 2. In contrast, our *Quality Gain* considers only one instance at each time point for the computation of $Q()$ and builds the arithmetical average incrementally, without considering past instances. This ensures that the instances are statistically independent from each other. The *Quality Gain* metric uses those independent instances to compute the goodness of a split. The result of this computation depends, however, on the performance of the classifier. Since, we consider a single node in a decision tree, the classifier and the entire path to the given node remain constant during the computation of the Hoeffding Bound. Consequently, all instances that fall into that node are conditionally independent given the classifier. This conditional independence of instances given the classifier allows us to use the Hoeffding Bound upon our split function.

5.2   Do We Need to Correct for Multiple Testing?

As explained in subsection 4.1, the split decision of `correctedVFDT` requires two tests on the same data sample: we compute ε for the best and second-best attributes. Since the likelihood of witnessing a rare event increases as the number of tests increases, it is possible that the α-errors (errors of first type) accumulate. To verify whether a correction for multiple tests (e.g. Bonferroni correction) is

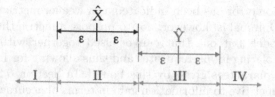

Fig. 2. When stating that Y is superior to X with confidence $1-\delta$, the error likelihood is δ; error and non-error areas are represented by numbers I - IV

necessary, we consider the different possible areas of value combinations separately. The areas, enumerated as I-IV, are depicted in Figure 2.

Figure 2 depicts a situation where the Hoeffding Bounds of attributes X and Y are separable, and allow us to state with confidence $1 - \delta$ that Y is superior to X. There is a chance of δ that this statement is wrong. We distinguish three cases for variable X (and equivalently for Y):

Case (1): the true average $\overline{X}$ is indeed in the ε-vicinity of $\widehat{X}$: $\widehat{X}-\varepsilon \le \overline{X} \le \widehat{X}+\varepsilon$
    (area represented by II in Figure 2)
Case (2): $\overline{X}$ is left to the ε-vicinity of $\widehat{X}$: $\overline{X} < \widehat{X} - \varepsilon$ (area I)
Case (3): $\overline{X}$ is right to the ε-vicinity of $\widehat{X}$: $\overline{X} > \widehat{X} + \varepsilon$ (areas III and IV)

According to the Hoeffding Inequality, the likelihood of the Case (1) is $1 - \delta$; we denote this case as `normal` or (n). We assume that the likelihood of error δ is distributed symmetrically around the ε-vicinity of $\widehat{X}$, hence the likelihood of Case (2) and of Case (3) is equal to $\delta/2$. In Case (2), the real average $\overline{X}$ is at the left of the ε-vicinity, hence the split decision would be the same as in Case (1). Therefore, we mark Case (2) as `not_harmful` (nh). In contrast, Case (3) for variable X may lead to a different split decision, because we would incorrectly assume that $\overline{X}$ is higher than it truly is. This is represented by areas III and IV in Figure 2. We mark Case (3) as `harmful` (h).

In Figure 3 we show all possible combinations of cases and their likelihoods. This tree depicts the likelihood of the outcome of each combination; the middle level corresponds to the first test, the leaf-level contains the outcomes after the first and the second test. For instance, the left node on the middle level denotes the `not_harmful` (nh) error of the first test. At its right we see the `normal` case (n) with likelihood $1 - \delta$. The leaf nodes of the tree represent the likelihood of outcomes after performing two tests: green nodes correspond to the `not_harmful` outcomes (n), (nh); red ones are potentially `harmful` (h); the blue ones contain both `harmful` and `not_harmful` outcomes.

Even if we consider all blue solid nodes as `harmful`, the sum of the likelihoods of `harmful` outcomes (cf. Eq. 9) is still smaller than δ, hence a correction for multiple tests (e.g. Bonferroni correction) is not necessary.

$$\frac{\delta^2}{4} + \frac{\delta}{2}(1 - \delta) + \frac{\delta^2}{4} + \frac{\delta}{2}(1 - \delta) + \frac{\delta^2}{4} = \delta - \frac{\delta^2}{4} \tag{9}$$

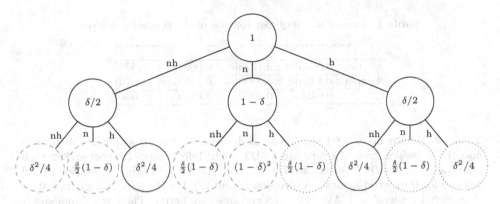

Fig. 3. Likelihood of all possible test outcomes. The middle level of the tree stands for outcomes of the first test. The leafs correspond to likelihood of outcomes after performing two tests. Green dashed leafs stand for **no error** (n) or **not_harmful** error (nh). Red dotted ones denote **harmful** error (h). Blue solid leafs combine **harmful** and **not_harmful** errors, so they have no label.

6 Experiments

We evaluate our `correctedVFDT` with $oq()$ implemented as *isCorrect*() (Eq. 7), i.e. with *AccuracyGain* as our split function (cf. end of Section 4). We measure the impact of the modifications to VFDT [1] on classifier performance.

To quantify the impact of the improper use of the Hoeffdingin Inequality we use two indicators: the number of *Incorrect Decisions* and the average number of instances (*Average n*) considered before taking a split decision. For this experiment, a dataset with known ground truth is necessary. The artificial dataset and the experiment are described in 6.1.

When experimenting on real data, we quantify the performance of the stream classifier as *Avg. Accuracy* and the tree size as *Avg. # Nodes*. For this experiment, presented in subsection 6.2, we use the Adult dataset from the UCI repository[2].

6.1 Experimenting under Controlled Conditions

For the experiment under controlled conditions, we generate a dataset with a discrete multivariate distribution, as described in Table 1. The dataset has two attributes: A_1 with three discrete values in $\{A, B, C\}$, and A_2 with two discrete values in $\{D, E\}$. The target variable takes values from $\{c_1, c_2\}$.

In this experiment, we simulate a decision tree node and observe what split decision is taken in it. Since the distribution of the dataset is known , the attribute that each split function should choose at each moment is known. As we show in Table 2, we consider VFDT with IG - denoted as 'InfoGain' (cf. first two rows of Table 2 below the legend) for a decision bound of 1ε and 2ε, and we compare with our `correctedVFDT` with *Accuracy Gain* - denoted as 'AccuracyGain' (cf. last two rows of Table 2), again for 1ε and 2ε. This means that

Table 1. Joint probability distribution of the synthetic dataset

A_2		D		E	
A	$c_1 : 0.0675$	$c_2 : 0.1575$	$c_1 : 0.0675$	$c_2 : 0.1575$
A_1 B	$c_1 : 0.1350$	$c_2 : 0.0900$	$c_1 : 0.1575$	$c_2 : 0.0675$
C	$c_1 : 0.0450$	$c_2 : 0.0050$	$c_1 : 0.0450$	$c_2 : 0.0050$

Table 2. Results of 100 000 repetitions of decision process on a split attribute at a node in a decision tree. We compare VFDT with 'InfoGain' to `correctedVFDT` with 'AccuracyGain' for the incorrect invocation of the Hoeffding Inequality (decision bound 1ε) and for the correct invocation (decision bound 2ε). For the performance indicators 'Incorrect Decisions' and 'Average n' lower values are better. The last column shows the results of the significance test on the deviation of the measured error from the theoretically permitted one, depicted in the 'Alternative Hypothesis' column, where the error-likelihood δ of the Hoeffding Bound is set to 0.05.

Setup	Incorrect Decisions	Average n	Alternative Hypothesis	p-value
InfoGain, 1ε	25738	117.31	P(incorrect decision) > 0.05	$< 2.2e - 16$
InfoGain, 2ε	1931	1671.53	P(incorrect decision) < 0.05	$< 2.2e - 16$
AccuracyGain, 1ε	3612	17.68	P(incorrect decision) < 0.05	$< 2.2e - 16$
AccuracyGain, 2ε	22	37.45	P(incorrect decision) < 0.05	$< 2.2e - 16$

we consider both the erroneous decision bound 1ε and the corrected invocation of the Inequality with 2ε (cf. 4.1) for both VFDT and `correctedVFDT`.

In Table 2 we show the results, aggregated over 100,000 runs. In the second column, we count the 'Incorrect Decisions' over a total of 100,000 decisions. The third column 'Average n' counts the number of instances seen before deciding to split a node. The confidence level of the Hoeffding Inequality was set to $1 - \delta = 0.95$, hence only 5,000 (5%) incorrect split decisions are theoretically permitted. We run a binomial test to check whether the difference between the observed error and the expected one is significant (column before last) and return the computed p-value (last column of Table 2).

The original VFDT (1st row below legend in Table 2) exceeds the theoretical threshold of 5000 Incorrect Decisions by far. The corrected invocation of the Hoeffding Inequality (subsection 4.1) reduces the number of incorrect decisions by 92.497% (cf. 2nd row in Table 2), but at the cost of increasing the number of instances required to make a split from 117.31 to 1671.53. This means that the learner would wait approximatively 10 times longer to take a split decision and would abstain from possibly good split decisions. In contrast, `correctedVFDT` makes less incorrect decisions and decides sooner, as seen in the last two rows of Table 2. The 3rd row shows that even with the incorrect decision bound, `correctedVFDT` makes less incorrect decisions than the theoretic threshold. Best results are achieved for the correct decision bound of course (4th row): only 22

Table 3. Results analogous to those in Table 2, but with a confidence level of 0.99

Setup	Incorrect Decisions	Average n	Alternative Hypothesis	p-value
InfoGain, 1ε	14034	347.55	P(incorrect decision) > 0.01	$< 2.2e - 16$
InfoGain, 2ε	339	2872.24	P(incorrect decision) < 0.01	$< 2.2e - 16$
AccuracyGain, 1ε	1062	22.42	P(incorrect decision) < 0.01	0.9757
			P(incorrect decision) > 0.01	0.02617
AccuracyGain, 2ε	2	49.7	P(incorrect decision) < 0.01	$< 2.2e - 16$

of the total 100,000 decisions are wrong, corresponding to an improvement of 99.915 %. At the same time, our method for 2ε needs only 2.24% of the instances needed by VFDT, i.e. `correctedVFDT` converges much sooner than VFDT.

To ensure that these results are statistically significant we present the results of the binomial tests. The alternative hypothesis in the 4th column in Table 2 differs from row to row. In the first row, the alternative hypothesis says that the number of incorrect decisions will be higher than the theoretic bound (by the Hoeffding Inequality); the p-value in the last column states that the alternative hypothesis should be accepted already at a confidence level lower then $2.2e - 16$. Hence, the theoretical bound *is clearly* violated by the original VFDT. The alternative hypothesis in the other three rows states that the number of incorrect decisions will stay within bound; this hypothesis is accepted.

In Table 3, we compare VFDT to `correctedVFDT` at a confidence level $1 - \delta = 99\%$. The results are similar to Table 2, except for the `correctedVFDT` with incorrect decision bound: the theoretic bound is violated (significantly, see last column), i.e. even a good method will ultimatively fail if the Hoeffding Inequality is invoked erroneously: both the corrected decision bound and a proper split function are necessary for good performance (see last row).

6.2   Experiments on a Real Dataset

We have shown that the `correctedVFDT` with *Accuracy Gain* and correct decision bound (2ε) leads to an essential reduction of incorrect split decisions and that the decisions are taken much sooner. We now investigate how these new components influence the classification performance and size of created models on a real dataset. We use the dataset "Adults" from the UCI repository [2].

Stream mining algorithms are sensitive to the order of data instances and to concept drift. To minimize the effect of concept drift in the dataset, we created 10 permutations of it and repeated our tests on each permutation, setting the grace period of each run to 1. Therefore, the results presented in this section are averages over ten runs. This also increases the stability of the measures and lowers the effect of random anomalies.

For the two algorithms, we used the parameter settings that lead to best performance. For VFDT, these were $1-\delta = 0.97$ and decision bound $\varepsilon = 0.05$, i.e.

Table 4. Performance of VFDT and `correctedVFDT` of it on the "Adult dataset". The columns "Avg. Accuracy" and "Avg. # Nodes" denote the accuracy and the number of nodes of the decision trees, as averaged over ten runs.

Algorithm	Avg. Accuracy	Avg. # Nodes
VFDT	81,992	863.7
correctedVFDT	80,784	547.2

the invocation of the Hoeffding Inequality is incorrect. According to subsection 4.1, the true confidence is therefore much lower. For `correctedVFDT`, the correct decision bound 2ε was used, the confidence level was set to $1 - \delta = 0.6$. The second column of Table 4 shows the average accuracy over the 10 runs, the third columns shows the average number of nodes of the models built in all runs.

According to the results in Table 4, VFDT reached a high accuracy, but it also created very large models with 863.7 nodes on average. That high amount of nodes not only consumes a lot of memory, but it also requires much computation time to create such models. Furthermore, such extensive models often tend to overfit the data distribution. In the second row of the table we see that `correctedVFDT` maintained almost the same accuracy, but needed only 63.36% of the nodes that were created by VFDT.

Our `correctedVFDT` does not only have the advantage of lower computation costs regarding time and memory usage, but also a split confidence that is interpretable. As we have shown in the previous subsection, the Hoeffding Bound of the VFDT cannot be trusted, for it does not bound the error the way it is expected. Consequently, setting the split confidence to 0.97 does not mean that the split decisions are correct with this level of confidence. In contrast to that, our method does not violate the requirements for using the Hoeffding Bound and thus, we can rely on the split decisions with the confidence that we have set.

For this particular amount of data and concept contained in this dataset (approximatively) optimal results have been achieved using the confidence of 0.6. This is much lower than 0.97 used with the VFDT, but this is only an illusory disproportion. In fact, the confidence guaranteed by the VFDT was much lower due to the violations of the requirements of the Hoeffding bound and it is probably not possible to estimate it. Usage of our method allows to interpret the results. We can see that it is necessary to give up the high confidence to achieve the best result on a so small dataset.

7   Conclusions

We have shown that the prerequisites for the use of the Hoeffding Inequality in stream classification are not satisfied by the VFDT core [1] and its successors. In a controlled experiment, we have demonstrated that the prerequisite violations do have an impact in classifier performance.

To alleviate this problem, we have first shown that the Hoeffding Inequality must be invoked differently. We have adjusted the decision bound accordingly.

We have further specified a family of split functions that satisfies the Inequality's prerequisites and incorporated them to our new core `correctedVFDT`. Our experiments on synthetic data show that `correctedVFDT` has significantly more correct split decisions and needs less instances to make a decision than the original VFDT. Our experiments on real data show that `correctedVFDT` produces smaller models, converges faster and maintains a similar level of accuracy. More importantly, the performance results of `correctedVFDT` are reliable, while those of the original VFDT are not guaranteed by the Hoeffding Inequality.

We are currently extending `correctedVFDT` to deal with concept drift. Further, we want to explicate the premises under which arithmetical averages and more elaborate computations on the arriving stream (as in [10] for McDiarmid's Inequality) satisfy the prerequisite of independence. In our future work we are also going to investigate the performance of Accuracy Gain, its robustness to noise and concept drift on further datasets.

References

1. Domingos, P., Hulten, G.: Mining High Speed Data Streams. In: ACM SIGKDD Conference on Knowledge Discovery and Data Mining (2000)
2. Frank, A., Asuncion, A.: UCI Machine Learning Repository (2010)
3. Gama, J., Rocha, R., Medas, P.: Accurate decision trees for mining high-speed data streams. In: KDD 2003: Proceedings of the Ninth ACM SIGKDD International Conference on Knowledge Discovery and Data Mining, pp. 523–528. ACM, New York (2003)
4. Gama, J., Kosina, P.: Learning Decision Rules from Data Streams. In: Walsh, T. (ed.) IJCAI, pp. 1255–1260. IJCAI/AAAI (2011)
5. Hastie, T., Tibshirani, R., Friedman, J., Ebooks Corporation: The Elements of Statistical Learning, ch. 9.2.3, pp. 324–329. Springer, Dordrecht (2009)
6. Hoeffding, W.: Probability inequalities for sums of bounded random variables. J. Amer. Statist. Assoc. 58, 13–30 (1963)
7. Hulten, G., Spencer, L., Domingos, P.: Mining Time-Changing Data Streams. In: ACM SIGKDD (2001)
8. Kosina, P., Gama, J.: Very Fast Decision Rules for multi-class problems. In: Ossowski, S., Lecca, P. (eds.) SAC, pp. 795–800. ACM (2012)
9. Pfahringer, B., Holmes, G., Kirkby, R.: New Options for Hoeffding Trees. In: Orgun, M.A., Thornton, J. (eds.) AI 2007. LNCS (LNAI), vol. 4830, pp. 90–99. Springer, Heidelberg (2007)
10. Rutkowski, L., Pietruczuk, L., Duda, P., Jaworski, M.: Decision Trees for Mining Data Streams Based on the McDiarmid's Bound. IEEE Trans. on Knowledge and Data Engineering (accepted in 2012)
11. Xu, W., Qin, Z., Hu, H., Zhao, N.: Mining Uncertain Data Streams Using Clustering Feature Decision Trees. In: Tang, J., King, I., Chen, L., Wang, J. (eds.) ADMA 2011, Part II. LNCS, vol. 7121, pp. 195–208. Springer, Heidelberg (2011)

Exploratory Data Analysis through the Inspection of the Probability Density Function of the Number of Neighbors

Antonio Neme[1,2] and Antonio Nido[2]

[1] Network Medicine Group, Institute of Molecular Medicine, Finland (FIMM),
Tukholmankatu 8, Helsinki, Finland
neme@nolineal.org.mx
[2] Complex Systems Group, Universidad Autónoma de la Ciudad de México
San Lorenzo 290, México, D.F. México

Abstract. Exploratory data analysis is a fundamental stage in data mining of high-dimensional datasets. Several algorithms have been implemented to grasp a general idea of the geometry and patterns present in high-dimensional data. Here, we present a methodology based on the distance matrix of the input data. The algorithm is based in the number of points considered to be neighbors of each input vector. Neighborhood is defined in terms of an hypersphere of varying radius, and from the distance matrix the probability density function of the number of neighbor vectors is computed. We show that when the radius of the hypersphere is systematically increased, a detailed analysis of the probability density function of the number of neighbors unfolds relevant aspects of the overall features that describe the high-dimensional data. The algorithm is tested with several datasets and we show its pertinence as an exploratory data analysis tool.

Keywords: exploratory data analysis, high-dimensional data, probability density function, neighborhood analysis.

1 Introduction

Exploratory data analysis (EDA) is a set of methodologies and techniques that aid in the identification of general patterns present in data. In general, it is based in visual inspection of some aspects of data, and it is considered to be a preliminary stage within the process of data mining [1].

Since the seminal works of Tukey strong attention has been paid to computational tools to extract as much information from data as possible. The original idea stands on the fact that there is no need of applying high-order statistics, which in general is computationally expensive, to grasp some relevant aspects of data [2]. For a panoramic revision of EDA, see [3,4].

The exploration of high-dimensional data presents several challenges. Among them, the construction of algorithms that are able to capture not only the most relevant, but also some subtle relationships between vectors as well, is of major

A. Tucker et al. (Eds.): IDA 2013, LNCS 8207, pp. 310–321, 2013.

relevance. In [5,6] can be found a wide discussion about the open problems and some of the drawbacks of many of the most common high-dimensional data analysis techniques.

Of particular relevance is the distance matrix (DM) of the vectors practitioners are interested to explore. The information encoded in the DM is exploited in different ways by several algorithms. For example, hierarchical clustering finds the nearest pair of vectors and constructs a new element that replaces both of them, and iteratively constructs a tree [7]. Multidimensional scaling and non-linear projections such as self-organizing maps can also construct a low-dimensional representation of the distribution of high-dimensional data by processing the distance matrix [8,9].

In an application to information retrieval, the structure of data in high-dimensional spaces is grasped by computing the relative probability of the distance between pairs of randomly chosen vectors [10]. There, the distribution of pairwise distances reveals the existence of some structure of the data when the distribution deviates from the expected distribution of a random collection of vectors. We continue into that direction but with some modifications.

In this contribution we apply distance-based concepts to construct the probability density function of the number of neighbors in a high-dimensional space. Neighbors are defined as those points or vectors separated by a distance lower than a given threshold, and we compute the histogram for the number of possible neighbors each vector has. By varying the neighborhood threshold, we can obtain relevant information about the distribution followed by data in the high-dimensional feature spaces.

The vast majority of tools to explore high-dimensional data are static in the sense that they present a unique map, a tree or a set of rules that describes the distribution followed by data in a high-dimensional space. We follow here a somehow different route: we present a simple methodology that presents some general properties of the geometry of data in a high-dimensional space and, at the time that allows us to track individual vectors and in their relationships with other vectors. The rest of the contribution goes as follows. In section 2 we describe the methodology to explore data by means of probability density distribution, whereas in section 3 we present the results of analyzing several datasets. Finally, in section 4 we present some conclusions and discussion.

2   The Proposed Algorithm

It is a common interest among practitioners to detect general properties of the distribution followed by data embedded in high-dimensional spaces. The algorithm we describe is based on the probability density function of the number of neighbors. A vector or data has as its neighbors all other vectors within a certain distance. The methodology is called Exploratory Data Analysis through systematic inspection of the Probability density Function of the number of Neighbors, hereafter *EDAPFuN*.

Let H represent the high-dimensional data, and N be the number of vectors in H ($N = |H|$). First, the distance matrix M is computed from H. The metric can capture different properties of data [11]. Let r be the radius of the hypersphere that defines the neighborhood. Next, the probability density function of the number of neighbors (PDF for short) is obtained by counting the number of vectors that have within a distance r all possible number of vectors as neighbors. It is better to express the number of neighbors as a probability, so hereafter we will refer to the fraction of the total number of analyzed vectors that a given vector has as its neighbors. Let γ_r be the PDF observed when neighborhood is defined with an hypersphere of radius r, whereas γ_r^i refers to the probability of having a fraction of i neighbors when the hypersphere is of radius r. Fig. 1 describes this stage of the algorithm.

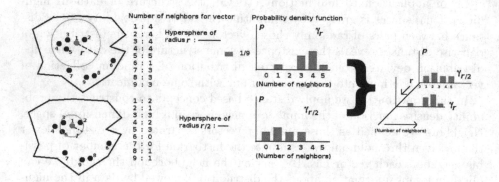

Fig. 1. The first stage of the algorithm *EDAPFuN*. PDF are obtained for several neighborhood values (r)

When we vary r from zero to the maximum distance in H, we obtain different γ_r. The idea of the methodology here described is to study these PDF's and how they vary with r. Note that if $r = 0$, then γ_r will be described as a spike for zero value since all vectors will have zero neighbors, and thus, the probability that a vector chosen at random will have zero neighbors is 1. In this case, the probability that a vector has one or more neighbors is zero. On the other hand, if r is maximal, then we observe a spike for the case of $i = 1$ since the neighborhood is large enough so that every vector has the maximum number of neighbors.

With EDAPFuN, instead of analyzing data in a high-dimensional space, we can analyze it in the three-dimensional space defined by r and γ_r. γ_r is two-dimensional, since in one axis it maintains the fraction of neighbors (n), and in the other, the probability of finding a vector with that fraction of neigbors (P_n). Since the computation of γ_r can be very demanding, and as it is not efficient to compute it from scratch for each value of r, we followed the algorithm depicted in 1:

Besides the visual inspection of PDF, several attributes from the obtained γ_r can be computed so as to unveil relevant aspects from data in H. We present some of those quantities in the next list.

Algorithm 1. EDAPFuN algorithm

Let N be the list of high-dimensional vectors to be analyzed.
Let R be the list of values r can take.
Let DM be the distance matrix for vectors in N.
Let NN be the matrix of R rows and N columns that maintains the
number of vectors that have the specified number of neighbors i for a
specific neighborhood r: NN_i^r
for all $i \in N$ **do**
  for all $j \in N, i \neq j$ **do**
    Let $L = fit - radius(r, R, DM_i^j)$
    for all $k \in L$ **do**
      $nv_k \leftarrow nv_k + 1$
    end for
  end for
  for all $r \in R$ **do**
    $NN_r^{nv[r]} \leftarrow NN_r^{nv[r]} + 1$
  end for
end for
γ_r is the normalization of NN_r

- γ_r is the PDF when a neighborhood is defined by an hypersphere of radius r. Since the number of possible neighbors is discrete, γ_r is represented by a list with elements $[i, p_i]$ where i is the number of neighbors and p_i is the probability that a vector, chosen at random, will have i neighbors.
- r_N the lowest r for which the probability of having as neighbors the maximum possible number of vectors $(N - 1)$ is greater than zero. That is, r_N refers to the value of r at which at least one vector has as its neighbors all other vectors. We just have to check the lowest r such that $N - 1 \in n_r$, in γ_r. r_N is then a measure of centrality. In a random distribution, it is expected that p_n is half the maximum possible distance in H.
- r_l is the smallest value of r for which the probability of having zero neighbors is zero for all vectors. That is, r_l is the r for which all vectors have at least one neighbor $(0 \notin n_r)$.
- r_z is the lowest value of r for which at least one vector has at least one neighbor. That is, $\exists i \gamma_r^i > 0$.
- C_r is the smallest r with the largest continuous n_r.
- D_r is be the smallest r with the largest n_r (not necessarily continuous).

It is time to observe what this representation can offer. First, note in fig. 2 how the PDF looks like for several benchmarks from the UCI repository [13]. γ_r is shown for $0 <= r < max$, where max is the maximum distance between two vectors in the dataset (diameter). Several metrics were considered and they can be contrasted there. The interpretation of the figures goes as follows. For a fixed r, γ_r represents the probability (in the z axis) of finding a vector with a fraction n of neighbors. We will refer to the z axis simply as p, the probability that a vector has the specified fraction of neighbors n for a radius r. Also, the

---

Algorithm 2. fit radius function

```
function fit-radius(r, R, d):
L = [], x = |R|, cont = 1
while cont == 1: do
   if d ≤ R_x then
      L ← L + x
   else
      return L
   end if
   x ← x − 1
   if x == 0 then
      cont = 0
   end if
end while
```

---

projections to the space (r, n) (neighborhood of radius r and fraction of vectors as neighbors n), (r, p), and (n, p) are shown.

We can observe in the three datasets that the spectrum of γ_r is deviated from what it is expected for random distributions. For the Iris dataset, $r_N > max/2$. Also, we can observe that r_l and r_z are larger than expected. There can be seen large discontinuities for values of $3 < r < 6$, which express the existence of clusters, since there are several vectors with a large number of neighbors, and then, several vectors with a much lower number of neighbors, but no vectors with an intermediate number of vectors.

For the red wine dataset, it is observed that r_N is very low, which says data is compacted within a subregion of the 11-dimensional space. There are, however, some vectors very isolated from the rest. There are also several discontinuities which unveils the existence of clusters.

Deviations from expected results for random distributions are a clue to grasp general properties in H. In fig. 3 a three-dimensional dataset is presented. The dataset consists of three clusters, each one with 350 points. It is observed that there are deviations from the expected PDF if vectors were randomly distributed. The minimum distance between points in different clusters is 0.29, and the maximal distance between points in the same cluster is also 0.29. It is observed that $\bar{\gamma}_{0.29}$ presents a very different behavior that the one expected for randomly distributed vectors. Also, $n_{0.29}$ is smaller that the corresponding number of possible neighbors for the random dataset (see fig. 2).

The deviations tell some relevant facts. First, at a given neighborhood, the maximum probability for some r is greater than expected. This is related to the existence of clusters in data. EDAPFuN finds patterns in data, that is, non-random distributions in data.

Table 1 shows the listed quantities for the two-dimensional datasets shown in figs. 2-3. In the next section, we apply the algorithm to several datasets.

r_N is expected to be $max/2$ for random distributions. If $r_N < max/2$, then it reflects that data is compacted, whereas if $r_N > max/2$, then data is disperse.

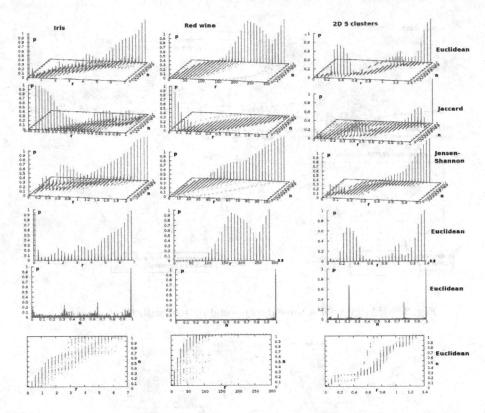

Fig. 2. PDF for several neighborhood values r (γ_r). Three datasets are included: Iris, red wine and a five-clustered two-dimensional data set. Besides the PDF for $0 <= r < max$, it is also shown the projection over (r,n), (n,p), and (r,n).

r_l takes into account isolated vectors. If there are isolated vectors, r_l will be higher. It can be observed in table 1 that this parameters are different for all two-dimensional datasets. C_r is a measure of continuity in data: if there are clusters, then C_r will tend to be smaller than expected for random distributions. D_r is also a measure of

Let $\gamma_r(v)$ be the fraction of neighbors that vector v has when considering a neighborhood of radius r. Let $\kappa_r(v)$ be the list of vectors that are neighbors of vector v for a radius r. Let $W_r(v)$ be the list of vectors with the same fraction i of neighbors as that of vector v when radius is r. With these three quantities and γ_r we can apply EDAPFuN in a different way. EDAPFuN allows not only to grasp some properties of data, it also allows a dynamic track of each vector and observe for each radius r the list of vectors that are its neighbors. This property will be explored further in the next section, by showing several examples.

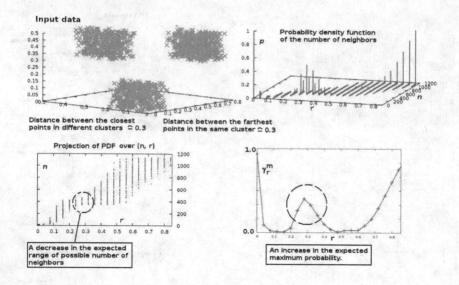

Fig. 3. EDAPFuN for a three-dimensional dataset with three clusters

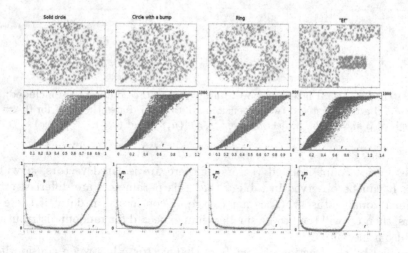

Fig. 4. EDAPFuN for some 2D artificial dataset, including a random distribution

Table 1. Attributes computed from γ_r for the datasets shown in previous figures

	2D square	2 clusters	3 clusters	4 clusters	2D circle	2D circle bump	ring	"Ef"
max	1.36	1.36	1.36	1.36	1.0	1.1	1.0	1.38
r_N	0.66	1.07	1.05	1.12	0.5	0.5	0.62	0.68
r_l	0.04	0.04	0.04	0.03	0.05	0.05	0.04	0.06
r_z	0.001	0.001	0.001	0.001	0.001	0.001	0.001	0.001
C_r	0.6	0.1	0.1	1.16	0.5	0.52	0.51	0.47
D_r	0.6	1.05	0.7	0.8	0.5	0.52	0.51	0.55

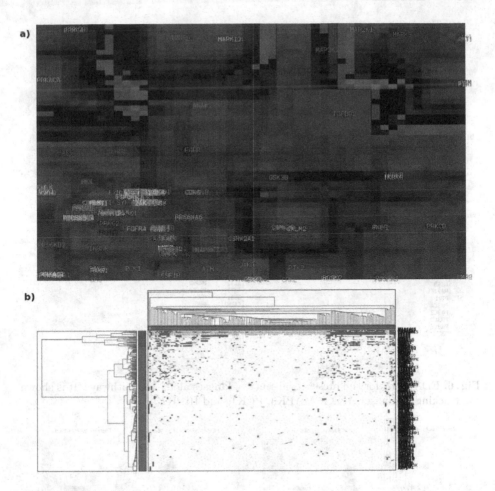

Fig. 5. SOM U-matrix and HC for the presence/absence of 715 kinases along 160 pathways

3   Results

We present in this section the results of analyzing two dataset from the molecular biology community. We contrast the analysis obtained by EDAPFuN with the more traditional tools of hierarchical clustering (HC) and self-organizing maps (SOM).

The first dataset consists of the presence/absence of 715 kinases in 168 pathways, derived from [12]. If we are interested in the general distribution of the 715 vectors in the 168-dimensional space, we will need to construct a low-dimensional representation. In fig. 5-b) it is presented a HC, whereas a SOM is shown in 5-a). In contrast, we applied EDAPFuN to the dataset and some outputs of the system are displayed in fig. 6.

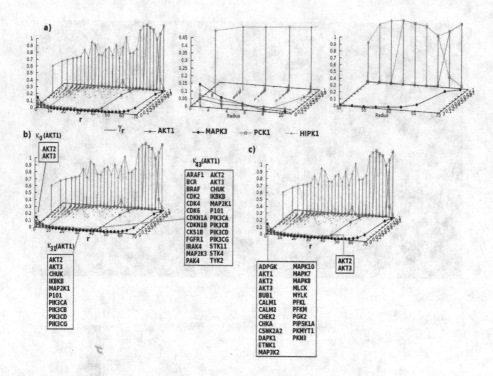

Fig. 6. EDAPFuN for the presence/absence of kinases along 160 pathways. It is shown the tracking of kinases AKT1, MAPK3, PCK1, and HIPK1

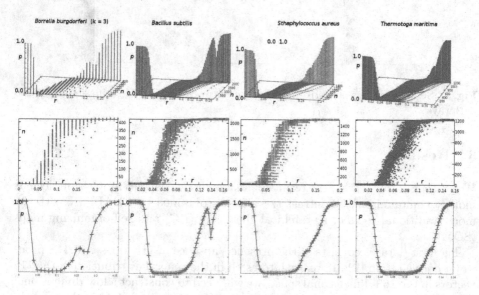

Fig. 7. EDAPFuN for different organisms

In the SOM and HC it is clear that some of the vector are located in a cluster very different to the rest of the vectors (upper right corner for the SOM, kinases MAPK3 and AKT1). However, in the EDAPFuN analysis it is even clearer this property, as it is shown by the steady probability of having a small and fixed fraction of neigbors for a large range of values for r. It is only when the radius r is large enough that the vectors in the isolated cluster start to have a larger fraction of neighbors (black line for MAPK3, and blue line for AKT1).

The relationship between any number of vectors can be traced. In fig. 6 four vectors are followed and marked by lines. The lines join the probability of finding vectors with a fraction of n neighbors for a radius r. If we are interested to know what are actually the neighbors of a relevant vector we can display them as $\kappa_r(v)$ maintains that information. In fig. 6-b it is shown the neighbors for one of the vectors. Finally, if we want to know what other vectors present the same fraction n of neighbors as the relevant vector v, we can also display it from the information contained in $W_r(v)$. Fig. 6-c shows those neighbors for a fixed vector.

In the second dataset, several genomes were analyzed using EDAPFuN. Each genome was divided in genes, and those genes were mapped to a high-dimensional space defined by the relative abundance of each of the 4^k possible sequences of length k (there are four possible bases: A, C, T, G). Each gene was then assigned to a position in the 4^k-dimensional space. We show in fig. 7 some results from EDAPFuN the case for $k = 3$ (64 dimensions), and four genomes where the number of genes varied from 3035 to 1986. It is observed that there are variations for what should have been expected for random distributions.

Also, it allowed us to identify genes atypically isolated in the high-dimensional kmer-abundance spaces. Fig 8 shows the cases for one of the genomes in which genes isolated genes in the 64-dimensional space were discovered. These isolated genes may be related to genes acquired by an horizontal transference event [14].

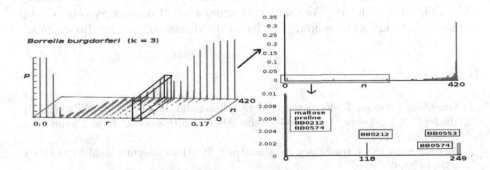

Fig. 8. EDAPFuN for *B. Burgdorferi*

4   Conclusions and Discussion

We have presented in this contribution a general overview of an exploratory data analysis methodology. It is based on the probability density function of the number of neighbors. The methodology, EDAPFuN from exploratory data analysis based on the probability density function of the number of neighbors, can aid in the task of finding general patterns of the distribution followed by data in high-dimensional spaces.

The algorithm obtains the PDF for several neighborhood thresholds. The variation of that PDF as a function of the neighborhood threshold is a tool that allows us to have a better idea of data distribution in high-dimensional spaces. EDAPFuN is based on the matrix distance of vectors, but it presents the information on it in a special format that allows that relevant information about the spatial distribution of high-dimensional data to be detected. That format is the PDF, and from a single distance matrix, several PDF's can be generated.

We have applied EDAPFuN to several datasets and it was able to unveil general attributes, such as the existence of clusters, and the sparseness of data. Also, several quantities that can help to abstract general properties of data can be computed from the obtained PDF, such as the minimum neighborhood threshold for the existence of a vector that has all the remaining vectors as neighbors.

We suggest that EDAPFuN may accompany existing high-dimensional data analysis tools, since it offers complementary information. It is not the intention of this paper to bring the drawbacks of HC or SOM, but just to contrast the information that can be retrieved from EDAPFuN with the one that can be inferred from the mentioned methods. Finally, the software was developed in Python and is available from authors webpage.

Acknowledgments. A. Neme thanks Secretaria de Ciencia y Tecnología del Distrito Federal, México (SCyTDF) for finantial support. A. Neme is also in the SNI - CONACYT México. We thank colleagues at the Complex Systems Group at the Universidad Autónoma de la Ciudad de México for fruitful discussions.

References

1. Dasu, T., Johnson, T.: Exploratory data mining and data cleaning. Wiley (2003)
2. Basford, K.E., Tukey, J.: Graphical analysis of multiresponse data. Chapman & Hall/CRC (1998)
3. Morgenthaler, S.: Exploratory data analysis. WIREs Computational Statistics 1, 33–44 (2009)
4. Martinez, W., Martinez, W.: Exploratory data analysis with Matlab. Chapman & Hall / CRC (2005)
5. Steinbach, M., Ertöz, L., Kumar, V.: The challenges of clustering high-dimensional data. In: New Vistas in Statistical Physics: Applications in Econophysics, Bioinformatics, and Pattern Recognition (2003)

6. Kriegel, H.P., Kröger, P., Zimek, A.: Clustering high-dimensional data: A survey on subspace clustering, pattern-based clustering, and correlation clustering. ACM Trans. on Knowledge Discovery from Data 3(1), Article 1 (2009)

7. Berthold, M., Wiswedel, B., Patterson, D.: Interactive exploration of fuzzy clusters using Neighborgrams Fuzzy Sets and Systems, vol. 149, pp. 21–37 (2005)

8. Borg, I., Groenen, P.: Modern Multidimensional Scaling: Theory and applications, 2nd edn. Springer (2005)

9. Vesanto, J., Sulkava, M.: Distance Matrix Based Clustering of the Self-Organizing Map. In: Dorronsoro, J.R. (ed.) ICANN 2002. LNCS, vol. 2415, pp. 951–956. Springer, Heidelberg (2002)

10. Brim, S.: Near neighbor search in large metric spaces. In: Proc. 21st VLDB Conf., Zürich, Switzerland, pp. 574–584 (1995)

11. Cha, S.H.: Comprehensive Survey on Distance/Similarity Measures between Probability Density Functions. Int. J. of Mathematical Models and Methods in Applied Sciences 4(1), 300–307 (2007)

12. Brough, R., Frankum, J., Sims, D.: Functional viability profiles of breast cancer. Cancer Discovery 1, 260–273 (2011)

13. Blake, C.L., Merz, C.U.: Repository of machine learning databases University of California, Irvine, Dept. of Information and Computer Sciences (1998), http://www.ics.uci.edu/mlearn/MLRepository.html

14. Garcia-Vallve, S., Romeu, A., Palau, J.: Horizontal Gene Transfer in Bacterial and Archaeal Complete Genomes. Genome Res. 10, 1719–1725 (2000)

The Modelling of Glaucoma Progression
through the Use of Cellular Automata

Stelios Pavlidis[*], Stephen Swift, Allan Tucker, and Steve Counsell

Department of Information Systems and Computing, Brunel University, London, UK
{Stelios.Pavlidis,Stephen.Swift,Allan.Tucker,
Steve.Counsell}@brunel.ac.uk

Abstract. We propose a model of glaucoma progression based on the application of Cellular Automata (CA) to visual field (VF) data, obtained through automated perimetry. VF sensitivities are converted into ganglion cell loss and CA are utilised to model the gradual deterioration of vision, mimicking degeneration of the actual ganglia. First we discuss the construction of a grid that approximates the VF map and the corresponding layer of ganglia in terms of cell counts in individual fields. The grid is populated with dead cells in accordance with patients' tests, and then we run a CA, utilising a majority and a probabilistic rule. Preliminary results are presented, showing that during its evolution, the CA often converges to configurations where the death of cells resembles VF data of the same patients, at later time. That is, the percentage loss of cells in VF fields observed in the CA resembles the real VF data.

Keywords: Predictive Modelling, Cellular Automata, Genetic Algorithms, Visual Field Testing.

1 Introduction

Glaucoma is an optic neuropathy accompanied by the progressive deterioration of vision with structural defects such as evident optic disc damage [1]. It constitutes the second major cause of blindness (8%) after cataracts, affecting millions of people worldwide [2]. Glaucoma progression is characterized by a gradually increasing functional deficit of the eye, with loss of VF, that is, the proportion of the space through which light can enter the eye and reach the retina. It has been established that early treatment can be extremely beneficial, to prevent irreversible loss of vision and permanent damage to the optic apparatus, hence prompt diagnosis is essential [3].

Understanding the mechanisms underlying glaucoma onset and progression is an active area of research, with substantial room left for investigation. The increasing availability of VF data has facilitated the application of machine learning methodologies to expand our knowledge and elucidate the various features of this ophthalmic condition. To date, research dealing with learning models of glaucoma progression from longitudinal data is somewhat limited. A few earlier studies have concentrated on basic methodologies, such as trend and event analysis [4,5]. More recent research

---

[*] Corresponding author.

A. Tucker et al. (Eds.): IDA 2013, LNCS 8207, pp. 322–332, 2013.
© Springer-Verlag Berlin Heidelberg 2013

has utilized multivariate time series statistical modelling and Bayesian Network classifiers [6,7]. Additionally, in [8], the authors examine the application of temporal abstraction, producing qualitative interval-based patterns from temporal data, in conjunction with association rule mining.

Here, we propose the use of CA to produce models of glaucoma progression. Our paper is organised as follows; the background section elaborates on glaucoma and establishes the conceptual relevance of a CA model to ganglion cell death. Then the construction of an appropriate grid system that models the VF and maps cells to VF points is discussed. The results produced by the CA model utilising the grid are presented and commented on, followed by conclusions and directions for future work.

2   Background

While the deeper underlying cause of glaucoma is still to be deciphered, it is believed that it is due to a blockage in part of the eye, preventing fluid draining out. This in turn leads to increased intraocular pressure, causing damage to the optic nerve, connecting the eye to the brain, and the nerve fibres from the retina. Hence, excluding low and normal tension glaucoma, the advance of the condition is accompanied by morphologic changes at the retinal nerve fibre layer and the optic nerve head [3,9].

Importantly, it has been established that the progressive loss of vision in glaucoma is due to the death of retinal ganglion cells and the degree of deterioration of vision is related to the extent of the loss [10,11]. More recently, a number of studies have provided quantitative relationships between the loss of ganglion cells and VF sensitivity [12-15]. These findings support the application of perimetry tests, widely used in clinical practice to acquire an estimate of the severity/stage of glaucoma.

In brief, tests consist of light stimuli production, which the patient is asked to confirm seeing. The spots of lights are located at several locations on a VF map, with different tests using a variety of maps and test programs. Currently, computerized automated perimetry is the test of choice in clinical practice, commonly utilizing Humphrey Field Analyzers[TM] under the 24-2 program (HFA, Carl Zeiss Meditec, Inc., Dublin, CA). In this case, the intensity of stimuli that the subject is able to see, at 52 distinct locations, excluding the 2 blind spots, serves as means of estimating the sensitivity of retina to light. A fovea-centred coordinates system, using a unique identification number for each area, is utilized, as shown in Fig. 1.

More precisely, for each VF location, a subject is shown a stimulus of adequate intensity, in terms of age and other parameters. If seen, intensity is decreased by a certain amount, and the process repeated until the stimulus cannot be perceived. The process is then reversed, with intensity gradually increased until the subject reports seeing the stimulus again. The two thresholds, known as first and second reversal respectively, serve to estimate a mean value of sensitivity per spot [16].

What is of particular interest for the research presented here is that the retinal layer this map represents consists of an estimated 1.2 and 1.5 million ganglion cells [17]. They connect to sensitive photoreceptors known as rods and cones, forming a layer above them, constituting their receptive field [18]. The 24-4 Humphrey visual field test, samples about 650,000, roughly 60% of the total number of cells [15]. Importantly, it has been suggested that the loss of ganglion cells may follow particular spatial patterns. In [19] the authors examine the hypothesis that ganglion cell death has an

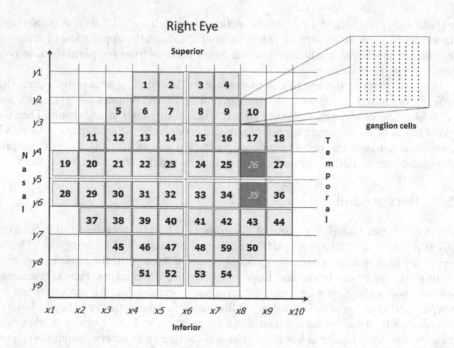

Fig. 1. Map of the visual field for the right eye, according to the Humphrey Field Analyzer™ 24-2 program. The darker squares correspond to the blind spot. The coordinates of threshold lines that define VF locations boundaries on this 2 dimensional representation of the VF are indicated as x_1 to x_{10} and y_1 to y_9. In our approach each VF holds a number of cells, modelling the actual layer of ganglion cells.

effect on the survival of neighbouring cells. Indeed, using nearest neighbour analysis, they show that loss appears in clustered patterns strengthening this hypothesis. It should be noted that there are plausible biological reasons accounting for this observation, such as the release of chemical agents by the dying cells, which influence surrounding cells [20]. In addition, further evidence has been found showing increased mortality of cells surrounding damaged retinal areas [21].

Therefore, given the availability of temporal clinical data from Humphrey 24-2 VF tests, the establishment of quantitative relationship between VF sensitivity and the loss of ganglion cells in a particular location, along with evidence that cell death follows spatial patterns, the utilization of CA seems an ideal candidate to study and model Glaucoma progression. In brief, a CA can be visualized as a grid of cells, each one having a finite number of possible states, two in the simplest case. The CA operates by iteratively scanning through the grid and changing the state of each cell depending on the state of a "neighbourhood" of cells and a predefined set of rules. A single instance of the grid is known as a generation [22]. In the implementation discussed here a 2 dimensional CA grid represents the layer of retina, or nerve cells on the back of the eye. That is, each cell in the CA grid corresponds to a ganglion cell in the retina, where 0 indicates a healthy cell, while 1 a damaged/dead cell.

Superior

2711 2840   2904 3292

3679 4712 5034   5099 5164 5422

3808 6261 10069 12070   12120 10392 7745 7100   T

N   e
      2840 4582 9294 20461 54930   56479 21494   8004   m
a   p
s   o
a   2904 4712 9424 21171 65258   66807 21494   8036   r
l   a
      4002 6906 12199 18331 18461   12586 8327 7358   l

4583 6455 8069   8069 6971 6326

4067 4454   4518 4647

Inferior

Fig. 2. Estimated number of ganglion cells within the Humphrey 24-2 test field, for the right eye according to [15]

However, the implementation of a CA representing cells, underlying the 24-2 Humphrey test VF, is not straightforward due to the fact that the density of ganglion cells varies widely at different locations, along with retinal eccentricity, that is, degrees of visual angle from the centre of the eye [15,23]. Generally speaking, density is much higher in centrally located VF locations and decreases abruptly towards the periphery. For example, VF location 32, on Figure 1, contains over 60,000 ganglion cells, whereas VF location 19 contains less than 3000 cells. Figure 2 exhibits the number of cells contained in individual VF locations according to [15].

Consequentially, to represent the topography of these cells on a standard, equally spaced 2 dimensional grid, areas need to be expanded accordingly, to contain the correct number of cells. This can be thought of as stretching a rubber band with cells on it so that dense areas expand in size in a fashion that allows the spacing to be equal, while preserving their relative locations and spatial relationship to neighbouring cells. While the mapping between low and high dimensional space may bring Kernels into mind, it should be noted that they are not relevant. Here, there is an attempt to represent the actual ganglion cells by a grid and at the same time be aware of which of the 52 locations on the VF map each cell belongs to. In the next section, we elaborate on the methodology that has been applied to approximate this spatial relationship before proceeding with the CA application.

3   Methods

3.1   Grid Preparation

A model of estimating ganglion cells density (D) in a retinal location based on its eccentricity was utilised [24]. The aim is to approximate the number of cells contained in each region, as given by [15], as closely as possible. Starting from the centre of the visual field, we estimate ganglion cell density (D) in degrees, according to [24], then converting it into distance (Dist) between adjacent cells, using equation (1). We store this value and move to an adjacent cell on the horizontal axis. Based on the new eccentricity of this cell, we perform a new calculation of cell distance. The process is repeated until the Humphrey field is covered, reaching the eccentricity of the corner most VF location on the horizontal axis. The process is then repeated to obtain the eccentricity (e) of all cells on the vertical axis, crossing the centre of the VF map. Once the distance of cells on the horizontal and vertical axis is computed, using equation (2), we estimate the eccentricities of diagonally located cells, and again using equation (1) the distances between them. Since the eccentricity of VF locations on the 24-4 Humphrey test is known, the above information allows us to map each cell on the grid to a VF location.

$$Dist = \frac{1}{\sqrt{3D}}$$

(1)

$$e = \tan^{-1} \sqrt{\left(\left(\sqrt{\tan(h)}\right)^2 + \left(\sqrt{\tan(v)}\right)^2 \right)}$$

(2)

In an attempt to further improve the mapping of cells on a square grid to VF fields, a heuristic search was also employed [26]. Given a grid (e.g. Figure 1) we move the threshold lines, which separate distinct VF locations. In particular, we implemented a genetic algorithm (GA) [27], where a vector represents the x and y coordinates of the threshold lines indicated on Figure 1. To avoid the grid moving around, the central coordinates (y_5, x_6) were set and not subjected to change. Hence, we work with a vector $G \in \{0\text{-}1000\}^{1\times17}$ where g_i represents a coordinate of a threshold line. An upper value of 1000 corresponds to a grid of 1,000,000 cells, allowing it to contain 650,000 cells in the area corresponding to a Humphrey 24-2 VF map. Here, g_1 to g_8 represent threshold coordinates y_1 to y_9, excluding y_5, while g_9 to g_{17} coordinates x_1 to x_{10}, excluding x_6. Naturally, the GA population consists of list of such vectors. A mutation consists of an increase or decrease of a value within a chromosome, thus, a movement of the threshold line. Uniform crossover was also utilised, allowing for chromosomes to exchange parts. Parameters were tested and judged according to the produced grid.

Additionally, we work with $CA \in \{0,1\}^{1000\times1000}$, representing a CA grid. The membership of each cell to a particular VF location as well as the total number of cells in a VF is readily available to us, as it is defined by the coordinates of the threshold lines. For example, the number of cells within VF location 1, is equal to $(y_2 - y_1) \times (x_5 - x_4)$, which, for our vector, is equal to $(g_2 - g_1) \times (g_{14} - g_{13})$. Hence, we obtain, for a set of coordinates, the number of cells in the matrix corresponding to each VF location V_i. For example the vector [1 10 20 30 40 50 60 70 80 1 10 20 30 40 50 60 70 80 90], with centre thresholds included for visualization purposes, would produce a regularly spaced grid, by ten cells, similar to the one in Figure 1.

The fitness of a vector/solution is estimated using the mean of percentage absolute differences between $\hat{V}$ and V, for each location in the VF map, using equation (3). V represents a vector, where v_i equals the number of cells in VF field i, according to [15], $\hat{V}$ a vector of estimated values produced by the GA, while N the number of VF fields, i.e. 52. Importantly, the algorithm was seeded with vectors corresponding to the grid produced with equations (1) and (2), which represents a configuration more similar in terms of cells contained in VF fields than a random or regularly spaced vector.

$$F = \frac{\sum_{i=1}^{N} \frac{|\hat{v}_i - v_i|}{\hat{v}_i}}{|V|} \tag{3}$$

3.2   CA Implementation

Having acquired a grid that approximates the Humphrey 24-2 visual field, we proceeded with the implementation of a CA according to the following rationale. First, from a data-set of VF tests from Moorfields Eye Hospital, pairs of tests were extracted, corresponding to patients' subsequent tests, taken one year apart, resulting in 1125 pairs of tests. This is a reasonable time interval to observe progression given the nature of glaucoma [28]. The total deviations (*td*) from normal sensitivity to light, for each VF location, were transformed into percentage of loss of ganglia with equation (4) [15].

$$\%Loss = 100 \times \left(1 - 10^{\left(\frac{td}{10}\right)}\right) \tag{4}$$

Based on the percentage of dead cells, from the first test of a patient, we seed VF locations on the CA grid (produced by the GA) with ones, representing these dead cells. Naturally, zeros represent healthy ganglia. As the CA evolves we examine the similarity of each generation to the second test of a patient, in terms of average percentage difference in cell loss.

At this early stage of the project, to study the merit of the discussed modelling approach, the CA was tested utilising some basic rules. First, the majority rule, where a cell surrounded by four or more dead neighbours (out of 8) dies (turns to "1") was used. Second, a probabilistic model was utilised, according to equation (5). Here, P is the probability of a cell dying and c the number of dead neighbours surrounding it, taking into account two layers of surrounding cells. Using two layers and thus 24 neighbours allows the probability value to vary more substantially. With no dead neighbours ($c=0$) $P_0 = a$, while with all 24 neighbours dead the probability becomes $P_1 = a \times e^{-\lambda \times 24}$. Hence λ can be estimated according to equation (6). By setting the probability of a cell dying, even if surrounded by healthy cells, to a value above zero, we account for random death. This is in agreement with medical rationale, as ganglia do gradually die throughout a person's lifetime [28]. In addition, having an increasing probability of a cell dying in a damaged area of the retina, that follows the degree of the damage, rather than a simple deterministic rule, seems more sensible from a biological point of view.

$$P = a \times e^{-\lambda \times c} \tag{5}$$

$$\lambda = -\frac{\ln\left(\frac{P_1}{P_0}\right)}{24} \tag{6}$$

4   Results and Discussion

4.1   Grid Construction

The grid obtained utilising equations (1) and (2) is a better approximation of the retinal layer represented by the Humphrey 24-2 map, than a regularly spaced grid, containing the same number of cells in each VF location. Still cells contained in each VF location differ about 50% on average, in terms of numbers, from actual estimations. Plausible reasons for this discrepancy include the fact that retinal ganglion cells are located on an elliptical layer rather than a flat surface. Additionally, equations (1) and (2) are only estimates and their repeated use exacerbates any existent error. Furthermore, cell configuration is not exactly similar to a square grid and it has been suggested that it resembles a hexagonal grid [25]. Nevertheless, for practical purposes of running an efficient CA, it is convenient to use a square grid as an approximation.

Importantly, the GA produced a grid where numbers of cells in VF locations are on average 87% similar to those reported in [15] (Figure 2). Hence a much better approximation of ganglion numbers in the VF points than the one produced with equations (1) and (2). Figure 3 displays the difference in terms of cell numbers between actual observations ([15]) and the initial method, utilising equations (1) and (2), as well as the seeded GA.

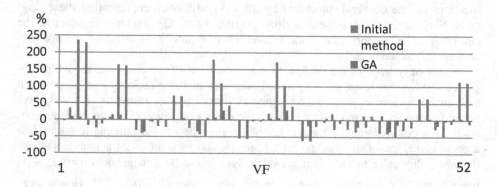

Fig. 3. Percentage difference between actual cell counts VF map and cells in the constructed grid, for the initial and GA method

4.2   CA Evolution

At this early stage of research, initial CA rule exploration was performed based on a set of pairs of VF tests of the same patient, taken a year apart. The grid was seeded with dead cells according to the first test of each patient and the CA was implemented, using the majority rule and the probabilistic model (section 3.2) with two sets of P_o and P_1 probabilities (table 1).

We observed that in a substantial number of cases subsequent generations of the CA converged towards a configuration that resembles the test taken a year later, before diverging. That is, the percentage of dead cells in the 52 VF fields resembled the one estimated with equation (4), for the second VF test. In each case the CA was run 10 times and the similarity between runs was found to be very high, with standard deviation of less than 0.1%. In particular, for the majority rule the CA converged towards a state more similar to the second VF test of a patient in 346 instances, in each of the 10 runs. Cases of convergence were significantly higher for the probabilistic model. The generation at which the most similar state was reached showed some variability for different patients, with some requiring only a few generations to reach the maximum similarity and others over a hundred. However, it was remarkably close for runs on the same patient, often reaching the same state at the same number of generations in each run.

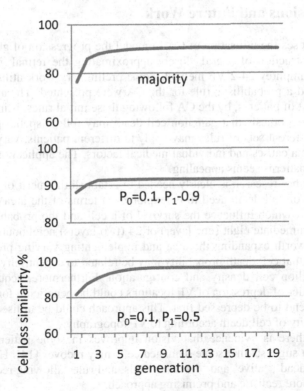

Fig. 4. Similarity in cell loss % between the evolving CA and the actual cell loss according to the VF data transformed with equation (4)

Table 1 summarises the results for the three CA rules, while Figure 4 displays an example of convergence of the CA to a subsequent VF test for each of the rules on table 1. The observation that incorporating the probability of random death leads to increased similarity between the CA and expected cell death, according to the data, is biologically sensible, as ganglion cells do degenerate with time. Finally, it should be mentioned that the blind spot area is not taken into account, as it is empty of ganglion cells.

Table 1. Comparison of results produced by the CA for different spatial rules

Rules	Number of experiments converging towards VF test	Mean similarity reached/ Standard deviation	
Majority	346	92.3%	5.3%
$P_0=0.1$ $P_1=0.9$	508	92.9%	5.1%
$P_0=0.1$ $P_1=0.5$	610	93.1%	4.9%

5   Conclusions and Future Work

This paper proposes the utilisation of CA to model the progression of glaucoma, presenting the construction of a grid closely approximating the retinal ganglion cells tested by the Humphrey 24-2 VF method. Some preliminary work utilising a simple majority rule and a probabilistic rule for the CA were presented. The approximation of the second test of patients, by the CA following these initial rules, is interesting and strengthens the hypothesis that ganglion cell death may follow spatial patterns. It is plausible that different sets of rules may apply to different patients, varying with underlying glaucoma causes and individual medical factors. The application of a CA for modelling such patterns seems appealing.

Naturally, further research is clearly needed to establish the merit of the proposed approach. Rules of cell death need to be explored, in terms of the area of the neighbourhood of cells which influence the survival of a cell and the probabilistic model. Here, only the immediate eight (one layer) or 24 (two layers) neighbours were considered, but it is worth expanding the area and implementing varying probabilities. It should be noted that cell death probability may be related to eccentricity, in the sense of varying ganglion cell density and configuration. Furthermore, sequence pattern mining of the order of depression of VF locations could be beneficial, for establishing if certain areas tend to be depressed first. This approach could be utilised to establish varying probability of cell death according to VF topography.

In addition, there is evidence that vision improves in some patients following treatment, which suggests that some damaged cells may recover [14]. Hence, a three state CA, with 'dead', 'alive' and 'diseased' cells, and rules allowing recovery of the latter, seems a more realistic and promising approach.

Given the large dataset that we have at our disposal, reverse engineering of CA rules based on patients that have taken a number of VF test is worth exploring. In addition, a CA can be seeded with a patient's test and allowed to evolve as proposed here. Then mapping of all subsequent tests of the same patient to later CA generations, in terms of similarity, can be examined. For this, a parallel CA, utilising GPU programming to massively increase speed, is worth consideration [29].

Finally, at later stage, when CA rule exploration has been performed, the predictive value of the model will be considered in regard to other work [6-8].

Acknowledgments. This work is funded by the EPSRC grant "Data Integrity and Intelligent Data Analysis Techniques Applied to a Glaucoma Progression Dataset" (EP/H019685/1) in collaboration with Moorfields Eye Hospital.

References

1. Quingley, H.A.: Glaucoma. Lancet 377(9774), 1367–1377 (2011), doi:10.1016/S0140-6736(10)61423-7
2. Pascolini, D., Mariotti, S.P.: Global estimates of visual impairment: 2010. Br. J. Ophthalmol (December 2011), doi:10.1136/bjophthalmol-2011-300539
3. Sharma, P., Sample, P.A., Zangwill, L.M., Schuman, J.S.: Diagnostic tools for glaucoma detection and management. Surv. Ophthalmol. 53(suppl. 1), S17–S7 (2008), doi:10.1016/j.survophthal.2008.08.003
4. Fitzke, F.W., Hitchings, R.A., Poinoosawmy, D., McNaught, A.I., Crabb, D.P.: Analysis of visual field progression in glaucoma. Br. J. Ophthalmol. 80(1), 40–48 (1996)
5. Heijl, A., Lindgren, G., Lindgren, A.: Extended empirical statistical package for evaluation of single and multiple fields in glaucoma: Statpac 2. In: Mills, R., Heijl, A. (eds.) Perimetry Update 1990/1, pp. 303–315. Kluger Publications (1990)
6. Swift, S., Liu, X.: Predicting glaucomatous visual field deterioration through short multivariate time series modelling. Artif. Intell. Med. 24(1), 5–24 (2002)
7. Tucker, A., Vinciotti, V., Liu, X., Garway-Heat, D.: A spatio-temporal Bayesian network classifier for understanding visual field deterioration. Artif. Intell. Med 34(2), 163–177 (2005)
8. Sacchi, L., Tucker, A., Counsell, S., Garway-Heath, D., Swift, S.: Understanding Glaucoma Progression Using Temporal Abstractions and Association Rules. In: Proceedings of the Annual Workshop on Intelligent Data Analysis in Biomedicine and Pharmacology, IDAMAP, Lake Bled, Slovenia (July 06, 2011)
9. Lan, Y.N., Henson, D.B., Kwartz, A.J.: The correlation between optic nerve head topographic measurements, peripapillary nerve fibre layer thickness, and visual field indices in glaucoma. Br. J. Ophthalmol. 87(9), 1135–1141 (2003)
10. Quigley, H.A.: Neuronal death in glaucoma. Prog. Retin. Eye. Res. 18(1), 39–57 (1999)
11. Quigley, H.A., Dunkelberger, G.R., Green, W.R.: Retinal ganglion cell atrophy correlated with automated perimetry in human eyes with glaucoma. American Journal of Ophthalmology 107(5), 453–464 (1989)
12. Harwerth, R.S., Carter-Dawson, L., Smith III, E.L., Barnes, G., Holt, W.F., Crawford, M.L.: Neural losses correlated with visual losses inclinical perimetry. Invest. Ophthalmol. Vis. Sci. 45(9), 3152–3160 (2004)
13. Harwerth, R.S., Quigley, H.A.: Visual field defects and retinal ganglion cell losses in patients with glaucoma. Arch. Ophthalmol. 124(6), 853–859 (2006)

14. Swanson, W.H., Felius, J., Pan, F.: Perimetric defects and ganglion cell damage: interpreting linear relations using a two-stage neural model. Invest. OphthalmolVis. Sci. 45(2), 466–472 (2004)
15. Drasdo, N., Mortlock, K.E., North, R.V.: Ganglion Cell Loss and Dysfunction: Relationship to Perimetric Sensitivity. Optom Vis. Sci 85(11), 1036–1042 (2008), doi:10.1097/OPX.0b013e31818b94af
16. Shaarawy, T., Sherwood, M.B., Crowston, J.G., Hitchings, R.A.: Glaucoma: Medical Diagnosis & Therapy, vol. 1. Saunders Ltd. (2009)
17. Hecht, E.: Optics, 3rd edn. Addison Wesley (1997)
18. Sjöstrand, J., Olsson, V., Popovic, Z., Conradi, N.: Quantitative estimations of foveal and extra-foveal retinal circuitry in humans. Vision Res. 39(18), 2987–2998 (1999)
19. Lei, Y., Garrahan, N., Hermann, B., Fautsch, M.P., Johnson, D.H., Hernandez, M.R., Boulton, M., Morgan, J.E.: Topography of neuron loss in the retinal ganglion cell layer in. Br. J. Ophthalmol. 93, 1676–1679 (2009), doi:10.1136/bjo.2009.159210
20. Neufeld, A.H.: Nitric oxide: a potential mediator of retinal ganglion cell damage in glaucoma. Surv. Ophthalmol. 43(suppl. 1), S129–S135 (1999)
21. Levkovitch-Verbin, H., Quigley, H.A., Martin, K.R., Zack, D.J., Pease, M.E., Valenta, D.F.: A model to study differences between primary and secondary degeneration of retinal ganglion cells in rats by partial optic nerve transection. Invest Ophthalmol. Vis. Sci. 44, 3388–3393 (2003)
22. Bandini, S.: Cellular Automata. Future Generation Computer Systems, vol. 18 (2002)
23. Drasdo, N., Millican, C.L., Katholim, C.R., Curcio, C.A.: The length of Henle fibers in the human retina and a model of ganglion receptive field density in the visual field. Vision Res. 47(22), 2901–2911 (2007)
24. Drasdo, N.: Receptive field densities of the ganglion cells of the human retina. Vision Res. 29(8), 985–988 (1989)
25. Snyder, A.W., Miller, W.H.: Photoreceptor diameter and spacing for highest resolving of gratings. J. Opt. Soc. Am. 67, 696–698 (1977)
26. Michalewicz, Z., Fogel, D.B.: How to solve it: Modern heuristics. Springer, Berlin (1998)
27. Holland, J.H.: Adaptation in Natural and Artificial Systems: An Introductory Analysis with Applications to Biology, Control, and Artificial Intelligence. University of Michigan Press, Ann. Arbor (1975)
28. Heijl, A.: Concept and Importance of Visual Field Measurements to Detect Glaucoma Progression. Glaucoma Now 2, 2–4 (2010)
29. Skrobanski, S., Pavlidis, S., Ismail, W., Hassan, R., Counsell, S., Swift, S.: Use of General Purpose GPU Programming to Enhance the Classification of Leukaemia Blast Cells in Blood Smear Images. In: Hollmén, J., Klawonn, F., Tucker, A. (eds.) IDA 2012. LNCS, vol. 7619, pp. 369–380. Springer, Heidelberg (2012)

Towards Narrative Ideation
via Cross-Context Link Discovery
Using Banded Matrices

Matic Perovšek[1,2], Bojan Cestnik[3,1], Tanja Urbančič[4,1],
Simon Colton[5], and Nada Lavrač[1,2,4]

[1] Department of Knowledge Technologies, Jožef Stefan Institute, Ljubljana, Slovenia
[2] International Postgraduate School Jožef Stefan, Ljubljana, Slovenia
[3] Temida d.o.o., Ljubljana, Slovenia
[4] University of Nova Gorica, Nova Gorica, Slovenia
[5] Department of Computing, Goldsmiths College, London, UK

Abstract. Knowledge discovery and computational creativity have until lately been investigated by two separate research communities. However, research in bisociative, cross-context knowledge discovery has recently started addressing creative tasks, including creative literature mining. This paper contributes to this effort by investigating an approach to cross-context link discovery based on banded matrices, aimed at identifying meaningful bridging terms (b-terms) at the intersection of two different domains. The proposed approach was applied to a simplified computational creativity task of narrative ideation from pairs of short sentences. As input, we took sentences from two different contexts: what-if sentences retrieved from Twitter, and morals from Aesop's fables, respectively. The approach resulted in a list of linked pairs of sentences from these two domains, illustrating the potential of the proposed approach to cross-context narrative ideation.

1 Introduction

The rate at which novelties are introduced in modern life has been so rapid that it invites changes in the way humans cope with creativity, especially how we gather information and how we make new connections between pieces of information. In the emerging field of computational creativity, Wiggins [1] has proposed the following definition: "computational creativity refers to performance of tasks (by a computer) which, if performed by a human, would be deemed creative".

This paper addresses a creative task of bisociatively connecting narratives from different domains. According to Koestler [2], a *bisociation* can be defined as (sets of) concepts that bridge two otherwise not (or very sparsely) connected domains. In his view, bisociative thinking occurs when a problem, idea, event or situation is perceived simultaneously in two or more different "matrices of thought" or domains. When two matrices of thought interact with each other, the result is either their fusion in a novel intellectual synthesis or their confrontation in a new aesthetic experience. Koestler regarded many different mental phenomena which are based on comparison (such as analogies, metaphors, jokes, identification, anthropomorphism, and so on), as special

A. Tucker et al. (Eds.): IDA 2013, LNCS 8207, pp. 333–344, 2013.

cases of bisociation. Following Koestler's ideas, the goal of this paper is the development of a computational system, which blends elements drawn from two previously unrelated matrices of thought into a new matrix of meaning.

In the area of knowledge discovery and literature mining, cross-domain connections have been the topic of recent research. In the area of bisociative knowledge discovery, Berthold [3] defines that two concepts are bisociated if there is no direct, obvious evidence linking them and if one has to cross different domains to find the link. Furthermore, a new link must provide some novel insight into the problem addressed. Several approaches to cross-domain knowledge discovery have been recently developed and reported [3]. In literature mining, on the other hand, cross-domain links in medical literature has been a topic of extensive research since early 1980s. For example, Smalheiser and Swanson [4] developed an online system ARROWSMITH which is supported by Swanson's ABC model approach. ARROWSMITH takes as an input two sets of titles from disjoint domains A and C and lists terms that are common to A and C. The resulting bridging terms (b-$terms$, forming set B) are further investigated for their potential to generate new scientific hypotheses.

Recently, the developers of the CrossBee system [5] investigated a specific form of bisociation, by exploring terms that appear in documents which represent bisociative links between concepts of different domains. Their methodology is based on an ensemble of specially tailored text mining heuristics which assign the candidate bridging concepts a bisociation score. The resulting ranked list of potential domain bridging terms enables the user to start inspecting b-terms with top-ranked terms, which should result in higher probability of finding observations that may lead to the discovery of new bridges between different domains. As described, the creative act is to find links which cross two or more different domains, leading out of the original "matrix of thought" or "out-of-the-plane" in Koestler's terms [2].

In our work we study heuristics for b-term ranking, resulting in an ordered list of potential bridging terms. The novel methodology introduced in this paper uses banded matrices [6] to discover structures which reveal the relations between the rows (representing documents) and columns (representing words/terms) of a given data matrix (representing a set of documents). We use this information in developing new heuristics for evaluating words/terms according to their bridging term (b-term) potential. In addition, the method enables the identification of document outliers, but this is out of the scope for this paper. While this methodology is mainly targeting cross-context knowledge discovery, this paper focuses on its use in the context of computational creativity for bisociative, cross-context narrative ideation. In the simplified narrative ideation scenario addressed in this paper, we use the proposed methodology for b-term ranking on documents from two domains to discover the bridging terms with the aim of combining sentences from the two domains. In our illustrative example, the first domain consists of 94 What-if sentences taken from Twitter, while the second domain consists of 277 morals from Aesop's fables. The presented approach resulted in a list of interesting linked pairs of sentences from both contexts. An example of such a pair of sentences is: *"What if humans could actually breathe in space and the government says we can't so we don't try to escape? Nothing escapes the master's eye."*

The paper is structured as follows. In the next two sections, we describe the methodology of discovering links between two unrelated contexts by using banded matrices. In the experimental section, we present the results of our methodology in a simplified narrative ideation scenario, to combine statements from two different domains. We conclude the paper by highlighting the most important findings and providing some plans for further work.

2 Banded Matrices: Definition and a Motivating Example

Our research aims at finding cross-domain links by exploring the bridging terms (b-terms) at the intersection of domains that establish links between two domains of interest. The proposed approach follows the work of Juršič et al. [5], which already contributed to b-term detection in cross-context literature mining. The approach to cross-context link discovery that we investigate in this paper is new: it is based on banded matrices [6].

Our methodology works by first encoding the documents from the two domains into the standard Bag-Of-Words (BOW) vector representation and then transforming the binary matrix of BOW vectors to its banded structure. The proposed banded matrices methodology is based on the assumption that similar documents, as well as the words that appear in the same document, will appear closer to each other in the matrix and will therefore form "clusters" along the main diagonal of the matrix in its banded form[1]. Our work is based on the intuition that terms that connect different domains will be positioned at the edges of clusters from different domains: we have developed different heuristics that should be able to identify these b-terms by ranking them high in the ranked list of terms with high potential for cross-context link discovery. We introduce below the banded matrices, and follow this with a toy example illustrating the approach.

2.1 Definition of Banded Matrices

Uncovering structures that reveal the nature of relations between rows and columns of data matrices is an important step towards solving real-world problems, as binary data occur in numerous real-world applications. Recent research in social networks, bioinformatics, and human genomics has shown the benefits of banded representations of matrices [7]. These representations have contributed to bringing huge performance boosts in various mathematical operations, including matrix multiplication.

To explain the algorithm that transforms a matrix into its banded structure we first need to define the basic concepts. A binary matrix has a banded structure if we can find a permutation of rows and columns such that the 1s exhibit a staircase pattern down the rows along the leading diagonal, as illustrated in Figure 1.

A binary matrix M is fully banded if there exists a permutation of rows κ and a permutation of columns π such that:

1. *for every row i in M_κ^π the entries with 1s occur in consecutive column indices $a_i, a_i + 1, \ldots, b_i$, and*
2. *the values of starting indices for 1s in successive rows (i and i+1) satisfy $a_i \leq a_{i+1}$ and $b_i \leq b_{i+1}$.*

---

[1] A correspondence between bi-clustering and banded structures has been shown in [7].

A necessary precondition for (1) to hold is that matrix M satisfies the *consecutive-ones property*: a binary matrix satisfies this property if it is possible to order the columns so that in every row the non-zero entries occur in consecutive positions.

As banded structured matrices cannot be expected to arise in noisy real-world environments, we need to redefine the problem in the sense that it is applicable to a wider range of real-world situations. We aim to minimize the number of transformations one needs to perform on a binary matrix to unveil a banded structure. The number of such transformations will measure how far the matrix is from being fully banded. The algorithm presented in the next section (following the motivating example) aims to solve this optimization problem.

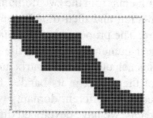

Fig. 1. An example of a fully banded matrix

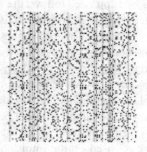

Fig. 2. Documents (rows) and words (columns) in an ideal-world domain. The color of the row indicates the domain of the document (blue for domain A and red for domain C).

2.2   A Motivating Example

Let us have two sets of documents A and C. For the purpose of explaining the methodology we constructed a small ideal-world dataset, which consists of 6 clusters of documents, 3 of which belong to domain A, while the others belong to domain C. Initially, we took a set of 120 different randomly selected words and randomly divided them into 6 clusters, so that there were no intersections, i.e., each word belonged to one cluster only. The number of possible words per cluster was 20, while each document in the cluster was randomly assigned only 15 of these words. Using a banded matrix algorithm presented in the next section this document set would first be transformed into the structure shown in Figure 3 and finally into a fully banded matrix form shown in Figure 4.

In order to illustrate our methodology, we randomly chose 8 words from each of the two domains A and C to act as artificially defined, preselected bridging terms. This effect was achieved by inserting the preselected terms into every document in every cluster with a 10% chance, thus spoiling the original clean separation of words within documents of different clusters. The resulting matrix showing documents as rows, and words as columns, is depicted in Figure 2, where the green vertical lines represent the artificially inserted b-terms. As the aim of our method is to identify the bridging terms,

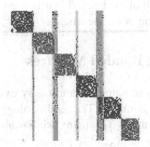

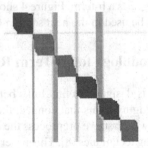

Fig. 3. Final result of our methodology: matrix of documents from Figure 2 permuted using row and column permutations obtained from the transformation of the matrix into its fully banded structure

Fig. 4. Matrix of documents shown after the transformation of documents in Figure 2 into a fully banded matrix structure. Rows represent the documents, while columns represent the terms. The green vertical lines represent the terms which were inserted as potential bridging terms to the documents.

we conducted experiments to check how the designed preselected terms will be ranked by our heuristics.

Having used our methodology on the ideal-world toy domain of Figure 2, we got the result shown in Figure 3. The green vertical lines represent the terms which were deliberately acting as bridging terms in this experiment. As can be seen from Figure 3, similar documents (documents from the same cluster) and similar terms (terms that are contained in the same document cluster) are located close to each other. As a result, the "clusters" along the matrix leading diagonal are clearly visible. Note that the preselected bridging terms occur mainly on the transitions between the clusters. All of our heuristics (explained in the next section) correctly assigned a b-term score greater than 0 only to the preselected bridging terms, which served as a proof-of-concept for the toy experiment.

Let us now consider a single document only. A document from domain A (represented with a horizontal yellow line on Figures 2, 3 and 4 consist of the following words: *magnesium blood cell prophylaxi relationship lithium red calcium effect sodium control membrane measurement potential perfuse* **ophthalmoplegic simultaneous**, where the first 15 words were randomly selected from the document's cluster term set, while the two words in bold were randomly inserted from the preselected set of bridging terms. The blue dots on of the horizontal yellow line in Figure 3 consequently symbolize the above words. According to the banded structure of the matrix (see Figure 4) the words **simultaneous** and **ophthalmoplegic** belong to word clusters of domains A and C, respectively. While the observed document belongs to domain A, the term **ophthalmoplegic** is representative for the documents from domain C. Therefore, our methodology should be able to identify this term as a potential b-term. In contrast, as the word **simultaneous** is used in the documents from the same domain A, it should not be considered as a b-term. Indeed, our heuristics (presented in the next section) have only identified *ophthalmoplegic* as a b-term. Figure 4 shows the final result of the banded matrix algorithm and will be used in the next section for the explanation of our heuristics.

3   A Methodology for B-Term Ranking Using Banded Matrices

Our approach is designed to find links between two domains, named A and C, by exploring the bridging terms that connect these two separate domains. The methodology works as follows. First, we preprocess the documents from the two domains using standard text mining techniques [8]. This is performed through a number of steps: stop-word removal, stemming or lemmatization, usage of synonym dictionaries, construction of n-grams of words and, finally, transformation to a Bag-Of-Words representation. Next, the result of the preprocessing step, i.e., the binary matrix of "Bag-Of-Words" vectors (the BOW matrix), is transformed to its banded matrix structure. Finally, we sort the terms according to their scores representing their bridging term potential, computed according to the new heuristics described below. In the following subsections, each step of the proposed bridging term detection and ranking methodology is described in detail.

3.1   Constructing a Banded Structure Using a Bidirectional MBA Algorithm

The optimization problem addressed to make a banded matrix is labeled Bidirectional Minimum Banded Augmentation (bidirectional MBA) [6] and is defined as follows:

Given a binary matrix M, find the minimum number of bidirectional flips (flips from both 1s to 0s and 0s to 1s) so that M becomes fully banded.

Algorithm 1: Bidirectional MBA algorithm

    1. Find fixed permutation of columns π.
    2. Solve the consecutive-ones property on the column permuted matrix M^π.
    3. Resolve Sperner conflicts (defined later in this section) between rows in M^π.
    4. Sort the rows in M^π and return the row permutation.

The presented MBA algorithm discovers a single band by first fixing the column permutations of the data matrix before proceeding with the rest of the algorithm. A good permutation of columns tends to put similar columns (i.e., terms) closer to each other. We use the Jaccard coefficient as a column similarity measure: $J(M^a, M^b) = \frac{M^a \cap M^b}{M^a \cup M^b}$. In our example, this similarity measure returns the highest value of 1 when two terms occur in the same set of documents. We used the spectral ordering algorithm [9] to find the fixed column permutation π of matrix M.

Next, the algorithm deals with solving the consecutive-ones properties on rows of matrix M^π, which is an essential step in finding the row permutation κ. Solving the consecutive-ones problem for row M_i^π with bidirectional flips corresponds to solving the maximum sub-array problem on matrix W_i^j [6], defined as follows:

$$W_i^j := \begin{cases} 1 & \text{if } M_i^j = 1 \\ -1 & \text{if } M_i^j = 0 \end{cases}.$$

The objective of solving the maximum sub-array problem is to find the sub-array of the matrix that has the maximum sum of numbers. This problem can be solved in linear time with respect to the size of the array using the scan-line algorithm [10]. This method returns the interval boundaries which we use to solve the consecutive-ones problem in M_i^π by setting the fields in M_i^π between the boundaries to 1 and others to 0.

Next, the algorithm deals with removing the Sperner conflicts between the rows of matrix M^π. A matrix has Sperner conflicts if its rows do not form a Sperner family of intervals:

Two rows $M_i = [a, b]$ and $M_j = [a', b']$ with consecutive-ones property, where $i \neq j$, form a Sperner family of intervals if they are overlapping such that $(a \geq a' \vee b' \geq b) \wedge (a' \geq a \vee b \geq b')$.

Additional flips on rows of M^π need to be made in order to ensure that rows have the Sperner family of intervals property.

Let $\hat{M}$ be the binary matrix M augmented with $M_{ij} = M_i \backslash M_j$ for every two rows $M_i \subset M_j$. Note that M is fully banded if and only if $\hat{M}$ has the consecutive-ones property (proof in [6]).

To eliminate all Sperner conflicts between row intervals of M^π, the algorithm has to go through all extra rows described in $\hat{M}$ and make sure that they have the consecutive-ones property. This can be done by solving the maximum sub-array problem on the extra rows of $\hat{M}$. We perform additional flips in order for the rows to obtain consecutive-ones property. Lastly, we update the rows in M^π according to the changes made over $\hat{M}$ in order to get a banded matrix.

Finally, the algorithm sorts rows $[a, b]$ of M^π in an ascending order of as, while deciding ties with the ascending order of their bs. The result of the algorithm is a banded matrix M_{band}, along with details of the row and column permutations that were performed. We use these permutations on our original matrix M, as the objective is to produce a matrix without distorting the data (i.e. without making the bidirectional flips). In the next section, we present the heuristics for calculating the b-terms potential scores.

3.2 New Heuristics for B-Terms Potential Evaluation

Here we describe the details of the four heuristics which we propose for computing the bridging term potential scores. After completing the step of term score computation, we sort the terms according to the values of one of the heuristics and present the top-ranked terms (hopefully representing the most interesting b-term candidates) to the expert. The designed heuristics should favor b-terms over non-b-terms by pushing interesting b-term candidates higher to the top of the ranked term list. For easier definition of the proposed heuristics we define variable d_{idx} to represent the row index of document d in the banded matrix M_{band} and t_{idx} to represent the column index of term t in M_{band}. Note that in order to compute the score of the proposed heuristics, we distinctively take into account the document-term matrix in two forms, banded (as shown in Figure 4) and full (as shown in Figure 3).

Heuristic 1: This is a frequency based heuristic for computing the b-term potential. If all document-term pairs in the t_{idx}-th column of matrix M_{Band}, which equal to 1, belong to the same domain, we denote this domain as D_1. Note that in such a case, the t_{idx}-th column, which represents term t in the banded matrix, should be "single-colored" in the matrix visualization in Figure 4. If the documents in the M_{Band} for term t do not belong to the same domain, the heuristic returns score 0 for this term ($h1score(t) := 0$). Otherwise, the score of Heuristic 1 for term t is defined as:

$$h1score(t) := countDoc_{D_2}(t),$$

where $countDoc_{D_2}(t)$ is the number of documents that contain term t and belong to domain D_2 (do not belong to domain D_1) in the matrix shown in Figure 3. This heuristic is based on the assumption that terms which strongly represent one domain (the single-colored column in the banded matrix of Figure 4), and at the same time there are many documents from the other domain that contain these terms, have a higher chance of being the bridging terms between the two domains.

Heuristic 2: This is also a frequency based heuristic. Similarly as described in Heuristic 1, if all documents for which the t_{idx}-th column of matrix M_{Band} equals to 1 belong to the same domain, we label this domain as D_1. Otherwise, the heuristic returns score 0 for term t ($h2score(t) := 0$). Heuristic 2 score for term t is defined as:

$$h2score(t) = \frac{countDoc_{D_2}(t)}{countOnDiagDoc_{D_1}(t)},$$

where $countDoc_{D_2}(t)$ is the number of documents from domain D_2 that contain term t, and $countOnDiagDoc_{D_1}(t)$ is the count of document-term pairs equaling 1 in the t_{idx}-th column of the banded matrix M_{Band}: $countOnDiagDoc_{D_1}(t) := |\{d_{idx}; d \in D_1 \wedge M_{Band}(d_{idx}, t_{idx}) = 1\}|$. Therefore, for term t $h2score(t)$ is the ratio between the count of documents that belong to domain D_2 and the documents in the M_{Band} for term t which belong to domain D_1. The intuition behind this heuristic is that a term that strongly represents a given domain according to the banded matrix, and is at the same time also contained in many documents of the other domain, should also have a high b-term potential score.

Heuristic 3: This is an inverse of Heuristic 2, defined as:

$$h3score(t) := \begin{cases} 0 & \text{if } h2score(t) = 0 \\ \frac{1}{h2score(t)} & \text{otherwise} \end{cases}$$

The intuition behind this heuristic is that a term that strongly represents a domain in banded matrix M_{band} should get a higher score compared to the other terms. The number of documents on the diagonal should be as high as possible: column t_{idx} should contain as many document-term pairs from the same domain.

Heuristic 4: If the documents in M_{Band} for term t do not belong to the same domain, this heuristic returns a score of 0 ($h2score := 0$). Otherwise, we label this domain as D_1 and define the Heuristic 4 score for term t as follows:

$$h4score(t) = \frac{countOnDiagDoc_{D_1}(t)}{countDoc_{D_1}(t)} * countDoc_{D_2}(t),$$

where $countOnDiagDoc_{D_1}(t)$ is the count of document-term pairs in the t_{idx}-th column of banded matrix M_{Band}, and where documents belong to domain D_1: $countOnDiagDoc_{D_1}(t) = |\{d_{idx}; d \in D_1 \wedge M_{Band}(d_{idx}, t_{idx}) = 1\}|$; $countDoc_{D_1}(t)$ denotes the number of documents from domain D_1 that contain term t, while $countDoc_{D_2}(t)$ is the number of documents from domain D_2 that contain term t. The bridging term potential score for term t is the ratio of documents from domain D_1 that lay on the diagonal multiplied by the number of documents from the other domain. The intuition behind this heuristic is that for term t, the more the term represents a domain (has a large proportion of document on the diagonal of the banded matrix) and also the more documents from the other domain that contain t exist, the higher the potential of term t to be a bridging term between the two domains.

All the heuristic scores are normalized by dividing the term scores with the highest score. The result of our methodology is a list of terms sorted by their b-term potential scores. It should be left to the domain expert to check whether the discovered bridging term suggests a valid, new and interesting relation. While the methodology has been currently applied to bridging two domains only, the method can be generalized to connecting several different domains, which is the topic of our further research.

4 Using Banded Matrix-Based B-Term Ranking for Bisociative Morals Ideation

Although not primarily designed for this task, our methodology can be used for creating pairs of sentences from different domains, which combine into surprising, funny or even insightful pieces of text when put together and considered as a whole. Bridging terms, appearing in both sentences, detected by our methodology function as a kind of glue, contributing to the coherency and increasing the potential for combinations to be meaningful.

To illustrate the potential of the proposed approach for narrative ideation, we chose two domains: What-if sentences and Aesop's fables' morals. The first domain consists of 94 What-if sentences, retrieved from Twitter (using hash tag #whatif) and the UKWaC British English web corpus. For instance, here is example of such a sentence: *"What if Google someday went down and we couldn't Google what happened to Google?"* The Aesop's fables[2] morals dataset is a collection of 277 fable morals. We created this dataset by crawling the Aesop's fables online collection.

---

[2] Aesop was a Greek fabulist, known as author of numerous fables; these are characterized by animals, which solve problems and have human characteristics. See
http://www.aesopfables.com/

Table 1. Results of our methodology: combinations of what-if sentences with Aesop's fable morals. In the brackets are the fable titles, which were not part of the document's contents, but are given here as an additional piece of information. The bridging terms are shown in bold.

What if **life** is one big dream, and when we die, we wake up. Evil tendencies are shown in early life. (The Man, the Boy and the Donkey)
What if we woke up, as a baby, and our whole life had been a dream? Evil tendencies are shown in early **life**. (The Man, the Boy and the Donkey)
What if, like Bhutan, we gauged our **life**'s success by how happy we are, not by how big the house is, the number. Evil tendencies are shown in early **life**. (The Man, the Boy and the Donkey)
What if someone you **love** dearly gave you a surprise bday party and when you arrived every1 was exchanging gifts but not nary a 1 was for you! Misery **loves** company. (The Fox Who Had Lost His Tail)
What if someone you **love** dearly gave you a surprise bday party and when you arrived every1 was exchanging gifts but not nary a 1 was for you! Even the wildest can be tamed by **love**. (The Lion in Love)
What if there are other beings in the same room as us but we can't **see** them and they can't see us. Not everything you **see** is what it appears to be. (The Dancing Monkeys)
What if there are other beings in the same room as us but we can't **see** them and they can't see us. Gossips are to be **seen** and not heard. (The Eagle the Cat and the Wild Sow)
What if we don't **see** tomorrow and everything you said today couldn't be undone. Would you be proud? Not everything you **see** is what it appears to be. (The Dancing Monkeys)
What if we don't **see** tomorrow and everything you said today couldn't be undone. Would you be proud? Gossips are to be **seen** and not heard. (The Eagle the Cat and the Wild Sow)
What if eyeballs had butts but we couldn't **see** them because they're hidden in our skulls? Not everything you **see** is what it appears to be. (The Dancing Monkeys)
What if eyeballs had butts but we couldn't **see** them because they're hidden in our skulls? Gossips are to be **seen** and not heard. (The Eagle the Cat and the Wild Sow)
What if humans could actually breathe in space? And the government says we can't so we don't try to **escape**? Nothing **escapes** the master's eye. (The Stag in the Ox-Stall)
What if humans could actually breathe in space? And the government says we can't so we don't try to **escape**? We had better bear our troubles bravely than try to **escape** them. (The Kings Son and the Painted Lion)
What if you simply stopped doing whatever it is that isn't a part of your **success**? The best intentions will not always ensure **success**. (The Monkeys and Their Mother)
What if, like Bhutan, we gauged our life's **success** by how happy we are, not by how big the house is, the number. The best intentions will not always ensure **success**. (The Monkeys and Their Mother)

Each what-if sentence as well as each Aesop's fables' moral was treated as a separate document. The documents from both domains were preprocessed using standard text mining techniques, described in the methodology section. This resulted in 383 distinct terms from all the obtained documents. We applied our methodology to find terms with the highest b-term potential. For simplicity, the terms were sorted using the scores of Heuristic 4, which we consider the most complete among the presented heuristics. The five terms with the highest b-term potentials were used to create all pairs of sentences sharing the selected bridging term (with the first sentence being from the what-if domain and the second from the Aesop's fables domain). The 15 highest scoring b-term based concatenated pairs of sentences which resulted are shown in Table 1. These results show–subjectively–that using the terms with the highest b-term potential resulted in several meaningful, creative combinations of sentences. Moreover, it is clear that a large proportion of the sentence pairs in Table 1 have meaningful relations which could form the basis of artefacts such as poems or stories. We can argue that it would be quite a laborious task to find similarly valuable combinations from all possible pairs (143) of sentences from the given domains without guidance provided by bridging terms. We plan to use crowd-sourcing to test the hypothesis that our approach can reliably produce such valuable combinations.

5 Conclusions and Further Work

The experimental evidence above indicates that the methodology presented in this paper has the potential for supporting the users in the task of bisociative, cross-domain narrative ideation. Banded matrices help us to discover the structures which reveal the nature of relations between terms and documents. We have shown that the approach can be used to construct creative combinations of sentences from different domains, coupled with bridging terms with the highest b-term potential. The results confirm the potential of the proposed approach to identify meaningful bridging concepts in the intersection of texts from different domains.

In further work, we will upgrade the methodology to combine not only pairs of sentences from two domains, but also to compose longer chains of sentences, resulting in narrative ideation. Another line of research will address more subtle connections between sentences. For instance, there is a semantic connection between *baby* and *early life* in *"What if we woke up, as a baby, and our whole life had been a dream? Evil tendencies are shown in early life."* However, this connection arises purely by coincidence: it was not detected by the system. Furthermore, introducing more semantic understanding into the ranking could substantially improve the performance. We further plan to improve to the set of heuristics, e.g., by introducing a more global view taking into consideration a term's local neighbourhood, and exploring the potential of outlier documents in guiding search. We plan to apply this narrative ideation approach to knowledge discovery in different domains, working with experts from different fields to address real-life artistic and scientific domains and getting valuable feedback from them.

An important next step will be to crowd-source opinions about how reliable the process is at producing sentence pairs (and larger constructs) which can be meaningfully interpreted in such a way that intelligence and possibly creativity are projected onto the software producing them. After the analysis of the results and further refinement of the techniques, we plan to embed fictional ideation processes and idea expansion via bisociation into software for generating artefacts of cultural value such as poems, stories and scientific hypotheses. We hope to show that the kinds of cross-context link discovery methods presented here can be used generically in Computational Creativity projects across domains.

References

1. Wiggins, G.: A preliminary framework for description, analysis and comparison of creative systems. Knowledge-Based Systems 19(7), 449–458 (2006)
2. Koestler, A.: The Act of Creation, vol. 13 (1964)
3. Berthold, M. (ed.): Bisociative Knowledge Discovery. Springer (2012)
4. Smalheiser, N., Swanson, D., et al.: Using ARROWSMITH: A computer-assisted approach to formulating and assessing scientific hypotheses. Computer Methods and Programs in Biomedicine 57(3), 149–154 (1998)
5. Juršič, M., Cestnik, B., Urbančič, T., Lavrač, N.: Cross-domain literature mining: Finding bridging concepts with CrossBee. In: Proceedings of the 3rd International Conference on Computational Creativity, pp. 33–40 (2012)
6. Garriga, G., Junttila, E., Mannila, H.: Banded structure in binary matrices. Knowledge and Information Systems 28(1), 197–226 (2011)
7. Alqadah, F., Bhatnagar, R., Jegga, A.: Mining maximally banded matrices in binary data. In: Proceedings of the 10th SIAM International Conference on Data Mining (SDM 2010), pp. 942–953 (2010)
8. Feldman, R., Sanger, J.: The Text Mining Handbook: Advanced Approaches in Analyzing Unstructured Data. Cambridge University Press (2006)
9. Atkins, J., Boman, E., Hendrickson, B.: A spectral algorithm for seriation and the consecutive ones problem. SIAM Journal on Computing 28(1), 297–310 (1998)
10. Cormen, T., Leiserson, C., Rivest, R., Stein, C.: Introduction to Algorithms. MIT Press (2001)

Gaussian Topographic Co-clustering Model

Rodolphe Priam[1], Mohamed Nadif[2], and Gérard Govaert[3]

[1] S3RI, University of Southampton, University Road, SO17 1BJ, Southampton, U.K.
[2] LIPADE, Université Paris Descartes, 45 rue des Saints Pères, 75006 Paris, France
[3] HEUDIASYC, CNRS 7253, University of Technology of Compiègne, France

Abstract. The visualization of the clusters obtained by a partitioning procedure is very informative as this helps to a better overview of the contents of a data table. For co-clustering, the latent block mixture model is very effective. We propose to define generative self-organizing maps with this model for Gaussian blocks. A perspective is the analysis and the visualization of continuous data.

Keywords: latent block mixture model, generative topographic mapping, block expectation maximization, variational expectation maximization.

1 Introduction

In data analysis, the reduction of the dimensions of a numerical data matrix leads to synthetic and understandable representations in a low dimensional space. The variables of a numerical matrix can be continuous, binary or discrete. For parametric modeling, one of the most studied data table is for continuous variables. When the data matrix is large, a clustering may lead to a quicker and easier access to the hidden contents of the data in comparison with a method which only reduces the dimensionality by projection. Moreover, combining clustering and reduction is often more informative. For instance, it is possible to show not only each datum by a point on a two dimensional map, but also their clusters which are obtained from a given algorithm. It is also possible to model the projection of clusters of rows (resp. columns) rather than the rows (resp. columns). This may be an interesting approach for data analysis because this can induce a reduction of the number of variables before a clustering. But most of the time, mapping and clustering operate separately [2], and one has to decide which method must be devoted to each problem. On the contrary, it exists a family of methods which solves them simultaneously.

The Kohonen's maps and more generally self-organizing maps (SOM) [1] are modified clustering methods for continuous variables most often. They are able to learn a relation of vicinity between the clusters and can induce a visualization by post-processing. Some modified versions are able to provide more suitable results in particular situations like the analysis of discrete data. Moreover, a parametric model is very flexible and even scalable when it is defined and learned properly. Hence, a probabilistic model for SOM is interesting for these diverse reasons.

A. Tucker et al. (Eds.): IDA 2013, LNCS 8207, pp. 345–356, 2013.

The Generative Topographic Mapping (GTM) [3] is a generative SOM with a more restricted set of parameter values than the Kohonen's map. It is extensively studied and improved ([4,5,6]) recently for data analysis purposes. It formulates a self-organizing map by adding the constraints of vicinity between the clusters at the level of the expectations of a Gaussian mixture model (GMM) [7].

These methods can be useful for many domains which need the analysis of a large amount of data such as often met today. In bioinformatics, Kohonen's maps are very useful [8,9,10] for the analysis of microarrays. The clustering of their variables is also a main concern because for instance only a few genes are expressed, activated or switch on in a particular biological situation. Here, the clustering of the two dimensions of a data matrix is therefore essential [11]. For all these reasons, we propose an extension of GTM with the help of a co-clustering model, the latent block mixture model (LBM) [12] for data matrices which have a block structure. An example in the presented experimental part is a preliminary test of the method with a real dataset of gene expression in a microarray.

The paper is organized as follows. In section 2, we review LBM with a Gaussian setting, and the related criterion to optimize for the parameters estimation. In section 3, we add the constraints in this model and we propose a learning algorithm. Two different approaches for the regularization of the parameters are considered: L_2 for a first method named GBGTM and L_1 for a second one named GBGTM-S. In section 4 we present the numerical experiments with the proposed methods. Finally, in section 5 we summarize the contribution and the perspectives.

2 LBM with Gaussian Blocks

Let us denote $\mathbf{x} = (x_{11}, x_{12}, \ldots, x_{ij}, \ldots, x_{nd})$, the data matrix in a latent block model. I is the set of the rows and J is the set of the columns. A possible assignment of I is modeled with the binary classification matrix $\mathbf{z} = (z_{ik})_{n \times g}$, it is such that $z_{ik} = 1$ indicates the component of the row i, and $\sum_{k=1}^{g} z_{ik} = 1$. Similarly for the assignments of J it is denoted the binary matrix $\mathbf{w} = (w_{j\ell})_{d \times m}$. The two sets of possible assignments $\mathbf{w}$ and $\mathbf{z}$ partition the cells of the matrix $\mathbf{x}$ into a number of contiguous, non-overlapping blocks. A block $k\ell$ is defined as the set of cells $\{x_{ij}; z_i = k, w_j = \ell\}$. For a latent block model, the $n \times d$ random variables which generate the observed cells x_{ij} of the data matrix are assumed to be independent, once $\mathbf{z}$ and $\mathbf{w}$ are fixed, they permit to define a co-clustering model.

2.1 Latent Block Model

The probability density function (p.d.f.) of a latent block model is defined as the following decomposition. It is obtained by independence of $\mathbf{z}$ and $\mathbf{w}$, by summing over all the assignments [12]:

$$f_{LBM}(\mathbf{x}; \boldsymbol{\theta}) = \sum_{(\mathbf{z},\mathbf{w}) \in \mathcal{Z} \times \mathcal{W}} P(\mathbf{z})P(\mathbf{w})P_{\boldsymbol{\theta}}(\mathbf{x}|\mathbf{z}, \mathbf{w})$$
$$= \sum_{(\mathbf{z},\mathbf{w}) \in \mathcal{Z} \times \mathcal{W}} \prod_i p_{z_i} \prod_j q_{w_j} \prod_{i,j} \varphi(x_{ij}; \alpha_{z_i w_j}), \tag{1}$$

where the set of all the possible assignments is denoted $\mathcal{Z}$ for I and $\mathcal{W}$ for J, while $\varphi(.; \alpha_{k\ell})$ is a probability density function defined for the cell (ij) on the set of reals $\mathbb{R}$ and $\{\alpha_{k\ell}\}$ are unknown parameters. The vectors of the probabilities p_k and q_ℓ that a row (resp. a column) belongs to the k-th component (resp. ℓ-th component) are denoted $\mathbf{p} = (p_1, \ldots, p_g)$ (resp. $\mathbf{q} = (q_1, \ldots, q_m)$). The set of parameters is denoted $\boldsymbol{\theta}$. It is compound of $\mathbf{p}$, $\mathbf{q}$ and $\boldsymbol{\alpha} = (\alpha_{k\ell})$ which aggregates the parameters from all the p.d.f. of the blocks, $\boldsymbol{\theta} = (\mathbf{p}, \mathbf{q}, \boldsymbol{\alpha})$. Hereafter, to simplify the notation, the sums and the products relating to rows, columns or clusters will be subscripted respectively by the letters i, j, k, or ℓ without indicating the limits of variation, which are implicit. The set of parameters $\boldsymbol{\theta}$ of the model can be estimated by maximizing the log-likelihood:

$$L(\mathbf{x}; \boldsymbol{\theta}) = \log f_{LBM}(\mathbf{x}; \boldsymbol{\theta}). \tag{2}$$

The block model is dramatically more parsimonious than the usual mixture model where each dimension of the data matrix is modeled separately. Next, we describe the latent block model where φ is a Gaussian density.

2.2   Univariate Gaussian Density Function

When the cells are generated with a Gaussian distribution, the density function for the block $(k\ell)$ is written:

$$\varphi(x_{ij}; \alpha_{k\ell}) = \frac{exp(-|x_{ij} - \mu_{k\ell}|^2 / 2\sigma_{k\ell}^2)}{\sqrt{2\pi}\sigma_{k\ell}}, \tag{3}$$

where the mean is $\mu_{k\ell}$ and the variance $\sigma_{k\ell}^2$ such that $\alpha_{k\ell} = (\mu_{k\ell}, \sigma_{k\ell}^2)$. With this distribution the model is called GLBM (see [13]). Next paragraphs, we review the criterion and the algorithm for an estimation of $\boldsymbol{\alpha}$.

2.3   Objective Function and EM

For the co-clustering model which is considered, we aim to address the problem of parameters estimation by a maximum likelihood (ML) approach such that:

$$\hat{\boldsymbol{\theta}} = argmax_{\boldsymbol{\theta}} L(\mathbf{x}; \boldsymbol{\theta}). \tag{4}$$

Let us focus on the estimation of a value of $\boldsymbol{\theta}$ by the maximum likelihood approach associated to the block mixture model. For this model, the completed data are taken to be the vector $(\mathbf{x}, \mathbf{z}, \mathbf{w})$ where the latent vectors $\mathbf{z}$ and $\mathbf{w}$ are the random labels for the rows and the columns respectively. The classification log-likelihood can then be written:

$$L_C(\mathbf{z}, \mathbf{w}; \mathbf{x}, \boldsymbol{\theta}) = \log P_{\boldsymbol{\theta}}(\mathbf{x}|\mathbf{z}, \mathbf{w}) + \log P(\mathbf{z}) + \log P(\mathbf{w}).$$

Here, without loss of generality, it is supposed that the mixing coefficients are equiproportional, $p_k = g^{-1}$ and $q_\ell = m^{-1}$. From the definition of the model, it is clear that the probabilities involving the vector $(\mathbf{x}, \mathbf{z}, \mathbf{w})$ can be written explicitly. The log-likelihood of the completed data is considered in the next paragraphs for the inference by taking benefit of the introduced latent variables. The EM algorithm [14] maximizes the log-likelihood w. r. to $\boldsymbol{\theta}$ iteratively by maximizing the conditional expectation of the log-likelihood of the completed data w. r. to $\boldsymbol{\theta}$ given a previous current estimate $\boldsymbol{\theta}^{(t)}$ and the observed data $\mathbf{x}$,

$$Q(\boldsymbol{\theta}, \boldsymbol{\theta}^{(t)}) = \sum_{\mathbf{z}, \mathbf{w}} P_{\boldsymbol{\theta}^{(t)}}(\mathbf{z}, \mathbf{w}|\mathbf{x}) L_C(\mathbf{z}, \mathbf{w}; \mathbf{x}, \boldsymbol{\theta}), \tag{5}$$

with the probabilities that the cell (i, j) belongs to the cluster (k, ℓ) conditionally to the whole table. The index (t) permits to denote a current estimation of a parameter or a function of the parameters. The probability mass function involved in the expectation (5) may be obtained by the Bayes rule but the structure of dependence among the random cells (taking for values x_{ij}) of the data table makes intractable their exact computation and an alternative is required.

2.4   Block EM Algorithm

The approach based on a generalized EM and a variational approximation has been proposed in [15] and named *Block EM* (BEM). Roughly speaking, the approximation used consists in replacing $P(\mathbf{z}, \mathbf{w}|\mathbf{x})$ by $P(\mathbf{z}|\mathbf{x})P(\mathbf{w}|\mathbf{x})$. The algorithm proceeds by defining a lower bound of the log-likelihood (see [15]) and repeats the two following steps:

- **E-step:** The posterior probabilities are found with the fixed point relations:

$$c_{ik}^{(t)} \propto e^{\sum_{j,\ell} d_{j\ell}^{(t)} \log \varphi(x_{ij}; \alpha_{k\ell}^{(t)})} \quad \text{and} \quad d_{j\ell}^{(t)} \propto e^{\sum_{i,k} c_{ik}^{(t)} \log \varphi(x_{ij}; \alpha_{k\ell}^{(t)})}. \tag{6}$$

- **M-step:** A temporary value of the parameters is found by solving:

$$\theta^{(t+1)} = argmax_{\boldsymbol{\theta}} \sum_{i,j,k,\ell} c_{ik}^{(t)} d_{j\ell}^{(t)} \log \varphi(x_{ij}; \alpha_{k\ell}). \tag{7}$$

Here the objective function is denoted $\tilde{Q}_{LBM}(\boldsymbol{\theta}|\boldsymbol{\theta}^{(t)})$. It is denoted the variational probabilities c_{ik} such that $\sum_k c_{ik} = 1$, and also $d_{j\ell}$ such that $\sum_\ell d_{j\ell} = 1$. For $1 \le k \le g$ and $1 \le \ell \le m$, it is denoted the aggregated statistics, $y_{k\ell}^{(t)} = \sum_{i,j} c_{ik}^{(t)} d_{j\ell}^{(t)} x_{ij}$, and $c_k^{(t)} = \sum_i c_{ik}^{(t)}$, $d_\ell^{(t)} = \sum_j d_{j\ell}^{(t)}$.

Hence, the parameters are estimated by an iterative way. The BEM algorithm proceeds by an alternated maximization of $\tilde{Q}$ and converges to a final solution which maximizes (locally) the log-likelihood of the latent block model. The models, the learning algorithms implementing BEM and the corresponding M-step have been proposed previously in the literature (see [13]). In the following, we explain how to adapt the model in order to make it capable of mapping directly the data to the plane.

3 Parsimonious Topographic Mapping

The parameters $\alpha_{k\ell}$ are re-parameterized with two sets of vectors, one for the dimension k, and one for the dimension ℓ. A particular choice for the mixing probabilities is also decided, these prior are chosen fixed and equiproportional. By this way, we define a general model for a parsimonious parametric SOM in order to map the rows of a numerical tables with a large number of columns into the plane. This family of methods is named Topographic latent block model or Block GTM. In the Gaussian case and with a bayesian setting, the log-likelihood can be written:

$$L_{GBGTM}(\mathbf{x}; \bar{\theta}) = \log \mathbb{E}_{\mathbf{z},\mathbf{w},\Omega,\beta} \left[P(\mathbf{z}, \mathbf{w}, \Omega, \beta; \mathbf{x}, \bar{\theta}) \right]$$
$$= \log \mathbb{E}_{\mathbf{z},\mathbf{w},\Omega,\beta} \left[P(x|\mathbf{z}, \mathbf{w}, \Omega, \beta; \bar{\theta}) P(\Omega|\beta) P(\beta) P(\mathbf{z}) P(\mathbf{w})) \right] .$$

The hidden matrix Ω is defined hereafter in sections 3.1 and 3.2 while β is a parameter for its hierarchical prior. The general quantity $\bar{\theta}$ is the covariance matrix $\Sigma = (\sigma_{k\ell}^2)$ which aggregates the variance parameters and is kept non random. Note that the distribution of the block model is recognized with a specific parameterization for handling the nonlinear mapping. In the section, the problem we address is to define more precisely the new log-likelihood and to estimate the parameters with a variational EM algorithm as explained next.

3.1 Parameter Transformation

It is defined a set of constant bivariate vectors s_k, the coordinates of the nodes of an had hoc regular rectangular planar mesh. By attaching the parameters of one cluster to the coordinates of one node, this may induce the wanted constraints for the self-organization of the row clusters. For nontrivial problems, the data cloud has a complex shape and higher dimensions than two are required for the modeling. Hence, the 2-dimensional coordinates s_k are transformed into higher dimensional vectors ξ_k. This leads to a discretization and modelization of the latent space where the data are projected, a square $[-1; 1] \times [-1; 1]$. The nodes of the rectangular mesh have a similar role than in the algorithm of Kohonen's maps. Finally, the discretized latent space is modeled as follows.

Nodes of a Mesh: The coordinates of these nodes can be written:

$$\mathcal{S} = \left\{ s_k = \begin{pmatrix} s_{k_1} \\ s_{k_2} \end{pmatrix} ; k = 1, ..., g \right\} .$$

This set is used for a ranking of the clusters of GLBM along the mesh such that it is attributed a unique coordinate s_k to each cluster k of the rows. Then the distance between the coordinates of two nodes from the mesh is related to the similarity of the contents of the two corresponding clusters. Note that we are mainly interested by a squared map in the experimental part, hence $k = k_1 + (k_2 - 1) \times \sqrt{g}$ where $(k_1, k_2) \in \{1, \cdots, \sqrt{g}\} \times \{1, \cdots, \sqrt{g}\}$.

Transformed Nodes: Each coordinate s_k is projected into a space of h dimensions with the help of well defined basis functions ϕ. This is written for $1 \leq k \leq g$ and $1 \leq s \leq h$:

$$\xi_k = (\phi_1(s_k), \phi_2(s_k), \cdots, \phi_h(s_k))^T,$$

where

$$\phi_s(s_k) \propto exp[-||s_k - \mu_{\phi_s}||^2/2\nu_{\phi_s}^2].$$

These functions ϕ_s are typically Gaussian density functions with mean centers $\mu_{\phi_s} \in \mathbb{R}^2$ and standard deviation ν_{ϕ_s}. Note that the choice of h may have a consequence on the accuracy of the clustering and mapping. For the constraints on the row clusters, a matrix aggregates the basis functions for all the nodes as follows, $\Phi = [\xi_k| \cdots |\xi_g]^T$. Next paragraph, we introduce the parameters for the columns in order to model the probabilities $\alpha_{k\ell}$.

Transformed Parameters: A set of latent vectors related to the clustering of the columns is defined as $\{w_\ell \in \mathbb{R}^h, 1 \leq \ell \leq m, h \in \mathbb{N}_+^*\}$, and its estimation is required. For modeling the dependence of each parameter $\alpha_{k\ell}$ with ξ_k and w_ℓ, it is considered their inner product. To map an inner product ($w_\ell^T \xi_k \in \mathbb{R}$) onto its corresponding parameter ($\alpha_{k\ell} \in [0;1]$), it is then considered:

$$\alpha_{k\ell} = w_\ell^T \xi_k \text{ for } 1 \leq k \leq g \text{ and } 1 \leq \ell \leq m. \tag{8}$$

This function may be different for each model φ, and is here the identity matrix for a Gaussian law.

3.2   Parameters and Variational Criterion

The reduced $g \times m$ matrices $(\mu_{k\ell})$ and $(\sigma_{k\ell}^2)$ in the previous co-clustering model are replaced by one matrix $\Omega = [w_1|w_2| \cdots |w_m]^T$. The resulting model remains parsimonious because h is small, less than half of one hundred in general. But for an even more parsimonious setting, a bayesian L_1 penalization can be used in order to cancel out some components of each vector w_ℓ. Let's have $\mathcal{N}(y; \mu, \Sigma)$ for y distributed as a Gaussian p.d.f. of mean μ and variance Σ, while $\mathcal{G}(y; d0, c0)$ is for y distributed as a Gamma distribution with two real parameters $d0$ and $c0$. Let's have $\tilde{\theta} = \{z, w, \Omega, \beta\}$ for the random variables. Then, the new parameters which are involved in the E-step are $\tilde{\theta}$. In the M-step, this is $\bar{\theta}$. Then, a hierarchical prior is defined from Gaussian and Gamma distributions for a penalization in the new completed log-likelihood which can be written:

$$\tilde{L}_C(z, w, \Omega, \beta; x, \bar{\theta}) = L_C(z, w; x, \theta(\Omega))$$
$$+ \sum_s \log \mathcal{N}(w_{(s)}; 0_m, V_{(s)}) + \sum_{s\ell} \log \mathcal{G}(\beta_{s\ell}; d_{\beta_s^0}, c_{\beta_s^0}).$$

Here, $\beta = (\beta_{s\ell})$ is a $h \times m$ real matrix, $w_{(s)}$ is a h-dimensional column of Ω, and $V_{(s)} = \mathbf{diag}_{1\leq\ell\leq m}(\beta_{s\ell})$ while $0_m = (0)$ is the h-dimensional null vector.

The parameter $\theta(\Omega)$ stands for the transformed quantities $\alpha_{k\ell}$ with the new matrix Ω, as defined in (8). The components of θ defined for the regularization are independent and the hierarchical setting can induce parsimony [16], while the hyperparameters $(d_{\beta_s^0})$ and $(c_{\beta_s^0})$ remain constant (for instance 10^{-3}). Note that a related hierarchical prior has been proposed recently [17] for GTM, with dramatic improvements. Here, each column of the matrix defined for the non-linear mapping is independent and has a Gaussian distribution with random parameters for the regularization.

For the distributions defined for the variational bound, the law of $(\mathbf{z}, \mathbf{w})$ remains identical. The other parameters have related distributions as defined just below for Ω and for β,

$$Q(\Omega) = \prod_\ell \mathcal{N}(w_\ell; \mu_\ell, S_\ell) \text{ and } Q(\beta) = \prod_{\ell,s} \mathcal{G}(\beta_{s\ell}; d_{\beta_{s\ell}}, c_{\beta_{s\ell}}).$$

By this way it is possible to have a new function to optimize which handles the regularization. For large values of $\beta_{s\ell}$, the corresponding cells in Ω might cancel out, leading to a parsimonious matrix as expected. The different variables in $\tilde{\theta}$ are supposed independent such that the related density Q is a product. Let's also have $\mathcal{H}(Q)$ stands for the entropy of Q. Following the extensive literature on variational learning, it can be written with this approach:

$$L_{GBGTM} \geq \mathbb{E}^Q_{\mathbf{z},\mathbf{w},\Omega,\beta} \left[\tilde{L}_C(\mathbf{z}, \mathbf{w}, \Omega, \beta; \mathbf{x}, \bar{\theta}) \right] - \mathcal{H}(Q). \tag{9}$$

Here the right member is the new function involved in the variational maximization, $\mathcal{F}(Q, \bar{\theta})$ with an expectation w.r. the distribution Q. This is a surrogate criterion for the estimation of the parameters. An algorithm for learning the parameters is presented next paragraphes.

3.3   General Algorithm

The purpose is to iteratively update a current value of the parameters such that the function $\mathcal{F}$ is increased. The algorithm BEM is altered in order to handle the randomness of Ω (and β) and finally reach a local ML solution. In a variational EM, the two following steps are repeated:

- The distribution of each parameter for Q are found by maximizing the free energy $\mathcal{F}$ after marginalizing out other parameters conditionnally to a known value $\bar{\theta}^{(t)}$, in a E-step:

$$Q^{(t)}(\tilde{\theta}_r) \propto \exp\left\{ \mathbb{E}^{Q^{(t)}}_{\tilde{\theta}_{-r}} \left[\tilde{L}_C(\mathbf{z}, \mathbf{w}, \Omega, \beta; \mathbf{x}, \bar{\theta}^{(t)}) \right] \right\}. \tag{10}$$

Here, it is denoted $\tilde{\theta}_r \in \tilde{\theta}$ while $\tilde{\theta}_{-r}$ is the result of removing $\tilde{\theta}_r$ from $\tilde{\theta}$.
- The eventual remaining parameters $\bar{\theta}$ where no prior is assumed can be estimated for a new current value by maximizing $\mathcal{F}$ with respect to $\bar{\theta}$, after expectation with the obtained variational $Q^{(t)}$ distribution, in a M-step:

$$\bar{\theta}^{(t+1)} = argmax_{\bar{\theta}} \mathbb{E}^{Q^{(t)}}_{\tilde{\theta}} \left[\tilde{L}_C(\mathbf{z}, \mathbf{w}, \Omega, \beta; \mathbf{x}, \bar{\theta}) \right]. \tag{11}$$

This is solved as follows. Actually, for $\mathbf{z}$ and $\mathbf{w}$, the previous update formula for the posterior probabilities in (6) might be relevant at least approximatively.

3.4   Detail of the Algorithm

When $\Lambda_{\beta,\ell}^{(t)}$ stands for the regularizing matrix at the current time, it can be written:

$$\Lambda_{\beta,\ell}^{(t)} = \begin{cases} \mathbf{diag}_{1\leq s\leq h}\left\{\frac{d_{\beta^0}+\frac{1}{2}}{c_{\beta_s^0}+\frac{1}{2}(\mu_{s\ell}^2+\sigma_{S,s\ell}^2)}\right\} & \text{with parsimony by (10)} ; \\ \mathbf{diag}_{1\leq s\leq h}\left\{10^{-3}d_\ell^{(t)}\right\} & \text{without parsimony.} \end{cases} \quad (12)$$

For the usual L_2 normalization, a simple constant $10^{-3}d_\ell^{(t)}$ in the diagonal of the regularizing matrix has been generally useful instead of the bayesian estimation for the case L_1. This regularization does not induce parsimonious values for the components of μ_ℓ. Here, $\sigma_{S,s\ell}^2$ is the sth diagonal component of $S_\ell^{(t)}$ while $\mu_{s\ell}$ is the sth component of $\mu_\ell^{(t)}$. When $\beta_{s\ell} = \beta_s$ $\forall \ell$, then the regularizing matrix is denoted $\Lambda_\beta^{(t)}$, with in the numerator $\frac{m}{2}$ which replaces $\frac{1}{2}$, and in the denominator a sum $\sum_\ell(\mu_{s\ell}^2 + \sigma_{S,s\ell}^2)$ instead of the ℓ-st component.

Let's have the matrix $\Upsilon_\ell^{(t)} = \mathbf{diag}_{1\leq k\leq g}(\{\sigma_{k\ell}^{(t)}\}^{-2})$ where $\sigma_{k\ell}$ is updated as in the unconstrained case. Morever, for the variational parameters, the updates may be obtained by a Laplace approximation with the particular new values for $1 \leq \ell \leq m$:

$$S_\ell^{(t)} = \left\{d_\ell^{(t)}\Phi^T\Upsilon_\ell^{(t)}G^{(t)}\Phi + \Lambda_\beta^{(t)}\right\}^{-1}, \text{ and } \mu_\ell^{(t)} = S_\ell^{(t)}\Phi^T\Upsilon_\ell^{(t)}y_\ell^{(t)}. \quad (13)$$

Here, $G^{(t)} = \mathbf{diag}_{1\leq k\leq g}(c_k^{(t)})$ is a diagonal matrix, and $y_\ell^{(t)} = (y_{k\ell}^{(t)})$ is a g-dimensional vector. In this version, for the variance, it is kept the fuzzy memberships for the columns, such that the update formula is:

$$\Sigma^{(t+1)} = \left(\frac{\sum_{ij}c_{ik}^{(t)}d_{j\ell}^{(t)}(x_{ij}-a_{k\ell})^2}{c_k^{(t)}d_\ell^{(t)}}\right). \quad (14)$$

Note that it is clear that the proposed model is a generalization of the generative topographic mapping [3]. Finally, the algorithm repeats the computational steps:

1. Update $c_{ik}^{(t)}$ or $d_{j\ell}^{(t)}$ with (6)
2. Update $\Lambda_{\beta,\ell}^{(t)}$ with (12)
3. Update $S_\ell^{(t)}$ with (13)
4. Update $\mu_\ell^{(t)}$ with (13)
5. Update $\Sigma^{(t+1)}$ with (14)
6. Return to 1. until stop.

Note that when instead of (12), this is $\Lambda_\beta^{(t)}$ which is in stake, then the aggregated version of the update replaces the step (2.) in the algorithm. The stopping criterium may be a maximal number of steps, typically less than 200 for instance.

Finally, the procedure converges towards $(\hat{\boldsymbol{\mu}}_1, \cdots, \hat{\boldsymbol{\mu}}_m, \hat{S}_1, \cdots, \hat{S}_m, \hat{\Lambda}_{\beta,1}, \cdots,$ $\hat{\Lambda}_{\beta,m}, \hat{\boldsymbol{\Sigma}})$, the final value of the parameters. The final classification matrices $\hat{C} = (\hat{c}_{ik})$ for the rows and $\hat{D} = (\hat{d}_{j\ell})$ for the columns are also obtained.

4   Experiments

A first experiment with a biological dataset illustrates the interest of the block model and the normalization. Other results with simulated datasets have confirmed the better behaviour of the proposed method in comparison with GTM for data matrices with a block structure. We are interested in this section on the comparison of GTM and GBGTM when the data is the gene expression from the microarray of the Colon tissue samples [18].

4.1   Post-treatment and Experimental Settings

Clustering and Mapping: When a row $i \in I$ has a higher posterior probability $\hat{c}_{ik}$ for a cluster k then it belongs to this cluster and the label for the ith row is estimated by $\hat{z}_i = k$ such as, $\hat{z}_i = argmax_k \hat{c}_{ik}$. Moreover, the set of two-dimensional coordinates S of the g clusters leads to the projections of the rows on the latent space. A row i has a representative on the latent space, a point with the coordinates, $\hat{\boldsymbol{s}}_i^{MAP} = \boldsymbol{s}_{\hat{z}_i}$. By performing this procedure for each row i, the model builds a reduced view of the n elements of I. When two nodes have their coordinates s_k and $s_{k'}$ near in the latent space, their corresponding clusters might have their parameters $\alpha_{k\ell}$ and $\alpha_{k'\ell}$ similar, so their corresponding contents should be also similar. A projection can be obtained by computing an average position of each row i with the posterior probabilities, $\hat{\boldsymbol{s}}_i = \sum_{k=1}^{g} \hat{c}_{ik} s_k$. If the vector of probabilities $(\hat{c}_{i1}, \hat{c}_{i2}, \cdots, \hat{c}_{ig})$ is binary, then the row i is in the cluster $\hat{z}_i$, and $\hat{\boldsymbol{s}}_i = \hat{\boldsymbol{s}}_i^{MAP}$.

Experimental Settings: After the estimation of the parameters, several indicators are computed for the comparison. By using the estimated label $\hat{z}_i$, the error rate is the percentage of missclassified when each node is labelled by majority vote, denoted Error-rate. When the mapping is continuous by using the coordinates $\{\hat{s}_i\}$, it can be computed other indicators than related to the classification error rate in order to reveal the quality of the projection onto the plane, the DB-Index [19] and also the average of the Silhouettes [20] denoted S-Index. The quality of the mapping and the accuracy of the clustering, which are obtained from the topographic latent block model are compared with the original GTM. Two different approaches for the regularization of the parameters are considered: L_2 for a first method named GBGTM and L_1 for a second one named GBGTM-S. In the following subsection, we consider a real dataset. The maps are obtained for the following sizes:

$$g \in \{5^2, 6^2\} \text{ and } m = 40.$$

For these settings, the resulting three indicators and the final maps from the points $\hat{s}_i$ are presented herefater.

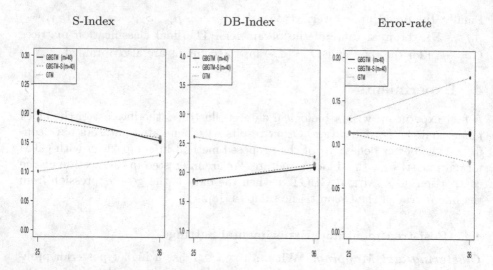

Fig. 1. Indicators obtained in the experiments for the methods GTM, GBGTM and GBGTM-S for several values of g and m

4.2   Output for a Biological 62 × 2000 Table

In this paragraph, we present the output for a real biological dataset. Among the samples, there 40 are tumour and 22 are normal. Initially there were 6500 genes for all the tissues on the Affymetrix array. Following previous authors, the genes are selected, filtered and the values are scaled with a logarithmic transformation plus a standardisation of the rows. This leads to finally about 1000 genes for the same number of samples 62. Note that this choice is random because the normalisation is decided from a training subsample. As the sample size is small, the clustering error is decreasing with the number of clusters for our method because the underlying metrics is well suited while on the contrary GTM has a higher error. For this sample the two versions of GBGTM have a similar behaviour from the point of view of the visual indicators. According to Figure 1, our new method is clearly able to outperform the usual GTM with an enhancement of both the clustering and the mapping. For an overview of the visualization obtained by the different approaches with this data, the maps which have been constructed from the final parameters of the three methods are also proposed in Figure 2 in order to help their comparisons. If the two versions of GBGTM gives almost similar maps, the available implementation of GTM leads to a map where a first part of the clusters have fuzzier posterior probabilities while another large part is empty. Finally, the consequence of the L_1 regularisation is more parsimony because useless cells of the matrix Ω may cancel out. The number of parameters is dramatically reduced in comparison with GTM, the reduction factor is about $\frac{3m}{2d}$.

GTM   GBGTM   GBGTM-S

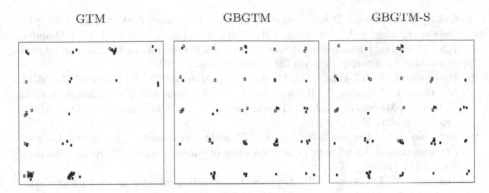

Fig. 2. Final maps from the experiments for the methods GTM, GBGTM and GBGTM-S for $g = 5^2$ and $m = 40$, after same jittering

5 Conclusion and Discussion

Herein, we have proposed a new model for the projection of continuous data tables with a block structure. The approach is parsimonious and flexible. In the experiments, it is observed that our model is able to outperform GTM. Several perspectives are possible. The exact variational posterior probabilities for the clustering of the rows and columns should be preferred for the L_1 regularization. A penalization for the variance parameters can be introduced. The column clusters offer additional information on the relations between the variables and can be discussed further. Additional simulations for the comparisons can be considered also in future. Unbalanced mixing coefficients are also interesting.

References

1. Kohonen, T.: Self-organizing maps. Springer (1997)
2. Jain, A.K.: Data clustering: 50 years beyond K-means. Pattern Recognition Letters 3(8), 651–666 (2010)
3. Bishop, C.M., Svensén, M., Williams, C.K.I.: GTM: A principled alternative to the self-organizing map. In: Mozer, M.C., Jordan, M.I., Petsche, T. (eds.) Advances in Neural Information Processing Systems 9, pp. 354–360. The MIT Press, Cambridge (1997)
4. Kabán, A., Girolami, M.: A combined latent class and trait model for analysis and visualisation of discrete data. IEEE Trans. Pattern Anal. and Mach. Intell., 859–872 (2001)
5. Tino, P., Nabney, I.: Hierarchical gtm: Constructing localized nonlinear projection manifolds in a principled way. IEEE Trans. Pattern Anal. Mach. Intell. 24(5), 639–656 (2002)
6. Vellido, A.: Selective smoothing of the generative topographic mapping. IEEE Transactions on Neural Networks 14(3), 847–852 (2003)
7. McLachlan, G.J., Peel, D.: Finite Mixture Models. John Wiley and Sons, New York (2000)

 8. Golub, T.R., Slonim, D.K., Tamayo, P., Huard, C., Gaasenbeek, M., Mesirov, J.P., Coller, H., Loh, M.L., Downing, J.R., Caligiuri, M.A., Bloomfield, C.D., Lander, E.S.: Molecular classification of cancer: class discovery and class prediction by gene expression monitoring. Science 286(5439), 531–537 (1999)
 9. Hautaniemi, S., Yli-Harja, O., Astola, J., Kauraniemi, P., Kallioniemi, A., Wolf, M., Ruiz, J., Mousses, S., Kallioniemi, O.-P.: Analysis and Visualization of Gene Expression Microarray Data in Human Cancer Using Self-Organizing Maps. Mach. Learn. 52(1-2), 45–66 (2003)
10. Newman, A.M., Cooper, J.B.: AutoSOME: a clustering method for identifying gene expression modules without prior knowledge of cluster number. BMC Bioinformatics 11, 117 (2010)
11. Shannon, W., Culverhouse, R., Duncan, J.: Analyzing microarray data using cluster analysis. Pharmacogenomics 4(1), 41–52 (2003)
12. Govaert, G., Nadif, M.: Clustering with block mixture models. Pattern Recognition 36(2), 463–473 (2003)
13. Nadif, M., Govaert, G.: Model-Based Co-clustering for Continuous Data. In: ICMLA, pp. 175–180. IEEE Computer Society (2010)
14. Dempster, A., Laird, N., Rubin, D.: Maximum-likelihood from incomplete data via the EM algorithm. J. Royal Statist. Soc. Ser. B 39, 1–38 (1977)
15. Govaert, G., Nadif, M.: An EM algorithm for the block mixture model. IEEE Trans. Pattern Anal. Mach. Intell. 27(4), 643–647 (2005)
16. Neal, R.M.: Bayesian Learning for Neural Networks. Lecture Notes in Statistics, vol. 118. Springer (1996)
17. Yamaguchi, N.: Variational bayesian inference with automatic relevance determination for generative topographic mapping. In: SCIS-ISIS 2012, pp. 2124–2129 (2012)
18. Alon, U., Barkai, N., Notterman, D.A., Gish, K., Ybarra, S., Mack, D., Levine, A.J.: Broad patterns of gene expression revealed by clustering analysis of tumor and normal colon tissues probed by oligonucleotide arrays. Proc. Natl. Acad. Sci. USA 96(12), 6745–6750 (1999)
19. Davies, D.L., Bouldin, D.W.: A cluster separation measure. IEEE Transactions on Pattern Analysis and Machine Intelligence PAMI-1(2), 224–227 (1979)
20. Rousseeuw, P.: Silhouettes: A graphical aid to the interpretation and validation of cluster analysis. J. Comput. Appl. Math. 20, 53–65 (1987)

Preventing Churn in Telecommunications:
The Forgotten Network

Dejan Radosavljevik and Peter van der Putten

LIACS, Leiden University, P.O. Box 9512, 2300 RA Leiden, The Netherlands
{dradosav,putten}@liacs.nl

Abstract. This paper outlines an approach developed as a part of a company-wide churn management initiative of a major European telecom operator. We are focusing on explanatory churn model for the postpaid segment, assuming that the mobile telecom network, the key resource of operators, is also a churn driver in case it under delivers to customers' expectations. Typically, insights generated by churn models are deployed in marketing campaigns; our model's insights are used in network optimization in order to remove the key network related churn drivers and therefore prevent churn, rather than cure it. The insights generated by the model have caused a paradigm shift in managing the network with the operator where the research was conducted.

Keywords: Mobile Network, Churn Prevention, Postpaid, Explanatory Model, Customer Centricity.

1   Introduction

The phenomenon of churn, which denotes loss of a client to competitors, is a key problem across industries. New customers are difficult to find, especially in saturated markets, such as the European mobile communications market. Furthermore, it is far less expensive to retain existing customers than to acquire new ones. Retention is usually a process that identifies customers that are likely to churn, using various predictive modeling techniques, followed by approaching these customers with suitable offers that would persuade the customer into extending the contract. But, can the customer be prevented from even wanting to churn? Can the main churn drivers be mitigated beforehand?

This paper is focused on a company-wide churn reduction initiative conducted in one of the largest European telecom operators. As explained above, churn/customer retention is typically a marketing based process. But, despite of the involvement of predictive analytics, this process is in its nature reactive, because the customer has already decided to churn and an action is being taken to stop this.

In this research we are taking a completely different approach: the model generated here is not to be used for campaigning. Our method attempts to tackle churn by identifying the key reasons why customers decide to churn in order to alleviate them, rather than identify prospective churners. This approach is even more justified taking into

A. Tucker et al. (Eds.): IDA 2013, LNCS 8207, pp. 357–368, 2013.

account the current and future stringent European Data Privacy regulations, which limit operators use of customers' data for campaigning purposes. This is especially the case with Internet usage data.

The mobile telecommunications network is a key resource for telecom operators. It is the means of service delivery as well as the most frequent touch-point with the customers. Problems with ability to use the network (services) have been identified by surveys internal to the company, as well as in literature (section 2), as one of the key reasons to churn. But, most of the time, customers are not experts and cannot pinpoint what exactly is going wrong. Most of the time, this is generalized as "coverage problems". This research is taking a deep dive into various network problems and their relation to customer churn. The main objective here is to identify the problems that customers that have churned were experiencing, so that they can be corrected for the current customer base and reduce their likelihood of churn. In other words, rather than treating symptoms, we are treating the cause of the disease. This research and its outcome have caused a paradigm shift in managing the network with the operator where the research was conducted.

In this research we are focusing on the post-paid customer segment. Even though these customers are bound by contract, which makes the task of churn prediction slightly less challenging, the revenues that are typically generated here are much higher than in the prepaid segment. Furthermore, postpaid customers' service usage is much higher, compared to the prepaid segment; therefore they would be more prone to experiencing network related issues which can potentially lead to churn. The combination of higher usage and revenues makes it easier to justify the network investments needed to remedy their problems.

The rest of the paper is structured as follows. Section 2 describes the related work on telecom churn. Section 3 discusses the dataset and methodology we used. Section 4 contains the results, their application. Limitations and future work are discussed in section 5. Finally, we present our conclusions in section 6.

2   Telecom Churn in Literature

Churn in various industries has been a growing topic of research for the last 15 years [1]. According to [2], churn management consists of predicting which customers are going to churn and evaluating which action is most effective in retaining these customers. Retention strategies are in the focus of [3]. However, most often churn prediction and improving model performance is analyzed following one of these two strategies: adding/improving the data to mine and inventing new algorithms or improving the existing ones [2].

The remark above is certainly valid in the case of telecom churn literature. Many papers are trying to find the best algorithm that would outperform all others: Logistic Regression, Decision Trees, Neural Networks, evolutionary learning, discriminant analysis, Bayesian approaches are examined in [4,5,6,7]; Support Vector Machines, Random Forest, Rotation Forest, Bagging and Boosting are analyzed in [1,8,9,10]. In our view, the value of this research is somewhat limited, at least for real world data mining, given the No Free Lunch theorem [11]. In recent years the overwhelming theme in (telecom) churn research is Social Networks Analysis (SNA), claiming to largely improve on existing churn models [12,13,14,15,16,17,18,19]. However, some of our recent work has demonstrated that this claim is not generally applicable, at least not in prepaid churn prediction on a European market [20]. Most of the SNA research focuses on the Asian or US Markets.

Taking into account the data perspective, most of the literature, especially the one focusing on SNA, is using features extracted from Call Detail Records (CDRs). Contractual, demographic, billing, handset, customer service, market (competitor's offers), and customer survey data is used by [3,4,5,6,7,8] in addition to CDRs. Just a few of these papers take into account any network usage related problems as possible factors affecting churn. For instance, dropped calls are considered in [4,21] as potential churn influencers. Service quality in general and innovativeness is marked as churn detractor by [22].

Predictive models trying to explain churn have not received as much attention in literature [2,23]. Nevertheless, there are studies in industries other than telecom illustrating the need to gain insight into causes of churn [24,25]. Furthermore, research based on customer surveys claims that network coverage, mobile signal strength and voice call drops are reasons for customers to churn [21,22,26,27,28]. However, all these papers are based on survey data, thus perception of quality and not actual network counters.

It is apparent that in most recent telecom churn research the physical telecom network- the means of delivering telecom services, has been largely neglected. At best quality (or the lack of) of voice call usage is considered. To the best of our knowledge, there is little or no research on how Internet usage on a mobile network and its quality parameters might affect churn. This is one of the key reasons why the topic of our research is an explanatory churn model for telecommunications with actual network quality usage parameters as its focus, not just the customers' perception of network quality. In addition, this model, unlike the related work, is not focusing on retention; instead, it is concentrating on eliminating what we see as one of the crucial causes of telecom churn- poor experience using the services on the network.

3   Dataset and Methodology

In this section we will describe the process and the data set used in this research. As mentioned previously, this research was not started with retention campaigns in mind. It was a part of a cross departmental company-wide churn tackling initiative, executed in parallel with regular churn campaigns. Therefore, the objective of this research was not to compete with churn models created for campaigning, but to detect whether there are telecom network quality related factors influencing churn and identify potential remedies.

Table 1. List of contractual, demographic and CDR based features

Contractual and demographic features	Features Extracted from CDRs
Contract expiry List of services/ products used Subscription fee Monthly Bill for each of services Age, gender, zip code Handset	Amount of Voice Calls, SMS and Internet Volume (MB) used, both local and roaming Breakdowns of Voice Calls and SMS onto national-international, internal-external(competitors network)

Table 2. List of network quality features per category

General Network Quality	Voice and SMS quality	Internet quality
2G and 3G Coverage at home Provisioning Errors	Voice Call and SMS Dropped Voice Call Setup Failures Voice Call and SMS drop rate Voice Call Setup Duration (Maximum and Average)	3G and 2G Data Attempts 3G and 2G Data Errors 3G and 2G Success Rate Ratio of 3G usage vs. 2G usage

3.1   Dataset

The results presented here are based on a random sample of 150,000 consumer post-paid subscribers of the operator from September 2012. This is just a fraction of the overall base, the exact percentage is confidential. There was a limitation enforced on the dataset related to contract expiry date: the sample was limited to subscribers whose contracts were expiring in three months or have already expired; thus only customers at risk of churn were taken into account. Churn was measured for the following two months, October and November 2012, combined.

The final dataset consisted of 750 features, gathered by merging tables from CRM and Network databases. In addition to the attributes similar to what was described in section 2 (see Table 1) we added so called Network quality or usability features [29] (see Table 2). The features extracted from CDRs and the network quality features represent monthly aggregates. We also examined their respective three-month aggregates, as well as if there is a rising or declining trend in the past three months for any of these features and use these as potential predictors of churn.

3.2   Methodology

Our research setup is similar to what we have described in [29]. The data originally residing in various CRM and Network quality databases was collected into a single Oracle database, which allowed easier manipulation and data cleansing [30]. For Data analysis, Predictor Selection and Model Development and Assessment we used the commercial tool Pegasystems Predictive Analytics Director [31].

We divided the sample into training, validation and testing set using the ratio 50:25:25. The validation set is used during the data analysis stage as a "pre-test" set, in order to verify the univariate performance of each predicting variable with relation to churn, established on the training set.

The performance measure used to evaluate the performance of each individual predictor, as well as the models, was Coefficient of Concordance (CoC), a rank correlation measure related to Kendall's tau, suitable for evaluating scoring models [31,32]. The CoC (Figure 1) is a measure equivalent to the Area under the ROC (AUC). One interpretation of the CoC measure is that in a scoring model it gives the probability that a randomly chosen positive case will get a higher score than a randomly chosen negative case. The CoC measures the grey area in the graph depicted on Figure 1 and can thus be translated to the Gini coefficient. The CoC value ranges from 50 to 100. The random choice has a CoC value of 50.

All models developed are scoring models, i.e. we calculate probabilities that someone will churn, without setting a cutoff point. As mentioned above, these models are not to be used for campaigning, but for network improvements, therefore setting a cutoff point to strictly classify whether an instance is a churner or not is not necessary. For this reason, using measures such as recall and precision are not applicable in our case.

During the data analysis stage, the continuous variables are discretized into bins. Bins without significant performance difference are then grouped together. Basically, this is a supervised, bottom-up approach to discretization of continuous variables. One of the advantages of this approach is that it can address non-linear effects of variables onto churn: namely, each separate bin gets a score which is concordant to churn and this score is used for modeling. This process is similar for symbolic variables. Variables can be inspected via histograms and the discretization settings can be manually changed if deemed necessary. The next step in the process is predictor grouping which assists feature selection. Namely, variables that are correlated to each other are grouped together. A given predictor may have a high univariate performance, but also be correlated with other candidate predictors that are even stronger, hence not adding value to a model.

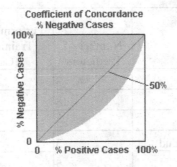

Fig. 1. Coefficient of Concordance

We first used the best predictor of each group and then selected/deselected variables manually to develop the models with a good performance, but also good explanatory value.

As explained previously, the topic of this research is not finding the next best algorithm. That is why we used standard algorithms, such as Logistic Regression and Decision Trees based on the CHAID splitting method [33]. These methods also perfectly fit the explanatory nature of our research, because they are easy to interpret. This is an advantage in commercial settings, where people that need to make investment decisions based on the model and implement its results are not data miners.

The modeling process results in scoring models: each instance is allocated a rank score concordant with the probability of being a churner. The CoC (AUC) measure is used to measure model quality. In addition, we use gain charts as visual representation of model performance. On the y-axis, these charts show the captured proportion of the desired class (i.e. churners in selection divided by total number of churners) with increasing selection sizes (x-axis, from highest scoring to lowest scoring) (see Figure 2).

4   Results, Application and Discussion

Even though optimizing model performance is not the topic of our paper, we deem it necessary to benchmark our network against the campaigning model. The performance (CoC) of the models we created is presented on Table 3.

It is worthwhile mentioning that all models presented here were built using Decision Trees with CHAID splitting criterion, which have an inherent characteristic of dealing with non-linear data. We also tested models using Logistic Regression, but they had somewhat worse performance (0.5 CoC points). Please note that due to the discretization process described in the methodology section, this implementation of logistic regression is able to handle non-linear dependencies too. It is worthwhile mentioning that in order to test for non-linear interaction effects between a combination of two variables and churn we created close to 280,000 new predictors using two way combinations of all of the 750 variables. However, no strong non-linear effects were noted.

Table 3. Model Performance

Model Description	Number of Predictors	Performance on Training set (CoC)	Performance on Test set (CoC)
Campaign	3	76.0	75.9
Campaign_PlusNetwork	6	76.8	76.7
ContractEnd_PlusNetwork	5	75.1	74.7
Campaign_MinusContractEnd	5	68.7	68.1
PurelyNetworkBased	5	66.6	66.5

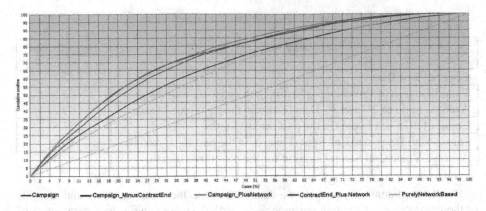

Fig. 2. Gain Charts of Models used

As can be seen on Table 3, adding network related features to a campaigning model (Model Campaign_PlusNetwork) only marginally increases performance (1 CoC point), visible on Figure 2 only after the 40th percentile of cases ranked by churn, which confirms our result from [29]. However, campaigning wise, this has no meaning because rarely do campaigns address more than 40% of the base that is at churn risk.

The PurelyNetworkBased model, which is the topic of our research, has the weakest performance. Nevertheless, just for comparison reasons, we built Campaign models without the strongest predictor - Contract End (Campaign_MinusContractEnd) and a model based on a combination of just the Contract End and Network Factors (ContractEnd_PlusNetwork). The Campaign_MinusContractEnd performs only somewhat better than the Pure Network Model (1.5 CoC on the test set in Table 3, or 5% more churners in the Top 20% of the scores on Figure 2), and the Model ContractEnd_PlusNetwork performs only marginally worse than the campaigning model (1.3 CoC on the test set in Table 3, or 4% less churners in the Top 20% of the scores on Figure 2). The conclusion here is that, less the Contract End variable, the network quality parameters from our Purely Network Based Model perform nearly as well as the other predictors.

However, performance was not the main topic of our research. The main aim was explanatory value of our model. On Figure 2 it is shown that Purely Network Based Model can address the 35% of churners in the top 20% of scores, while the Campaign model addresses nearly 55% of all churners in the top 20% of scores. This may be interpreted as the Network factors being "responsible" for the 35% out of 55% of churners in the Top 20% of all scores and that correcting these parameters would mitigate at least a part of them[1]. The rest of the churn (the other 20%) is due to other reasons, e.g. a better competitor offer. Having this in mind, it was worthwhile analyzing the parameters that constitute this Purely Network Based model.

---

[1] In retention campaigns too, one cannot expect 100% acceptance rate.

Table 4. Univariate performance of predictors (CoC)

Variable	Performance (CoC)
Contract End Date	73.1
Total Duration of Provisioning Errors in the past six months	62.5
Average Ratio of 2G and 3G Data Events in the past three Months	59.2
Count of 2G Data Events in the past three Months	57.5
Sum of Call Drops and Call Setup Failures in the past three months	56.8
Average Voice Call Setup Duration for the past three months	52.4

Due to confidentiality reasons we cannot disclose the exact numbers and weights of the parameters constituting our model. Nevertheless, we can disclose parameters of which our network model is consisted, ranked by their individual performance (CoC): The Total Duration of Provisioning Errors in the past six months; The Average Ratio of 2G and 3G Data Events in the Past three months; The Count of 2G Data Events in the past three Months; The Sum of Call Drops and Call Setup Failures in the past three months and The Average Voice Call Setup Duration for the past three months. The individual influence (CoC) of each of these parameters onto churn is presented on Table 4. Just for comparison, we also show the performance of the best predictor, the contract end date, which has a superior prediction power. However, the purpose of these models is to models is to investigate why customers churn from a network perspective and offer means of alleviating these reasons. In this case, the relationship with contract end date is secondary. When customers get closer to the end of their contract, there is a higher risk of churn. Moreover, customers out of contract for a longer period of time have proven to be loyal, as the other customers have left.

The influence of each of these parameters onto customer experience and therefore churn can be explained and is agreed upon by the company experts. First of all, it is interesting to note that the Sum of Call Drops and Call Setup Failures in the past three months is not a rate, but an absolute count. Namely, it is irrelevant if a customer dropped two calls out of 30 or out of a 100, the two dropped calls drive churn. The parameter Average Voice Call Setup Duration for the past three months implies that customers do not appreciate having to wait a long time to establish a voice call. Provisioning errors are errors where customers have not been enabled to use certain services on the network even though they have subscribed for them (e.g. not being provisioned to use Internet), or do not have the appropriate quality of service (e.g. being provisioned to used Internet at 1 Mbps when subscribed to 3 Mbps). These errors do not occur frequently but are deemed by experts to have a severely negative influence onto satisfaction even if they occur once during the contract duration; therefore we summed up six months of these errors' history. It is interesting to see the growing influence of mobile Internet services onto churn, especially the strong preference of customers to use the 3G network, which is by design much faster than the 2G network[2]. The low 2G speed is not deemed satisfactory, it can be in fact perceived by

---

[2] 3G networks reach speed of 21Mbps, while for 2G the maximum speed is only 64 Kbps.

the customers as not being connected at all. The influence of quality of Internet services onto churn is represented via the Number of 2G Data Events and the Ratio of 3G vs. 2G Data Events.

The added value of these parameters is that they denote clear actions for the technology department on which actions to take in order to prevent churn. In order to develop these actions, we went back to analyzing the predictors mentioned above. Namely, we were looking for thresholds in these parameters that, once crossed, point to higher churn probabilities (e.g. Customers having more than 5 dropped calls in 3 months are 2 times more likely to churn). Projects have been developed to maintain and correct these parameters and their respective critical values (increased churn risk thresholds). This also had a profound effect onto the mindset of the department maintaining the network: the focus has shifted from a network centric approach to a customer centric approach in managing the network. We will explain what this means using the example of Voice Call Drops. The network centric approach in managing this key performance indicator would be to just measure a network wide call drop rate and attempt to maintain it above a certain threshold by giving priority to fixing network cells with a large number of dropped calls. The customer centric approach in managing this parameter is to monitor the number of customers experiencing dropped calls and giving highest priority to network cells where *most customers* experience dropped calls. The customer centric approach allows addressing the problem of a higher number of customers, rather than focusing on network cells where only few customers experience a large number of dropped calls. It has already been implemented and has helped reduce the number of customers experiencing dropped calls in general, which resulted in improved satisfaction in customer surveys (internal to the operator), implicating that churn reduction should follow. Similar approaches are developed to address the other parameters from our model. Also, it is possible that the solution applied to a given network cell to reduce the number of customers experiencing dropped voice calls may also influence some of the other quality parameters, especially in a case of a 3G network cell (e.g. increasing the coverage area or adding extra capacity to a 3G cell might reduce both the number of customers experiencing dropped calls and prevent them from falling back to a neighboring 2G cell when using Internet). As an extension of this approach, it can be envisioned that cells where a high number of customers that are already at churn risk experience dropped calls are given priority, but this is subject to legal limitations with regard to data privacy[3].

To summarize, even though our churn model based entirely on network quality parameters has lesser performance compared to a normal campaigning models, it does have many other advantages: it addressed churn in a preventive manner, as it is not necessary to run retention campaigns with it; it provided guidance on what are the critical network parameters that need to be corrected in order to address churn from a network perspective; and it created a mind-shift in the department managing the network into a customer centric perspective, which already resulted in increased customer satisfaction.

---

[3] It involves storing the cells/locations of particular customers.

5   Limitations and Future Work

The first limitation we would like to address is the lack of coverage data per customer. We were only able to calculate (not measure) the coverage at home for each customer. Loss of coverage for each customer is impossible to measure from the network side. Having adequate coverage information could have improved our model. However, the Ratio between 2G and 3G data events does imply the influence of loss of 3G coverage or insufficient 3G capacity in certain areas onto churn.

Other limitations of this research are of legal nature. Namely, in most European countries stringent Data Privacy Acts or Net Neutrality Laws (will soon) exist. This makes it impossible to look into individual consumption of different types of Internet use (e.g. browsing, streaming, messaging, VoIP etc), which could provide even better insights into what type of service degradation leads to churn.

Next, as usage patterns change, so do the expectations from the service quality that the network provides. Therefore, in time we expect a change in the influence on churn of the various factors that we discussed which makes the model outdated. This will especially be the case after the introduction of 4G (LTE) networks, which allow much faster Internet speed (throughput). However, these issues can be addressed by re-modeling.

As future work, we would like to go one step further, and investigate the benefits network experience measured directly on the phone, via a preinstalled app, of course with customers' permission. We believe that this would provide a 360 degrees view of customers' network experience and close the gap created by the data that is difficult to obtain due to technical or legal limitations. Measurements taken directly on the phone are the ultimate determinant of customer's network experience.

6   Conclusions

In this paper we presented an atypical approach to churn management in commercial settings. We succeeded in explaining at least a part of churn via actual measurements of network quality. The main benefits of our approach are the following: First, we managed to build an explanatory churn model by sacrificing only a part of the performance. Second, our churn model is based on features that are extracted from actual network parameters rather than surveys (real network experience vs. perception). Third, this model generates insights on which network parameters are necessary to be corrected in order to reduce churn, which is a new way of churn reduction. The insights generated caused a shift from network centricity towards customer centricity in managing the telecom network. Using this process, the churn mitigation process is no longer just a retention campaign: the churn efforts are no longer the responsibility of just the CRM teams, Marketing and Customer service, but also the Technology department, managing the network is involved. Finally, our research is deployed and in use in one of the largest European telecom operators and has already contributed to increased customer satisfaction, implicating that churn reduction should follow.

Last but not least, we would like to point out the possibility of applying our research onto domains other than mobile telecom. Obviously, this approach can be mirrored onto fixed telecommunications and potentially into churn in other industries, but also in many other cases where prevention is more important than the cure, like certain medical research.

References

1. Verbeke, W., Martens, D., Mues, C., Baesens, B.: Building comprehensible customer churn prediction models with advanced rule induction techniques. Expert Systems with Applications (2010), doi:10.1016/j.eswa.2010.08.023
2. Ballings, M., Van den Poel, D.: The Relevant Length of Customer Event History for Churn Prediction: How long is long enough? Expert Systems with Applications (2012)
3. Hung, S., Yen, D.C., Wang, H.: Applying data mining to telecom churn management. Expert Systems with Applications 31(3), 515–524 (2006)
4. Mozer, M., Wolniewicz, R., Johnson, E., Kaushansky, H.: Churn reduction in the wireless industry. In: Proceedings of the Neural Information Systems Conference (1999)
5. Wei, C., Chiu, I.: Turning telecommunications call details to churn prediction: A data mining approach. Expert Systems with Applications 23, 103–112 (2002)
6. Au, W., Chan, K., Yao, X.: A novel evolutionary data mining algorithm with applications to churn prediction. IEEE Transactions on Evolutionary Computation 7(6), 532–545 (2003)
7. Neslin, S., Gupta, S., Kamakura, W., Lu, J., Mason, C.: Detection defection: Measuring and understanding the predictive accuracy of customer churn models. Journal of Marketing Research 43(2), 204–211 (2006)
8. Archaux, C., Martin, A., Khenchaf, A.: An SVM based churn detector in prepaid mobile telephony. In: International Conference on Information & Communication Technologies (ICTTA), pp. 19–23 (2004)
9. Lemmens, A., Croux, C.: Bagging and Boosting Classification Trees to Predict Churn. Journal of Marketing Research 43(2), 276–286 (2006)
10. Idris, A., Khan, A., Lee, Y.S.: Intelligent churn prediction in telecom: Employing mRMR feature selection and RotBoost based ensemble classification. Applied Intelligence, 1–14 (2013)
11. Wolpert, D.H., Macready, W.G.: No free lunch theorems for optimization. IEEE Transactions on Evolutionary Computation 1(1), 67–82 (1997)
12. Dasgupta, K., Singh, R., Viswanathan, B., Chakraborty, D., Mukherjea, S., Nanavati, A.A.: Social Ties and their Relevance to Churn in Mobile Telecom Networks. In: Proceedings of the 11th International Conference on Extending Database Technology, pp. 668–677 (2008)
13. Wang, Y., Cong, G., Song, G., Xie, K.: Community-based Greedy Algorithm for Mining Top-K Influential Nodes in Mobile Social Networks. In: Proceedings of the 16th ACM SIGKDD International Conference on Knowledge Discovery and Data Mining, pp. 1039–1048 (2010)
14. Richter, Y., Yom-Tov, E., Slonim, N.: Predicting customer churn in mobile networks through analysis of social groups. In: Proceedings of the SIAM International Conference on Data Mining, pp. 732–741 (2010)
15. Motahari, S., Mengshoel, O.J., Reuther, P., Appala, S., Zoia, L., Shah, J.: The Impact of Social Affinity on Phone Calling Patterns: Categorizing Social Ties from Call Data Records. In: Proc. of the Sixth Workshop on Social Network Mining and Analysis (2012)

16. Nanavati, A.A., Gurumurthy, S., Das, G., Chakraborty, D., Dasgupta, K., Mukherjea, S., Joshi, A.: On the structural properties of massive telecom call graphs: Findings and implications. In: Proceedings of the 15th ACM International Conference on Information and Knowledge Management, pp. 435–444. ACM (2006)

17. Ngonmang, B., Viennet, E., Tchuente, M.: Churn prediction in a real online social network using local community analysis. In: 2012 IEEE/ACM International Conference on Advances in Social Networks Analysis and Mining (ASONAM), pp. 282–288. IEEE (2012)

18. Polepally, A., Mohan, S.: Behavior Analysis of Telecom Data Using Social Networks Analysis. In: Behavior Computing, pp. 291–303. Springer, London (2012)

19. Saravanan, M., Raajaa, G.V.: A Graph-Based Churn Prediction Model for Mobile Telecom Networks. In: Zhou, S., Zhang, S., Karypis, G. (eds.) ADMA 2012. LNCS, vol. 7713, pp. 367–382. Springer, Heidelberg (2012)

20. Kusuma, P.D., Radosavljevik, D., Takes, F.W., van der Putten, P.: Combining Customer Attribute and Social Network Mining for Prepaid Mobile Churn Prediction. In: Proceedings of the 23rd Annual Belgian Dutch Conference on Machine Learning (Benelearn), pp. 50–58 (2013)

21. Ahn, J.H., Han, S.P., Lee, Y.S.: Customer churn analysis: Churn determinants and mediation effects of partial defection in the Korean mobile telecommunications service industry. Telecommunications Policy 30(10), 552–568 (2006)

22. Malhotra, A., Malhotra, C.K.: Exploring switching behavior of US mobile service customers. Journal of Services Marketing 27(1), 13–24 (2013)

23. Lima, E., Mues, C., Baesens, B.: Domain knowledge integration in data mining using decision tables: Case studies in churn prediction. Journal of the Operational Research Society 60, 1096–1106 (2009)

24. Buckinx, W., Van den Poel, D.: Customer base analysis: Partial defection of behaviourally loyal clients in a non-contractual FMCG retail setting. European Journal of Operational Research 164(1), 252–268 (2005)

25. Anil Kumar, D., Ravi, V.: Predicting credit card customer churn in banks using data mining. International Journal of Data Analysis Techniques and Strategies 1(1), 4–28 (2008)

26. Birke, D., Swann, G.P.: Network effects and the choice of mobile phone operator. Journal of Evolutionary Economics 16(1-2), 65–84 (2006)

27. Min, D., Wan, L.: Switching factors of mobile customers in Korea. Journal of Service Science 1(1), 105–120 (2009)

28. Seo, D., Ranganathan, C., Babad, Y.: Two-level model of customer retention in the US mobile telecommunications service market. Telecommunications Policy 32(3), 182–196 (2008)

29. Radosavljevik, D., van der Putten, P., Kyllesbech Larsen, K.: The impact of experimental setup in prepaid churn prediction for mobile telecommunications: what to predict, for whom and does the customer experience matter. Trans. Mach. Learn. Data Min. 3(2), 80–99 (2010)

30. Oracle.: Oracle Database Documentation Library, http://www.oracle.com/pls/db102/homepage

31. Pegasystems: Predictive Analytics Director (Version CDM 6.3) (Software). Pegasystems, Inc. (2008), http://www.pega.com/products/decision-management

32. Kendall, M.: A New Measure of Rank Correlation. Biometrika 30(1-2), 81–89 (1938)

33. Witten, I.H., Frank, E.: Data Mining: Practical Machine Learning Tools and Technique, 2nd edn. Morgan Kaufmann, San Francisco (2005)

Computational Properties
of Fiction Writing and Collaborative Work

Joseph Reddington[1], Fionn Murtagh[2], and Douglas Cowie[1]

[1] Royal Holloway, University of London, UK
[2] De Montfort University, Leicester, UK
{j.reddington,douglas.cowie}@rhul.ac.uk, fmurtagh@acm.org

Abstract. From the earliest days of computing, there have been tools to help shape narrative. Spell-checking, word counts, and readability analysis, give today's novelists tools that Dickens, Austen, and Shakespeare could only have dreamt of. However, such tools have focused on the word, or phrase levels. In the last decade, research focus has shifted to support for collaborative editing of documents. This work considers more sophisticated attempts to visualise the semantics, pace and rhythm within a narrative through data mining. We describe real life applications in two related domains.

Keywords: Visualisation, narrative, HCI, data mining.

1 Introduction

This work considers sophisticated attempts to visualise pace and rhythm within a narrative. The key insight of these techniques is not to replace a qualitative evaluation (the reading of the text) with a quantitative assessment, but, by means of a rigorous deterministic process, to extract relationships from input data and display them for interpretation. In essence, one qualitative evaluation (of the text) is augmented with another (of an image); however, the qualitative evaluation of the image has the advantage that it is not only vastly faster, but also independent of both language and reader familiarity.

Fiction writing is a competitive industry, and supports several sub-sectors in the form of writing classes, manuscript consultants, and networking events. Writers face challenges in getting feedback on their work, particularly in terms of rhythm and pace. Not only is quality subjective, the process is extremely time-consuming for the reader. Moreover, if the writer is to iterate through drafts of their work, then the feedback of any given reader becomes less and less useful as the reader becomes familiar with the text. There are also situational difficulties, such as if the writer simply doesn't accept aspects of the criticism as valid.

A naïve tool might split a narrative into chapters and then plot a chart showing how a measure like the Flesch reading index [10] changed between chapters. Such a chart would have limited general use; however, if a chapter had a significantly different index it would be sensible to conclude that the chapter was considerably

A. Tucker et al. (Eds.): IDA 2013, LNCS 8207, pp. 369–379, 2013.

different in style to the surrounding chapters and that the writer should be aware of this. It is key that the writer certainly shouldn't be expected to change the narrative simply because one chapter is somewhat unusual by some measure. There are many possible reasons for the anomaly, but it is our position that it is to the writer's advantage that they are aware of both the result and the tool, so they can reason about why the result occurred. If the writer has purposely caused the effect to further the narrative, then such a result would be a validation, otherwise, if the writer has accidentally caused the effect then they can consider the worth of the effect and potentially adjust or remove it.

This work uses a framework for narrative analysis proposed in [19] and applies such techniques to two example domains, with a view to evaluating the system to see if it can provide insights of value in literary research. One domain is in the traditional agent/consultant model, whereas the other is a group process, situated much closer to writing for TV or film scripts. Of course, our comparisons are not an adequate or complete way of assessing individual style; they are nonetheless an element that can be employed usefully for our specific purpose.

This paper first details work in related areas and places the techniques examined here in an insightful and innovative context. The following sections describe the operational use domains. Visualisations of the narrative mapping are described. The analysis of these mappings is accompanied by examples and notes on how the use was suited, or not, to particular aspects of each domain.

2   Previous Work

Previous work relating narrative and computer science tends to focus on creation – for example, designing systems that produce emergent narrative [3, 13, 14] or by modelling an existing narrative as a sequence of actions with pre and postconditions [21]. There are also many instances where media outlets have announced computer systems that can pick the next bestselling book, script, or music [23]. The failure of these systems to live up to the hype has led people to be naturally cautious about any analysis system in the creative domains.

The techniques examined in this work were first used in [19, 20] to distinguish the style and structure of film and TV scripts. Murtagh et al. focused on capturing the semantics of the data and the plausibility of taking text as a practical expression of underlying story. This work can be characterised as providing a platform to construct visual representations of semantics encoded in the data.

There is an overlap with the area of *sentiment analysis*, which analyses user-generated content: often by determining if the author of a blog comment or tweet is in favour of, or against, a product. Although visualisations have been constructed this way [4?], such approaches are based, thus far, in examining a small set of sentiment-bearing words, and they consider the source text as a single block, rather than a set of discrete scenes comprising a narrative arc.

3   Methodology and Evaluation Domain

This section presents details of domains of deployment. Later sections will evaluate how different information mapping methods are used to enhance the workflow of each. Our evaluation is based on testimonials and our observations.

Interviews conducted with experts in the publishing industry that made it clear that there was a large degree of resistance to what the industry might see as "replacement by robots". The two mapping through visualisation techniques we evaluate here are of interest because they require a level of interpretation from the user, and so may be much more acceptable to the industry.

The use of these techniques was evaluated in two domains, which were selected to represent the extremes of creative writing. The Writer's Desk is a consultancy offering a very traditional feedback mechanism to authors, whereas Project TooManyCooks models the deadline-driven high intensity creativity found in group writing for TV, film, or magazines.

3.1   Project TooManyCooks

Project TooManyCooks (TMC) (described briefly in [18]) is a creative writing project that runs camps of 8 to 10 student writers who collaboratively create a novel (depending on the age of the students this is normally in the 30,000 to 65,000 word range) over a period of five days. It has two core goals: to increase the contact time and feedback between students interested in fiction writing; and to give students experience of the lifecycle of the novel from inception to printing. Example outputs include [5–8]. In this domain, users were particularly interested in using the analysis techniques to quickly alert them to sections that in some sense didn't follow the overall voice of the rest of the novel. The project was also interested in mapping and visualisation of overall plot arcs: allowing them to reorder sections in such a way that particular scenes do not overshadow each other within the narrative.

3.2   The Writer's Desk

The Writer's Desk (TWD) is a commercial entity specialising in the review of manuscripts for authors [12]. TWD's role is in giving professional feedback to authors over the style and structure of their work. This study spent six months providing narrative analysis for a selection of the submissions they received. The analysis reports were either used internally for developing TWD feedback or passed on to authors as an appendix[1]. TWD and their writers were particularly interested in seeing the chapter-to-chapter flow and, as an extension of this, how an author's work sits as a whole. As a commercial enterprise, TWD was also interested in identifying target markets – and in grooming submissions to hit an area of particular interest to the public more precisely.

---

[1] An example of a report prepared for TWD can be found at
http://www.cs.rhul.ac.uk/home/joseph/hosted/angel.eps

4   Visualisations

This work reports experiences using two mappings to express the narrative arc. Firstly, quite general frequency of occurrence data is determined for word usage in context. Based on all interrelationships between words and text segments, a mapping is obtained that is Euclidean and hence easily visualised as a map-like representation. From that, and aided greatly by the Euclidean map – most often, of full inherent dimensionality and hence not suffering any loss of information – a tree or hierarchical visualisation is obtained. A further innovative development is to have such a hierarchy respect a given ordering of the input text related to narrative development or chronology.

Each input text is automatically divided into a number of segments, with chapter headings being used to delimit segments[2]. Given these segments and a list of unique words in the input text, a cross-tabulation is constructed which gives the count of the occurrences of a given word in a given segment. From a machine-learning perspective our data was semi-structured, in that it is organised into discrete chapters or segments.

One can use correspondence analysis to extract from a cross-tabulation some level of structure from the text in the form of an embedding in Euclidean space. Details of the construction are available in [18, 19]. We refer to the extracted structure as mapping the semantics of the text, because each word is a weighted average of text segments, and each text segment is a weighted average of the words it contains. Both the tree visualisations to be presented use Euclidean space embedding as a starting point. For each visualisation, a description is given with examples and then a detailed analysis is reported on of the advantages and also limitations of the visualisation in each domain.

4.1   Unordered or Geometric

The relationships in the data given by the set of all frequency of occurrence (including $0 =$ no presence) values can be projected into two dimensions to show the relative position of each chapter (text segment) in the projection. Figure 1 shows such a projection from *Owen Noone and the Marauder* [9], with each segment of the text represented by a point on the projection. Since the process used the relative word counts as its starting point, two segments in the novel will appear closer to each other in the projection if they have similar relative word frequencies. It is our position that when an author writes a segment in a distinctly different style or tone (examples might be moving to a different tense or a sudden change in the tension in the storyline) then these word frequencies will change significantly and be visible on the projection for interpretation.

For example, Figure 1 shows a tight core grouping over to the right hand side of the projection, with a number of outliers. Subjectively one might say that this grouping represents the "voice" of the author or novel – and it may be considered

---

[2] If there are no chapters in the text, but sections are divided by some distinct typo-graphical convention, then section boundaries may be used instead.

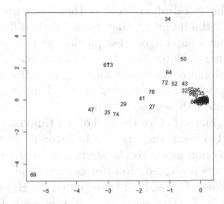

Fig. 1. *Owen Noone and the Marauder* as a geometric, best planar projection, visualisation

worthwhile to investigate the nature of those segments that did not fit in with this voice. If one examines segment 69, which is the most extreme outlier, it can been seen that it is written as a fictional extract from the newspaper *USA Today*, as opposed to the majority of the novel, which is written with a more conventional third-person narrator. The author very much intended to give this segment a different "voice". In this particular work, the majority of the other relative outliers are similar plot devices in the form of radio announcements, magazine articles and so on. Of course, the software makes no judgement here. It simply displays the information for an expert evaluation.

The example of *Owen Noone* is a static study of a published novel after a rigorous proofing and editing process. We shall shortly show how TWD used the visualisation to examine a snapshot of styles to position a novel in the market, while TMC used the visualisation to track the progress of construction over time.

TooManyCooks. One of the core goals within the TooManyCooks process was to give the appearance of having one single author with a clear style and "voice". The group originally relied on the "wikipedia effect" – that is that if enough different authors proofread and rewrite the same section repeatedly, then differences in style become invisible to the causal reader. However the 2-dimensional projection allowed users to visualise the style and see which sections might benefit from a stylistic rewrite.

It is tempting to assume that this "core style" was simply the average of the styles of the writers. In fact, this was the working model used in TwoManyCooks – this visualisation was introduced in the proofreading stages as a way of applying a consistent style across the novel. During the latter two days of a TooManyCooks project, the current draft of the novel is repeatedly printed out, proofread, and has changes made to it (generally on the order of three iterations per day). In early iterations, group coordinators would identify outliers – evaluate each of them to see if the outlier was an intentional outlier and, if not, paired it with another segment that was in some sense opposing the first. The group members who wrote the first drafts of each of these were instructed to copy-edit each other's draft with the intention that the stylistic differences would cancel out. One could imagine a similar process pairing writers and sub-editors on a magazine or a newspaper. In later iterations, this becomes much more a process of identifying unintentional outliers and focusing the stronger writers on those chapters for rewriting, while other writers polished more minor corrections in those chapters that hadn't shown as outliers.

Later work provided more grist for the mill of our thinking. A recent TooMany-Cooks group was selected from students who had won a short story competition. Figure 2 shows a projection in which the short stories are compared with both the novel that the writers produced, and (for context) the popular novels *Harry Potter and the Half Blood Prince* [22], represented by H, and *Pride and Prejudice* [1], represented by P. The short stories, represented by the I symbols, unexpectedly do not surround the novel that the authors later collaborated on (represented by S symbols). This suggests that in fact the core clustering is more a result of the group of writers improving the consistency of their prose with regard to an intended style, rather than being shackled to a literary fingerprint. Note also that the clustering of the TooManyCooks novel is much less tight than either of the two popular authors, which is probably to be expected from a small group of 6th-form students[3] writing over a five day period.

The major use of the unordered visualisation for the TooManyCooks project was in identifying sections of unusual style and being able to evaluate each for its role in the story. Being able to highlight those aspects of the story that did not have the same "voice" as the main narrative allowed the writers to streamline the feedback process and present to readers a more consistent narrative.

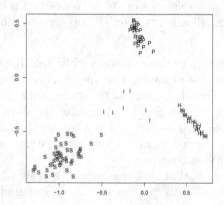

Fig. 2. Comparing author short stories (I) with their collaborative novel (S) and commercial examples (H and P)

---

[3] 16-17 years old.

The Writer's Desk (TWD). An attraction of the projections for TWD was the ability to quickly compare with other artists within the same genre. A regular complaint of publishers and agents is that they are sent manuscripts for genres in which they do not specialise and end up rejecting the vast majority of these out of hand. At the fine-grained level, editors have regularly commented that an author does not necessarily write in the style that they believe they do and, more crucially, they do not necessarily aim at a market segment that they are best suited for. By using the projection visualisation to compare a target manuscript with a selection of commercial novels one can compare explicitly.

For example, TWD had a commission to examine a particular target novel that was aimed at the style of romance novel exemplified by Danellie Steel. Figure 3 shows the the target novel text (T), compared with several other novels. These are: *Kaleidoscope*, by Danellie Steele (S); *Emma*, by Jane Austen (A); and *Eclipse*, by Stephanie Meyer (M) [25, 2, 15]. This allowed TWD to evaluate, to their own satisfaction, if the style and word choice in this instance was closer to the Steele-style romance than either the classic or teen styles of the other examples. Furthermore, the overall consistency of the text is similar to what would be expected from a published novel. There is, of course, a psychological component to some of this feedback. Some authors react viscerally to the idea of this sort of analysis, fearing that the approach reduces creativity, while some react very favourably, having more faith in their own interpretation of the visualisations than they necessarily have in their agents or editors (who they might see as sparing them hard truths).

The ability to highlight anomalous sections was also of great interest within the TWD domain as it provided a useful metric for working one-on-one with authors, and to invite them to interpret the results in relation to their work. This allowed the conversation to be more about the guiding of the author and not about a difference in personal tastes between people.

Feedback from the company was universally positive, particularly in the area of how comfortable they were in interpreting the visualisation for themselves, and in helping with the more commercial aspects of the business.

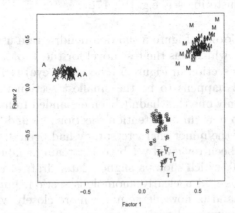

Fig. 3. Comparing a novel (T) with its commercial competitors

4.2 Ordered or Hierarchical

Although the planar presentations are useful, they do not address the fact that the narrative is consumed linearly, and so they reflect only those differences that we are referring to as style or mood between any given successive pair of segments. To gain more insight into the actual structure of the narrative, a visualisation is used that respects the sequentiality of the segments. This section evaluates this hierarchical arrangement of the information, again starting from quite generic text/word association data with relatively minimal pre-processing..

The hierarchical clustering algorithm used here is detailed in [16], and was used as a device to deconstruct the film *Casablanca* in [19]. Briefly, the algorithm repeatedly merges the least dissimilar pair of adjacent scenes to form a tree-like structure that shows how segments of a narrative cluster together.

This sequential ordering allows the viewer to notice how, although a chapter or set of chapters may fit within the overall 'style' of a novel, they may not necessarily match with their immediate neighbours. Once again, there can be outliers, and a human can decide if an outlier is there to intentionally shape the narrative or not.

For example, Figure 4 shows the ordered visualisation of *Harry Potter and the Half-Blood Prince* by J.K. Rowling [22], in which each segment is a chapter in the novel. Viewing the structure, one can see that the cluster comprising only the first chapter is rated as being remarkably dissimilar to the cluster containing all other chapters. The opening chapter of the novel is a conversation between the Prime Minister of the UK, and the Minister for Magic; the chapter is used mainly for setting up the narrative and the mood, and neither character features significantly in the remainder of the text. A subjective reading of the novel may support that the first chapter was separate structurally from the text. Although the comparative deconstructing of such works to a much lower level of detail is a fascinating subject in its own right, it is outside the scope of this work. In particular, our two target domains focus much more heavily on the use of this ordered visualisation for examining novels as works-in-progress. For more information on this clustering see, e.g. [16, 17].

Project TooManyCooks. Figure 5 shows a dendrogram using an early draft of *The Shadow Hours*, which was the test novel for the TooManyCooks Project. The major anomalous section in Figure 5 (chosen by eye) is 44, followed by 26, 27, and 6. Section 44 happens to be the smallest section in the narrative in that draft – and the only one that hadn't been expanded from a skeletal outline into a draft section so it required attention. Sections 26 and 27 were character development of one of the minor characters; they had been drafted by one team member and had not been reviewed yet by other team members.

Examining Figure 6, which shows a slightly later draft of the same novel, in this draft section 44 is still a clearly anomolous section, but by much less of a degree, and 26, 27, and 6, now merge much more closely with the surrounding chapters. This is compared with contemporaneous notes from the project showing that 44 had now been drafted, and the other scenes were going though

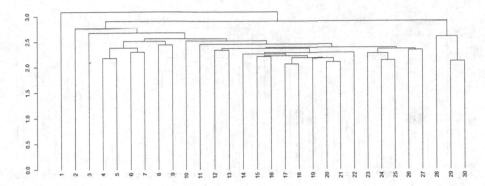

Fig. 4. Ordered visualisation of chapters in *Harry Potter and the Half-Blood Prince*

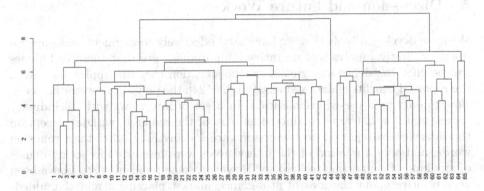

Fig. 5. Chapter structure of *The Shadow Hours*, first snapshot from drafts

second drafts. In this case the dendrogram allowed an "at a glance" notification of areas that required particular attention and revealed that a section had been missed due to a communication error in the team.

The Writer's Desk (TWD). Given the much greater amount of time that staff at TWD had to examine a manuscript, the ability to "immediately evaluate" the structure of a document was less important. Instead the structural diagrams were used to validate, and later guide, the reviewer's own evaluation. During the early stage of the project, staff reviewed documents as normal, and then examined the structural diagrams to see how much their interpretation of the diagram agreed with their interpretation of the text. As trust built, this progressed to reviewing documents before using the diagrams to check that no obviously anomoulous sections were missing, and then to reviewing both the text and the diagram at the same time, allowing the reviewer to re-examine text on the fly and get a much stronger impression of not only where the current section of text is going but how it slots into the overall narrative.

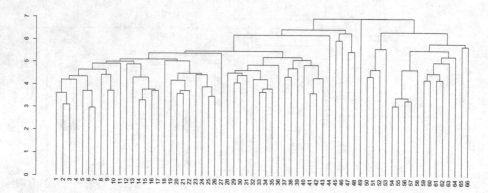

Fig. 6. Structure of *The Shadow Hours*, second snapshot from drafts

5   Discussion and Future Work

We have developed tools that we have used effectively to augment and improve upon qualitative analyses of narrative. Our findings are that these techniques can be effective, depending greatly on the situation they are applied in. Given the reported benefits of data visualisation [26, 24], the publishing sector has been slow to engage in use of visualisation. In a set of interviews with 14 industry representatives that were conducted as part of the research, without exception the interviewees reported no use of software for anything other than counting words, and only a fraction of the interviewees were interested in seeing demos of any kind of supporting technology. However, some publishing staff have been very positive about the idea of at-a-glance market placement and the added-value of being able to check that the section of the book that one has read is typical of the author's voice. Those who have made use of the technology are positive, and provided us with testimonials.

Acknowledgements. We would like to thank especially Adam Ganz for his guiding expertise and long association along with the staff at TWD, in particular Jacqueline Kibby. Thanks are also due to all participants on the TooManyCooks projects, and to the varied expertise provided by David Wells, Tony Greenwood, Adam Roberts, Meg Mitchell, Lucy Yeomans, Emm Johnstone, Patrick Leman, Yvonne Skipper, Peter Dunsmuir, John Vines, and Mark Dorling.

References

1. Austen, J.: Pride and Prejudice (1812)
2. Austen, J.: Emma (1815)
3. Callaway, C.B., Lester, J.C.: Narrative prose generation. Artificial Intelligence 139(2), 213–252 (2002)

4. Chen, C., Ibekwe-sanjuan, F., Sanjuan, E., Weaver, C.: Visual analysis of conflicting opinions. In: IEEE Symposium on Visual Analytics Science and Technology (VAST) (2006)
5. Cooks, T.: Delivery (June 2009), http://amazon.co.uk/o/ASIN/B004PLN008/
6. Cooks, T.: The Deception of Success (December 2011), http://amazon.com/o/ASIN/B006LLEY2M/
7. Cooks, T.: Playing with Controversy (February 2012), http://amazon.com/o/ASIN/B008I2LDG2/
8. Cooks, T.: Roadkill Casserole (February 2012), http://amazon.com/o/ASIN/B0076IJPT6/
9. Cowie, D.: Owen Noone and the Marauder: A Novel, Bloomsbury USA (2005)
10. Flesch, R.: A new readability yardstick. Journal of Applied Psychology 32(3), 221–233 (1948)
11. Gamon, M., Aue, A., Corston-Oliver, S., Ringger, E.: Pulse: Mining customer opinions from free text. In: Famili, A.F., Kok, J.N., Peña, J.M., Siebes, A., Feelders, A. (eds.) IDA 2005. LNCS, vol. 3646, pp. 121–132. Springer, Heidelberg (2005)
12. Kibby, J.: The Writer's Desk (2012), http://www.thewritersdesk.co.uk/ (accessed April 4, 2012)
13. Kriegel, M., Aylett, R.S.: Emergent narrative as a novel framework for massively collaborative authoring. In: Prendinger, H., Lester, J.C., Ishizuka, M. (eds.) IVA 2008. LNCS (LNAI), vol. 5208, pp. 73–80. Springer, Heidelberg (2008)
14. Louchart, S., Swartjes, I., Kriegel, M., Aylett, R.S.: Purposeful authoring for emergent narrative. In: Spierling, U., Szilas, N. (eds.) ICIDS 2008. LNCS, vol. 5334, pp. 273–284. Springer, Heidelberg (2008)
15. Meyer, S.: Eclipse (The Twilight Saga, Book 3) (August 2007)
16. Murtagh, F.: Multidimensional clustering algorithms. Physika Verlag, Würzburg (1985)
17. Murtagh, F.: Correspondence analysis and data coding with Java and R. Chapman & Hall/CRC Press, Boca Raton (2005)
18. Murtagh, F., Ganz, A., Reddington, J.: New methods of analysis of narrative and semantics in support of interactivity. Entertainment Computing 2, 115–121 (2011)
19. Murtagh, F., Ganz, A., McKiep, S.: The structure of narrative: The case of film scripts. Pattern Recognition 42, 302–312 (2009)
20. Murtagh, F., Ganz, A., McKie, S., Mothe, J., Englmeier, K.: Tag clouds for displaying semantics: The case of lmscripts. Information Visualization Journal 9, 253–262 (2010)
21. Porteous, J., Cavazza, M., Charles, F.: Applying planning to interactive storytelling: Narrative control using state constraints. ACM Trans. Intell. Syst. Technol. 1(2) , 10:1–10:21 (2010)
22. Rowling, J.K.: Harry Potter and the Half-Blood Prince (Book 6), 1st edn. Scholastic, Inc. (July 2005)
23. Sheridan, B.: Can computers pick the next big thing? Businessweek (August 12, 2010)
24. Spiegelhalter, D., Pearson, M., Short, I.: Visualizing uncertainty about the future. Science 333(6048), 1393–1400 (2011)
25. Steel, D.: Kaleidoscope. Dell (November 1988)
26. Tufte, E.R.: The Visual Display of Quantitative Information. Graphics Press, Cheshire (1986)

Classifier Evaluation
with Missing Negative Class Labels

Andrew K. Rider, Reid A. Johnson, Darcy A. Davis,
T. Ryan Hoens, and Nitesh V. Chawla

Department of Computer Science and Engineering
University of Notre Dame
Notre Dame IN, 46556, USA

Abstract. The concept of a negative class does not apply to many problems for which classification is increasingly utilized. In this study we investigate the reliability of evaluation metrics when the negative class contains an unknown proportion of mislabeled positive class instances. We examine how evaluation metrics can inform us about potential systematic biases in the data. We provide a motivating case study and a general framework for approaching evaluation when the negative class contains mislabeled positive class instances. We show that the behavior of evaluation metrics is unstable in the presence of uncertainty in class labels and that the stability of evaluation metrics depends on the kind of bias in the data. Finally, we show that the type and amount of bias present in data can have a significant effect on the ranking of evaluation metrics and the degree to which they over- or underestimate the true performance of classifiers.

Keywords: Evaluation, Classification, False Negatives.

1   Introduction

Classification is often applied in cases where only one class is well defined. In the biological domain, scientists can identify protein interactions with high confidence but negative interactions can never be measured. When training classifiers on such data, the classifier is trained on a positive class consisting of truly interacting proteins and a negative class consisting of proteins that have *not been observed interacting*. Thus the negative class exhibits bias, as it may—and often does—consist of many interacting proteins which have been mislabeled as not interacting. Similarly, in the medical field we can be confident that a patient who has been diagnosed with a disease has in fact contracted the disease, while a patient who has not been diagnosed may simply not yet have been tested for that disease. Knowing the reliability of a classifier's prediction in the presence of noise is essential in these fields.

Standard classifiers are often applied to data with a poorly defined negative class [1]. In many cases, there is an implicit assumption that data are mislabeled completely at random. This is common even among algorithms that are designed

A. Tucker et al. (Eds.): IDA 2013, LNCS 8207, pp. 380–391, 2013.

for mislabeled positive class data [2]. This assumption is unrealistic in real world scenarios where there may be multiple sources of different systematic biases in experimentation and data collection. Furthermore, the proportion of true negative class instances to mislabeled positive class instances is often expected to be overwhelmingly large. While this would seem to validate the assumption of completely random data bias, it has not been shown to be a safe assumption for an unknown proportion of mislabeled instances with unknown bias.

We motivate this study through the analysis of real world experiment that is used to try to address some of the most pressing issues in biology today. In performing the study we uncover additional critical questions that must be addressed in order to answer our motivating question, "How reliable are evaluation metrics when the negative class contains an unknown proportion of mislabeled positive class instances?"

2   Case Study

Physical interactions between proteins are one of the primary mechanisms by which a cell carries out its function. While there are high-throughput methods to measure protein-protein interactions (PPI), expense, noisy measurements, and the sheer number of possible interactions in even relatively simple organisms renders complete tests for all interactions infeasible. The identification of interacting proteins based on known interactions and related information is a common classification task in the biological domain [3].

The discovery of unknown protein interactions can have significant impact in pharmaceuticals and biology. With this in mind, we trained naive Bayes classifiers on incremental updates to known protein interactions in yeast. We collected these data from BIOGRID, a curated repository for protein interaction data sets from multiple organisms. These data provide a real world case in which mislabeled class instances (unknown protein interactions) were incrementally revealed to be positive class instances with each update of the system [4]. Features consisted of expression data, Gene Ontology information, and known pathways. Each of these types of data have been used in the classification of protein interactions [5].

Expression data measure the amount of gene product (i.e. RNA) produced by each gene. It is an indirect way to measure the amount of protein produced by a cell. We gathered two features from expression data: one from a line cross experiment, in which two strains of yeast were bred, and one from a compendium of treatments in which yeast were exposed to chemicals and given mutations before measurements were taken [6,7]. We collected a third feature based on the Gene Ontology (GO). The GO is a hierarchy of categories that describe the function, process, and biological components in which genes are involved. This feature was created by counting the number of GO slim terms (a high level set of GO terms) shared between each pair of genes. We used the number of shared pathways between genes as the fourth feature [8]. Pathways describe a series of interactions that lead to a product or change in a cell. Yeast has approximately 6,000 genes, translating to roughly 18 million unique protein interactions. We trained naive Bayes classifiers on this data set for five versions of BIOGRID. There was an average difference of about 20,000 interactions between each version of BIOGRID. Each data set contained 8 million instances with all positive protein interactions from that version of BIOGRID. The remainder of the instances were randomly under-sampled from the remaining potential protein interactions.

In order to evaluate classifier performance, we measured both the area under the receiver operating characteristic curve (AUROC) and area under the precision-recall curve (AUPR) of models trained on data using five versions of BIOGRID. We were interested in how accurate the evaluation metrics were in measuring classifier performance when many of the positive class instances were mislabeled. To this end, we measured the AUROC and AUPR based on the class labels from each given version of the BIOGRID database and the class labels from a more recent version of BIOGRID (version 3.1.85). We call the AUROC and AUPR that are based on the class labels from earlier versions of BIOGRID the "bias class AUROC and bias class AUPR" because of the presence of mislabeled instances. Similarly, we call the AUROC and AUPR that are based on the class labels from the most recent version of BIOGRID the "true class AUROC and true class AUPR" because of the additional positive class instances that are correctly labeled.

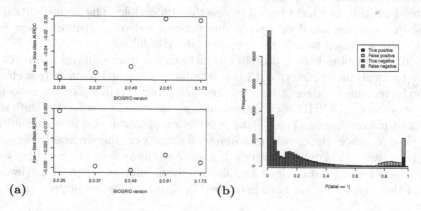

(a)   (b)

Fig. 1. (a) AUROC and AUPR of classifiers trained to predict protein interactions. The x-axis shows the BIOGRID update used to label positive interactions. (b) Histogram of positive class (interacting protein) probabilities based on known interactions from BIOGRID version 2.0.25. For clarity, only the 47 smallest bins are shown.

Figure 1a shows the difference between the true and bias class AUROC and the true and bias class AUPR of classifiers trained on the PPI data sets. Both the bias class AUROC and the bias class AUPR tend to overestimate classifier performance. The fact that the difference between the true class and bias class for both metrics does not reliably improve suggests that additional correctly labeled positive class instances are not giving the classifier enough information about the remaining mislabeled instances. In other words, the decision boundary remains noisy despite the smaller number of mislabeled positive class instances. Figure 1b further supports this assessment.

Figure 1b shows a bar chart of the instances colored according to their class labels. True positives, false positives, true negatives, and false negatives were identified by comparing the predicted class labels from classifiers trained on known interactions from the earliest BIOGRID version to the "true class" labels from the latest version of BIOGRID. For example, true positives are instances that are known to be positive protein interactions in the latest version of BIOGRID that were also predicted as positive class by classifiers trained on known protein interactions from the first version of BIOGRID. The distribution appears multimodal, indicating that there is information within the given features that clearly separates many protein interactions into distinct groups. False negatives appear randomly spread throughout the true negatives. This may indicate that protein interactions classified as false negatives are not related within the features in this data set to the protein interactions identified correctly as interacting.

Figures 1a and 1b demonstrate that our evaluation of the classifier is optimistic and that the addition of correctly labeled proteins does not seem to reliably affect classifier performance. This may indicate that the mislabeled positive class instances are mislabeled completely randomly. However, there is in fact at least one known systematic bias in the data used for this study. The Gene Ontology contains many more annotations for genes that are known to be related to heavily researched topics than for genes related to less interesting biological functions or processes [9]. Is there a latent variable that captures this notion of "interestingness?" Is the absence of a latent variable or the presence of sufficient information within our data suggested by the evaluation? Does the lack of reliable improvement as more mislabeled interactions were corrected suggest that the proportion of mislabeled instances does not affect the classifier, or does the slight improvement in the last BIOGRID version indicate that there is some important threshold? There may be specific answers to these questions for this data set, but we attempt to answer these questions more generally in the following sections.

3 Generalizing the Problem

Many of the questions brought up by the case study concern whether or not the mislabeled positive class instances are mislabeled systematically. In real world problems, we often know that there is bias in the data, but we do not know what kind of bias exists. In biology, there is a bias in the well studied protein interactions that is related to how interesting the protein's function is. As a result, the poorly understood proteins may be poorly characterized in the data, confounding attempts at classification. In such cases we often know *that* instances may be mislabeled, but are unable to ascertain *how* the data are mislabeled. Bias in the data may be systematic or random. Furthermore, it may be expressed as mislabeled instances or missing data. While bias in the data is a commonly studied problem in the literature, the focus has been on learning in biased data sets [10]. It is equally important to study the effect of bias on the performance metrics used to evaluate the performance of learning algorithms.

Generally speaking, data can be missing in three ways, mirroring the missingness mechanisms set out by Allison et al.: missing at random (MAR) when values are missing in a way that is explained within the data; missing not at random (MNAR) when values are missing in a way that could be explained by a latent variable to which a learner does not have access; and missing completely at random (MCAR) when values are missing and there is no variable, latent or observed, that explains the missing values [11]. In this work we consider an analogous problem in which the bias takes the form of mislabeled instances in the data rather than missing instances. We term these cases BAR, BCAR, and BNAR for this type of bias. These three cases may have marked effects on the evaluation of classifier performance.

In a typical supervised learning scenario, classifiers can be trained and ranked by any of a large number of evaluation metrics [12]. This situation is complicated by the presence of bias in the data. Not only can different evaluation metrics give conflicting rankings, but they may react to the presence of different types of bias in different ways. We focus on the AUROC and the AUPR. These measures are commonly used as a single representative number to describe classifier performance. AUROC and AUPR have been studied in the context of class imbalance and in comparison to each other [13]. However, AUROC and AUPR have not been studied in the context of mislabeled bias.

4   Systematic Bias in Class Labels

We consider class labels to be *poorly defined* if the *positive* class contains only correctly labeled instances, whereas the *negative* class contains both correctly labeled and incorrectly labeled instances. While many data sets can be considered poorly defined, the underlying cause can vary greatly between data sets. In particular, depending on how the data are collected, different types of biases may be injected into the mislabeling of instances in the data set (e.g., a positive class instance may not have a completely random chance of being mislabeled). Therefore, in this section we discuss the various types of biases that can be found in real world data sets, and the way in which we simulate each of the types of bias. Note that in each of the bias injection mechanisms, only one class (the positive class) can have its labels flipped.

4.1   Injecting Bias

We modeled each type of bias by injecting it into data sets. This approach may compound existing bias in the data sets, but our assumption is that the data sets are correctly labeled. Completely random bias (BCAR) was injected into data sets by changing the label of positive class instances uniformly at random. We injected random bias (BAR) into data sets by sorting the data by a single feature and flipping the class label of the first X% of the positive class instances. Data sets were made to be biased not at random (BNAR) by sorting the instances by a single feature, flipping the class label of the first X% of the positive class instances, and removing the feature that was used to sort the data.

In order to isolate the effect of correlated features on the bias, we injected bias into data sets based on the most independent feature f as defined in Equation 1.

$$f = \arg\min_i \sum_{j \in X, i \neq j} |corr(X_i, X_j)| \tag{1}$$

This equation minimizes the absolute value of the correlation between each pair of features, where X is the set of feature vectors and $corr(X_i, X_j)$ is the Pearson correlation coefficient computed between features i and j.

5 Experimental Design

It is difficult to separate the behavior of an evaluation metric from specific classifiers. To approach this problem we observe how AUROC and AUPR behave over multiple classifiers trained on the same data sets. To preserve the validity of comparisons, we trained classifiers on the same folds with the same randomly permuted data with precisely the same biased instances.

In order to highlight differences between the two evaluation metrics, we measure both using the true class labels and the flipped class labels. This allows us to measure the AUROC and AUPR under two common scenarios in the practice of data mining: one in which classifiers are trained on data with an unknown bias and one in which classifiers are trained on data with an unknown bias but true class labels are discovered afterwards.

We simulate these two scenarios by measuring the AUROC and AUPR with the flipped class labels (the first scenario) and the true class labels (the second scenario). Classifiers were trained on data with varying levels of bias. We used the probability estimates output by classifiers to rank the instances. We then used the ranking and the biased class labels to calculate the "bias class" AUC and the true class labels to calculate the "true class" AUC. This enables us to measure the effects of bias on the performance measures, and how robust each of the metrics and classifiers are to varying degrees of bias. If the performance on the "true" labels is much worse than that of the performance on the "biased" labels, the classifier metric combination is not effective at ascertaining the true performance of the classifier on the problem. Similarly, if the "true" class performance is much better than the "biased" class performance, then the metric is overly pessimistic, and not suitable for cases where there is noise in the negative class label.

5.1 Evaluation Metrics

ROC curves compare true positive rate and false positive rate while precision-recall curves compare precision to recall (or true positive rate). ROC curves measure the "completeness" of predictions as the amount of false positives increases while precision-recall curves measure the "purity" of predictions as the

Table 1. Data sets used in this study

Name	Features	Feature type	Instances	Name	Features	Feature type	Instances
letter	16	continuous	20000	credit-a	15	mixed	690
ism	6	continuous	11180	crx	15	mixed	690
page	10	continuous	5473	vote	16	discrete	435
estate	12	continuous	5322	vote1	15	discrete	435
krkp	36	discrete	3196	horse-colic	22	mixed	368
hypo	25	mixed	3163	ion	34	continuous	351
SVMguide1	4	continuous	3089	bupa	6	continuous	345
segment	19	continuous	2310	heart-c	12	mixed	303
artificial	8	continuous	2000	threenorm	19	continuous	300
splice	60	continuous	1000	twonorm	20	continuous	300
tic-tac-toe	9	discrete	958	heart-h	13	mixed	294
oil	49	continuous	937	breast-y	9	mixed	286
pima	7	continuous	768	sonar	59	continuous	208
breast-w	9	continuous	699	heart-v	13	mixed	200

amount of captured true positives increases. This difference underlies some of the observed strengths and weaknesses of using the area under these curves.

AUROC can be overly optimistic in cases of imbalanced data while making fewer assumptions about misclassification costs than other metrics such as accuracy [14]. This makes sense in the context of viewing ROC as a measurement of "completeness," as a model may have a low precision but a high recall. AUPR has been used to overcome this concern in highly skewed data sets [15]. It has been shown that AUPR and AUROC can give conflicting rankings for different classifiers trained on the same data [13]. We will demonstrate that this occurs across data sets and at various levels of bias.

5.2 Classifiers

To minimize the likelihood of sampling error, we trained classifiers on 100 random permutations of each data set in Table 1 using 10-fold cross validation. Classifiers included C4.5 decision trees (C4.5), naive Bayes (NB), 5-nearest neighbors (NN), support vector machines (SVM), and multilayer perceptrons (MLP). We used unpruned and uncollapsed C4.5 trees with Laplace smoothing at the leaves. These are common parameters for C4.5 when used in imbalanced problems [16]. Unspecified parameters remained as their default in WEKA [17]. These algorithms were chosen to provide a range of classification approaches. AUROC and AUPR calculations were averaged across folds and permutations of the data.

5.3 Data Sets

We selected 27 real data sets from the UCI repository, and generated one artificial data set [18]. The real data sets were selected to maximize diversity, allowing us to draw conclusions based on a wide range of evidence. These data sets were considered ground truth data, with accurately labeled instances, thereby allowing us to construct the "true" baseline performance. Regardless of the accuracy of this assumption, the availability of the original class labels allows us to calculate performance metrics with both true and biased data. Combined with the injection of different types of bias, this allows us to evaluate the stability of performance metrics. All data sets are listed in Table 1.

6   Results

In order to determine how AUROC and AUPR behave under different levels
and types of bias, we used signed rank tests to evaluate the hypothesis that
the mean rank of a classifier as given by the true class AUC was less than
or equal to the mean rank of the classifier as given by the bias class AUC.
Tied ranks corresponded to data sets. This test was done for each classifier and
with each type and level of bias. Significant values indicate that the bias class
AUC overestimates performance. We also tested the opposite hypothesis, that
the mean rank of a classifier as given by the true class AUC was greater than
or equal to the mean rank of the classifier as given by the bias class AUC.
This corresponds to the bias class AUC underestimating performance. P-values
shown in Tables 2a and 2b reflect tests of the first hypothesis, and numbers in
bold indicate significance at a level of $\alpha = 0.01$ for either test. Values in bold
that are greater than 0.01 indicate that the second hypothesis was rejected.

Most of the significant differences occur in data that are BAR, but some are
present in BNAR data sets. Some differences are consistent between BAR and
BNAR for C4.5, NB, and NN in both Tables 2a and 2b. Comparing the two
tables, we see that the bias class AUROC for C4.5 classifiers tends to overestimate performance, while the bias class AUPR underestimates performance.
NB classifiers show the opposite trend, where the bias class AUROC underestimates performance, while the bias class AUPR overestimates performance. It
is interesting to note this statistically significant difference in light of the fact
that AUROC and AUPR both overestimated classifier performance in the case
study.

7   Case Study Revisited

Now that we have observed how AUROC and AUPR behave with a variety of
classifiers trained on data with different systematic biases and different levels of
bias, we can make better-informed conclusions about where to look for bias and
what type of bias to expect. These observations may guide us to improve the
performance of classifiers on these data.

It is important to note that the ranking of classifiers given by AUROC and
AUPR are different. The fact that both overestimate classifier performance in
the case study indicates that the ranking is neither optimizing completeness nor
precision in the mislabeled positive class instances. Recall that there is a known
bias in the GO feature related to how interesting researchers find particular
genes or functions. Given the behavior of AUROC and AUPR for NB classifiers
in Table 2, if the bias in the data were BAR, we would expect the AUROC and
AUPR to under- and overestimate classifier performance, respectively. However,
both AUROC and AUPR overestimated performance in Figure 1a. This suggests
a few possibilities. First, the data may not be BAR. This is strongly suggested
by the results in Table 2 and by our use of a reduced set of GO terms. Second,
there may be a latent variable, either "interestingness" of particular proteins to

Table 2. True class versus bias class AUC. Signed rank tests compared the rank of classifiers across data sets to determine if the mean rank given by the true class AUC was less than or equal to the mean rank given by the bias class AUC.

(a) True class AUROC versus bias class AUROC. Signed rank tests compared the rank of classifiers across data sets to determine if the mean rank given by the true class AUROC was less than or equal to the mean rank given by the bias class AUROC.

Bias	Classifier	True class AUROC versus bias class AUROC									
		0%	10%	20%	30%	40%	50%	60%	70%	80%	90%
BCAR	C4.5	1.000	0.977	0.386	0.681	0.986	0.682	0.293	0.212	0.074	0.120
	MLP	1.000	1.000	0.681	0.807	0.044	0.386	0.807	0.981	0.978	0.681
	NB	1.000	1.000	0.681	0.977	0.977	0.977	0.825	0.117	0.963	0.979
	NN	1.000	0.977	0.681	0.074	0.579	0.383	0.425	0.579	0.579	0.960
	SVM	1.000	0.173	0.977	0.977	0.579	1.000	1.000	0.500	1.000	0.049
BAR	C4.5	1.000	0.026	**0.003**	**0.000**	**0.000**	**0.000**	**0.001**	**8e-05**	**0.000**	**8e-05**
	MLP	1.000	0.033	0.021	**0.004**	0.028	0.035	0.015	0.559	0.822	0.740
	NB	1.000	0.982	**0.999**	**1.000**	**0.999**	**0.999**	**0.998**	0.991	**0.997**	**0.992**
	NN	1.000	0.932	0.426	0.911	0.986	**0.999**	0.987	**0.999**	0.975	0.719
	SVM	1.000	0.977	0.978	0.991	0.975	0.956	**0.995**	0.954	0.912	0.918
BNAR	C4.5	1.000	0.388	0.579	0.152	0.133	0.196	0.297	0.755	0.951	0.519
	MLP	1.000	0.681	0.330	0.027	**0.003**	**0.009**	0.014	0.087	0.138	0.784
	NB	1.000	0.970	0.936	0.974	**0.998**	**0.998**	**0.997**	0.920	0.836	0.943
	NN	1.000	0.286	0.283	0.548	0.666	0.813	0.500	0.696	0.529	0.529
	SVM	1.000	0.977	0.500	0.579	0.500	0.087	0.500	0.173	0.060	0.153

(b) True class AUPR versus bias class AUPR. Signed rank tests compared the rank of classifiers across data sets to determine if the mean rank given by the true class AUPR was less than or equal to the mean rank given by the bias class AUPR.

Bias	Classifier	True class AUPR versus bias class AUPR									
		0%	10%	20%	30%	40%	50%	60%	70%	80%	90%
BCAR	C4.5	1.000	0.273	0.536	0.500	0.029	0.586	0.623	0.099	0.370	0.777
	MLP	1.000	0.035	0.546	0.304	0.932	0.372	0.793	0.589	0.537	0.682
	NB	1.000	0.133	0.060	0.120	0.286	0.867	0.286	0.536	0.030	0.021
	NN	1.000	0.967	0.669	0.931	0.396	0.010	**0.003**	**0.006**	0.039	0.231
	SVM	1.000	0.931	0.809	0.802	0.870	0.972	0.985	0.990	0.991	0.972
BAR	C4.5	1.000	0.952	**0.996**	**1.000**	**1.000**	**1.000**	**0.998**	**1.000**	**0.998**	**0.995**
	MLP	1.000	0.812	0.992	**0.994**	0.982	0.625	0.749	0.625	0.571	0.401
	NB	1.000	**0.003**	**0.001**	**0.000**	**0.000**	**0.001**	**0.002**	0.054	0.122	0.018
	NN	1.000	0.762	0.606	0.323	0.151	0.025	**0.003**	**0.005**	**0.001**	0.416
	SVM	1.000	0.204	0.627	0.404	0.518	0.580	0.658	0.102	0.292	0.187
BNAR	C4.5	1.000	0.647	0.897	0.792	0.860	0.853	0.329	0.240	0.554	0.918
	MLP	1.000	0.637	0.964	0.810	0.500	0.841	0.970	0.988	0.837	0.935
	NB	1.000	**0.007**	0.015	0.015	**0.006**	**0.004**	0.017	0.040	0.018	**0.008**
	NN	1.000	0.986	0.585	0.156	0.314	0.095	0.076	0.445	0.663	0.750
	SVM	1.000	0.411	0.420	0.981	**0.994**	0.980	0.993	0.963	0.862	0.802

researchers or something else that could provide the classifier vital information to improve the ranking. This is further suggested by the middle mode in Figure 1b. Third, and most likely of all, there may be a combination of systematic biases in the data. Each feature was drawn from data gathered through experiments with their own biases and may combine to create data that seem BCAR. From this analysis, we can conclude first, that the data are not simply BCAR, and second, that the first place to start looking for additional features that explain the mislabeled positive class instances is the middle mode in Figure 1b.

8   Discussion

An understanding of the strengths and limitations of evaluation metrics can allow us to use and interpret them more effectively. Knowing the expected behavior of a performance metric under specific conditions can facilitate the detection of anomalous behavior and help to more accurately measure performance. While the expected behavior of any combination of evaluation metric and classifier does not mean the same behavior will be observed on a specific data set, it can be used to guide investigation and identify potential sources of systematic bias.

The approach taken in this study can be used more generally as a framework to approach the analysis of data with a poorly defined negative class. If researchers have access to a data set with incremental updates as we did in our case study, then the ideas of "true class" and "bias class" can be used to make an educated guess about what kind of bias is being added to the data set. Additionally, the use of multiple evaluation metrics helped to identify anomalous behavior and their agreement in our case study allows us to more confidently assess the usefulness in the ranking of false negatives. Each figure gave us further insight into the data. Namely, how the evaluation metrics were over- or underestimating performance (Figure 1a), how the classifier grouped the data (Figure 1b), and how informative the ranking was about mislabeled positive class instances.

In this work we sought to address the question "How reliable are evaluation metrics when the negative class contains an unknown proportion of mislabeled positive class instances?" We showed that there is much that we can uncover about the nature of bias in the data and the reliability of evaluation. We addressed two key questions in this study. First, "how do AUROC and AUPR behave under varying levels of bias in the data set?" Our experiments show that the trend to over- or underestimate classifier performance (Tables 2a and 2b) is fairly stable across levels of bias. A second question addressed is, "What is the effect of different types of bias in the data on AUROC and AUPR?" Tables 2a and 2b indicate that the type of bias does have an effect on whether the class AUROC and class AUPR tend to under- or overestimate the performance of NB and C4.5 classifiers. Of course, it is difficult to observe the behavior of an evaluation metric outside of the context of classifiers. Indeed, we found that different combinations of classifier and evaluation metric have different behaviors.

One concern that arose while studying how the amount of mislabeled data affects evaluation was that the class imbalance rose with the proportion of mislabeled instances. A data set with evenly balanced classes would end up with a 19:1 class imbalance ratio when 90% of the class labels were flipped. The added effects of the imbalance problem could have a confounding effect on the evaluation metrics. Regardless, because we observed changes in AUROC and AUPR across all proportions of mislabeled instances, we feel that the effect of the class imbalance problem is controlled in our experiments.

This study relied on an idealized scenario in which only one type of bias affected a data set at a time through a single feature. The combinatorial problem of applying each type of bias to each feature was prohibitive both in terms of time as well as complexity of analysis. However, we showed that in many data sets, even if data are mislabeled with respect to the least dependent feature, AUROC and AUPR can over- or underestimate classifier performance.

We focused on AUROC and AUPR, but it is reasonable to expect still more different behaviors from additional evaluation metrics. One future direction might be to investigate the use of combinations of evaluation metrics to overcome individual biases. Perhaps the tendency of AUROC to overestimate performance and the tendency for AUPR to underestimate performance for C4.5 (and the opposite tendencies for NB) can be used together to get a measure that is more robust to mislabeled instances. By exploring these sorts of possibilities, future work may be able to provide principled methods for overcoming the problem of missing negative class labels.

References

1. Pandey, G., Zhang, B., Chang, A.N., Myers, C.L., Zhu, J., Kumar, V., Schadt, E.E.: An integrative multi-network and multi-classifier approach to predict genetic interactions. PLoS Comput. Biol. 6(9), e1000928+ (2010)
2. Elkan, C., Noto, K.: Learning classifiers from only positive and unlabeled data. In: Proceeding of the 14th ACM SIGKDD International Conference on Knowledge Discovery and Data Mining, pp. 213–220. ACM (2008)
3. Qi, Y., Bar-Joseph, Z., Klein-Seetharaman, J.: Evaluation of different biological data and computational classification methods for use in protein interaction prediction. Proteins 63(3), 490–500 (2006)
4. Breitkreutz, B.J., Stark, C., Reguly, T., Boucher, L., Breitkreutz, A., Livstone, M., Oughtred, R., Lackner, D.H., Bähler, J., Wood, V., Dolinski, K., Tyers, M.: The BioGRID Interaction Database: 2008 update. Nucleic Acids Research 36(suppl. 1), D637–D640 (2008)
5. Jansen, R., Yu, H., Greenbaum, D., Kluger, Y., Krogan, N.J., Chung, S., Emili, A., Snyder, M., Greenblatt, J.F., Gerstein, M.: A bayesian networks approach for predicting protein-protein interactions from genomic data. Science 302(5644), 449–453 (2003)
6. Brem, R.B., Kruglyak, L.: The landscape of genetic complexity across 5,700 gene expression traits in yeast. Proceedings of the National Academy of Sciences of the United States of America 102(5), 1572–1577 (2005)
7. Hughes, T.R., Marton, M.J., Jones, A.R., Roberts, C.J., Stoughton, R., Armour, C.D., Bennett, H.A., Coffey, E., Dai, H., He, Y.D., Kidd, M.J., King, A.M., Meyer, M.R., Slade, D., Lum, P.Y., Stepaniants, S.B., Shoemaker, D.D., Gachotte, D., Chakraburtty, K., Simon, J., Bard, M., Friend, S.H.: Functional discovery via a compendium of expression profiles. Cell 102(1), 109–126 (2000)
8. Christie, K.R., Hong, E.L., Cherry, J.M.: Functional annotations for the Saccharomyces cerevisiae genome: the knowns and the known unknowns. Trends in Microbiology 17(7), 286–294 (2009)
9. Myers, C., Barrett, D., Hibbs, M., Huttenhower, C., Troyanskaya, O.: Finding function: evaluation methods for functional genomic data. BMC Genomics 7(1), 187+ (2006)
10. Zhang, S., Zhang, C., Yang, Q.: Data preparation for data mining. Applied Artificial Intelligence 17(5-6), 375–381 (2003)
11. Allison, P.D.: Missing data: Quantitative applications in the social sciences. British Journal of Mathematical and Statistical Psychology 55, 193–196 (2002)

12. Forman, G.: An extensive empirical study of feature selection metrics for text classification. The Journal of Machine Learning Research 3, 1289–1305 (2003)
13. Davis, J., Goadrich, M.: The relationship between Precision-Recall and ROC curves. In: Proceedings of the 23rd International Conference on Machine Learning, ICML 2006, pp. 233–240. ACM, New York (2006)
14. Drummond, C., Holte, R.C.: Explicitly representing expected cost: an alternative to ROC representation. In: Knowledge Discovery and Data Mining, pp. 198–207 (2000)
15. Landgrebe, T.C.W., Paclik, P., Duin, R.P.W., Bradley, A.P.: Precision-recall operating characteristic (P-ROC) curves in imprecise environments. In: 18th International Conference on Pattern Recognition, ICPR 2006, vol. 4, pp. 123–127. IEEE (2006)
16. Cieslak, D.A., Hoens, T.R., Chawla, N.V., Kegelmeyer, W.P.: Hellinger distance decision trees are robust and skew-insensitive. In: Data Mining and Knowledge Discovery, pp. 1–23 (2012)
17. Hall, M., Frank, E., Holmes, G., Pfahringer, B., Reutemann, P., Witten, I.: The WEKA data mining software: an update. Special Interest Group on Knowledge Discovery and Data Mining Explorer Newsletter 11(1), 10–18 (2009)
18. Bache, K., Lichman, M.: UCI machine learning repository (2013)

Dynamic MMHC: A Local Search Algorithm for Dynamic Bayesian Network Structure Learning

Ghada Trabelsi[1,2], Philippe Leray[2], Mounir Ben Ayed[1], and Adel Mohamed Alimi[1]

[1] Research Group on Intelligent Machines,
National School of Engineers (ENIS) of Sfax, University of Sfax, Tunisia
[2] Knowledge and Decision Team, Laboratoire d'Informatique de Nantes
Atlantique (LINA), University of Nantes, France
{ghada.trabelsi,mounir.benayed,adel.alimi}@ieee.org,
philippe.leray@univ-nantes.fr

Abstract. Dynamic Bayesian networks (DBNs) are a class of probabilistic graphical models that has become a standard tool for modeling various stochastic time-varying phenomena. Probabilistic graphical models such as 2-Time slice BN (2T-BNs) are the most used and popular models for DBNs. Because of the complexity induced by adding the temporal dimension, DBN structure learning is a very complex task. Existing algorithms are adaptations of score-based BN structure learning algorithms but are often limited when the number of variables is high. We focus in this paper to DBN structure learning with another family of structure learning algorithms, local search methods, known for its scalability. We propose Dynamic MMHC, an adaptation of the "static" MMHC algorithm. We illustrate the interest of this method with some experimental results.

Keywords: Dynamic Bayesian networks, structure learning, scalability, local search methods.

1 Introduction

Bayesian networks (BNs) are one of the most complete and consistent formalisms for the acquisition and representation of knowledge and for reasoning from incomplete and/or uncertain data. Structure learning of these models from data is an NP-hard problem [1]. Many studies have been conducted on this subject, leading to three different families of approaches: (1) constraint-based methods, (2) score-based methods, and (3) hybrid methods combining the advantages of both previous families. These last methods deal with local structure identification and global model optimization constrained with these local information. These methods are able to scale to distributions with more than thousands of variables.

Dynamic Bayesian networks (DBNs) are a general and flexible model class for representing complex stochastic processes [9] and are used in several areas such as speech recognition, target tracking and identification or genetics. Because of the complexity induced by adding the temporal dimension, DBN structure learning is also a very complex task. Existing algorithms [6,16,17,20,8] are adaptations of score-based BN structure learning algorithms, but are often limited when the number of variables is high.

A. Tucker et al. (Eds.): IDA 2013, LNCS 8207, pp. 392–403, 2013.

Some others more scalable algorithms [5,21,18] have been proposed for a subclass of DBNs.

We focus in this paper to DBN structure learning with local search methods, by adapting the MMHC algorithm proposed by Tsamardinos and al. [14], one of the "state of the art" algorithms of this family. We claim that these local search algorithms can easily take into account the temporal dimension.

Section 2 provides the background of our work with a brief introduction to DBN structure learning and to MMHC algorithm. In section 3, our proposed algorithm *Dynamic MMHC* is explained in three sub-algorithms. We present the related works in section 4. Section 5 describes our experimental results. Finally, section 6 presents conclusions and perspectives.

2 Background

2.1 Dynamic Bayesian Networks

A DBN is a probabilistic graphical model devoted to represent sequential systems [9]. More precisely, a DBN defines the probability distribution over $\mathbf{X}[t]$ where $\mathbf{X} = \{X_1 \ldots X_n\}$ are the n variables observed along discrete time t.

In this work, we consider a special class of DBNs, namely the 2-Time slice BN (2T-BN). A 2T-BN is a DBN which satisfies the Markov property of order 1 $\mathbf{X}[t-1] \perp \mathbf{X}[t+1] \mid \mathbf{X}[t]$. As a consequence, a 2T-BN is described by a pair $(M_0, M_\rightarrow)$.

M_0 (**initial model**) is a BN representing the initial joint distribution of the process $P(\mathbf{X}[t=0])$ and consisting of a direct acyclic graph (DAG) G_0 containing the variables $\mathbf{X}[t=0]$ and a set of conditional distributions $P(X_i[t=0] \mid pa_{G_0}(X_i))$ where $pa_{G_0}(X_i)$ are the parents of variable $X_i[t=0]$ in G_0.

$M_\rightarrow$ (**transition model**) is another BN representing the distribution $P(\mathbf{X}[t+1] \mid \mathbf{X}[t])$ and consisting of a DAG $G_\rightarrow$ containing the variables in $\mathbf{X}[t] \cup \mathbf{X}[t+1]$ and a set of conditional distributions $P(X_i[t+1] \mid pa_{G_\rightarrow}(X_i))$ where $pa_{G_\rightarrow}(X_i)$ are the parents of variable $X_i[t+1]$ in $G_\rightarrow$, parents which can belong to time t or $t+1$.

2.2 Local Search Algorithms

Local search algorithms are hybrid BN structure learning methods dealing with local structure identification and global model optimization constrained with these local information.

Several local structure identifications for static BNs have been proposed, dedicated to discover the candidate Parent-Children (PC) set of a target node algorithm [13,10] or the Markov Blanket (MB) i.e. parents, children and spouses, of the target node [15,11,10]. If the global structure identification is the final goal, Parent-Children identification is sufficient in order to generate a global undirected graph which can be used as a set of constraints in the global model identification. For instance, the recent Max-Min Hill-Climbing algorithm (MMHC) (cf. algorithm 1) proposed by Tsamardinos and al. [14] combines the local identidication provided by Max-Min Parent Children (MMPC) algorithm [13] and a global greedy search (GS) where the neighborhood of a given graph

Algorithm 1. MMHC(D)

Require: Data (D)
Ensure: BN structure (DAG)

1: $G_c \leftarrow \phi, G \leftarrow \phi, S \leftarrow 0$
   % Local identification
2: **for all** $X \in \mathbf{X}$ **do**
3:     CPC_X=MMPC(X, D)
4: **end for**
5: **for all** $X \in \mathbf{X}$ And $Y \in CPC_X$ **do**
6:     $G_c \leftarrow G_c \bigcup (X,Y)$
7: **end for**
   % Greedy search (GS) optimizing score function in DAG space
8: $Test \leftarrow$ True, $S \leftarrow$ Score(G,D)
9: **while** $Test$=True **do**
10:     $N \leftarrow$ Generate_neighborhood(G, G_c)
11:     G_{max}= arg max$_{F \in N}$Score(F,D)
12:     **if** Score(G_{max},D) > S **then**
13:         $G \leftarrow G_{max}$
14:         $S \leftarrow$ Score(G_{max},D)
15:     **else**
16:         $Test \leftarrow$ False
17:     **end if**
18: **end while**
19: return the DAG G found

Algorithm 2. MMPC(T, D)

Require: target variable (T); Data (D)
Ensure: neighborhood of T (CPC)

1: $ListC =\mathbf{X} \setminus \{T\}$
2: $CPC = \overline{\text{MMPC}}(T, D, ListC)$
   % Symmetrical correction
3: **for all** $X \in CPC$ **do**
4:     **if** $T \notin \overline{\text{MMPC}}(X, D, \mathbf{X} \setminus \{X\})$ **then**
5:         $CPC = CPC \setminus \{X\}$
6:     **end if**
7: **end for**

---

is generate with the following operators: add_edge (if the edge belongs to the set of constraints and if the resulting is acyclic DAG), delete_edge and invert_edge (if the resulting is acyclic DAG) (this algorithm is not describe for lack of space).

The MMPC local structure identification, described in Algorithm 2, is decomposed into two tasks, the neighborhood identification itself ($\overline{\text{MMPC}}$), completed by a symmetrical AND correction (X belongs to the neighborhood of T if the opposite is also true). The neighborhood identification ($\overline{\text{MMPC}}$), described in Algorithm 3, uses the Max-Min Heuristic defined in Algorithm 4 in order to iteratively add (forward phase) in the candidate Parent-Children set (neighborhood) of a target variable T the variable the most di-

---

Algorithm 3. $\overline{\text{MMPC}}(T, D, ListC)$

Require: target variable (T); Data (D); List of potential candidates ($ListC$)
Ensure: neighborhood of T (CPC)

1:   $CPC = \emptyset$
    % Phase I: Forward
2: **repeat**
3:     $<F, assocF> = $ MaxMinHeuristic($T, CPC, ListC$)
4:     **if** $assocF \neq 0$ **then**
5:       $CPC = CPC \bigcup \{F\}$
6:       $ListC = ListC \setminus \{F\}$
7:     **end if**
8: **until** CPC has not changed or $assocF = 0$ or $ListC = \emptyset$
    % Phase II: Backward
9: **for all** $X \in CPC$ **do**
10:    **if** $\exists S \subseteq CPC$ and $assoc(X; T|S) = 0$ **then**
11:      $CPC \setminus \{X\}$
12:    **end if**
13: **end for**

Algorithm 4. MaxMinHeuristic($T, CPC, ListC$)

Require: target variable (T); current neighborhood (CPC); List of potential candidates ($ListC$)
Ensure: the candidate the most directly dependent to T given CPC (F) and its association measurement ($AssocF$)

1: $assocF = max_{X \in ListC} Min_{S \subseteq CPC} Assoc(X; T|S)$
2: $F = argmax_{X \in ListC} Min_{S \subseteq CPC} Assoc(X; T|S)$

---

rectly dependent on T conditionally to its current neighborhood (line 1 in algorithm 4). This procedure can potentially add some false positives which are then deleted in the backward phase. Dependency is measured with an association measurement function $Assoc$ like χ^2, mutual information or G^2.

3   Dynamic Max-Min Hill-Climbing

3.1   Principle

Local search methods have been proposed to solve the problem of the structure learning in high dimension for static BN. The dimensionality of the search space also increases for DBN, because of the temporel dimension. We think that these methods could be adapted and give relevant results for 2T-BN models.

In 2T-BN models, temporality is constrained by the first order Markov assumption. We claim that local search algorithms can easily take into account this temporal constraint. We propose this adaptation as a general principle of hybrid structure learning methods (local identification with global search). This paper proposes a new DBN structure learning algorithm inspired from local search methods, by adapting the MMHC algorithm described in the previous section. But an adaptation of other local identification methods is also possible.

Algorithm 5. $\overline{\mathrm{DMMPC}}(T, D)$

Require: target variable (T); Data (D)
Ensure: neighborhood of T in G_0 (Ne_0) and in $G_\rightarrow$ (Ne_+)

    % search Ne_0 of T in $t = 0$
1: $ListC_0 = \mathbf{X}[0]\backslash\{T\} \bigcup \mathbf{X}[1]$
2: $Ne_0 = \overline{\mathrm{MMPC}}(T, D, ListC_0)$
    % search Ne_+ of T in $t > 0$
3: $ListC = \mathbf{X}[t\text{-}1] \bigcup \mathbf{X}[t] \backslash\{T\} \bigcup \mathbf{X}[t+1]$
4: $Ne_+ = \overline{\mathrm{MMPC}}(T, D, ListC)$

Inspired from MMHC algorithm detailed in section 2.2, Dynamic MMHC algorithm proposes to identify independently these graphs by applying a GS algorithm (adapted by [6] for 2T-BN) (cf. Algorithm 3) constrained with local informations. These informations are provided by the identification of the neighborhood Ne_0 (resp. Ne_+) of each node in G_0 (resp. $G_\rightarrow$) (cf. Algorithm 5) .

By mimicking the decomposition procedure of MMHC, our local structure identification $\overline{\mathrm{DMMPC}}$ will be decomposed into two tasks: the neighborhood identification itself ($\overline{\mathrm{DMMPC}}$) completed by a symmetrical correction. We notice here that because of the non-symmetry of temporality, our local structure identification will be able to automatically detect some directed parent or children relationships if the corresponding variables do not belong to the same time slice.

3.2   Neighborhood Identification and Symmetrical Correction

$\overline{\mathrm{DMMPC}}$ algorithm consists of two phases detailed in Algorithm 5 respectively dedicated to the identification of the neighborhood Ne_0 (resp. Ne_+) of a target variable T in G_0 (resp. $G_\rightarrow$).

$\mathbf{X}[0]$ and $\mathbf{X}[1]$ respectively denote the variables $\mathbf{X}$ for $t = 0$ and $t = 1$. We recall that the 2T-BN model is first-order Markov. Hence, it is possible that the neighborhood Ne_0 of a variable T in $\mathbf{X}[0]$ can belong to $\mathbf{X}[0]$ and $\mathbf{X}[1]$.

Let us define CPC_0 the parents or children of T in slice 0 and CC_1 the children of T in slice 1.

In $\overline{\mathrm{DMMPC}}$, we propose using the static MMPC algorithm with the candidate variables $ListC_0 = \mathbf{X}[0] \cup \mathbf{X}[1] \backslash\{T\}$ in order to identify Ne_0. Because of the temporal information, we will then be able later to separate $Ne_0 = CPC_0 \cup CC_1$.

In the same way, $\mathbf{X}[t\text{-}1]$, $\mathbf{X}[t]$ and $\mathbf{X}[t + 1]$ respectively denote the variables $\mathbf{X}$ for times $t - 1, t$ and $t + 1$. Ne_+ of a variable T in $\mathbf{X}[t]$ can belong to $\mathbf{X}[t - 1]$, $\mathbf{X}[t]$ and $\mathbf{X}[t + 1]$.

So let us define CPC_t the parents or children of T in slice t, CC_{t+1} the children of T in slice $t + 1$ and CP_{t-1} the parents of T in slice $t - 1$.

We propose using the static MMPC algorithm with the candidate variables $ListC = \mathbf{X}[t - 1] \cup \mathbf{X}[t] \cup \mathbf{X}[t + 1]\backslash\{T\}$ in order to identify Ne_+. Because of the temporal information, we will then be able later to separate $Ne_+ = CPC_t \cup CC_{t+1} \cup CP_{t-1}$.

Algorithm 6. DMMPC(T, D)

Require: target variable (T); Data (D)
Ensure: neighborhood of T in G_0 (Ne_0) and $G_\rightarrow$ (Ne_+)

1: $Ne_0 = \overline{\text{DMMPC}}(T, D,).Ne_0$ % the set of all neighborhoods of T in G_0 returned by $\overline{\text{DMMPC}}$
2: $Ne_+ = \overline{\text{DMMPC}}(T, D).Ne_+$ % the set of all neighborhoods of T in $G_\rightarrow$ returned by $\overline{\text{DMMPC}}$
   % symmetrical correction Ne_0 of T in $t = 0$
3: $CPC_0 = Ne_0 \cap \mathbf{X}[0], CC_1 = Ne_0 \cap \mathbf{X}[1]$
4: **for all** $X \in CPC_0$ **do**
5:    **if** $T \notin \overline{\text{DMMPC}}(X, D).Ne_0$ **then**
6:       $CPC_0 = CPC_0 \setminus \{X\}$
7:    **end if**
8: **end for**
9: **for all** $X \in CC_1$ **do**
10:    **if** $T \notin \overline{\text{DMMPC}}(X, D).Ne_+$ **then**
11:       $CC_1 = CC_1 \setminus \{X\}$
12:    **end if**
13: **end for**
14: $Ne_0 = CPC_0 \bigcup CC_1$
    % symmetrical correction Ne_+ of T in $t > 0$
15: **for all** $X \in Ne_+$ **do**
16:    **if** $T \notin \overline{\text{DMMPC}}(X, D).Ne_+$ **then**
17:       $Ne_+ = Ne_+ \setminus \{X\}$
18:    **end if**
19: **end for**
20: $CPC = Ne_+ \cap \mathbf{X}[t] \; ; CC = Ne_+ \cap \mathbf{X}[t + 1] \; ; CP = Ne_+ \cap \mathbf{X}[t - 1]$

Algorithm 7. DMMHC(D)

Require: Data D
Ensure: G_0 and $G_\rightarrow$
   % Construction initial model G_0
1: **for all** $X \in \mathbf{X}[0]$ **do**
2:    $CPC_X = \text{DMMPC}(X, D).CPC_0$
3:    $CC_X = \text{DMMPC}(X, D).CC_1$
4: **end for**
   % Greedy search (GS)
5: Only try operator add_edge $Y \rightarrow X$ if $Y \in CPC_X$
   % Construction transition model $G_\rightarrow$
6: **for all** $X \in \mathbf{X}[t]$ **do**
7:    $CPC_X = \text{DMMPC}(X, D).CPC \; ; CC_X = \text{DMMPC}(X, D).CC \; ; CP_X = \text{DMMPC}(X, D).CP$
8: **end for**
   % Greedy search (GS)
9: Only try operator add_edge $Y \rightarrow X$ if $Y \in CPC_X$ and $\{X, Y\} \in \mathbf{X}[t]$
10: Only try operator add_edge $X \rightarrow Y$ if $Y \in CC_X$ and $X \in \mathbf{X}[t]$ and $Y \in \mathbf{X}[t + 1]$
11: Don't try operator reverse_edge $X \rightarrow Y$ if $Y \in CC_X$ and $X \in \mathbf{X}[t]$ and $Y \in \mathbf{X}[t + 1]$

As its static counterpart, DMMPC algorithm described in Algorithm 6 has to perform a symmetrical correction. Because of the non-symmetry of temporality, we have to adapt this correction. When $t = 0$, we have to apply separately the symmetrical correction on CPC_0 and CC_1 because for all $X \in CC_1$, X doesn't belong to slice $t = 0$ but to slice $t = 1$ and its temporal neighborhoods are given by $Ne_+(X)$.

3.3 Global Model Optimization

Our Dynamic MMHC algorithm described in Algorithm 7 proposes to identify independently these graphs by applying a greedy search algorithm constrained with local information. These information are provided by the identification of the neighborhood Ne_0 (resp. Ne_+) of each node in G_0 (resp. $G_\rightarrow$).

As its static counterpart, DMMHC will consider adding an edge during the greedy search, if and only if the starting node is in the neighborhood of the ending node. G_0 learning only concerns the variables in slice $t = 0$, so we can restrict the add_edge operator only to variables found in a CPC_0 of another variable.

$G_\rightarrow$ is a graph with variables in slices $t-1$ and t but this graph only describes temporal dependencies between $t - 1$ and t and "inner" dependencies in t. So we also restrict our operators in order to consider adding edges with these constraints and we don't authorize reversing temporal edges.

3.4 Time Complexity of the Algorithms

This section presents the time complexity of the algorithm. In the static case and according to the work from Tsamardinaos and al. [14] the number of independence tests for all variables with the target conditioned on all subsets of CPC (target parents and children set) is bound by $O(|V|.2^{|CPC|})$, where $|V|$ the number of all variables in the BN and $|CPC|$ is the number of all variables in the parents/children set of target. The overall cost of identifying the skeleton of the BN is $O(|V|^2 \, 2^{|PC|})$, where PC is the largest set of parents and children overall variables in V.

In our dynamic (temporal) case, as for the static, we first identify the number of tests in the $\overline{DMMPC}$. In this part we have two cases to present (i.e. t=0 and t>0). We start with t=0, $\overline{DMMPC}$ will calculate the association of all variables in time slices t=0 and t=1 with target in slice t=0 conditioned on all subsets of Ne_0 (in the worst case). Thus, when t=0, the number of tests is bounded by $O(|2V|.2^{|Ne_0|})$, where V is the set of all variables in a time slice. For t>0 case, the number of tests is bounded by $O(|3V|.2^{|Ne_+|})$, because $\overline{DMMPC}$ calculates the association of all variables in three time slices t, t-1, t+1 with the target in slice t conditioned on all subsets of Ne_+ (in the worst case).

The total number of tests in both phases at $t = 0$ and $t > 0$ is bounded respectively by $O(|2V|.2^{|Ne_0|})$ and $O(|3V|.2^{|Ne_+|})$. Thus, the total number of tests in both phases is bounded by $O(|3V|.2^{|Ne_+|})$. The overall cost of identifying the skeleton of initial and transition models in DBN (i.e., calling DMMPC with all targets in G_0 and $G_\rightarrow$) is $O(|3V|^2.2^{|Ne|})$, where Ne is the largest set of neighborhood over all variables in the time slice t.

4 Related Works

Daly and al. [2] propose a recent and interesting state of the art about BN and DBN. We focus here in 2T-BN structure learning. Friedman and al. [6] have shown that this task can be decomposed in two independent phases: learning the initial graph G_0 as a static BN structure with a static dataset corresponding to $\mathbf{X}[t = 0]$ and learning the transition graph $M_\rightarrow$ with another "static" dataset corresponding to all the transitions $\mathbf{X}[t] \cup \mathbf{X}[t + 1]$. Then they proposed to apply usual score-based algorithms such as greedy search (GS) in order to find both graphs.

Tucker and al. [17] propose an evolutionary programming framework in order to learn the structure of higher order kT-BN. Gao and al. [8] develop another evolutionary approach for 2T-BN structure learning. Wang and al. [20] also look at using evolutionary computation in 2T-BN structure learning by incorporating sampling methods.

All these approaches are validated on benchmark models with about 10 variables. In a more general context, due to inherent limitations of score-based structure learning methods, all these methods will have a very high complexity if the number of variables increases. With the help of local search, DMMHC is able to constraint the search space in the final global optimization (GS). By this way, we can theoritically work in high dimensions like MMHC with static BNs.

Another way to deal with scalability is to restrict the class of 2T-BN by only considering parents of $X_i[t+1]$ in time slice t (for 2T-BN) or any previous time slice (for kT-BN). Dojer [5] proposes a score based method named polynomial time algorithm for learning this class of DBN (with BDe and MDL scores). The implementation of this algorithm is given in [21]. An experimental study have been conducted with microarray data and about 23 000 variables. The time running is about 48 hours with the use of MDL score and 170 hours with the use of BDe score.

Vinh et al. [19] propose another polynomial time algorithm in the same context (equicardinality requirement) with other scoring function (MIT). Vinh et al. [18] propose another score based algorithm without equicardinality assumptions named MIT-global. Also, they propose another contribution in this work consist to an hybrid method with local Blanket identification (MIT-MMMB). An experimental study have been conducted with 1595 variables and show that the local search MIT-MMMB and global-MIT have similar results better than advanced score based algorithm (simulated annealing).

DMMHC and MIT-MMMB are both hybrid methods with local search and global optimization. When DMMHC try to identify the candidate parents/children CPC set of a target variable in a 2T-BN, MIT-MMMB try to identify the (more complex) Markov Blanket MB but in a restricted subclass of 2T-BNs. In one hand, this assumption permits to simplify both MB discovery and global optimization. In other hand, contrary to DMMHC, MMMB is not able to identify the intra-time dependencies.

5 Experimental Study

5.1 Algorithms

We have implemented the Greedy Search (GS) for 2T-BN such as described in section 4. This algorithm is considered as the reference algorithm for DBN structure learning. More complex evolutionary algorithms exist (cf. section 4) but there no available implementation for these specific algorithms.

We have also implemented our proposal, DMMHC, described in section 3. As a first step in our study, our algorithm uses a constrained greedy search, whereas the original

MMHC algorithm proposes using a Tabu search. Extending our approach by using this Tabu search is a simple task and will be one of our immediate perspectives.

We implemented these algorithms in our structure learning platform in C++ using Boost graph[1] and ProBT [2] libraries.

The greedy search used in GS and DMMHC optimizes the BIC score function. DMMPC also uses χ^2 independence test with $\alpha = 0.05$ as *Assoc* function.

Experiments were carried out on a dedicated PC with Intel(R) Core(TM) 2.20 Ghz, 64 bits architecture, 4 Gb RAM memory and under Windows 7.

5.2   Networks and Performance Indicators

Contrary to static BN, evaluating a DBN structure learning algorithm is more difficult. First reason is the unavailability of standard benchmarks, except for instance some reference networks with a small number of variables (less than 10). Second reason is the articles about DBN structure learning use different indicators to argue about the reliability of their proposals.

In [12], we provided tools for benchmarking DBN structure learning algorithms by proposing a 2T-BN generation algorithm, able to generate large and realistic 2T-BNs from existing static BNs. Our companion website[3] proposes some 2T-BNs generated from 6 well-known static BNs (Asia, Alarm, Hailfinder, Win95pts, Andes and Link) with a number of variables in $G_{\rightarrow}$ from 16 to 112 for the 3 first 2T-BNs and from 156 to 1448 for the 3 last ones. For each of these 2T-BNs, we have respectively sampled with Genie/Smile software [4] 2.000, 5.000 and 10.000 sequences of length equal to 6, which correspond to datasets of size 2.000, 5.000 and 10.000 for G_0 structure learning and 5x2.000, 5x5.000 and 5x10.000 for $G_{\rightarrow}$ structure learning. In [12], we also proposed a novel metric for evaluating performance of these structure learning algorithms, by correcting the existing Structural Hamming distance (SHD) in order to take into account temporal background information. As 2T-BNs are defined by two graphs G_0 and $G_{\rightarrow}$, the distance between one theoretical 2T-BN and the learnt one is defined as the pair of the SHD for initial and transition graphs.

Running time is also measured (in seconds). Experiments are canceled when computations did not complete within four days.

5.3   Empirical Results and Interpretations

Figure 1 presents the average results of SHD and running time obtained by GS and DMMHC algorithms with respect to sample size. Figure 1.(a) describes results for initial model corresponding to six (small and large) benchmarks (Asia, Alarm, Hailfinder, Win95pts, Andes, Link). Figure 1.(b) describes results for transition model corresponding to benchmarks used before. We can notice that for every benchmark, DMMHC algorithm obtains better SHD than GS. Also, we can observe than DMMHC overperforms GS running time. This situation is really significant even for benchmarks with a

---

[1] http://www.boost.org/

[2] http://www.probayes.com/index.php

[3] https://sites.google.com/site/dynamicbencmharking/

[4] http://genie.sis.pitt.edu/

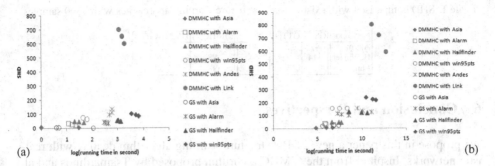

Fig. 1. Average SHD vs running time obtained by GS and DMMHC with respect to sample size. (a) Initial graph (b) Transition graph.

small number of variables. Unlike GS, DMMHC is able to provide results in a decent time for large benchmarks (Andes and Link).

From these results, we can see that DMMHC is an efficient algorithm for 2T-BN structure learning. The quality of the learnt structure is better for our reference algorithm (2T-BN greedy search) with a better scalability: results are obtained in a lower running time and DMMHC can manage high dimensional benchmarks such as 2T-BN generated from Andes and Link.

Comparison with existing works is a difficult task because existing works about 2T-BN structure learning deal with specific benchmarks or specific evaluation metric, such as reported in [12]. As a first comparison, learning the initial model is very similar to static structure learning. As we created our 2T-BN benchmarks from usual static BNs, we can compare the results of our implementations to static ones. [7] provides one comparative study involving GS and MMHC for Alarm, Hailfinder and Link benchmarks with 5000 samples, with BDeu score (and similar results for BIC score). [4] provides SHD results for Alarm and Hailfinder with 5000 samples for another structure learning algorithm, Greedy Thick Thinning (GTT) [3], a two-phases hill-climbing heuristic. Table 1 summarizes the SHD obtained for 3 different hill-climbing algorithms and 2 implementations of MMHC, in about the same contexts (5000 samples). We can observe than our implementation give similar results than concurrent ones for Alarm and Link benchmarks. Some strange results occur for Hailfinder benchmark. [7] reports SHD results equal to 114 (resp. 88) for GS (resp. MMHC) and our implementation obtains 54 (resp. 46) when [4] obtains 48. A deeper study is currently being conducted in order to understand this phenomenon.

It is the first time that existing static benchmarks have been extended for 2T-BN structure learning. It is also the first time that the SHD has been used for 2T-BN, with a temporal correction as described in [12]. For these reasons, comparisons with existing results obtained by concurrent 2T-BN structure learning algorithms are not possible. We intend to disseminate our benchmarks and performance indicators in order to propose a unified evaluation framework for 2T-BN structure learning.

Table 1. SHD comparison with existing static structure learning approaches with 5000 samples

Networks	GS[7]	GTT[4]	GS(G_0)	MMHC[7]	DMMHC(G_0)
Alarm	47	44	40	24	27
Hailfinder	**114**	48	54	**88**	46
Link				687	650

6   Conclusion and Perspectives

We propose in this paper a new 2T-BN structure learning algorithm dealing with realistic networks. Inspired from the MMHC algorithm proposed by Tsamardinos and al. [14] for static BNs, DMMHC algorithm is a local search algorithm dealing with local structure identification (DMMPC) and global model optimization constrained with these local information.

We have shown that the local structure identification can easily take into account the temporal dimension in order to provide some additional information about temporality that can then be used with a greedy search in order to learn the global structure of the initial model G_0 and the transition one $G_\rightarrow$. As far as we know, no method able to learn dynamic network models with such approaches has been proposed previously. We also tested DMMHC in high dimensional domains, with thousands of variables.

Our main immediate perspective concern local improvements of our algorithm. We think that our scalability can be improved during the local structure identification by using temporality constraints in $\overline{\text{DMMPC}}$ in order to decrease the number of *Assoc* calls. We also think that the quality of reconstruction can be improved during the global optimization by using a more evolved meta-heuristic such as a Tabu search. Moreover, we will extend our dynamic approach with the use of others local identification methods such as PCD algorithm [10] provably correct under faithfulness assuption. we have also seen that [18] local approch has good properties in a specific subclass of 2T-BNs. our last perspective is to adapt our algorithm in this context and compare it with this state of art scalable algorithms.

Acknowledgment. The first, third and fourth authors would like to acknowledge the financial support provided by grants from the General Department of Scientific Research and Technological Renovation (DGRST), Tunisia, under the ARUB program 01/UR/11/02.

References

1. Chickering, D., Geiger, D., Heckerman, D.: Learning bayesian networks is NP-hard. Tech. Rep. MSR-TR-94-17, Microsoft Research Technical Report (1994)
2. Daly, R., Shen, Q., Aitken, S.: Learning bayesian networks: approaches and issues. Knowledge Engineering Review 26(2), 99–157 (2011)
3. Dash, D., Druzdzel, M.: A robust independence test for constraint-based learning of causal structure. In: Proceedings of the Nineteenth Conference Annual Conference on Uncertainty in Artificial Intelligence (UAI 2003), pp. 167–174. Morgan Kaufmann (2003)

4. de Jongh, M., Druzdzel, M.: A comparison of structural distance measures for causal bayesian network models. In: Recent Advances in Intelligent Information Systems, Challenging Problems of Science. Computer Science series, pp. 443–456 (2009)
5. Dojer, N.: Learning bayesian networks does not have to be np-hard. In: Královič, R., Urzyczyn, P. (eds.) MFCS 2006. LNCS, vol. 4162, pp. 305–314. Springer, Heidelberg (2006)
6. Friedman, N., Murphy, K., Russell, S.: Learning the structure of dynamic probabilistic networks. In: UAI 1998, pp. 139–147 (1998)
7. Gámez, J.A., Mateo, J.L., Puerta, J.M.: Learning bayesian networks by hill climbing: efficient methods based on progressive restriction of the neighborhood. Data Mining and Knowledge Discovery 22, 106–148 (2011)
8. Gao, S., Xiao, Q., Pan, Q., Li, Q.: Learning dynamic bayesian networks structure based on bayesian optimization algorithm. In: Liu, D., Fei, S., Hou, Z., Zhang, H., Sun, C. (eds.) ISNN 2007, Part II. LNCS, vol. 4492, pp. 424–431. Springer, Heidelberg (2007)
9. Murphy, K.: Dynamic Bayesian Networks: Representation, Inference and Learning. Phd, University of California, Berkeley (2002)
10. Peña, J.M., Björkegren, J., Tegnér, J.: Scalable, efficient and correct learning of markov boundaries under the faithfulness assumption. In: Godo, L. (ed.) ECSQARU 2005. LNCS (LNAI), vol. 3571, pp. 136–147. Springer, Heidelberg (2005)
11. Rodrigues de Morais, S., Aussem, A.: A novel scalable and data efficient feature subset selection algorithm. In: Daelemans, W., Goethals, B., Morik, K. (eds.) ECML PKDD 2008, Part II. LNCS (LNAI), vol. 5212, pp. 298–312. Springer, Heidelberg (2008)
12. Trabelsi, G., Leray, P., Ben Ayeb, M., Alimi, A.: Benchmarking dynamic bayesian network structure learning algorithms. In: ICMSAO 2013, Hammamet, Tunisia, pp. 1–6 (2013)
13. Tsamardinos, I., Aliferis, C., Statnikov, A.: Time and sample efficient discovery of markov blankets and direct causal relations. In: ACM-KDD 2003, vol. 36(4), pp. 673–678 (2003)
14. Tsamardinos, I., Brown, L.E., Constantin, F., Aliferis, C.F.: The max-min hill-climbing bayesian network structure learning algorithm. Mach. Learn. 65(1), 31–78 (2006)
15. Tsamardinos, I., Constantin, F.A., Statnikov, A.: Algorithms for Large Scale Markov Blanket Discovery. In: FLAIRS, pp. 376–380 (2003)
16. Tucker, A., Liu, X.: Extending evolutionary programming methods to the learning of dynamic bayesian networks. In: The Genetic and Evolutionary Computation Conference, pp. 923–929 (1999)
17. Tucker, A., Liu, X., Ogden-Swift, A.: Evolutionary learning of dynamic probabilistic models with large time lags. International Journal of Intelligent Systems 16(5), 621–646 (2001)
18. Vinh, N., Chetty, M., Coppel, R., Wangikar, P.: Local and global algorithms for learning dynamic bayesian networks. In: ICDM 2012, pp. 685–694 (2012)
19. Vinh, N., Chetty, M., Coppel, R., Wangikar, P.: Polynomial time algorithm for learning globally optimal dynamic bayesian network. In: Lu, B.-L., Zhang, L., Kwok, J. (eds.) ICONIP 2011, Part III. LNCS, vol. 7064, pp. 719–729. Springer, Heidelberg (2011), http://dx.doi.org/10.1007/978-3-642-24965-5_81
20. Wang, H., Yu, K., Yao, H.: Learning dynamic bayesian networks using evolutionary mcmc. In: ICCIS 2006, pp. 45–50 (2006)
21. Wilczynski, B., Dojer, N.: Bnfinder: exact and efficient method for learning bayesian networks. Bioinformatics 25(2), 286–287 (2009)

Accurate Visual Features
for Automatic Tag Correction in Videos

Hoang-Tung Tran, Elisa Fromont,
François Jacquenet, and Baptiste Jeudy

Université de Lyon, F-42023 St-Etienne, France,
CNRS UMR 5516, Laboratoire Hubert-Curien,
Université de Saint-Etienne Jean Monnet, F-42023 St-Etienne, France

Abstract. We present a new system for video auto tagging which aims at correcting the tags provided by users for videos uploaded on the Internet. Unlike most existing systems, in our proposal, we do not use the questionable textual information nor any supervised learning system to perform a tag propagation. We propose to compare directly the visual content of the videos described by different sets of features such as Bag-Of-visual-Words or frequent patterns built from them. We then propose an original tag correction strategy based on the frequency of the tags in the visual neighborhood of the videos. Experiments on a Youtube corpus show that our method can effectively improve the existing tags and that frequent patterns are useful to construct accurate visual features.

1 Introduction

As a result of the recent explosion of online multimedia content, it is more and more important to index all forms of web content for various search and retrieval tasks [13]. Classic text-based search engines already offer a good access to multimedia contents in the online world. However, these search engines cannot accurately index the extensive resource of online videos unless these videos are carefully annotated (mostly by hand) while being uploaded on the web. However, user-provided annotations are often incorrect (i.e. irrelevant to the video) and incomplete. The reason for the former is because uploaders might want to rapidly increase the video's number-of-view by tagging it with a popular tag such as "Harry Potter", even though that video has no relationship with this famous book series. Incompleteness means that a given list of correct tags might not be sufficient to describe the video. Because of these two issues, a lot of online videos are hidden to text-based search engines (i.e. to users).

To overcome these drawbacks, we will focus on the task of improving annotations of web video data. Our aim is to set up a system which would be able to handle the two above drawbacks. There have been many efforts to automatically annotate videos (e.g [10], [13]). However, most of the current proposed systems use limited concepts (tags) and some supervised information to learn one or multiple classifiers to tag a video dataset. These approaches seems inappropriate to correct the tags of any video on a large website such as Youtube where the

A. Tucker et al. (Eds.): IDA 2013, LNCS 8207, pp. 404–415, 2013.

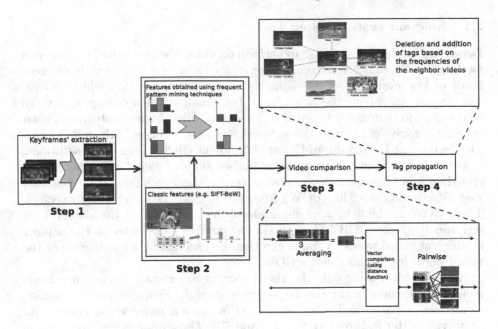

Fig. 1. Video tag propagation process

number of possible tags is infinite and where the ground truth (true labels) is inaccessible a priori. Thus, we would like to propose an unsupervised approach based on the comparison of the visual content of the videos to propagate the tags of the most similar videos based on their textual frequency. In this approach the major scientific issues lie in: i) the choice of the features that are used to make relevant unsupervised comparisons, ii) the comparison method itself, iii) the propagation process and iv) the evaluation of the entire system.

The remainder of this paper is organized as follows. A review of related works concerning the above mentioned problems is given in Section 2. In Section 3, we describe in details how to apply data mining techniques as well as our proposed method to compare videos. Experiments on a real Youtube dataset converted into a huge transactional dataset are presented in Section 4 and show that our system can correct relevant tags. We conclude in Section 5.

2 Related Works on the General Framework

The first step of our system (described in Fig. 1) is to decompose a video into a sequence of keyframes (using for example [24]). In the following, the words frame and keyframe will be used interchangeably. The related works concerning the subsequent steps taken in our tag correction approach are presented below.

2.1   Relevant Features (step 2)

Depending on the task you wish to perform on video, the best suited features can be different. So, the current trend in computer vision is to concatenate different kinds of low level features into a high dimensional vector that will be subsequently used for solving the vision tasks. For example, when dealing with video comparison for automatic tagging, [11] uses edge distribution histograms, color moments or wavelet texture color autocorrelograms. [21] uses both audio and visual features or Histograms of Oriented Gradient (HOG) from [4] as additional features. In [10], frame features include other kinds of global color histograms, and Haar and Gabor wavelets. Another very popular technique is to construct *Bag Of visual Words* [20] (BOW) from the original low-level feature vectors. These BOW are built by applying a clustering algorithm on the low-level descriptors (e.g. color RGB 3-D vectors). The number k of clusters is the number of different visual words. A frame can then be encoded by an histogram of the visual words it contains called a BOW.

However, when using only the visual content to compare videos, the above-mentioned features might not be accurate enough. Frequent pattern mining techniques are more and more often used in the computer vision community to provide better features (see e.g. [22] and [7]). Those approaches often rely on class information to select, in a post-processing step, a compact set of relevant features from the output of the mining algorithms. Without this selection step, this output would not be usable in practice to describe images.

2.2   Video Similarity (step 3)

Even though a video is considered as a sequence of images, variation in the videos duration or in the number of keyframes makes them more difficult to compare. We describe three categories of methods to compare videos. The first one consists in considering the average of the features of the keyframes. For example [21] consider the average of all frame histograms to produce a single histogram for the whole video. The histogram can be thresholded to remove some potential noise. Here classical distance functions such as $L1$ or $L2$ or histogram intersection can be used to estimate the similarity between videos. Even if this method is computationally efficient, one loses a lot of the available information by averaging all the frames. The second approach consists in comparing pairs of keyframes. For example in [11], the authors measure the similarity between two videos as the similarity between the two most similar frames of the videos. The comparison of the two videos is made using a unique pair of frames and no sequential information is taken into account. The last approach makes use of common identical frames called *near duplicate* to compare videos (see e.g. [23]). These frames are visually similar but different in terms of formatting, viewpoint, change in camera parameters, etc. but their common parts can still be used to compute a similarity score. Even though near-duplicate phenomenon appears quite often on video sharing sites, this approach can not be applied to all videos and especially not for the large set of videos found, e.g., on YouTube.

2.3   Tag Propagation (step 4)

The problem of automatic tag corrections of videos has often been tackled in the literature especially during the TRECViD [12] competition. However, it is often treated as a multi-label classification problem [16] or as a tag ranking problem [8,5]. The latter consists in finding a list of the most relevant tags for a new video given information about its neighborhood (this information can be visual as in our case or, for example, social in the context of social networks). Even if they are close to our problem, these two methods assume that the number of tags is fixed and known in advance and that they can have access to a perfectly tagged set of videos to learn a good model for each tag.

Since most video auto tagging systems use a supervised approach, the tag propagation step is not needed. However [23] uses such propagation procedure on which we base ours. For each video v, a list of possible-relevant tags is obtained from the k most similar videos (using a k-nearest neighbor algorithm). After that, a score function is applied for each tag to estimate its relevance according to the video v. This score function depends on the tag frequency (the higher the frequency, the higher the score), the number of tags associated with a video (the higher the number, the smaller the score), and the video similarity (the higher the similarity, the higher the score). Finally, all scores that are larger than a predefined threshold will be considered as suitable tags for the video v. Other tags (with smaller scores) will be deleted if they appear in Video v tag list.

This approach is similar to the collaborative filtering (CF) approach [15] which is a successful method to build recommender systems. In CF, some user's attributes (here, the attributes are tags and the user is a video) are predicted using the information about these particular attributes for similar users. In CF, the similarities between users would typically be computed on the other common tags of the videos (the one that are not being predicted). Our approach is different since neighborhood videos do no necessarily contain similar tags and, in particular, their distribution can be completely different. Besides, contrarily to CF, we do not assume that the tags already present are correct, and we even allow the system to remove some.

3   Our Auto-tagging System

3.1   Proposed Features

As explained in the introduction, we can use many possible features to describe a video and this is a crucial point to work on to get a relevant tag propagation at the end of the process. As our low level features, we propose, as often suggested in the literature to describe images, to use 1,000-dimensional (1,000-D) BOW constructed from SIFT descriptors [9] (128-D descriptors) obtained regularly on a grid. However, as suggested in [7], we propose to extract those BOW locally from each frame to obtain more relevant frequent patterns later on. More precisely, for each point on the grid, we will create a BOW by counting the visual words which corresponds to that point and to its 18 nearest neighbors.

This arbitrary number depends on the resolution of the video but roughly corresponds to a local description of half overlapping windows around each point. Each keyframe is thus described by a large number of BOW (in practice around 250 per frame).

Data Mining Techniques to Find More Discriminative Features. As mentioned in section 2, the data mining techniques used in the literature to obtain better features for video processing (for example APRIORI [1] or LCM [18]) output a huge number of patterns (exponential in the number of dimension of the binary vectors). Those patterns can be filtered out using supervised information as, e.g, shown in [3]. However, in our case, no supervised information is available thus different criteria have to be proposed.

Both KRIMP [19] and SLIM [14] algorithms have been proposed to reduce the number of output patterns without relying on supervised information but by optimizing a criterion based on the Minimum Description Length principle. Both algorithms solve a minimal coding set problem but they differ in the way they choose the collection of candidate patterns. KRIMP follows a straightforward two-phases approach: it first mines a collection of frequent itemsets, then it considers these candidates in static order, accepting a pattern if it improves a compression criterion. However, mining candidates is expensive and by considering candidates only once, and in a fixed order, KRIMP sometimes rejects candidates that could have been used latter on. SLIM greedily constructs pattern sets in a bottom-up fashion, iteratively joining co-occurring patterns such that compression is maximized. It employs a simple yet accurate heuristic to estimate the gain or cost of adding a candidate. For this reason, SLIM is faster and can handle larger datasets than KRIMP. In conclusion, SLIM seems a good candidate algorithm to filter out our patterns.

Converting Features into Binary Form. Most frequent pattern mining techniques use binary or transactional data. Therefore, the BOW must be converted into binary vectors. The most simple (and classical) method to do so is to transform all non-zero values into one. A lot of information is lost during this conversion if the original histogram is dense with many different values for each bin of the histogram. However, in our case, the 1,000-D histogram contains at most 19 non null values (corresponding to the 18 neighbors of the current visual word + itself) and this is also the maximum value for a bin. This simple procedure thus seems appropriate to avoid unnecessary large binary vectors.

Encoding videos with BOW and FP. If F is the set of frequent patterns obtained using SLIM, we build a binary vector V of size $|F|$ for each keyframe. In this vector, $V(i)$ is set to 1 if the ith pattern of F appears in this keyframe and 0 otherwise.

In our experiments, we encode our videos using BOW vectors, frequent pattern vectors (FP) built from them or with both of them (BOW+FP). For this last case, we need to normalize the feature vectors since both types of features have different distributions. Let N_{BOW}, σ^2_{BOW} be the number and the variance of the values in the BOW sub vector; and N_{FP}, σ_{FP}^2 the number and variance of the values in the FP sub vector. We modified all the values of the FP vector as follows:

$$FP_{new}[i] = FP[i] * \frac{\sigma_{BOW}}{\sigma_{FP}} * \frac{N_{BOW}}{N_{FP}}$$

The new BOW+FP feature vector is the concatenation of BOW and FP_{new}.

3.2 Proposed Asymmetrical Video Similarity Measure

We propose an asymmetrical similarity measure inspired by the video pairwise comparison techniques to increase the relevance of the video comparison. The first step consists in calculating all the pairwise similarities between all the keyframes of two videos. After that, instead of taking the optimum value of all the pairwise similarity scores (as in [23]), we propose to take the average of all maximum similarities corresponding to one video. In other words, for each keyframe of a video A, we search in all the keyframes of video B for the highest pairwise matching score and we keep this value. Then, we take the average of all the computed values for all the keyframes of the video A to return the similarity score of video A towards video B. If we denote $A(i)$ the ith keyframe of A and $|A|$ the number of keyframes in A, then

$$sim(A, B) = \frac{1}{|A|} \sum_i \max_j sim(A(i), B(j)).$$

The similarity $sim(A(i), B(j))$ between frames is just the inverse of a distance between the vectors representing the frames (in the experimental section, we use the histogram intersection [17]). When the two frames are identical, this asymmetrical similarity is set to a maximal value.

3.3 Proposed Tag Propagation Algorithm

As explained in Section 2, to tag a given video $v \in V$, we rely on the tags $t \in T$ of the k most similar videos in its neighborhood. To propagate a given tag t to v, one need to set a threshold on the number of times t should appear in the neighbors. However, given the very different distribution of each tag, we decided to use two comparison statistical tests between the distribution of a tag in the entire dataset and its distribution in the k nearest neighbors. The first one states that the probability of a given tag should be significantly greater than 0 in the entire dataset to be propagated and the second one states that it should be significantly more present in the neighbors than in the entire dataset. Formally:

– (Global scale) A tag can be propagated if:

$$\hat{p} \geq u_{\frac{\alpha}{2}} \sqrt{\frac{\hat{p}(1 - \hat{p})}{N}}$$

where $\hat{p}$ is the proportion of a tag over the whole dataset, N is the total number of videos, $u_{\alpha/2}$ is $\frac{1-\alpha}{2}$ percentile of a standard normal distribution.

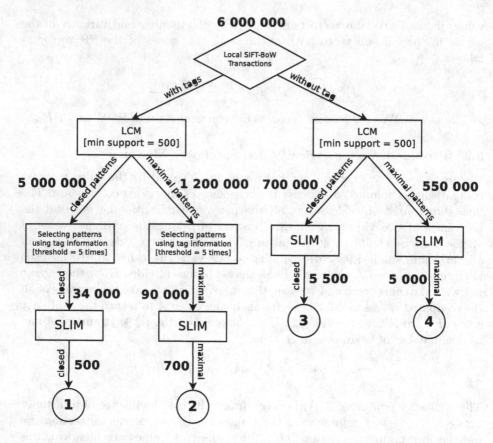

Fig. 2. Different strategies to obtain an accurate set of frequent patterns

- (Local scale) A tag is propagated if:

$$\hat{p}_1 \geq \hat{p} + u_{\frac{\alpha}{2}} \sqrt{\frac{\hat{p}_1(1 - \hat{p}_1)}{k} + \frac{\hat{p}(1 - \hat{p})}{N}}$$

where $\hat{p}_1$ is the proportion of a tag in the k neighbors.

We arbitrarily decide to remove a tag from a video if it is never present in its neighbors. Note that the central limit theorem applies whenever $k \geq 30$.

4   Experiments

4.1   Protocol

Dataset. We pre-processed a Youtube dataset [2] with more than 10, 000 videos already decomposed into shots and keyframes. There are about 18 shots per video and 1.5 keyframes per shot, i.e. about 27 keyframes for one video. We first decided to focus on videos with some common tags to obtain an interesting sample of the original dataset. For that, we focused on the 500 most frequent tags in the original set. In practice, the tail of the tag distribution will not be modified by our method. We then removed the articles, pronouns, prepositions, words with less than two letters and the 50 most common tags in the remainder list (those, such as the word "video", were considered too frequent to be informative). This gave us a list of 150 authorized tags. We then kept the videos that contains at least 5 of those 150 tags and more than 1 keyframe. That led us to consider a corpus of 668 videos. Note that this is smaller than the 10, 000 initial videos but enough to illustrate our method. From this set of videos and from the local SIFT-BOW feature vectors computed from the keyframes we created a binary dataset of about 6, 000, 000 1,000-D transactions.

Evaluation of the Results. We randomly chose 50 videos from the 688 videos, and tagged them by hand using the 150 authorized tags to obtain a ground truth. We ran our tag propagation method on the 688 videos and reported the accuracy results for these particular 50 videos. This accuracy is measured in terms of "percentage of good corrections" (PGC). Let $T_{add,correct}$ be the correctly added tags out of $T_{add,total}$ added tags and $T_{remove,correct}$ be the correctly removed tags out of $T_{remove,total}$ removed tags.

$$PGC = \frac{T_{add,correct} + T_{remove,correct}}{T_{add,total} + T_{remove,total}}$$

If PGC is larger than 0.5, our system improves the tags in the videos. Note that since most existing tag correction systems use some supervised information, we do not compare our system to them. The following experiments stand for a proof of concept of our system.

Frequent Patterns Mining. As explained in Sect. 3, we first decided to use SLIM on the original transactional dataset. However, SLIM did not provide any fixed results after more than a week running. We then decided to use the well known and fast LCM algorithm [18] to mine closed or maximal patterns first and use SLIM as a post-processing step to select a less redundant number of patterns. The different strategies are shown in Figure 2.

Without tag information. We use LCM to find closed and maximal frequent patterns from the dataset made up of the 6, 000, 000 original 1,000-D transactions. The support threshold was set as low as possible which corresponded for us to 500 (less than 1%). LCM produces 700, 000 closed and 550, 000 maximal patterns as shown in Fig. 2 (outputs 3 and 4). We then used SLIM to select a smaller set of around 5, 000 non redundant patterns (final output of SLIM).

With tag information. For this experiment, we concatenate all these 1,000-D vectors to 150-D vectors which describe for each transaction belonging to a

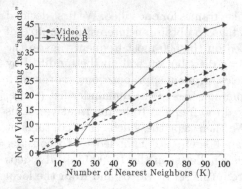
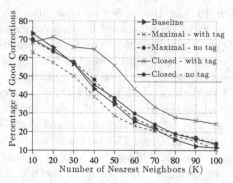

Fig. 3. Number of nearest neighbor videos that contain (plain line) or should contain to trigger the propagation step (dashed line) the tag "amanda" according to the number of nearest neighbors for 2 different videos.

Fig. 4. of good corrections according to the number of neighbors in 50 videos represented with 1) SIFT-BOW (baseline) and frequent patterns obtained with 2-3) LCM (closed and max) with a post-processing using SLIM and 4-5) LCM (closed and max) with SLIM and some tag information.

frame of a given video, the list of tags that were associated with this video. Each of the $6,000,000$ transactions is thus described by a sparse 1,150-D binary vector. As shown in Fig. 2 (outputs 1 and 2), LCM is first used to produce around $5,000,000$ closed and $1,200,000$ maximal frequent patterns in a couple of hours. In this experiment, we make use of the existing tag information to filter out relevant patterns. A pattern is considered relevant for a certain tag if it is five times more frequent in videos that contain that tag than in videos that do not contain it. It is kept if it is relevant for at least one tag (note that this is similar to the concept of emerging patterns [6]). After going through this filtering process, the number of pattern is reduced to about $90,000$ for the maximal and $34,000$ for the closed. Then, SLIM is used to produce (also in a couple of hours) around 600 non redundant patterns.

4.2   Tag Propagation Results

Test and Neighbors. As explained in Sect. 3, the number of neighbors considered for the tag propagation and the statistical tests are directly responsible for the propagation (or the removal) of a tag. To evaluate our choices experimentally, we selected a frequent tag ("amanda") in our video and 2 videos that should not be tagged with this particular tag. Fig. 3 shows the number of nearest neighbor videos that contain the tag "amanda" for the 2 different videos (plain lines). It also shows how many nearest neighbor videos should contain the tag "amanda" to trigger the propagation step (dashed line). Since the tag should not be propagated to these videos, the plain lines should stay below their corresponding dashed line. This is correct for 30 nearest neighbors. It is not correct anymore

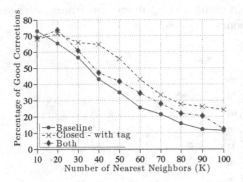 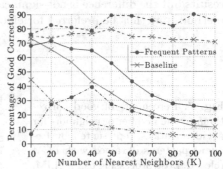

Fig. 5. Percentage of good corrections according to the number of neighbors for a video dataset represented with 3 different features: 1) SIFT-BOW (baseline), 2) frequent patterns obtained with LCM (closed) with a post-processing using SLIM and some tag information, and 3) the concatenation of both feature vectors.

Fig. 6. Percentage of good corrections when counting only the deletions (dashed lines), the additions (dashed-dotted lines) and both (solid lines) according to the number of neighbors for propagation results obtained with 1) SIFT-BOW (baseline) and 2) frequent patterns obtained with LCM (closed), a post-processing step with SLIM and some tag information.

when increasing the number of neighbors. This means that for this particular tag, increasing the number of neighbors will actually degrade the propagating system. Other similar experiments tend to confirm the relevance of choosing 30 neighbors.

Figure 4 shows the results for the 5 experimental settings described above. Note that, as explained in Sect. 3, using less than 30 neighbors questions our statistical tests. On the contrary, using 100 videos out of 668 clearly introduces a lot of noise. The best results are obtained using the combination of LCM (closed), SLIM and the tag information (output 1 of Fig. 2). In this case for 30 neighbors, the system is able to produce around 65% of good corrections which means that 54 correct tags were added or deleted in the process (28 were wrong propagations). For the same setting, the baseline only allows us to produce 57% good corrections. Figure 5 and 6 focus on the baseline and the best case. It also shows the results when both vectors are concatenated. The concatenation produces worse results than the best case which means that the information given by the baseline is not complementary to the information given by the frequent patterns.

Figure 7 shows a comparison between our proposed asymmetrical similarity measure and the basic method which consists in simply averaging for one video all the keyframe features and computing an histogram intersection between the feature vectors representing two videos. Our proposed method does give better results but the difference is not significant using 30 neighbors. The frame average method may thus be preferred for efficiency reasons.

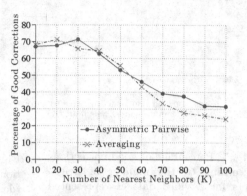

Fig. 7. Percentage of good corrections to the number of neighbors for a video dataset represented by frequent patterns obtained with LCM (closed, post-processing using SLIM, tag information), using the asymmetrical similarity measure and simply averaging the keyframes features and computing an histogram intersection.

5   Conclusion

We have presented a complete tag correction system which corrects and completes original tags on videos without learning any model. We have proposed a new high level video feature vector to describe our videos based on frequent patterns and decided to compare the videos directly using an histogram intersection distance function. We have evaluated our method on a real Youtube dataset and shown that our system can effectively be used to correct tags.

However, the new proposed feature vector and the pairwise video comparison procedure do not always make a significant improvement compared to naive methods which use simple BOW features (from SIFT) and averaging over the videos. As future work, we thus propose to take into account the sequential information in the video to create better high level features (such as frequent sub sequences). Besides, our proposed method should clearly be used off-line since the mining part takes a substantial amount of time. Efficient algorithms (for example able to deal with streams) should also be designed to tackle a Youtube-scale dataset.

References

1. Agrawal, R., Srikant, R.: Fast algorithms for mining association rules in large databases. In: 20th Int. Conf. on Very Large Data Bases, pp. 478–499 (1994)
2. Cao, J., Zhang, Y., Song, Y., Chen, Z., Zhang, X., Li, J.: MCG-WEBV: A benchmark dataset for web video analysis. Tech. rep., ICT-MCG-09-001, Institute of Computing Technology (2009)
3. Cheng, H., Yan, X., Han, J., Yu, P.S.: Direct discriminative pattern mining for effective classification. In: 24th Int. Conf. on Data Engineering, pp. 169–178 (2008)

4. Dalal, N., Triggs, B.: Histograms of oriented gradients for human detection. In: Conf. on Computer Vision and Pattern Recognition, vol. 1, pp. 886–893 (2005)
5. Denoyer, L., Gallinari, P.: A ranking based model for automatic image annotation in a social network. In: 4th Int. Conf. on Weblogs Social Media, pp. 231–234 (2010)
6. Dong, G., Li, J.: Efficient mining of emerging patterns: discovering trends and differences. In: Int. Conf. on Knowledge Disc. and Data Mining, pp. 43–52 (1999)
7. Fernando, B., Fromont, E., Tuytelaars, T.: Effective use of frequent itemset mining for image classification. In: Fitzgibbon, A., Lazebnik, S., Perona, P., Sato, Y., Schmid, C. (eds.) ECCV 2012, Part I. LNCS, vol. 7572, pp. 214–227. Springer, Heidelberg (2012)
8. Liu, D., Hua, X., Yang, L., Wang, M., Zhang, H.: Tag ranking. In: 18th Int. Conf. on World Wide Web, pp. 351–360. ACM (2009)
9. Lowe, D.G.: Distinctive image features from scale-invariant keypoints. Int. Journal of Computer Vision 60(2), 91–110 (2004)
10. Morsillo, N., Mann, G., Pal, C.: YouTube scale, large vocabulary video annotation. In: Schonfeld, D., Shan, C., Tao, D., Wang, L. (eds.) Video Search and Mining. SCI, vol. 287, pp. 357–386. Springer, Heidelberg (2010)
11. Moxley, E., Mei, T., Manjunath, B.: Video annotation through search and graph reinforcement mining. IEEE Transactions on Multimedia 12(3), 184–193 (2010)
12. Over, P., Awad, G., Michel, M., Fiscus, J., Sanders, G., Shaw, B., Kraaij, W., Smeaton, A.F., Quénot, G.: An overview of the goals, tasks, data, evaluation mechanisms and metrics. In: TRECVID 2012. NIST, USA (2012)
13. Shen, J., Wang, M., Yan, S., Hua, X.S.: Multimedia tagging: past, present and future. In: 19th ACM Int. Conf. on Multimedia, pp. 639–640 (2011)
14. Smets, K., Vreeken, J.: SLIM: Directly mining descriptive patterns. In: SIAM Int. Conf. on Data Mining, pp. 236–247 (2012)
15. Su, X., Khoshgoftaar, T.M.: A survey of collaborative filtering techniques. In: Advances in Artificial Intelligence 2009, vol. 4 (2009)
16. Sun, Y.Y., Zhang, Y., Zhou, Z.H.: Multi-label learning with weak label. In: 24th AAAI Conf. on Artificial Intelligence, pp. 593–598 (2010)
17. Swain, M.J., Ballard, D.H.: Color indexing. Int. J. Comp. Vision 7(1), 11–32 (1991)
18. Uno, T., Kiyomi, M., Arimura, H.: LCM ver. 3: Collaboration of array, bitmap and prefix tree for frequent itemset mining. In: 1st Int. Workshop on Open Source Data Mining: Frequent Pattern Mining Implementations, pp. 77–86. ACM (2005)
19. Vreeken, J., van Leeuwen, M., Siebes, A.: KRIMP: mining itemsets that compress. In: Data Mining and Knowledge Discovery, pp. 1–46 (2011)
20. Yang, J., Jiang, Y., Hauptmann, A., Ngo, C.: Evaluating bag-of-visual-words representations in scene classification. In: Int. Workshop on Multimedia Information Retrieval, pp. 197–206. ACM (2007)
21. Yang, W., Toderici, G.: Discriminative tag learning on youtube videos with latent sub-tags. In: Computer Vision and Pattern Recognition, pp. 3217–3224 (2011)
22. Yuan, J., Yang, M., Wu, Y.: Mining discriminative co-occurrence patterns for visual recognition. In: Conf. on Comp. Vision and Pat. Recognition, pp. 2777–2784 (2011)
23. Zhao, W., Wu, X., Ngo, C.: On the annotation of web videos by efficient near-duplicate search. IEEE Transactions on Multimedia 12(5), 448–461 (2010)
24. Zhuang, Y., Rui, Y., Huang, T., Mehrotra, S.: Adaptive key frame extraction using unsupervised clustering. In: Int. Conf. on Image Processing, pp. 866–870 (1998)

Ontology Database System and Triggers

Angelina A. Tzacheva, Tyrone S. Toland, Peyton H. Poole, and Daniel J. Barnes

University of South Carolina Upstate, Department of Informatics
800 University Way, Spartanburg, SC 29303, USA
atzacheva@uscupstate.edu

Abstract. An ontology database system is a basic relational database management system that models an ontology plus its instances. To reason over the transitive closure of instances in the subsumption hierarchy, an ontology database can either unfold views at query time or propagate assertions using triggers at load time. In this paper, we present a method to embed ontology knowledge into a relational database through triggers. We demonstrate that by forward computing inferences, we improve query time. We find that: first, ontology database systems scale well for small and medium sized ontologies; and second, ontology database systems are able to answer ontology-based queries deductively; We apply this method to a Glass Identification Ontology, and discuss applications in Neuroscience.

1 Introduction

Researchers are using Semantic Web ontologies extensively in intelligent information systems to annotate their data, to drive decision-support systems, to integrate data, and to perform natural language processing and information extraction. Researchers in biomedicine use ontologies heavily to annotate data and to enhance decision support systems for applications in clinical practice [24] .

An ontology defines terms and specifies relationships among them, forming a logical specification of the semantics of some domain. Most ontologies contain a hierarchy of terms at minimum, but many have more complex relationships to consider. Ontologies provide a means of formally specifying complex descriptions and relationships about information in a way that is expressive yet amenable to automated processing and reasoning. As such, they offer the promise of facilitated information sharing, data fusion and exchange among many, distributed and possibly heterogeneous data sources.

However, the uninitiated often find that applying these technologies to existing data can be challenging and expensive. What makes this work challenging in part is: to have systems that handle basic reasoning over the relationships in an ontology in a simple manner, that scales well to large data sets. In other words, we need the capabilities of an efficient, large-scale knowledge-based system (such as Semantic Web Ontology), but we want a solution as simple as managing a regular relational database system, such as MySQL. We argue that: when used in the proper context, regular databases can behave like efficient, deductive systems.

In this paper, we present an easy method for users to manage an ontology plus its instances with an off-the-shelf database management system like MySQL, through triggers. We call these sorts of databases *ontology databases*.

A. Tucker et al. (Eds.): IDA 2013, LNCS 8207, pp. 416–426, 2013.

An *ontology database system* takes a Semantic Web ontology as input, and generates a database schema based on it. When individuals in the ontology are asserted in the input, the database tables are populated with corresponding records. Internally, the database management system processes the data and the ontology in a way that maintains the knowledge model, in the same way as a basic knowledge-base system.

After the database is bootstrapped in this way, users may pose SQL queries to the system declaratively, i.e. based on terms from the ontology, and they get answers in return that incorporate the term hierarchy or other logical features of the ontology. In this way, our proposed system is useful for handling ontology-based queries. That is, users get answers to queries that take the ontological subsumption hierarchy into account.

We find that ontology databases using our trigger-based method scale well.

The proposed method can be extended to perform integration across distributed, heterogeneous ontology databases using an inference-based framework [10].

The rest of the paper is organized as follows: in Section 2 we review related work and provide background information; in Section 3 we discuss the main ideas behind ontology databases and the proposed method; in Section 4 we present case studies; in Section 5 we suggest directions for the future and conclude.

2 Related Work

Knowledge-Based Systems. Knowledge-based systems (KBs) use a knowledge representation framework, having an underlying logical formalism (a language), together with inference engines to deductively reason over a given set of knowledge. Users can tell statements to the KB and ask it queries [20], expecting reasonable answers in return. An ontology, different from but related to the philosophical discipline of Ontology, is one such kind of knowledge representation framework [15]. In the Semantic Web [4], description logic (DL) [2] forms the underlying logic for ontologies encoded using the standard Web Ontology Language1 (OWL). One of the major problems with Semantic Web KBs is they do not scale to very large data sets [18].

Reasoning. Researchers in logic and databases have contributed to the rich theory of deductive database systems [12], [13]. For example, Datalog [25] famously uses views for reasoning in Horn Logic. We already mentioned EKS-V1 [27]. Reasoning over negations and integrity constraints has also been studied in the past [19]. Of particular note, one of the side-remarks in one of Reiters papers [22] formed an early motivation for building our system: Reiter saw a need to balance time and space in deductive systems by separating extensional from intensional processing. However33 years laterspace has become expendable. Other works move beyond Datalog views to incorporate active rules for reasoning. An active rule, like a trigger in a database, is a powerful mechanism using an event-condition-action model to perform certain actions whenever a detected event occurs within a system that satisfies the given condition. Researchers in object-oriented and deductive database systems use active technologies in carefully controlled ways to also manage integrity constraints and other logical features [7], [26]. Researchers are studying how to bring database theory into the Semantic Web [21], but more work is needed in that regard.

Scalability. Since reasoning generally poses scalability concerns, system designers have used the Lehigh University Benchmark (LUBM) [16] in the past, to evaluate and compare knoweldge-based (KB) systems. The Lehigh University authors have also proposed the Description Logic Database (DLDB) [17]. We would charactarize DLDB as an *ontology database* because it is similar to our proposed system. It mimics a KB system by using features of a basic relational database system, and it uses a decomposition storage model [1] and [6] to create the database schema.

Information Integration. Another important motivation for using ontologies is the promise they hold for integrating information. Researchers in biomedical informatics have taken to this idea with some fervor [14]. One system in particular, OntoGrate, offers an inferrential information integration framework using ontologies which integrates data by translating queries across ontologies to get data from target data sources using an inference engine [8]. The same logical framework can be extended to move data across a network of repositories.

3   Ontology Database System

In the following sections, we use a simple running example, the SisitersSiblings example, to illustrate how we implement an ontology database system. We begin with the basic idea, and then explain how we structure the database schema and implement each kind of logical feature using triggers and integrity constraints.

3.1   The Basic Idea

We can perform rudimentary, rule-based reasoning using either views or triggers. For example, suppose we assert the statement (a rule): *All sisters are siblings*. Then we assert the fact: *Mary and Jane are sisters*. Logically, we may deduce using modus ponens (MP) that Mary (M) and Jane (J) are siblings. The notation {x/M, y/J} denotes that the variable x gets substituted with M, y with J, and so on, as part of the unification process.

$$\frac{Sisters(x,y) \rightarrow Siblings(x,y) Sisters(M,J)}{Siblings(M,J)} MP\{x/M, y/J\}$$

If sibling and sister facts are stored in two-column tables (prefixed with $a\_$ to denote an asserted fact), then we can encode the rule as the following SQL view:

```
CREATE VIEW siblings(x, y) as
SELECT x,y FROM a_siblings
UNION
SELECT x,y FROM sisters
```

In the view-based method, every inferred set of data necessarily includes its asserted data (e.g., siblings contains *a_siblings* and sisters contains *a_sisters*). Note: when the view is executed, the subquery retrieving sisters will unfold to access all asserted sisters data. Recursively, if sisters subsumes any other predicate, it too will be unfolded. Database triggers can implement the same kind of thing:

```
CREATE TRIGGER subproperty_sisters_siblings
ON INSERT (x, y) INTO sisters
FIRST INSERT (x, y) INTO siblings
```

The deduction is reflected in the answer a query such as *Who are the siblings of Jane?* Of course, the answer returned, in both cases, is: *Mary*. We easily formulate the SQL query:

```
SELECT x FROM siblings WHERE y=Jane
```

What differentiates these two methods is that views are goal-driven: the inference is performed at query time by unfolding views. Whereas, triggers forward propagate facts along rules as they are asserted, i.e., at load time. We advocate a trigger-based approach as our preferred method of implementation. If we consider the spacetime tradeoff: essentially, triggers use more space to speed up query performance. The technique has some of the advantages of materialized views but it differs in some important ways, specifically: deletions. Assume we assert the following in order: $A \rightarrow B$, insert $A(a)$, delete $A(a)$. Next, ask the query $B(?x)$. A trigger-based implementation returns $\{x/a\}$. A view-based implementation returns *null*. In addition, triggers can differentiate negation from deletion. Finally, aside from rule-based reasoning, triggers support other logical features we find important, such as domain and range restrictions, and inconsistency detection. We describe the methods for handling each case in the following implementation details.

3.2   Implementation Details

Decomposition Storage Model. We use the decomposition storage model [1], [6] because it scales well and makes expressing queries easy. Arbitrary models result in expensive and complicated query rewriting, so we would not consider them. The two other suitable models in the literature are the horizontal and vertical models. Designers rarely use the horizontal model because it contains excessively many null values and is expensive to restructure: The administrator halts the system to add new columns to service new predicates. The vertical model is quite popular because it avoids those two drawbacks. Also, the vertical model affords fast inserts because records are merely appended to the end of the file. Sesame [5] and other RDF stores use the vertical storage model.

However, the vertical storage model is prone to slow query performance because queries require many joins against a single table, which gets expensive for very tall tables. Furthermore, type-membership queries are somewhat awkward. As a typical workaround, designers first partition the vertical table to better support type-membership queries, then they partition it further along other, selected predicates which optimize certain joins based on an informed heuristic. However, this leads back toward complicated query rewriting because the partitioning choices have to be recorded and unfolded in some way.

We view the decomposition storage model as a fully partitioned vertical storage model, where the single table is completely partitioned along every type and every predicate. In other words: each type and each predicate gets its own table. When taken to this extreme, query rewriting becomes simple, because each table corresponds directly to a query predicate. In this way, the decomposition storage model keeps the advantages of the vertical model while improving query performance (because of the partitions) without introducing complex query rewriting. Figure 1 illustrates the three different models using the Sisters - Siblings example.

Object	Type	Sister-of	Sibling-of
maryDoe	Female	janeDoe	janeDoe
johnDoe	Male	null	maryDoe

(a) Horizontal Model

Predicate	Subject	Object
Type	janeDoe	Female
Type	maryDoe	Female
Sister-of	maryDoe	janeDoe
Sibling-of	maryDoe	janeDoe

(b) Vertical Model

Female

ID
janeDoe
maryDoe

SisterOf

Subject	Object
maryDoe	janeDoe

SiblingOf

Subject	Object
maryDoe	janeDoe

(c) Decomposition Storage Model

Fig. 1. The SistersSiblings examples using the 1(a) horizontal, 1(b) vertical, and 1(c) decomposition storage model

Subsumption. Ontology engineers often specify subclass relationships in Semantic Web ontologies, which form a subsumption hierarchy. That constitutes the majority of reasoning for biomedical ontologies [2] as well. As we mentioned earlier (Section 3.1), we handle subclass relationships by using triggers.

The same can be handled by using Views in Datalog. However, Datalog views differ from inclusion axioms in description logic [3]. In other words, the semantics of these two logical formalisms differ:

$$Sisters \rightarrow Siblings \neq Siblings \subseteq Sisters$$

The literature suggests that these differences are formally captured using modal logic [3]. In our proposed method, we ensure that the contrapositive of the rule is enforced as an integrity constraint [23] (and not as a rule): if *Siblings(M,J)* is not true, then *Sisters(M,J)* cannot possibly be true (otherwise, raise an inconsistency error). We, therefore, implement the contrapositive as a foreign-key constraint as follows:

```
CREATE TABLE Siblings(
subject VARCHAR NOT NULL,
object VARCHAR NOT NULL,
CONSTRAINT fk-Sisters-Siblings
FOREIGN KEY subject, object
REFERENCES Sisters(subject, object) ...)
```

Figure 2(a) illustrates the two parts of an inclusion axiom graphically. The trigger rule event is indicated in the figure by the a star-like symbol, denoting that the detected assertion causes a trigger to fire. In this example, we enforce the following rule: $All females are person(s), i.e., Female \rightarrow Person$. Therefore, asserting Female(Mary) causes the trigger to actively assert Person(Mary).

Finally, the contrapositive is checked using the foreign key. Note: consistency requires that forward-propagations occur *before* integrity checking, which explains using the keywords *before* or *first* in our trigger definitions.

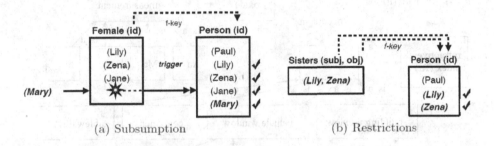

(a) Subsumption   (b) Restrictions

Fig. 2. The star-like symbol denotes an event fires a trigger rule. The checkmark symbol denotes an integrity check occurs. (a) Subsumption is implemented using a combination of triggers and integrity constraints. (b) Domain and range restrictions are implemented using foreign-key (f-key) constraints.

Domain and Range Restrictions. Another important feature of Semantic Web ontologies are domain and range restrictions. These restrict the possible set of instances that participate in property assertions. For example, *Only person(s) may participate in the sisters relationship*. Restrictions are formalized using modal logic. They correspond to integrity constraints. In our proposed method, we implement them as foreign key constraints on the subject or object (i.e. domain and range) of the property.

```
CREATE TABLE Sisters(
subject VARCHAR NOT NULL,
object VARCHAR NOT NULL,
CONSTRAINT fk-Sisters-Subject-Person
FOREIGN KEY subject
REFERENCES Person(id) ...)
```

4 Experiment

We apply our proposed trigger-based ontology database system method to a Glass Identification Ontology. Next, we discuss its application for a Neural ElectroMagnetic Ontology.

Glass Identification. We apply our proposed method to a Glass Identification Ontology - shown on Figure 3. The Glass Identification Dataset [11] is donated by the Central Research Establishment, Home Office Forensic Science Service, Reading, Berkshire, England. The study of classification of types of glass was originally motivated by criminological investigation. At the scene of the crime, the glass left can be used as evidence, if it is correctly identified.

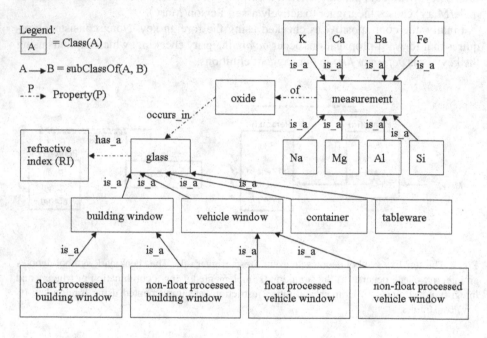

Fig. 3. Glass Identification Ontology

Applying the decomposition storage model [1], [6], we obtained a table for each glass type, i.e. FloatProcessedBuildingWindow, BuildingWindow, Glass, Container, and so on. We implement the subclass relationships in Glass Ontology, which form a subsumption hierarchy, through triggers. For instance:

```
CREATE TABLE FloatProcessedBldWin(
subject VARCHAR NOT NULL,
object VARCHAR NOT NULL,
CONSTRAINT fk-FloatProcessedBldWin-BuildingWindow-Glass
FOREIGN KEY subject, object
REFERENCES Sisters(subject, object) ...)
```

We implement domain and range restrictions as foreign key constraints. These restrict the possible set of instances that participate in property assertions. For example, *Only Windows may participate in the Glass - BuildingWindow relationship*. Restrictions correspond to integrity constraints.

We apply Lehigh University Benchmarks (LUBM) [16] to evaluate the performance of this system. It consists of a set of 14 queries for evaluating load time and query time of knowledge bases. A data generator generates assertins, i.e. *datainstances*, which can be saved as a set of Web Ontology Language (OWL) files and loaded into a knowledge base for evaluation. The main idea is to vary the size of the data instances to quantify the scalability of a Semantic Web Knowledge Base. Mesuring both load time and query time with respect to the imput parameters provides an evaluation of the total performance. We confirmed that by using triggers to materialize inferences, query performance improves by several orders of magnitude. That comes at the cost of load time. Load time is slower. The proposed trigger method increases the disk space required by roughly three times. In other words, our proposed method improves query speed, by costing more space. The proposed approach scales well to very large datasets, and medium-sized ontologies.

Neural ElectroMagentic Ontology. Our ontology database system can be applied to a Neural ElectroMagentic Ontology (NEMO) [9]. NEMO records experimental measurments from brainwave studies, which classify, label, and annotate event related potentials (ERP) using ontological terms. Brainwave activity is measured when certain event happens - such as a word or a sentence is read or heard. Information about scalp distribution, and neural activity during cognitive and behavioral tasks is included. A partial representation is shown on Figure 4.

We show that our propsoed method is useful for answering queries that take subsumption into account - answering queries deductively. In other words, we are able to answer Ontology-Based Queries. For example, the following query requires taking the submsumption hierarchy into account:

Return all data instances that belong to ERP patter classes,
which have a surface positivity over frontal regions of interst and
are earlier than the N400.

In this query, *frontal region* can be unfolded into constituent parts (ex. right frontal, left frontal) as shown in Figure 4. At higher abstraction level, the *N400* is a pattern class that is also associated with spatial, temporal, and functional properties. The patter class labels can be inferred by applying a set of conjunctive rules. This can be implemented by using Semantic Web Rule Language.

Preliminary results indicate that neurosientists are attracted by the ability to pose queries at the conceptual level, without having to formulate SQL queires; which, would require taking complex logical interactions and reasoning aspects into consideration. Those high-level, logical interactions are modeled only once by specifying the ontology. Other examples of conceptual level queries include:

Which patterns have a region of interest that is left-occipital
and manifests between 220 and 300ms?

What is the range of intensity mean for the region of interst
for N100?

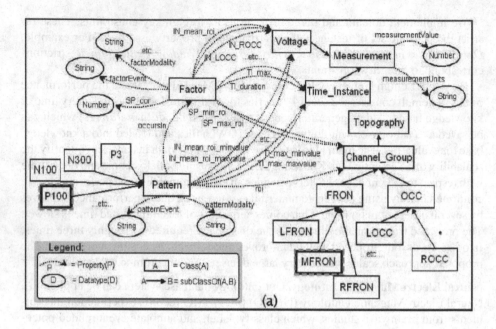

Fig. 4. Neural ElectroMagnetic Ontology (NEMO)

5 Conclusions and Directions for the Future

We present an ontology database system, a tool for modeling ontologies plus large numbers of instances using off-the-shelf database management systems such as MySQL. An ontology database system is useful for answering ontology-based, scientific queries that require taking the subsumption hierarchy and other constraints into account (answering queries deductively). Furthermore, our trigger implementation method scales well with small and medium-sized ontologies, used with very large datasets.

The proposed method pre-computes inferences for the subsumption hierarchy, so larger and deeper ontologies will incur more costly up-front penalties. However, the query answering time is significantly improved at the end.

Mapping rules between ontologies also can be implemented as trigger-rules, giving us an efficient and scalable way to exchange data among a distributed ontology database systems.

With Ontology Database Systems, it is possible to integrate two knowledge bases. The key idea is to map ontology terms together, then to reason over them as a whole, which comprizes - *a merged ontology*. We can use namespaces to distinguish terms from each knowledge base; next, we can map the ontologies together using *bridging axioms* [10]; and finally, we can reason over the entire merged ontology to achieve integration. As a direction for the future, we can adopt the Dou and LePendu [10] approach for ontology-based integration for relational databases.

Further future steps include studying ontology evolution and concept drift to propagate changes within an ontology database. Changes in the ontology affect the structure, rules and data for an ontology database, which makes efficiently managing the knowledge model a challenging problem.

References

1. Abadi, D.J., Marcus, A., Madden, S.R., Hollenbach, K.: SW-Store: A Vertically Partitioned DBMS for Semantic Web Data Management. VLDB Journal 18(2), 385–406 (2009)
2. Baader, F., Calvanese, D., McGuinness, D.L., Nardi, D., Patel-Schneider, P.F. (eds.): The Description Logic Handbook: Theory, Implementation, and Applications. Cambridge University Press (2003)
3. Baader, F., Nutt, W.: Basic Description Logics. In: Description Logic Handbook, pp. 43–95 (2003)
4. Berners-Lee, T., Hendler, J., Lassila, O.: The Semantic Web. Scientific American (May 2001)
5. Broekstra, J., Kampman, A., van Harmelen, F.: Sesame: A Generic Architecture for Storing and Querying RDF and RDF Schema. In: Horrocks, I., Hendler, J. (eds.) ISWC 2002. LNCS, vol. 2342, pp. 54–68. Springer, Heidelberg (2002)
6. Copeland, G.P., Khoshafian, S.N.: A decomposition storage model. In: Proceedings of the ACM SIGMOD International Conference on Management of Data, SIGMOD 1985, New York, NY, USA, pp. 268–279 (1985)
7. Curé, O., Squelbut, R.: A Database Trigger Strategy to Maintain Knowledge Bases Developed Via Data Migration. In: Bento, C., Cardoso, A., Dias, G. (eds.) EPIA 2005. LNCS (LNAI), vol. 3808, pp. 206–217. Springer, Heidelberg (2005)
8. Dou, D., Pan, J.Z., Qin, H., LePendu, P.: Towards Populating and Querying the Semantic Web. In: International Workshop on Scalable Semantic Web Knowledge Base Systems, SSWS, pp. 129–142 (2006), co-located with ISWC 2006
9. Dou, D., Frishkoff, G., Rong, J., Frank, R., Malony, A., Tucker, D.: Development of Neuro-ElectroMagnetic Ontologies (NEMO): A Framework for Mining Brainwave Ontologies. In: Proceedings of the 13th ACM International Conference on Knowledge Discovery and Data Mining, KDD, pp. 270–279 (2007)
10. Dou, D., LePendu, P.: Ontology-based Integration for Relational Databases. In: ACM Symposium on Applied Computing, SAC, pp. 461–466 (2006)
11. Frank, A., Asuncion, A.: UCI Machine Learning Repository. University of California, School of Information and Computer Science, Irvine, CA (2010),
    http://archive.ics.uci.edu/ml
12. Gallaire, H., Minker, J., Nicolas, J.-M.: Logic and Data Bases (1977)
13. Gallaire, H., Nicolas, J.-M.: Logic and Databases: An Assessment. In: Kanellakis, P.C., Abiteboul, S. (eds.) ICDT 1990. LNCS, vol. 470, pp. 177–186. Springer, Heidelberg (1990)
14. Goble, C., Stevens, R.: State of the nation in data integration for bioinformatics. Journal of Biomedical Informatics (February 2008)
15. Guarino, N.: Formal Ontology in Information Systems. In: International Conference on Formal Ontology in Information Systems
16. Guo, Y., Pan, Z., Heflin, J.: LUBM: A benchmark for OWL knowledge base systems. Journal of Web Semantics 3(2-3), 158–182 (2005)
17. Guo, Y., Pan, Z., Heflin, J.: An Evaluation of Knowledge Base Systems for Large OWL Datasets. In: McIlraith, S.A., Plexousakis, D., van Harmelen, F. (eds.) ISWC 2004. LNCS, vol. 3298, pp. 274–288. Springer, Heidelberg (2004)
18. Haarslev, V., Möller, R.: High Performance Reasoning with Very Large Knowledge Bases: A Practical Case Study. In: Proceedings of the International Joint Conferences on Artificial Intelligence, IJCAI 2001, pp. 161–168 (2001)

19. Kowalski, R.A., Sadri, F., Soper, P.: Integrity Checking in Deductive Databases. In: VLDB, pp. 61–69 (1987)
20. LePendu, P., Dou, D., Howe, D.: Detecting Inconsistencies in the Gene Ontology using Ontology Databases with Not-gadgets. In: Meersman, R., Dillon, T., Herrero, P. (eds.) OTM 2009, Part II. LNCS, vol. 5871, pp. 948–965. Springer, Heidelberg (2009)
21. Motik, B., Horrocks, I., Sattler, U.: Bridging the Gap Between OWL and Relational Databases. In: Proceedings of the 16th International Conference on World Wide Web, WWW 2007, pp. 807–816 (2007)
22. Reiter, R.: Deductive Question-Answering on Relational Data Bases. In: Logic and Data Bases, pp. 149–177 (1977)
23. Reiter, R.: What Should a Database Know? Journal of Logic Programming 14(1), 127–153 (1992)
24. Shah, N., Jonquet, C., Chiang, A., Butte, A., Chen, R., Musen, M.: Ontology-driven Indexing of Public Datasets for Translational Bioinformatics. BMC Bioinformatics 10 (2009)
25. Ullman, J.D.: Principles of Database and Knowledge-Base Systems, vol. I. Computer Science Press (1988)
26. Vasilecas, O., Bugaite, D.: An algorithm for the automatic transformation of ontology axioms into a rule model. In: Proceedings of the International Conference on Computer Systems and Technologies, CompSysTech 2007, pp. 1–6. ACM, New York (2007)
27. Vieille, L., Bayer, P., Küchenhoff, V., Lefebvre, A., Manthey, R.: The EKS-V1 System. In: Voronkov, A. (ed.) LPAR 1992. LNCS, vol. 624, pp. 504–506. Springer, Heidelberg (1992)

A Policy Iteration Algorithm for Learning from Preference-Based Feedback

Christian Wirth and Johannes Fürnkranz

Knowledge Engineering, Technische Universität Darmstadt, Germany
{cwirth, fuernkranz}@ke.tu-darmstadt.de

Abstract. Conventional approaches to reinforcement learning assume availability of a numerical feedback signal, but in many domains, this is difficult to define or not available at all. The recently proposed framework of preference-based reinforcement learning relaxes this condition by replacing the quantitative reward signal with qualitative preferences over trajectories. In this paper, we show how to estimate preferences over actions from preferences over trajectories. These action preferences can then be used to learn a preferred policy. The performance of this new approach is evaluated by a comparison with SARSA in three common reinforcement learning benchmark problems, namely mountain car, inverted pendulum, and acrobot. The results are showing convergence rates that are comparable, but achieved with a much less time consuming tuning of the setup.

1   Introduction

Most common methods for reinforcement learning are facilitating feedback in the form of numerical rewards. This form of feedback is very informative, which eases the learning process, but it is often not easy to define. Numeric reward definition often requires detailed knowledge of the task at hand. In some domains, there is no natural choice, so that certain outcomes have to be associated with arbitrary reward signals. As an example, consider the cancer therapy domain studied by Zhao et al. [15], in which a negative reward of -60 has been assigned to the death of the patient in a medical treatment. Even for classical benchmark problems such as the inverted pendulum or the mountain car, there are many choices for modeling the reinforcement signal. The most natural choice, giving a positive reinforcement if the problem has been solved (i.e., the car gets up the mountain) is not sufficient because solution length is a crucial factor (the quicker the car manages to get up the mountain, the better). Thus, one has to use step penalties, trace decay or a reinforcement signal that depends on solution length. Again, these choices result in different results and convergence speeds. Moreover, these choices interact with the definition of features representing a state, possibly the parameterizable policy, as well as the parametrization for the learning algorithm itself. Our main goal is to overcome those problems for easing the use of reinforcement learning and to expand its applicability to domains where numerical feedback is not readily available.

The growing field of preference learning [6] enables an intuitive form of feedback, which does not require numerical values. For reinforcement learning, preference-based feedback signals can be modeled as pairwise comparisons of trajectories [7]. This form of feedback does not require detailed knowledge about the problem domain and can

A. Tucker et al. (Eds.): IDA 2013, LNCS 8207, pp. 427–437, 2013.

often be given by non-expert users. We also only expect feedback based on complete trajectories. For keeping the process as simple as possible, states are only represented in a tabular way and the policy learned is not parameterized. Additionally, we are considering approaches requiring only a minimal amount of hyper-parameters for tuning the algorithm.

Our main contribution is hence an policy iteration algorithm which is able to learn from such preference-based feedback (Section 3). This step towards ready-to-use black-box algorithms is evaluated on three common reinforcement learning benchmark problems: mountain car, inverted pendulum and acrobot. (Section 4 & 5)

2   Problem Definition

In this paper, we formally define a problem setting, which we call *preference-based sequential decision processes* (PSDP). While this setting shares many similarities with Markov decision processes, the basic scenario in which most reinforcement learning algorithms operate, there are some important differences, most notably the absence of a numerical reward signal.

A PSDP $\{S, A, \delta, \succ\}$ is defined by a state space $S = \{s_i\}, i = 1 \ldots |S|$, a finite action space $A = \{a_j\}, j = 1 \ldots |A|$, a stochastic state transition function $\delta : S \times A \times S \rightarrow [0, 1]$, and a preference relation $\succ$ that is defined over trajectories through this state space. Each state s is associated with a set of available actions $A(s)$. A state is an absorbing state, if there are no actions available, i.e., the set of all absorbing states is $S^F = \{s \in S | A(s) = \emptyset\}$. A (stochastic) policy π is a probability distribution over the actions in each states, i.e., $\pi : S \times A \rightarrow [0, 1]$, where $\sum_{a' \in A(s)} \pi(s, a') = 1$.

A trajectory is a state/action sequence $T = (s_0, a_0, s_1, ..., a_{n-1}, s_n)$ where $s_i \in S, a_i \in A(s_i)$, and $s_n \in S^F$ is an absorbing (final) state. The set of states occurring in a trajecory T is denoted as $S(T)$, the set of actions as $A(T)$. We also write $(s, a) \in T$ when we want to denote that in trajctory T, action a has been taken in state s. Note that the same state may occur multiple times in a trajectory, so for two actions $a \neq a'$, it can be the case that both $(s, a) \in T$ and $(s, a') \in T$.

The associated learning problem assumes that the learner is able to observe a set P of trajectory preferences, i.e., $P = \{T_i^1 \succ T_i^2\}, i = 1 \ldots N$. The goal of the learner is to find a policy, which respects these observed preferences.

In this paper, we assume that the underlying preference relation $\succ$ forms a total order. Admitting partial orders (i.e., pairs of trajectories that cannot be compared), could yield a set of optimal policies, as opposed to a single one.

3   The Policy Iteration Cycle

Our approach is part of the class of Policy Iteration algorithms [12]. In each iteration, a small amount of preferences is requested, based on trajectory samples of the current policy. Based on those samples, we are determining the probability that an action choice (s, a) is belonging to a set of preferred actions. Because we are restricted to feedback in form of pairwise preferences, we are only calculating this probability for a pairwise comparison to another action. It is not easily possible to determine by "how much" an action should be preferred over another one, hence we are assuming that an action is maximal preferred, if its part of the set of preferred actions. (See sec. 3.2) The probability of membership to this set is then used to update a probabilistic value function (Section 3.1, 3.4). Section 3.3 is concerned with the exploration/exploitation dilemma that has to be considered in reinforcement learning scenarios.

3.1 Probabilistic Value Function

In this paper, we make the assumption that the order between trajectories is total. In fact, we assume that we can represent the policy with a value function $V(s, a)$ that induces an ordering on the actions a in each state s. However, the semantics of this function is not the expected reward of selecting action a in state s (which we cannot compute because we do not have a reward signal), but instead the value function should estimate the probability $\Pr(a|s)$ that action a is the most appropriate action in state s.

For computing $V(s, a) = \Pr(a|s)$ we rely on comparisons between pairs of actions. Let $\Pr(a \succ a'|s)$ denote the probability that action a is preferred over action a' in state s. As we assume a total order, $\Pr(a \succ a'|s) = 1 - \Pr(a' \succ a|s)$. The problem of reconstructing $\Pr(a|s)$ from such probabilities for pairwise comparisons is known as *pairwise coupling* [8]. Several approaches have been proposed in the literature [14]; we rely on the suggestion of Price et al. [9], who estimate $\Pr(a|s)$ as

$$\Pr(a|s) = \frac{1}{\sum\limits_{a' \in A(s), a' \neq a} \frac{1}{\Pr(a \succ a'|s)} - (|A(s)| - 2)}. \tag{1}$$

3.2 Policy Evaluation

For estimating $\Pr(a \succ a'|s)$, which is needed in Eq. (1), we use the current policy to sample a fixed number of K trajectories, starting in the same state s_0, for which we obtain their total order as a feedback signal. In the simplest case $K = 2$, i.e., we only compare a pair of trajectories, for larger K, we will obtain $K(K-1)/2$ pairwise preferences for these K trajectories.

Given $T_1 \succ T_2$, we want to estimate the probability $\Pr(a \succ a'|s, T_1 \succ T_2)$ that an action a, $(s, a) \in T_1$ is preferred over an action a', $(s, a') \in T_2, a \neq a'$ in a particular state $s \in S(T_1) \cap S(T_2)$. For doing so, we make the following assumptions:

1. The reason for $T_1 \succ T_2$ can be found in the overlap of states $N = S(T_1) \cap S(T_2), n = |N|$ if we are following an optimal policy in all other states.
2. For each state s in an unknown subset M, the action a taken in T_1 is preferable to the action a' taken in T_2, i.e.

$$\Pr(a \succ a'|s, T_1 \succ T_2, s \in M) = 1.$$

We call M the set of *decisive states*.

3. The probability $T_1 \succ T_2$ is proportional to the ratio of *decisive states* M to all overlapping states N, meaning the more often we did pick a preferred action, the more probable is the resulting trajectory preferred.

$$\Pr(T_1 \succ T_2 | m = p) = \frac{p}{n} \tag{2}$$

$m = |M|$ denotes the number of *decisive states* and $n = |N|$ is the amount of overlapping states. p is the number of decisive states assumed to exist. Note that a state may occur multiple times in each trajectory and therefore also in M and N.

For estimating $\Pr(a \succ a' | s, T_1 \succ T_2)$, we require $\Pr(s \in M | T_1 \succ T_2)$ to be able to resolve equation (3).

$$\Pr(a \succ a' | s, T_1 \succ T_2) = \Pr(a \succ a' | s, T_1 \succ T_2, s \in M) \cdot \Pr(s \in M | T_1 \succ T_2) \tag{3}$$

This is reduced to $\Pr(a \succ a' | s, T_1 \succ T_2) = \Pr(s \in M | T_1 \succ T_2)$, because of assumption 2, which is in-line with our approach (Sec. 3), where we assume that it is not possible to determine the exact value of $\Pr(a \succ a' | s, T_1 \succ T_2, s \in M)$ easily. $\Pr(s \in M | T_1 \succ T_2)$ can be calculated directly for specific m, using the a priori probability $\Pr(s \in M) = \frac{1}{2}$ as well as $m = |M|$ and $n = |N|$. (Equation 4)

$$\Pr(s \in M | m = p, T1 \succ T2) = \frac{p}{n} \tag{4}$$

The observed probabilities $\Pr(s \in M)$ could not be used because of computational reasons, namely time and numerical stability problems.

Assuming a binomial distribution induced by $\Pr(s \in M) = \frac{1}{2}$, we obtain $\Pr(m = p) = \binom{n}{p} \cdot \frac{1}{2}^n$. Using Bayes theorem, eq. (2) and the prior probability $\Pr(T_1 \succ T_2) = \frac{1}{2}$, yields

$$\Pr(m = p | T_1 \succ T_2) = \frac{\Pr(T_1 \succ T_2 | m = p) \Pr(m = p)}{\Pr(T_1 \succ T_2)} = \frac{p}{n} \binom{n}{p} \frac{1}{2}^{n-1}$$

We can now determine $\Pr(s \in M | T_1 \succ T_2)$ by calculating the combined probability and summing over all possible cases for p:

$$\Pr(s \in M \cap m = p | T_1 \succ T_2) = \Pr(s \in M | m = p, T_1 \succ T_2) \Pr(m = p | T_1 \succ T_2)$$

$$\Pr(s \in M | T_1 \succ T_2) = \sum_{p=0}^{n} \Pr(s \in M \cap m = p | T_1 \succ T_2) = \frac{n+1}{2n} \tag{5}$$

Equation (5) is now used to determine the value of eq. (3). The case $T_2 \succ T_1$ can be mapped to $\Pr(a \succ a' | s, T_1 \succ T_2)$ by using $\Pr(a \succ a' | s, T_1 \succ T_2) = 1 - \Pr(a' \succ a | s, T_2 \succ T_1)$, as mentioned in sec. 3.1.

Of course, we have to be able to assume that the decisive action(s) did really occur in a state $s \in S(T_1) \cap S(T_2)$. This can be achieved by always applying a deterministic, optimal policy π^* in all states $s \in S(T_1)$, $s \notin S(T_2)$ and vice versa. This will yield the same, optimal outcome for all non-overlapping states. Of course, π^* is unknown, but we can use the current best approximation $\tilde{\pi}^*(s, a)$:

$$\tilde{\pi}^*(s, a) = \begin{cases} 1 & \text{if } a = \arg\max_{a'}(\Pr(s|a')) \\ 0 & \text{else} \end{cases} \tag{6}$$

This means we are applying $\tilde{\pi}^*(s, a)$ to all states $s \notin S(T(\pi))$ with $T(\pi)$ as the set of trajectories already sampled for evaluating policy π. It should be noted, that we are sampling the first trajectory of each policy evaluation with the stochastic policy mentioned in section 3.3, because we need to guarantee, that every (s, a) can be sampled infinitely often [12]. Due to acting always deterministically for all $s \notin S(T(\pi))$, we do need to ensure that each state (s, a) can be part of the first trajectory $T \in T(\pi)$. This can be resolved by using a stochastic policy, that guarantees $\pi(s, a) > 0$.

It should also be noted, that all samples for a policy π are starting in the same state s_0, because defining preferences over trajectories starting in different states is non trivial. Of course, s_0 is randomly picked for each iteration of π.

3.3 The Exploration/Exploitation Dilemma

It is still required to solve the exploration/exploitation dilemma [12] in a way that guarantees $\pi(s, a) > 0$ which can be achieved by applying algorithms like EXP3 [4]. Our method was inspired by this policy, because it is an stochastic action selector, that can be applied to adversarial bandit problems. We chose to respect adversarial bandits, for reducing the amount of prequirements that have to be considered.

$$\tilde{g}_{s,a,i} = \frac{\Pr^{\pi_i}(a|s)}{\text{EXP3}(s,a)_i} \mathbf{1}_{s \in P_s^{\pi_i}} \qquad \tilde{G}_{s,a,i} = \sum_{t=1}^{i} \tilde{g}_{s,a,t}$$

$$\text{EXP3}(s, a)_{i+1} = \frac{\eta}{|A|} + (1 - |A(s)|\frac{\eta}{|A|}) \frac{\exp(\frac{\eta}{|A|}\tilde{G}_{s,a,i})}{\sum_{b \in A(s)} \exp(\frac{\eta}{|A|}\tilde{G}_{s,b,i})} \tag{7}$$

This formulation (Eq. 7) is equivalent to Audibert and Bubeck [2], with $g_{i,t} = \Pr^{\pi_i}(a|s)$. $\Pr^{\pi_i}(a|s)$ is an action preference sample obtained from the i-th policy evaluation, which is comparable to a reward sample $r(s, a)$.

But $\Pr(a|s)$ is not 0 for state action pairs (s, a) that have not been evaluated, because we have to assume $\Pr(a \succ a'|s) = \frac{1}{2}$ without further information. Hence we are updating all pairwise action probabilities, concerning $\Pr(a|s)$, $s \in P_s^{\pi_i}$. Additionally, we did replace η with $\frac{\eta}{|A|}$ to be able to define $\eta \in (0, 1]$ instead of $\eta \in (0, \frac{1}{|A|}]$.

3.4 Policy Improvement

We are now updating $\Pr(a \succ a'|s)$ by applying eq. 8. Any update on $\Pr(a \succ a'|s)$ will result in a divergence from $\Pr(a \succ a'|s) = \frac{1}{2}$, because eq. (5) can never become $\frac{1}{2}$ in a finite horizon scenario. This increases the probability to pick the action a in s.

$$\Pr(a \succ a'|s) = (1 - \alpha)\Pr(a \succ a'|s) + \alpha \tilde{\Pr}^\pi(a \succ a'|s) \tag{8}$$

$\tilde{\mathrm{Pr}}^{\pi}(a \succ a'|s)$ is the probability obtained from evaluating all samples from policy π and α defines a learning rate. $\mathrm{Pr}_j^{\pi}(a \succ a'|s, T_1 \succ T_2) \in P_s^{\pi}$ is the j-th action preference sample, obtained by evaluating policy π. Note that a single trajectory preference can result in multiple $\mathrm{Pr}^{\pi}(a \succ a'|s, T_1 \succ T_2)$, concerning the same state. P_s^{π} is the subset of action preferences for state s only. For determining the best approximation $\tilde{\mathrm{Pr}}^{\pi}(a \succ a'|s, T_1 \succ T_2)$ available, those samples are averaged ($\tilde{\mathrm{Pr}}^{\pi}(a \succ a'|s) = \frac{1}{|P_s^{\pi}|} \sum_{P_s^{\pi}} \mathrm{Pr}_j^{\pi}(a \succ a'|s, T_1 \succ T_2)$).

4   Experiments

We compare our algorithm to SARSA(λ) [12] in two test domains. We chose SARSA(λ) because it is a well-known, widely used, but not specialized algorithm. It is required to decrease the SARSA ϵ parameter over time for guaranteeing convergence [11], hence we decided to use an ϵ_n-greedy decay scheme ($\epsilon_n = \min\{1, \frac{c|A|}{d^2t}\}$), presenting a generic solution to the problem [3]. For keeping our comparison as fair as possible, we are only using terminal rewards, and are setting the step reward to 0. This is in-line with our preference approach, which is also only having access to feedback based on complete solutions (pairs). It should be noted, that even terminal-only numeric rewards are still much more informative than preferences, because they define a relation to all other solutions, as opposed to a singular, binary preference relation.

Our first testing domain is the well known mountain car problem, parameterized as defined by Sutton and Barto [12]. The state is defined by creating a tabular representation based on a 40 equal-width bins discretization of the position and 20 bins for the velocity. The horizon was set to 500 and start states are random. As a second testing domain, we are using the inverted pendulum problem [5]. The parametrization is mostly the same as defined by Dimitrakakis et al., namely: $F = -50N/0N/50N \pm 10, m = 2kg, M = 8kg, l = 0.5m, \tau = 0.05, horizon = 500$. The discretization was performed with 10 equal-width bins each for θ and $\dot{\theta}$ and the setup is using a fixed start state. The third testing domain, acrobot, was again parameterized as defined by Sutton and Barto [12], also using a 10 equal-width bin discretization. We introduced a transition noise of 20% for all scenarios, in order to have a bit more realistic problems.

SARSA was run in two settings: With $\lambda = 0.\bar{9}$ (exact: $1 - 10^{-15}$), because our own algorithm is also not using any decay and with an optimized λ value for showing a comparison to the best possible setting. The terminal reward was set to *horizon-stepcount* for mountain-car and *-(horizon-stepcount)* for inverted-pendulum, because it is required to define a reward (or decay) that allows distinguishing solutions by length. The decaying version is using a terminal reward of 1 (-1 for inverted-pendulum).

The optimal hyper-parameter set for SARSA has been determined by 250 random search trials, followed by 625 (5^4) grid search trials within the boundaries of the min/max values found for the best (by terminal policy) 20 random trials, except for λ which was fixed to the best value found within the random trials. This two step approach was chosen for reducing the amount of required experiments without using any assumptions. For our own approach, we only used 100 grid search trials, because of the lower parameter count. We only need to tune EXP3-η and α, opposed to the parameters $\lambda, \alpha, \epsilon$ and the ϵ-decay scheme (which contributes two additional parameters in our case) required for applying SARSA. All values have been uniformly selected within the given $[0, 1]$ range, except for $\epsilon.c$ which was uniformly sampled within $[1, 1000]$. A single trial is an average over 100 repeated runs in all cases.

5   Results

Figure 1 shows the results for the best configuration found for SARSA, SARSA($0.\bar{9}$), and our preference-based approach. The best result was picked out of all available results by comparing the terminal policy quality. Note that the values reported for mountain car and acrobot are negative, because of the step-penalty, as opposed to the step-reward used in the inverted pendulum problem. We can see that preference-based reinforcement learning converges more slowly than SARSA(λ) for the first steps, but that was to be expected due to the lower information content available within the feedback for each solution.

Additionally, it is clearly visible that decay is better suited for learning by solution length than the more sophisticated reward structure used for the SARSA($0.\bar{9}$) scenario. The results for the inverted pendulum, shown in Figure 1(b), are similar. The advantage of a well tuned decay is even more significant, supporting our claim that a good setup is essential for a successful application of SARSA. Especially the progress made in the first iterations seems to greatly depend on the parameterization. The acrobot domain (Figure 1(c)) seems to be quite a bit more challenging and our algorithm is not able to reach the quality of the SARSA(λ) solution within the given episode count, but it is still outperforming the SARSA($0.\bar{9}$) configuration. The differences between the SARSA(λ) and the SARSA($0.\bar{9}$) are also much greater than for the other testing domains.

Figure 2 shows the results obtained with different amounts of trajectory samples per iteration ($K = 2$, 4, and 10, cf. section 3.2). The setup with $K = 2$ is worse than the other parameterizations in both domains, but the differences between $K = 4$ and $K = 10$ are insignificant. The advantage of higher pairings is especially prominent in the inverted pendulum domain. Better results of the higher pairings are probably due to the higher amount of information available per evaluation. It should be noted, all configurations are generating the same amount of trajectories in total, but a higher K value results in a higher amount of preferences that need to be requested (see section 3.2).

6   Related Work

A preliminary version of this paper appeared as [13]. We improved this version by not using an index function but instead replacing the fixed scoring of action preferences with the probabilistic version of Section 3.2, by using a better founded pairwise coupling function, and by introducing some improvements concerning the application of EXP3 to preference learning.

The work of Fürnkranz et al. [7] is also comparable. They are also identifying action preferences, but based on a roll-out sampling method that is not considering trajectory overlaps, as well as requiring evaluations for multiple states for each policy iteration cycle. Each trajectory preference is only used to update a single state. This means each

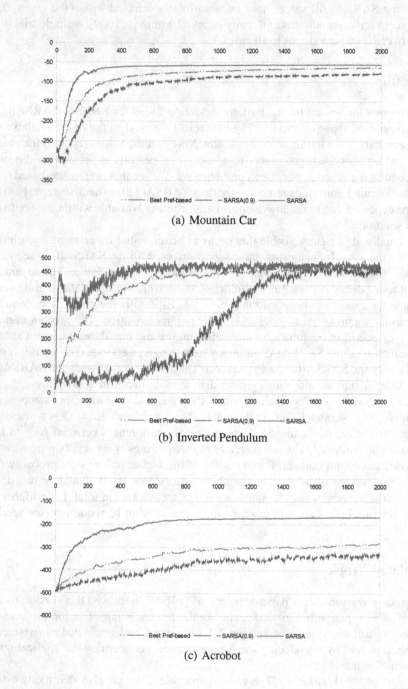

(a) Mountain Car

(b) Inverted Pendulum

(c) Acrobot

Fig. 1. Average step-penalty for the best result compared with SARSA for 2000 episodes

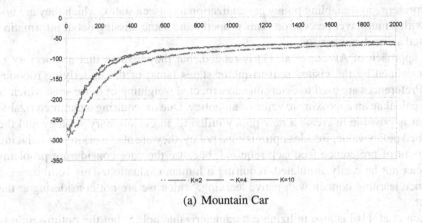

(a) Mountain Car

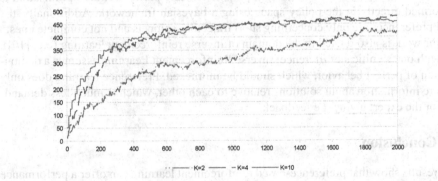

(b) Inverted Pendulum

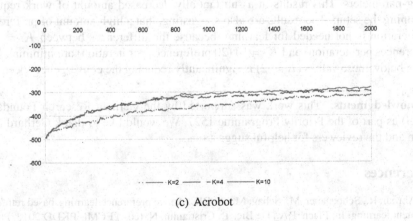

(c) Acrobot

Fig. 2. Average step-penalty for our new algorithm with different K for 2000 episodes

policy update is more reliable, but the intermediate state information of the trajectory is not exploited, resulting in less training data. Additionally, they are using a non-tabular state representation, enabling policy generalization to unseen states, which they are utilizing with a multilayer perceptron. This property and some missing detail information prevented a direct comparison.

The approach of Akrour et al. [1] is related, but they assume that a trajectory is already defined by the visited sensori-motor states (sms) of the underlying robotics agent. Preferences are used to determine an expected weighting of those sms, which is used to calculate an approximate value of an policy. Due to requiring a parameterizable policy, it is possible to create a new policy utilizing an evolutionary strategy and the mentioned policy value. Besides optimizing a policy, they are also working on reducing the amount of preference feedback required, because they are considering problems which can not be easily simulated, requiring a human evaluation. This combines the preference learning domain with active learning, which we are not considering at the moment.

Wilson et al. [16] is also utilizing a parameterizable policy, but the optimisation is preformed directly in the policy space using a bayesian framework. Additionally, the used preferences are only concerning short trajectory snippets and not complete ones.

Our work is also related to the domain of inverse reinforcement learning (e.g. [10]), but with one significant difference: Inverse Reinforcement Learning assumes a demonstration of perfect behavior, which should be mimicked. Preference learning does only require information about solutions relative to each other, which is much less demanding for the expert giving the feedback.

7   Conclusion

Our results show that preference-based reinforcement learning can offer a performance comparable to SARSA, but with a simpler feedback structure and a lower amount of hyper-parameters. This results in a substantially decreased amount of work required for tuning the setup. The results are also suggesting, that a high amount of preferences per iteration is not needed for learning, because the differences between $K = 4$ (6 preferences per iteration) and $K = 10$ (20 preferences per iteration) are minimal. Only going below these values ($K = 2$) is significantly reducing the convergence speed.

Acknowledgments. This work was supported by the German Research Foundation (DFG) as part of the Priority Programme 1527. We would like to thank Gerhard Neumann and the reviewers for helpful suggestions.

References

1. Akrour, R., Schoenauer, M., Sebag, M.: APRIL: Active preference learning-based reinforcement learning. In: Flach, P.A., De Bie, T., Cristianini, N. (eds.) ECML PKDD 2012, Part II. LNCS, vol. 7524, pp. 116–131. Springer, Heidelberg (2012)
2. Audibert, J.Y., Bubeck, S.: Minimax policies for adversarial and stochastic bandits. In: Proceedings of the 22nd Conference on Learning Theory (COLT 2009), Montreal, Quebec, Canada, pp. 773–818 (2009)

3. Auer, P., Cesa-Bianchi, N., Fischer, P.: Finite-time analysis of the multiarmed bandit problem. Machine Learning 47(2-3), 235–256 (2002)
4. Auer, P., Cesa-Bianchi, N., Freund, Y., Schapire, R.E.: Gambling in a rigged casino: The adversarial multi-arm bandit problem. In: Proceedings of the 36th Annual Symposium on Foundations of Computer Science, pp. 322–331 (1995)
5. Dimitrakakis, C., Lagoudakis, M.G.: Rollout sampling approximate policy iteration. Machine Learning 72(3), 157–171 (2008)
6. Fürnkranz, J., Hüllermeier, E. (eds.): Preference Learning. Springer (2010)
7. Fürnkranz, J., Hüllermeier, E., Cheng, W., Park, S.H.: Preference-based reinforcement learning: a formal framework and a policy iteration algorithm. Machine Learning 89(1-2), 123–156 (2012), special Issue of Selected Papers from ECML PKDD 2011
8. Hastie, T., Tibshirani, R.: Classification by pairwise coupling. The Annals of Statistics 26, 451–471 (1998)
9. Price, D., Knerr, S., Personnaz, L., Dreyfus, G.: Pairwise neural network classifiers with probabilistic outputs. In: Proceedings of the 7th Conference Advances in Neural Information Processing Systems (NIPS 1994), vol. 7, pp. 1109–1116. MIT Press (1994)
10. Rothkopf, C.A., Dimitrakakis, C.: Preference elicitation and inverse reinforcement learning. In: Gunopulos, D., Hofmann, T., Malerba, D., Vazirgiannis, M. (eds.) ECML PKDD 2011, Part III. LNCS, vol. 6913, pp. 34–48. Springer, Heidelberg (2011)
11. Singh, S.P., Jaakkola, T., Littman, M.L., Szepesvári, C.: Convergence results for single-step on-policy reinforcement-learning algorithms. Machine Learning 38(3), 287–308 (2000)
12. Sutton, R.S., Barto, A.: Reinforcement Learning: An Introduction. MIT Press, Cambridge (1998)
13. Wirth, C., Fürnkranz, J.: Learning from trajectory-based action preferences. In: Proceedings of the ICRA 2013 Workshop on Autonomous Learning (to appear, May 2013)
14. Wu, T.F., Lin, C.J., Weng, R.C.: Probability estimates for multi-class classification by pairwise coupling. Journal of Machine Learning Research 5, 975–1005 (2004)
15. Zhao, Y., Kosorok, M., Zeng, D.: Reinforcement learning design for cancer clinical trials. Statistics in Medicine 28, 3295–3315 (2009)
16. Wilson, A., Fern, A., Tadepalli, P.: A Bayesian Approach for Policy Learning from Trajectory Preference Queries. Advances in Neural Information Processing Systems 25, 1142–1150 (2012)

Multiclass Learning
from Multiple Uncertain Annotations

Chirine Wolley and Mohamed Quafafou

Aix-Marseille University, CNRS, LSIS UMR 7296
13397, Marseille, France
chirine.wolley@etu.univ-amu.fr,
mohamed.quafafou@univ-amu.fr

Abstract. Annotating a dataset is one of the major bottlenecks in supervised learning tasks, as it can be expensive and time-consuming. Instead, with the development of crowdsourcing services, it has become easy and fast to collect labels from multiple annotators. Our contribution in this paper is to propose a Bayesian probabilistic approach integrating annotator's uncertainty in the task of learning from multiple noisy annotators (annotators who generate errors). Furthermore, unlike previous work, our proposed approach is directly formulated to handle categorical labels. This is an important point as real-world datasets often have multiple classes available. Extensive experiments on datasets validate the effectiveness of our approach against previous efficient algorithms.

Keywords: supervised learning, multiple annotators, uncertainty.

1 Introduction

A classical supervised classification problem aims at assigning a class label for each input example. Let us consider a training dataset $\{(x_i, y_i)\}_{i=1}^{N}$, where $x_i \in \Re^D$ is an instance described by D descriptors, and $y_i \in \{0, 1, 2, ..., K\}$ is the class label. The goal is to learn a function f that classify a new unseen example x as $f(x) \approx y$. However, in many application fields, obtaining the ground truth label y_i for each instance x_i of the dataset is a major bottleneck: it is often time-consuming, expensive and even sometimes impossible. For example, in medical images, only a biopsy can detect the presence of cancer. Even so, this is clearly a risky and costly procedure.One approach to tackle this issue is to make use of the vast human resources available on the Internet such as Amazon's Mechanical Turk[1]. It provides an ideal solution to collect labels from a large number of annotators in a very short amount of time: the unknown ground truth label y for an instance x is replaced by multiple judgments $y^1, y^2, ..., y^T$ given by T annotators.

Combining the knowledge from different sources is far from being a solved problem. Indeed, one major drawback of most crowdsourcing services is that we do not have control over the quality of each labeler. Consequently, crowdsourced

---

[1] http://www.mturk.com

A. Tucker et al. (Eds.): IDA 2013, LNCS 8207, pp. 438–449, 2013.

labels tend to suffer from poor quality control, as they are typically not free of errors and exhibit high variance. Hence, an important research question occurs: how to learn a good classifier when we have multiple annotators providing noisy labels instead of one absolute gold standard?

Very recently, a few approaches that aim at addressing the challenges provided by this new setting have been proposed, including [4,7,6]. The common target of this family of work is both to learn a good classifier and to evaluate the trusthworthiness of each annotator.

Although various successful techniques have been proposed to solve the problem of learning from multiple annotators, a significant obstacle remains: they all focus on the knowledge of the annotators without trying to learn from their uncertainty. However, uncertainty may be widely present in this type of data. Therefore, machine learning methods which deal with annotations from differents sources should overcome both the inconsistency and unreliability of annotators. In this paper, we are interested in the situation where the annotations are dominated by uncertain annotators, that is, annotators with a significant lack of knowledge. We develop an effective probabilistic Bayesian approach to learning from multiple noisy labelers, and we show that our method is clearly more robust than other efficient algorithms when confronted with high levels of uncertainty. Another strength of our proposed algorithm is that it is directly formulated in order to handle multi-class classification. Indeed, most of the previous work focused on the solution in the two class case, where the labels y_i can only take the modalities 0 and 1. We believe this is a major drawback as many real-world applications have more than two classes to deal with. Typical examples include text categorization, object and gesture recognition, and microarray data analysis.

This paper is organized as follows; the next section reviews related work in the literature. Section 3 explains in detail the proposed framework, followed by the empirical evaluation in Section 4. Finally, Section 5 presents our conclusions.

2   Background

The problem of building classifiers in the presence of several labelers has been receiving increasing attention over the past years in many areas. For example, in the computer vision community [9] and in the natural language processing [8], authors use Amazon's Mechanical Turk to collect annotations from multiple labelers. In [5], authors show that building a classifier using annotations from many people can be potentially as good as employing one annotation provided one single expert.

In a more supervised learning context, a very recent and important work has been developed by Raykar and al. in [4]: their EM-based algorithm estimates the actual hidden labels and evaluates the reliability of each annotator. Many other papers aim at addressing the same problem. In [7,6], a probabilistic model is developed for learning a classifier from multiple annotators, where the effectiveness of an annotator may vary depending on both his reliability, and the data instance presented.

All the above works aim to address the problem of learning from multiple noisy annotators, i.e, annotators that generate errors when labeling. However, they all focus on the annotators' knowledge without trying to learn from their uncertainty. We here believe this is a major drawback of all these previous works as uncertainty and fuzziness are widely present in this kind of study.

Introduced in [10], uncertainty is widely present in many application areas and several different approaches have been developed dealing with this issue. In a context of multi-class classification, we are not aware of any approach integrating annotator's uncertainty in order to generate a classifier. We address the issue by simultaneously generating a classifier and estimating the performance of each annotator. Annotators may have a significant lack of knowledge and may feel uncomfortable with labeling an instance in unsure situations. Therefore, unlike [4,7,6], we let annotators express their uncertainty by adding the flag '?' to the label they propose. Our classifier is then generated taking both annotators knowledge and uncertainty into consideration. Finally, an entire spectrum of different levels of uncertainty exists, ranging from the unachievable ideal of complete understanding at one end of the scale to total ignorance at the other. For seek of simplicity, we consider in this paper the case where uncertainty is equivalent to total ignorance. From this point, the closest approach to our work has been studied in [11], where a classification method integrating annotator's ignorance has been developed. However, their model only handles binary classification, and we here believe this is a major drawback as real datasets have often multiple classes available.

In this paper, we present a probabilistic Bayesian method to learning from multiple uncertain annotators. Extensive experiments on several datasets validate the effectiveness of our approach against the baseline method of Raykar et al. [4].

3   Problem and Solution Overview

Given N instances $\{x_1, ..., x_N\}$, each data point x_i is described by D descriptors and labeled by T noisy annotators. In order to obtain a classifier dealing with the uncertainty issue, the labelers are asked to mark '?' in unsure situations, in addition to the label they give. From this point, for (K+1) possible classes we have $y_i^t \in \mathcal{Y} = \{0, 1, ..., K\} \cup \{(k, ?)\}_{k=0}^K$. Let z_i be the true label for the i-th instance, z_i is unknown and z_i belongs to $\mathcal{Z} = \{0, 1, ..., K\}$. For compactness, we define the matrices $X = [x_1^T; ...; x_N^T] \in R^{N \times D}$ is the matrix transpose, $Y = \left[y_1^{(1)}, ..., y_1^{(T)}; ...; y_N^{(1)}, ..., y_N^{(T)}\right] \in R^{N \times T}$ and $Z = [z_1, ..., z_N]^T$.

One question occurs: how to integrate all uncertain labels $\{(k, ?)\}_{k=0}^K$ in our approach? In other word, how uncertainty should be modeled ? For the sake of simplicity, we consider in this paper uncertainty as a state of total ignorance (cf Section 2), and we define the matrix of Ignorance $H \in R^{N \times T}$ as follows:

$$h_i^t = \begin{cases} 1 \text{ if } y_i^t = '?' \\ 0 \text{ otherwise} \end{cases} \tag{1}$$

H has been introduced in [11] in case of binary classification. We here extend it to handle categorical variables. From this point, H can be seen as the traceability of the ignorance of the annotators. For compactness, we set $H = \left[h_1^{(1)}, ..., h_1^{(T)}; ...; h_N^{(1)}, ..., h_N^{(T)}\right] \in R^{N \times T}$.

Conditionally to the uncertainty (i.e ignorance) of each labeler, our goal is to:

(1) Estimate the ground truth labels Z,
(2) Produce a classifier to predict the label z for a new instance x,
(3) Estimate the reliability of each annotator.

The joint conditional distribution can be expressed as $P(X, Y, Z|H, \Theta)$, where Θ is the set of parameters to be estimated (defined later in the paper). In this model, the annotation provided by labeler t depends on the unknown ground truth label z, but also on the domain of ignorance h of each annotator.

$P(X, Y, Z|H, \Theta)$ is estimated using a Bayesian approach: we set different prior distributions on the parameters depending on whether the annotator is sure or not of his response. If $h_i^t = 0$, we assume labeler t is almost sure of his label for the instance x_i. Therefore, a high prior is considered in this case. On the contrary, if $h_i^t = 1$, labeler t is uncertain about the label to assign, and a less informative prior is set in this case.

All the parameters Θ are estimated by maximising the log-posterior with the maximum-a-posteriori estimator (MAP):

$$\hat{\Theta}_{MAP} = argmax_{\Theta} \{ lnPr[X, Y, Z|H, \Theta] + lnPr[\Theta] \} \qquad (2)$$

These estimates are obtained using a combination of the Expectation-Maximization (EM) Algorithm [2] and the Newton-Raphson update.

4   The Model

4.1   Maximum a Posteriori Estimator

In this work, we make the assumption that the labelers (resp. the instances) are independent from each other. As the ground truth labels z are supposed unknown for all instances, we first estimate the probability $P(X, Y|H, \Theta)$. z will be integrated later in the model as missing variables. The likelihood of $P(X, Y|H, \Theta)$ can be written as follows:

$$Pr(X, Y|H, \Theta) = \prod_{i=1}^{N} \prod_{t=1}^{T} Pr[y_i^t|x_i, h_i^t, \Theta] Pr[x_i|h_i^t, \Theta] \qquad (3)$$

$$\propto \prod_{i=1}^{N} \prod_{t=1}^{T} Pr[y_i^t|x_i, h_i^t, \Theta] \qquad (4)$$

where the term $Pr(X|H, \Theta)$ is dropped as we are more interested in the other conditional distribution. Conditionning on the ground true label z_i, and also making the assumption y_i^t is conditionally independent (of everything else) given Θ and z_i, the likelihood of $P(X, Y, Z|H, \Theta)$ can be decomposed as:

$$Pr[X, Y, Z|H, \Theta] \propto \prod_{i=1}^{N} \sum_{k=0}^{K} Pr(z_i = k|x_i, \Theta_{z_k}) \prod_{t=1}^{T} Pr(y_i^t|z_i = k, h_i^t, \Theta_{y_k}^t) \quad (5)$$

We estimate each conditional-distribution individually.

Concerning $Pr(z_i = k|x_i, \Theta_{z_k})$, while the proposed method can use any other multi-class classifier, we set for simplicity a multinomial logistic regression distribution, i.e:

$$Pr(z_i = k|x_i, \Theta_{z_k}) = \begin{cases} \frac{e^{w_k^T x_i}}{1 + \sum_{k=0}^{K-1} e^{w_k^T x_i}} & \forall k \neq K \\ \frac{1}{1 + \sum_{k=0}^{K-1} e^{w_k^T x_i}} & \text{otherwise} \end{cases} \quad (6)$$

Hence, we have $\Theta_{z_k} = \{w_k\}, \forall k \in \{0, 1, ..., K\}$

Concerning $Pr(y_i^t|z_i = k, h_i^t, \Theta_{y_k}^t)$, we model it as a mixture of two multinomial distributions, depending on the value of h. We introduce two vectors of multinomial parameters $\alpha_k^t = (\alpha_{k0}^t, ..., \alpha_{kK}^t)$ and $\beta_k^t = (\beta_{k0}^t, ..., \beta_{kK}^t)$ such as:

$$y_i^t \sim \begin{cases} Multinomial(\alpha_k^t) & \text{if } h_i^t = 0 \\ Multinomial(\beta_k^t) & \text{otherwise} \end{cases}$$

with $\sum_{c=0}^{K} \alpha_{kc}^t = 1$ and $\sum_{c=0}^{K} \beta_{kc}^t = 1$. Here, α_{kc}^t (resp. β_{kc}^t) denotes the probability that the annotator t assigns class c to an instance given the ground truth label k in case of knowledge (resp. uncertainty).

Our motivation to model y_i^t as a mixture of two multinomial distributions comes from the fact that later in the paper, we will set different prior distributions on the parameters depending on the value of h.

From this point, we have $\Theta_{y_k}^t = \{\alpha_k^t, \beta_k^t\}, \forall k \in \{0, 1, ..., K\}, \forall t \in \{1, 2, ..., T\}$, and:

$$P(y_i^t|z_i = k, h_i^t, \Theta_{y_k}^t) = (1 - h_i^t) * \prod_{c=0}^{K} (\alpha_{kc}^t)^{\delta(y_i^t, c)} + h_i^t * \prod_{c=0}^{K} (\beta_{kc}^t)^{\delta(y_i^t, c)} \quad (7)$$

where $\delta(u, v) = 1$ if u = v, 0 otherwise.

We denote:

$$p_i^k = P(z_i = k|x_i, \Theta_{z_k}) \quad (8)$$

$$a_i^k = \prod_{t=1}^{T} Pr(y_i^t|z_i = k, h_i^t, \Theta_{y_k}^t) \quad (9)$$

Finally, the likelihood can be written:

$$Pr(X, Y, Z|H, \Theta) \propto \prod_{i=1}^{N} \sum_{k=0}^{K} p_i^k a_i^k \tag{10}$$

where $\Theta = \left\{ \Theta_{z_k}, \Theta_{y_k}^t \right\} = \{w_k, \alpha_k^t, \beta_k^t\}$.
Next section describes which prior is assumed on each parameter.

4.2   Prior Distributions

Different prior distributions are set on the parameters, depending on the behavior of the annotator when labeling. More precisely, we assume that an annotator is almost sure of his response when he labels an instance. Hence, a high prior is fixed in this case. On the contrary, he labels '?' in case of doubt, and a less informative prior is fixed.

Prior on Certain Annotations. In case of knowledge, the parameters to take into consideration are the parameters $\{\alpha_{kc}^t\}$ of the Multinomial distribution. A prior often used for this distribution is the Dirichlet prior, as it represents its conjugate prior. Hence, we set a Dirichlet distribution with parameters $\{\gamma_{kc}^t\}$ on all parameters $\{\alpha_{kc}^t\}$, and we have:

$$\alpha_k^t = (\alpha_{k0}^t, \alpha_{k1}^t, ..., \alpha_{kK}^t) \sim Dirichlet(\gamma_{k0}^t, \gamma_{k1}^t, ..., \gamma_{kK}^t)$$

Parameters $\{\gamma_{kc}^t\}$ are estimated as follow: Let $\lambda = \{\lambda_1, \lambda_2, ..., \lambda_N\}$ be the parameters of a Dirichlet distribution $X = \{X_1, X_2, ..., X_N\}$. λ is estimated by fixing the mean and the variance, and solving:

$$E[X_i] = \frac{\lambda_i}{\sum_{i=1}^{N} \lambda_i} \tag{11}$$

$$Var[X_i] = \frac{\lambda_i(\sum_{i=1}^{N} \lambda_i - \lambda_i)}{(\sum_{i=1}^{N} \lambda_i)^2(\sum_{i=1}^{N} \lambda_i + 1)} \tag{12}$$

Solving these equations according to λ_i, we obtain $\lambda_i = E[X_i] \sum_{i=1}^{N} \lambda_i$ with $\sum_{i=1}^{N} \lambda_i = \frac{E[X_i](1-E[X_i])}{Var[X_i]} - 1$. As $h^t = 0$, we assume annotator t is sure when labeling. Hence, we set $E[\alpha_{kk}^t] = 0.9$, and as $\sum_{c=0}^{K} \alpha_{kc}^t = 1$, we have $E[\alpha_{kc}^t] = 0.1/K \; \forall \; k \neq c$. Concerning the variance, we set $Var[\alpha_{kc}^t] = 0.01$ to reflect a high confidence on the prior. Finally, equations (11) and (12) are used to estimate the parameters γ_{kc}^t.

Prior on Uncertain Annotations. In case of uncertainty, the parameters to be estimated in the model are the parameters $\{\beta_{kc}^t\}$ of the multinomial distribution. We consider and analyse three different less informative prior distributions:

Dirichlet Prior: As for α_k^t, a Dirichlet prior can also be fixed on β_k^t. Hence, we have:

$$\beta_k^t = (\beta_{k0}^t, \beta_{k1}^t, ..., \beta_{kK}^t) \sim Dirichlet(\phi_{k0}^t, \phi_{k1}^t, ..., \phi_{kK}^t)$$

However, as $h^t = 1$, we assume that annotators ignore the label to give. Hence, we set $E\left[\beta_{kc}^t\right] = 1/(K+1)$ and $Var\left[\beta_{kc}^t\right] = 0.01, \forall \ \{k,c\} \in \{0,1,2,...,K\}$. Equations (11) and (12) are used again to estimate the parameters $\{\phi_k^t\}$.

Uniform Prior: Non-informative priors are often used to model ignorance, as they reflect a situation where there is a lack of knowledge about a parameter. A non-informative prior often used is the uniform distribution, which can be written with a dirichlet distribution as follows: $\beta_k^t \sim Dirichlet(\mathbf{1})$.

Jeffreys Prior: Another non-informative prior widely used is Jeffreys prior [3]. For β_k^t, we obtain: $\beta_k^t \sim Dirichlet(\mathbf{0.5})$.

Prior on the Weights $\{w_k\}$. For the sake of completeness, we assume a zero mean Gaussian prior on the weights w with inverse covariance matrix Γ: $w_k \sim N(w_k|0, \Gamma_k^{-1})$.

4.3   Computing Issues

We want to estimate the parameters $\hat{\Theta}_{MAP}$ by calculating (2). This maximisation problem can be simplified a lot using the Expectation-Maximisation (EM) algorithm [2]. Indeed, the EM algorithm is an iterative method for finding maximum likelihood or maximum a posteriori (MAP) estimates of parameters in statistical models, when the model depends on unobserved variables. In our case, the ground truth label z are unknown and will be considered as the unobserved variables. The EM algorithm alternates between two steps: the expectation step (E-Step), which creates a function for the expectation of the log-likelihood, and the Maximisation step (M-Step), which computes parameters maximizing the expected log-likelihood found on the E-Step.

Applying the EM algorithm in our model we obtain:

E-Step: The conditional expectation is computed as

$$E\left[ln \, Pr\left[Y, Z, X|H, \Theta\right]\right] \quad \propto \quad \sum_{i=1}^{N}\sum_{k=0}^{K}\mu_i^k ln\left[p_i^k a_i^k\right] \quad + \quad ln[Pr(\Theta)] \quad (13)$$

where $\mu_i^k = Pr\left[z_i = k|y_i^1, ..., y_i^T, x_i, h_i^1, ..., h_i^T, \Theta\right]$.

Using Bayes theorem and equations (8) and (9), we obtain the following expression for μ_i^k :

$$\mu_i^k \quad \propto \quad P(z_i = k|x_i, \Theta_{z_k}) \prod_{t=1}^{T} Pr(y_i^t|z_i = k, h_i^t, \Theta_{y_k}^t) \qquad (14)$$

$$= \quad \frac{p_i^k * a_i^k}{\sum_{k=0}^{K} p_i^k a_i^k} \qquad (15)$$

M-Step: Parameters Θ are estimated by equating, according to each parameter, the gradient of (13) to zero. In our case, if $\alpha_k^t \sim Dirichlet(\gamma_{k0}^t, \gamma_{k1}^t, ..., \gamma_{kK}^t)$ and $\beta_k^t \sim Dirichlet(\phi_{k0}^t, \phi_{k1}^t, ..., \phi_{kK}^t)$, we obtain:

$$\alpha_{kc}^t = \frac{\gamma_{kc}^t - 1 + \sum_{i \in \{i|h_i^t=0\}} \mu_i^k \delta(y_i^t, c)}{\sum_{c=0}^K \gamma_{kc}^t - (K+1) + \sum_{i \in \{i|h_i^t=0\}} \mu_i^k} \tag{16}$$

$$\beta_{kc}^t = \frac{\phi_{kc}^t - 1 + \sum_{i \in \{i|h_i^t=1\}} \mu_i^k \delta(y_i^t, c)}{\sum_{c=0}^K \phi_{kc}^t - (K+1) + \sum_{i \in \{i|h_i^t=1\}} \mu_i^k} \tag{17}$$

Concerning w_k, $\forall k \in \{0, 1, .., K\}$ we use the Newton-Raphson method given by:

$$w_k^{q+1} = w_k^q - \eta \Omega_k^{-1} g_k \tag{18}$$

where g_k is the gradient vector, Ω_k the Hessian matrix and η the step length. The gradient vector is given by:

$$g_k(w_k) = \sum_{i=1}^N \left[\mu_i^k - p_i^k \right] x_i - \Gamma_k w_k \tag{19}$$

and the Hessian matrix is given by

$$\Omega_k(w_k) = -\sum_{i=1}^N \left[p_i^k \right] \left[1 - p_i^k \right] x_i x_i^T - \Gamma_k \tag{20}$$

The steps E and M are iterated until convergence.. μ_i^k initialized using $\frac{1}{T}\sum_{t=1}^T \delta(y_i^t, k)$. Once the parameters are estimated in the EM algorithm, a new instance x_i is classified by calculating $p(z_i = k|x_i) = (1 + exp(-w_k^T x_i))^{-1}$, $\forall k \in \{0, 1, ..., K\}$, the probability that x_i has the ground truth label k.

5   Experiments

5.1   Data Description and Preprocessing

Simulations have been performed on twelve datasets from the UCI Machine Learning Repository [1] for which categorical labels are available. The number of classes ranges from 3 to 10, the number of samples ranges from 106 to 6435 and the number of attributes ranges from 4 to 36. Each data has been randomly divided into two folds: a training and a testing set, representing respectively 80% and 20% of the data. See Table 1 for more details. For each training dataset D, we simulate T=5 annotators as follows:

1. D is divided into T folds $\{d_1, d_2, ..., d_T\}$ using K-means.

Table 1. The description of datasets

Dataset name	Id	# trainings	# tests	# attributes	# classes
Breast Tissue	1	85	21	10	6
Cardiotography	2	1701	425	23	3
Ecoli	3	269	67	8	8
Iris	4	120	30	4	3
Seeds	5	168	42	19	7
Image Segmentation	6	1848	462	19	7
Vertebral Column	7	248	62	6	3
Wine	8	143	35	13	3
Yeast	9	9	1188	496	8
Dermatology	10	293	73	34	6
Satimage	11	5148	1287	36	6
Vehicle	12	757	189	18	4

Table 2. Experimental results (error mean ± std) of Multinomial Logistic Regression, Baseline, Dirichlet prior, Jeffreys prior and Uniform prior for all datasets

Id	MLR	Baseline	Error rate across different ignorance levels Dirichlet	Jeffreys	Uniform
1	0.622 ± 0.006	0.561 ± 0.003	0.419 ± 0.004	0.414 ± 0.004	0.404 ± 0.003
2	0.567 ± 0.005	0.541 ± 0.004	0.495 ± 0.001	0.493 ± 0.001	0.493 ± 0.001
3	0.247 ± 0.013	0.210 ± 0.008	$0.163 \pm 8 \times 10^{-4}$	$0.166 \pm 8 \times 10^{-4}$	$0.171 \pm 8 \times 10^{-4}$
4	0.228 ± 0.002	0.192 ± 0.001	$0.174 \pm 1 \times 10^{-4}$	$0.176 \pm 1 \times 10^{-4}$	$0.173 \pm 1 \times 10^{-4}$
5	0.119 ± 0.003	0.071 ± 0.001	$0.066 \pm 5 \times 10^{-4}$	$0.067 \pm 2 \times 10^{-4}$	$0.057 \pm 4 \times 10^{-4}$
6	0.354 ± 0.003	$0.298 \pm 3 \times 10^{-4}$	$0.096 \pm 6 \times 10^{-4}$	$0.100 \pm 9 \times 10^{-4}$	0.107 ± 0.001
7	0.613 ± 0.003	0.587 ± 0.003	$0.530 \pm 3 \times 10^{-4}$	$0.539 \pm 6 \times 10^{-4}$	$0.535 \pm 6 \times 10^{-4}$
8	0.247 ± 0.013	0.210 ± 0.008	$0.166 \pm 8 \times 10^{-4}$	$0.166 \pm 8 \times 10^{-4}$	$0.171 \pm 9 \times 10^{-4}$
9	0.539 ± 0.005	0.461 ± 0.002	$0.448 \pm 1 \times 10^{-4}$	$0.448 \pm 1 \times 10^{-4}$	$0.449 \pm 2 \times 10^{-4}$
10	0.216 ± 0.020	0.187 ± 0.014	$0.081 \pm 3 \times 10^{-4}$	$0.081 \pm 3 \times 10^{-4}$	$0.082 \pm 3 \times 10^{-4}$
11	0.236 ± 0.008	0.180 ± 0.001	$0.156 \pm 5 \times 10^{-4}$	$0.145 \pm 4 \times 10^{-4}$	$0.172 \pm 4 \times 10^{-4}$
12	0.294 ± 0.005	0.268 ± 0.002	$0.222 \pm 1 \times 10^{-4}$	$0.232 \pm 2 \times 10^{-4}$	$0.225 \pm 1 \times 10^{-4}$
Mean	$\mathbf{0.356 \pm 0.007}$	$\mathbf{0.314 \pm 0.004}$	$\mathbf{0.252 \pm 8 \times 10^{-4}}$	$\mathbf{0.252 \pm 7 \times 10^{-4}}$	$\mathbf{0.253 \pm 6 \times 10^{-4}}$

2. We assume annotator t is an expert for the set d_t, i.e:

$$\begin{cases} Error(t, d_t) = 10\% \\ Uncertainty(t, d_t) = 0\% \end{cases}$$

In other word, annotator t always gives the correct label and is always confident for the set d_t.

3. On the rest of the data $\bar{d}_t$, we assume annotator t makes 10% of errors and U% of uncertainty, $U \in \{0\%, 10\%, ..., 80\%, 90\%\}$:

$$\begin{cases} Error(t, \bar{d}_t) = 10\% \\ Uncertainty(t, \bar{d}_t) = U\% \end{cases}$$

Let $\mathcal{E}$ be the set of instances with errors, $\mathcal{U}$ the set of instances with uncertainty and $\mathcal{R}$ the rest of the instances. We have:

$$\begin{cases} Label(t, \mathcal{E}_i) = \{0, ..., K\} - z_i \\ Label(t, \mathcal{U}_i) = Random(0, ..., K) \\ Label(t, \mathcal{R}_i) = z_i \end{cases}$$

In our experiments, uncertain annotations '?' are replaced by simulating randomly any label. This choice is motivated by the fact that in our approach, we considered uncertainty as a form of total ignorance, which means that

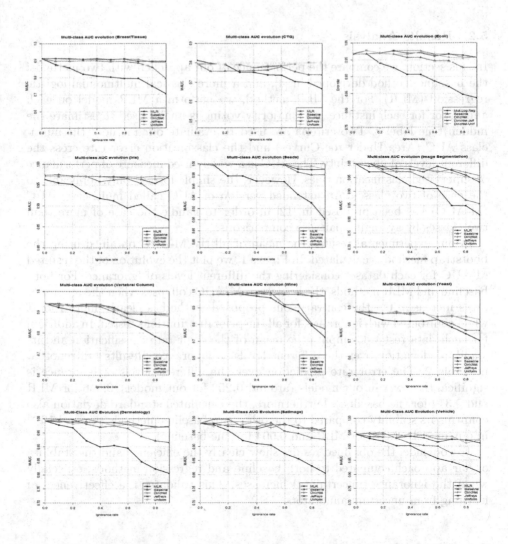

Fig. 1. Comparison of Multi-class AUC between Multinomial Logistic Regression, Baseline, Dirichlet prior, Jeffreys prior, Uniform prior

annotators give randomly any label in unsure situations. We simulate our method for different levels of uncertainty in order to test how robust is our method when confronted with ignorance. Each model has been simulated a hundred times using the bootstrap method.

5.2   Results Analysis

In this section, we compare the performance of our approach with two methods: the baseline method developed in [4], and a more classical multinomial logistic regression (MLR). For the MLR method, we perform a MLR model on each annotator for each instance, and majority voting is finally used to estimate the hidden true label z. Two criteria are used to evaluate our method: the multi-class AUC (Area Under roc Curves) and the classification error rate cross the different levels of uncertainty. The AUC is a widely used measure of performance of supervised classification rules. However, the simple form is only applicable to the case of two classes. An extended version of AUC called multi-class AUC (M-AUC) has been proposed in [12] in order to handle the case of more than two classes by averaging pairwise comparisons.

In our experimental results, the mean of all the M-AUC obtained using the bootstrap method is calculated. In Figure 1, we plot the evolution of the resulted M-AUC for each dataset considering the different levels of ignorance. For both baseline and MLR methods, the bootstrap M-AUC collapses when the ignorance rates increases. On the contrary, our proposed method is clearly more robust when confronted with ignorance, for all the prior distributions tested. In addition, for each data test we compute the mean of the error rate classification and its standard deviation over the different levels of ignorance. Results are reported in Table 2. The error rate prediction over the different levels of ignorance is significantly lower in our models (around 0.252 for our models, 0.356 for MLR and 0.314 for the baseline). Furthermore, the calculated standard deviation also confirms its stability compared to the other two baseline methods (around 10^{-4} in our models, 0.007 for MLR and 0.004 for the baseline).

To conclude, the obtained results show clearly the efficiency and the stability of our approach compared to both baseline and regression methods, especially when the ignorance (uncertainty) increases. This validates the effectiveness of learning from uncertain annotators.

6   Conclusion

We presented in this paper a novel probabilistic approach to learning from multiple doubtful annotators in a context of multi-class classification. Our main contribution was a comprehensive algorithm that lets the labelers express their uncertainty using a specific "uncertain" flag, and further integrates these labels using different prior distributions. Our model goes beyond current state of the art algorithms as it is formulated to handle categorical labels. Indeed, this is an important property of our model because real-world datasets usually have multiple classes available. Results of our experiments over many datasets validate our approach and prove that it outperforms previous efficient algorithms.

References

1. Asuncion, A., Newman, D.: Uci machine learning repository (2007)
2. Dempster, A.P., Laird, N.M., Rubin, D.B.: Maximum likelihood from incomplete data via the em algorithm. Journal of the Royal Statistical Society, Series B 39(1), 1–38 (1977)
3. Jeffreys, H.: An invariant form for the prior probability in estimation problems. Proceedings of the Royal Society of London. Mathematical and Physical Sciences, 453–461 (1946)
4. Raykar, V.C., Yu, S., Zhao, L.H., Valadez, G.H., Florin, C., Bogoni, L., Moy, L.: Learning from crowds. Journal of Machine Learning Research 11, 1297–1322 (2010)
5. Snow, R., O'Connor, B., Jurafsky, D., Ng, A.Y.: Cheap and fast - but is it good? evaluating non-expert annotations for natural language tasks. In: Proceedings of the Conference on Empirical Methods in Natural Language Processing, pp. 254–263. Association for Computational Linguistics, Stroudsburg (2008)
6. Whitehill, J., Ruvolo, P., Wu, T., Bergsma, J., Movellan, J.R.: Whose vote should count more: Optimal integration of labels from labelers of unknown expertise. In: NIPS, pp. 2035–2043 (2009)
7. Yan, Y., Rosales, R., Fung, G., Schmidt, M.W., Valadez, G.H., Bogoni, L., Moy, L., Dy, J.G.: Modeling annotator expertise: Learning when everybody knows a bit of something. Journal of Machine Learning Research - Proceedings Track, 932–939 (2010)
8. Sheng, V.S., Provost, F., Ipeirotis, P.G.: Get another label? Improving data quality and data mining using multiple, noisy labelers. In: Proceedings of the 14th ACM SIGKDD International Conference on Knowledge Discovery and Data Mining, pp. 614–622 (2008)
9. Sorokin, A., Forsyth, D.: Utility data annotation with Amazon Mechanical Turk. In: Proceedings of the First IEEE Worshop on Internet Vision at CVPR 2008, pp. 254–263 (2008)
10. Goetghebeur, E., Molenberghs, G., Kenward, M.G.: Sense and sensitivity when intended data are missing. Kwantitatieve Technieken 62, 79–94 (1999)
11. Wolley, C., Quafafou, M.: Learning from multiple naive annotators. In: Zhou, S., Zhang, S., Karypis, G. (eds.) ADMA 2012. LNCS, vol. 7713, pp. 173–185. Springer, Heidelberg (2012)
12. Hand, D.J., Till, R.J.: A simple generalisation of the Area Under the ROC Curve for multiple class classification problems. Mach. Learn. 45, 171–186 (1995)

Learning Compositional Hierarchies
of a Sensorimotor System

Jure Žabkar[1,*] and Aleš Leonardis[2,3,**]

[1] AI Lab, University of Ljubljana, Faculty of Computer and Information Science,
Tržaška 25, SI-1000 Ljubljana, Slovenia
jure.zabkar@fri.uni-lj.si
[2] Intelligent Robotics Lab, School of Computer Science,
University of Birmingham, UK
[3] Centre for Computational Neuroscience and Cognitive Robotics,
University of Birmingham, UK

Abstract. We address the problem of learning static spatial representation of a robot motor system and the environment to solve a general forward/inverse kinematics problem. The latter proves complex for high degree-of-freedom systems. The proposed architecture relates to a recent research in cognitive science, which provides a solid evidence that perception and action share common neural architectures. We propose to model both a motor system and an environment with *compositional hierarchies* and develop an algorithm for learning them together with a mapping between the two. We show that such a representation enables efficient learning and inference of robot states. We present our experiments in a simulated environment and with a humanoid robot Nao.

Keywords: compositional hierarchy, sensorimotor representation, computational modeling.

1 Introduction

Learning to play a cello and improving your skills in a video game surprisingly have a lot in common. In both cases, the interaction between perceptual and motor learning is a crucial term of success [35]. As the pitch helps a cello player to improve his/her grip on the strings, vision guides a game player to shorten the reaction time or improve on the accuracy. In the early stage of learning, the motor system samples the sensory space, which iteratively helps to improve the skills of the motor system [11,10]. This intertwined learning process is building our inner sensorimotor representation [20,15], which has developed a plethora of

* This work was partially supported by ARRS grant J2–4222, Machine learning for building intelligent tutoring systems (basic research project).
** This work was partially supported by ARRS grants: J2–4284, J2–3607, Programme P2–0214 and EU project PaCMan, Probabilistic and Compositional Representations of Objects for Robotic Manipulation.

A. Tucker et al. (Eds.): IDA 2013, LNCS 8207, pp. 450–461, 2013.

remarkable capabilities through evolution, such as efficient learning and generalization, and the ability of abstraction that enables reasoning.

Although computational models of sensorimotor systems [22,4,33] have made great progress in the last decade, the gap between the performance of the artificial sensorimotor systems and a human is still huge. The advantages of the state-of-the-art robotic systems are in sensor accuracy and high speed processing while they are significantly lacking in the efficient architecture of the sensorimotor systems.

In this paper, we present a computational representation inspired by the architecture of human brain. The properties of the brain that we try to mimic in the suggested representation are the following [2,9]:

- deep architecture: mammalian brain is organized in a deep architecture where perceptual input presents at multiple levels of abstraction. Learning deep architectures was introduced in [12].
- sparse distributed representation: many neurons get activated simultaneously. In each level of abstraction, the concepts are compositions of lower-level features sharing the features across different domains/tasks [26,19,2].
- unsupervised learning: learning deep architectures is carried out layer by layer in an unsupervised manner [12].

Specifically, we address the problem of learning a static spatial representation of robot/human motor system, representation of the environment and the connection between the two. We focus on Compositional Hierarchical Abstract Representation of Motor System (CHARMS), which is based on motor primitives [7]. In our setting, motor primitives are joint angles, i.e. the angles formed at the joints of the robot. CHARMS combines and abstracts motor primitives into higher layer concepts. The proposed architecture relates to recent research in cognitive science [33,16,36], which suggests that hierarchical representations are best suited for computational models of human cognition [6,3,33]. Similarly, in computer vision, the principle of hierarchical compositionality has recently proved very successful for visual object categorization [8].

CHARMS can generalize learning from motor primitives across different types of behavior. In robotics, several impressive examples of motor learning have been presented [14,21,23,32], yet only recently [34,17,1], the authors addressed the problem of generalization without re-learning the task. The approach we propose here, fundamentally differs from the above since it is biologically inspired and relates to memory models rather than differential equations. Our work also relates to the field of *learning robot control* [28], more specifically to learning sensorimotor models based on motor primitives [7,30,18]. To our knowledge, the existing work in this area uses dynamic movement primitives [29] either in learning by imitation or reinforcement learning. The models are predominantly expressed in the forms of differential equations [14,5,17].

The organization of the paper is as follows: first, we propose our computational model of sensorimotor system and introduce the architecture through a simple example. We continue by describing the process of building a hierarchical structure from sensorimotor data. We present the experiments in a simulated environment and humanoid robot Nao. In the simulator, we observed a n-degrees of freedom (DOF) manipulator acting in a planar task-space. We modeled the forward and inverse kinematics for different degrees-of-freedom and different sizes of training data.

Forward kinematics traditionally refers to using kinematic equations of a robot to compute the position of its end-effector given positions of its joints (i.e. joint configuration). Similarly, inverse kinematics refers to the mapping in the opposite direction: compute the positions of the joints given the position of the robot's end-effector. The mappings between robot's inner coordinate system (configuration space) and its environment (work space) are important because motor commands control the robot in the configuration space while the tasks for the robot are usually defined in its work space. In practice, kinematic models are either too complex, inaccurate or uncalibrated with robot's sensors perceiving the environment (e.g. cameras). In these situations, learning forward/inverse kinematics [7] provides an alternative way to analytical solutions.

In our setting, the training data were collected by motor babbling - randomly generated motor commands. First, we analyzed the generalization property of our representation by measuring root mean squared error (RMSE) in 10-fold cross-validation and varying the parameters: the size of the learning set and DOF. We show that the error decreases exponentially with the increasing size of the training set. Second, we studied the spatial complexity of the representation by measuring the number of nodes in each layer of the compositional hierarchy. While the convergence is quite slow when the compositional hierarchy (CH) is learned from motor babbling data, we show that learning from traces of intended actions induces rather small compositional hierarchies. We illustrate this by capturing the whole body motion of humanoid robot Nao (26 DOF). We conlude with the discussion of the results and possible future work.

2   Sensorimotor Representation

A human or a robot operating in a physical environment needs to combine the sensory information about the environment and the inner representation of its own embodiment. The sensory information typicaly comes from several different sources/sensors, e.g. visual [27], audio [25], touch sensors [35]. By the inner representation of the embodiment we refer to the coordinate system of the joints. In the following we propose a sensorimotor representation that combines the representation of the environment and the representation of the motor system in the form of compositional hierarchies.

2.1   The Architecture: Compositional Hierarchy

Compositional hierarchy (CH) is a hierarchical AND-OR graph, where AND nodes represent compositions and OR nodes represent alternative options. The bottom layer of a CH, $\mathcal{L}^0$, is a set of primitive nodes that represent the features of the given domain. The layers of the CH represent different levels of abstraction of the domain. The representation is built bottom-up, starting with singletons, which we combine and abstract in the subsequent layers. The process of composition/abstraction is recursively repeated until the level of abstraction covers the entire domain.

For example, let us represent a vector in a given domain by a compositional hierarchy. Let x, y, z be variables defined over the same domain $I = [0, 8]$ and let $\{1, 2, \ldots, 8\}$ be the finest quantization (resolution) of I. Our goal is to represent the point $P = (2, 7, 4)$ in the domain $I \times I \times I$.

In the primitive layer $\mathcal{L}^0$, P is represented by nodes $x = 2$, $y = 7$ and $z = 4$. The resolution of $\mathcal{L}^0$ is 1, i.e. variables x, y and z are quantized into intervals $[i, i+1)$ for $i = 0, \ldots, 7$. Layer 0 nodes are combined pairwise into three $\mathcal{L}^1$ nodes. Note that in $\mathcal{L}^1$ we also reduce the resolution from 1 to 2 units, so the layer $\mathcal{L}^1$ is an abstraction of the primitive layer $\mathcal{L}^0$. In this simple example, there is only a single $\mathcal{L}^2$ node that can be constructed in three different ways. In layer 2, the dimensionality of the node equalizes with the dimensionality of the domain. The layer $\mathcal{L}^3$ only abstracts $\mathcal{L}^2$ but can not do node compositions. Since $\mathcal{L}^3$ node covers the entire domain, no further abstractions are possible.

Our representation of the domain $I \times I \times I$ has 4 layers. The top layer is a single (root) node that covers the entire domain. The root node is the most abstract node in the hierarchy, i.e. it has the lowest possible resolution. Figure 1 shows the formal AND-OR graph representation of the compositional hierarchy.

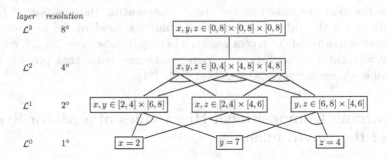

Fig. 1. An example representation of point $P = (2, 7, 4)$

To illustrate the benefits of compositionality and abstraction, let us examine the complexity of the above representation and compare it with a flat representation of the domain. We measure the complexity by counting the number of nodes in the representation. Flat representation of our domain has $8 \times 8 \times 8 = 512$ three-dimensional features.

In layer $\mathcal{L}^0$, our representation has 8 features per variable, i.e. $3 \times 8 = 24$ nodes. We combine $\mathcal{L}^0$ nodes into $\mathcal{L}^1$ nodes, but we reduce the resolution by half. So in $\mathcal{L}^1$, the quantization of each variable is $\{2, 4, 6, 8\}$, which gives $4 \times$

$4 \times 4 = 64$ possible nodes in $\mathcal{L}^1$. The $\mathcal{L}^1$ nodes are two-dimensional. There are 3 different combinations of variables x, y, z, each having $4 \times 4 = 16$ possible features. Compositional hierarchy therefore contains $3 \times 16 = 48$ nodes in layer $\mathcal{L}^1$.

The resolution is again decreased in $\mathcal{L}^2$, where two-dimensional nodes from $\mathcal{L}^1$ are combined into three-dimensional nodes. The quantization in $\mathcal{L}^2$ is $\{4, 8\}$, which gives $2 \times 2 \times 2 = 8$ possible concepts. Altogether, spatial compositional hierarchy needs $24 + 48 + 8 = 80$ nodes instead of $512 + 64 + 8 = 584$ as a non-compositional hierarchical structure would. The savings increase by increasing dimensionality and the complexity can be controlled by changing the resolution in each layer.

The representation of the environment. Our representation of the environment is a compositional hierarchy similar to the one from the introductory example - in a way it is even simpler as we only deal with two-dimensional environment in our experiments. Our experimental domain consists of 10^6 primitive features, covering the space of 1000×1000 points.

Motor-system representation. Compositional hierarchy of robot's motor system ($CHARMS$) builds upon the motor primitives that reside in the primitive layer. A motor primitive is a one-dimensional feature representing the state of the joint/motor. For example, a node $\varphi_3 = 30°$ represents the joint φ_3 at position $30°$.

The nodes of both hierarchies are linked during the learning process so that the corresponding body postures match the positions in the coordinate system of the environment, for example, the nodes describing the positions of all the joints of the arm are linked to the nodes representing the position of the tip of the finger on the table. The proposed computational model of perception-action cycle corresponds to recent research in cognitive science [24,13], cognitive and developmental psychology, and cognitive neuroscience that provide strong empirical evidence for perception-action cycle [31].

3   Learning Compositional Hierarchies of a Motor System and the Environment

Learning a sensorimotor model refers to learning the compositional hierarchies of the motor system and the environment from traces of sensorimotor data and linking the corresponding nodes of the hierarchies.

Input data is a trace of learning examples $\mathbf{e} = (\varphi_1, \ldots, \varphi_n, x, y)$, where φ_i represents a state of the motor i (i.e. the angle of the joint i), while sensory variables x and y are the coordinates of the end-effector. Motor and sensory variables are used separately to learn each of the two hierarchies. The corresponding nodes of the hierarchies are linked regarding the relationship between the motor and the sensory variables as given by $\mathbf{e}$.

Besides learning examples, the abstraction structure is also given, specifying the resolution at each layer. For example, the abstraction structure may state that the resolution is decreasing exponentially from the primitive layer to the root node of the hierarchy. In all the experiments described below the quantization in layer $\mathcal{L}^i$ is 2^i, $i = 0, \ldots, 9$, so $\mathcal{L}^0$ has the highest and $\mathcal{L}^9$ the lowest resolution.

3.1 Learning the Structure

The learning is layer-wise incremental. Given learning example $\mathbf{e}$, the joint angles φ_i $(i = 1, \ldots, n)$ form n primitive nodes. Primitive nodes in $\mathcal{L}^0$ are abstracted and combined into $\mathcal{L}^1$ nodes. The process of abstracting and combining nodes is recursively repeated until the node combines all n features. We call such nodes *full compositions*. Full compositions end the composition process while the abstraction proceeds in the higher layers until the root node of the hierarchy.

During the process described above, many nodes are reused again and again, possibly forming different combinations, which leads to one node having many alternative pairs of children in a layer below. Shareability of nodes is one of the key properties of the proposed representation.

The way the nodes are combined greatly influences the space complexity and thus the scalability of the representation. In our experiments, we combine the adjacent joints (1 and 2, 3 and 4, etc.) as given in the learning data. We ignore other potentialy interesting combinations of joints for the sake of simplicity and leave this for future work.

The CH of the environment is learned in the same way and is similar to our introductory example. Since the workspace in our case is two dimensional, we find the full compositions already in layer $\mathcal{L}^1$ of CH of the environment. Altogether, the CH of the environemnt in our example has eleven layers, with quantization: 1 $(\mathcal{L}^0), 2, 4, 8, \ldots 1024$ $(\mathcal{L}^{10})$. The top layer covers the entire domain, 1000×1000.

3.2 Linking and Inference

The main objective of the sensorimotor representation is to efficiently deal with forward and inverse kinematics by establishing the mapping between the inner coordinate system and the environment. Here we present our approach to solving forward/inverse kinematics with compositional hierarchies. We start with a formal definition and conclude with an illustrative example.

To enable the interaction between the motor system and environment, we have to properly link both representations. Given joints' states and the corresponding position of the end-effector, $\mathbf{e} = (\varphi_1, \ldots, \varphi_n, x, y)$, the joint vector $(\varphi_1, \ldots, \varphi_n)$ should relate to the corresponding workspace point (x, y) in the highest resolution. In order to do so for a large data set of (joints, point) pairs and to enable generalization, the mapping should connect the nodes across all the layers of both compositional hierarchies. More specifically, all the nodes covering $(\varphi_1, \ldots, \varphi_n)$ in CHARMS should be linked to all the nodes covering (x, y) in CH of the environment.

Note that both forward and inverse mappings are of type many-to-many. Whereas this makes perfect sense for the inverse kinematics the forward kinematics should only yield a single solution in the workspace. Many solutions are due to the fact that a great deal of information is lost because of the abstraction and space complexity reduction. We propose the following way of mapping reconstruction that solves the mapping in both directions at the same time.

Let $\mathcal{C}$ be a configuration space hierarchy, i.e. an AND-OR graph representing robot configurations, and similarly let $\mathcal{W}$ be the workspace hierarchy, representing positions of robot's end-effector in the work space. Let $f : \mathcal{C} \to \mathcal{W}$ and $g : \mathcal{W} \to \mathcal{C}$ be multi-valued functions. We observe the composite mappings of f and g and define:

- Forward kinematics: given configuration $c \in \mathcal{C}$, the corresponding position of the end-effector w is a fixed point of mapping $f \circ g$, $f(g(w)) = w$.
- Inverse kinematics: given the position of the end-effector $w \in \mathcal{W}$, the corresponding configuration c is a fixed point of $g \circ f$, $g(f(c)) = c$.

Let us illustrate the mapping mechanism through the following example of inverse kinematics of 5-DOF manipulator operating in two-dimensional workspace. Suppose the end-effector is at point $w = (w_x, w_y)$ and for the sake of simplicity, suppose w has been visited before - we examine a general case in our experiments in section 4. Inverse kinematics in our sensorimotor representation starts by activating the primitive nodes w_x and w_y in $\mathcal{W}$. The signals propagate from w_x and w_y to their common parent in $\mathcal{L}^1$ and subsequent parent nodes in higher layers. Eventually, the tree $\mathcal{T}_e$ representing the current position of the end-effector in all resolutions is activated. At the same time, the links from $\mathcal{T}_e$ to $\mathcal{C}$ are activated. The end-nodes of these links reside in different layers of $\mathcal{C}$. We proceed in $\mathcal{L}^3$ of $\mathcal{C}$, which in this case is the lowest layer with *full composition* nodes. These abstract nodes have many possible realizations in the primitive layer, $C \subset \mathcal{C}$, i.e. several different combinations of primitive nodes exist that share these abstract higher layer nodes. We map each $c_i \in C$ back to $\mathcal{W}$. Each mapping proceeds analogously as described above but in the opposite direction, yielding a set W_i of possible positions of the end-effector. The configuration c_i corresponds to end-effector position $w = (w_x, w_y)$ if $w \in W_i$.

4   Experimental Results

To study the proposed approach, we designed two types of experiments. First, we run a set of experiments in which we observed the generalization property of CHARMS. We measured the accuracy of the model for a different set of parameters in simulated motor babbling. Second, we observed the space complexity of the proposed representation. By training the models from traces of intended actions of humanoid robot Nao, we show that compositional hierarchies can scale well when learned from real, high DOF data.

4.1   Generalization Property of CHARMS

In this set of experiments our goal was to find out how well CHARMS can generalize. To this end we measured the accuracy of our sensorimotor representation learned from motor babbling data in a kinematic simulator. Our goal was to learn the forward and inverse kinematics of the robot. The accuracy refers to the precision of the end-effector.

The input data was a set of learning examples, where a learning example is a vector $(\varphi_1, \ldots, \varphi_n, x, y)$, as defined above. In our experiments, we varied the number of learning examples (N) in the learning set as well as the number of degrees of freedom of the robot, n. The domain of joint angles, φ_i, was the interval $[-180°, 180°]$, while the position of the end-effector was $(x, y) \in [0, 1000] \times [0, 1000]$, which defined the workspace. Neither learning nor test data had duplicate examples and there was no intersection between the learning and the test set.

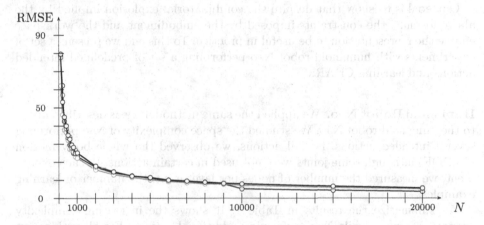

Fig. 2. Results of 10-fold cross-validation for different sizes of the learning sets. Colors indicate experiments for different values of degrees of freedom: blue (3 DOF), green (5 DOF), red (10 DOF), black (20 DOF). The overlap of the curves suggests that degrees of freedom do not affect the accuracy of sensorimotor model.

We observed RMSE in 10-fold cross-validation for different values of N and robots with different DOF (n). The experiments showed that RMSE converges with increasing N (see Fig. 2), which means that the larger learning data set implies higher accuracy of sensorimotor model. To achieve better precision of the end-effector, we can increase the resolution of the primitive layer. Note that the actual points have resolution 2, since a point only emerges in the first layer. The error does not depend on the number of degrees of freedom as indicated by the overlapping lines in Fig.2 but only on the number of learning data. The results also show that the error of sensorimotor model decreases very quickly and is negligible after $N = 15000$. Even for low sampling frequencies, e.g. 1 s, this means that after 6 hours of babbling, sensorimotor model would be sufficiently well trained. Since babbling samples the workspace relatively uniformly, generalization is good enough even with relatively small amount of learning data.

4.2   Space Complexity of Motor Memory

We measure space complexity of the representation by the number of nodes per layer. The latter is very difficult to compute analytically. We can estimate the absolute upper bound of the number of nodes per layer $\mathcal{L}^i$ with resolution r_i, $i = 1, 2, \ldots$, by

$$\binom{n}{i} \left(\frac{D}{r_i} \right)^i$$

where n is the degree of freedom of the robot and D is the quantization of the domain in the primitive layer, e.g. robot's joints are quantized into $D = 360$ angles.

Our goal is to show that despite the combinatorial explosion implied by the above formula, the constraints imposed by the embodiment and the workspace enable the representation to be useful in practice. To this end we present a set of experiments with humanoid robot Nao performing a set of predefined intended actions and learning CHARMS.

Humanoid Robot Nao. We applied the same methodology as described above to the humanoid robot Nao. We studied the space complexity of Nao performing several intended actions. For all actions, we observed the whole body motion (26 DOF) although some joints were not used in certain actions. In each experiment, we measured the number of nodes per layer w.r.t. the number of learning examples.

We summarize the results in Table 1. It shows the increasing complexity of motor memory while Nao performing additional actions. For all actions, we observe whole body motion, i.e. the values of all 26 joints. The actions are as follows. Nao started a straight 10 m walk from its initial standing posture after which it performed the left and the right turn respectively. During walking, Nao is also moving its arms, each arm following the opposite leg. When it stops walking, Nao makes a bow with its head, joining both hands in front of its head and returns to the initial pose after the bow. Then it performs three actions with its right arm: a wave, a hand shake and pushing of the ball. Finally, it uses the whole body again while leaning forward and picking up an object from the floor. Training data is collected while Nao is performing each action and CHARMS is updated after each action is completed. Column N in Table 1 indicates the number of training examples in the execution trace, i.e. the total number of samples collected during the execution of all the actions so far (e.g. Nao collected 6501 samples while walking straight ahead and additional 1003 while making a left turn, which sums to 7504; altogether Nao collected 10587 samples after performing all the actions). The columns $\mathcal{L}^0, \ldots, \mathcal{L}^9$ denote the number of nodes at each layer of CHARMS trained on the samples collected up to the current action. Note that the compositions are performed up to layer $\mathcal{L}^5$ while from $\mathcal{L}^6$ to $\mathcal{L}^9$ only the abstractions take place; compositionality stops when the nodes cover a full configuration of joints, e.g. all 26 joints in this experiment. After the whole repertoire of actions, the number of nodes per layer is still small compared to the results from motor babbling experiments with similar DOF manipulator.

Table 1. The number of nodes per layer in compositional hierarchy representing Nao's whole body configurations (26 DOF) while performing different actions

Robot's action	N	$\mathcal{L}^0$	$\mathcal{L}^1$	$\mathcal{L}^2$	$\mathcal{L}^3$	$\mathcal{L}^4$	$\mathcal{L}^5$	$\mathcal{L}^6$	$\mathcal{L}^7$	$\mathcal{L}^8$	$\mathcal{L}^9$
Straight walk	6501	508	781	738	409	195	69	9	4	3	1
+ Left turn	7504	562	996	1025	550	267	92	10	4	3	1
+ Right turn	8507	915	1382	1386	730	336	131	23	6	4	1
+ Bow	9008	1116	1581	1513	830	367	154	32	12	9	1
+ Wave	9509	1217	1724	1646	946	398	171	38	18	13	1
+ Hand shake	10010	1281	1824	1736	1030	416	183	39	18	13	1
+ Push	10511	1405	2138	1981	1211	486	211	52	30	17	1
+ Pick up	10587	1526	2435	2185	1341	558	250	79	50	24	1

5 Conclusions

We proposed a computational model of sensorimotor system that consists of compositional hierarchies of motor system and the environment. It provides an alternative approach to mathematical models that have been used traditionally for modeling robot and human motion. The proposed representation of the motor system, CHARMS, is learned from the training data and does not require a kinematic model of the robot. Additionally, the compositional hierarchy of the environment is learned and linked to CHARMS to form the sensorimotor model. This model enables forward and inverse kinematics, is immune to singularities and can cope with many degrees of freedom. In the ongoing work we focus on including dynamics in the proposed representation.

References

1. Ben Amor, H., Kroemer, O., Hillenbrand, U., Neumann, G., Peters, J.: Generalization of human grasping for multi-fingered robot hands. In: Proc. of 25th International Conference on Intelligent Robots and Systems, IROS 2012 (2012)
2. Bengio, Y.: Learning deep architectures for AI. Foundations and Trends in Machine Learning 2(1), 1–127 (2009)
3. Braun, D.A., Aertsen, A., Wolpert, D.M., Mehring, C.: Motor task variation induces structural learning. Current Biology 19(4), 352–357 (2009)
4. Braun, D.A., Waldert, S., Aertsen, A., Wolpert, D.M., Mehring, C.: Structure learning in a sensorimotor association task. PLoS ONE 5(1) (January 2010)
5. Degallier, S., Righetti, L., Gay, S., Ijspeert, A.: Toward simple control for complex, autonomous robotic applications: combining discrete and rhythmic motor primitives. Auton. Robots 31(2-3), 155–181 (2011)
6. Demiris, Y., Simmons, G.: Perceiving the unusual: Temporal properties of hierarchical motor representations for action perception. Neural Networks 19(3), 272–284 (2006)

7. D'Souza, A., Vijayakumar, S., Schaal, S.: Learning inverse kinematics. In: Proceedings of the 2001 IEEE/RSJ International Conference on Intelligent Robots and Systems, vol. 1, pp. 298–303 (2001)
8. Fidler, S., Boben, M., Leonardis, A.: Similarity-based cross-layered hierarchical representation for object categorization. In: Proc. of CVPR (2008)
9. Fuster, J.M.: Cortex and memory: Emergence of a new paradigm. J. Cognitive Neuroscience 21(11), 2047–2072 (2009)
10. Green, C., Bavelier, D.: Enumeration versus multiple object tracking: the case of action video game players. Cognition 101(1), 217–245 (2006)
11. Green, S.C., Bavelier, D.: Action video game modifies visual selective attention. Nature 423(6939) (2003)
12. Hinton, G.E., Osindero, S., Teh, Y.W.: A fast learning algorithm for deep belief nets. Neural Comput. 18(7), 1527–1554 (2006)
13. Hommel, B., Müsseler, J., Aschersleben, G., Prinz, W.: The Theory of Event Coding (TEC): A framework for perception and action planning. Behavioral and Brain Sciences 24(05), 849–878 (2002)
14. Ijspeert, A., Nakanishi, J., Schaal, S.: Learning Attractor Landscapes for Learning Motor Primitives. In: Advances in Neural Information Processing Systems 15, NIPS 2002, pp. 1547–1554 (2002)
15. Jackson, P.L., Decety, J.: Motor cognition: a new paradigm to study self-other interactions. Curr. Opin. Neurobiol. 14(2), 259–263 (2004)
16. Knoblich, G., Flach, R.: Predicting the effects of actions: interactions of perception and action. Psychol. Sci. 12(6), 467–472 (2001)
17. Kober, J., Wilhelm, A., Oztop, E., Peters, J.: Reinforcement learning to adjust parametrized motor primitives to new situations. Autonomous Robots 33, 361–379 (2012)
18. Krüger, V., Herzog, D., Baby, S., Ude, A., Kragic, D.: Learning actions from observations. IEEE Robot. Automat. Mag. 17(2), 30–43 (2010)
19. McClelland, J.L., Rogers, T.T.: The parallel distributed processing approach to semantic cognition. Nat. Rev. Neurosci. 4(4), 310–322 (2003)
20. Miall, R.C., Wolpert, D.M.: Forward models for physiological motor control. Neural Networks 9(8), 1265–1279 (1996)
21. Nakanishi, J., Morimoto, J., Endo, G., Cheng, G., Schaal, S., Kawato, M.: Learning from demonstration and adaptation of biped locomotion. Robotics and Autonomous Systems 47, 79–91 (2004)
22. Nishimoto, R., Tani, J.: Development process of functional hierarchy for actions and motor imagery. In: Proceedings of the 2009 IEEE 8th International Conference on Development and Learning, DEVLRN 2009, pp. 1–6. IEEE Computer Society, Washington, DC (2009)
23. Peters, J., Schaal, S.: Reinforcement learning of motor skills with policy gradients. Neural Networks 21(4), 682–697 (2008)
24. Prinz, W.: Perception and Action Planning. European Journal of Cognitive Psychology 9(2), 129–154 (1997)
25. Rauschecker, J.P.: An expanded role for the dorsal auditory pathway in sensorimotor control and integration. Hearing Research 271(1-2), 16–25 (2011)
26. Rumelhart, D.E., McClelland, J.L.: Parallel Distributed Processing: Explorations in the Microstructure of Cognition. Foundations, vol. 1. MIT Press (1986)
27. Sailer, U., Flanagan, J.R., Johansson, R.S.: Eye-hand coordination during learning of a novel visuomotor task. J. Neurosci. 25(39), 8833–8842 (2005)
28. Schaal, S.: Learning robot control, 2nd edn., pp. 983–987. MIT press (2002)

29. Schaal, S.: Dynamic movement primitives - a framework for motor control in humans and humanoid robots. In: The International Symposium on Adaptive Motion of Animals and Machines (2003)

30. Schaal, S., Mohajerian, P., Ijspeert, A.: A.j.: Dynamics systems vs. optimal control — a unifying view. In: Progress in Brain Research, pp. 425–445 (2007)

31. Sperry, R.W.: Neurology and the mind-brain problem. American Scientist 40(2) (1952)

32. Tamosiunaite, M., Nemec, B., Ude, A., Wörgötter, F.: Learning to pour with a robot arm combining goal and shape learning for dynamic movement primitives. Robot. Auton. Syst. 59(11), 910–922 (2011)

33. Tenenbaum, J.B., Kemp, C., Griffiths, T.L., Goodman, N.D.: How to grow a mind: statistics, structure, and abstraction. Science (New York) 331(6022), 1279–1285 (2011)

34. Ude, A., Gams, A., Asfour, T., Morimoto, J.: Task-specific generalization of discrete and periodic dynamic movement primitives. Trans. Rob. 26(5), 800–815 (2010)

35. Wolpert, D.M., Diedrichsen, J.A., Flanagan, J.R.: Principles of sensorimotor learning. Nat. Rev. Neurosci. 12(12), 739–751 (2011)

36. Wolpert, D.M., Flanagan, J.R.: Motor learning. Current Biology 20(11), 467–472 (2010)

Author Index